DL

Determination of
Organic
Structures by
Physical Methods

VOLUME 6

Contributors

Patrick J. Arpino

J. M. Bellama

John W. Easton

L. Lunazzi

F. W. McLafferty

Gail M. Pesyna

John K. Saunders

J. Schraml

Robert L. Strong

Determination of Organic Structures by Physical Methods

VOLUME 6

Edited by

F. C. NACHOD
Sterling-Winthrop Research Institute
Rensselaer, New York

J. J. ZUCKERMAN
Department of Chemistry
State University of New York
  at Albany
Albany, New York

EDWARD W. RANDALL
Department of Chemistry
Queen Mary College
London, England

1976
ACADEMIC PRESS
New York   London   San Francisco
A Subsidiary of Harcourt Brace Jovanovich, Publishers

ACADEMIC PRESS, INC.
111 Fifth Avenue, New York, New York 10003

United Kingdom Edition published by
ACADEMIC PRESS, INC. (LONDON) LTD.
24/28 Oval Road, London NW1

Library of Congress Cataloging in Publication Data
Main entry under title:

Determination of organic structures by physical
methods.

Includes bibliographies.
1.   Chemistry, Organic.   2.   Chemistry, Analytic.
I.   Braude, Ernest Alexander,   (date)   ed.
II.   Nachod, Frederick C., ed.   III.   Phillips,
William Dale,   (date)   ed.   IV.   Zuckerman, Jer-
old J.,   (date)   ed.
QD271.D46   547'.3   54-11057
ISBN 0−12−513406−1

Contents

List of Contributors

Numbers in parentheses indicate the pages on which the authors' contributions begin.

PATRICK J. ARPINO (1), Department of Chemistry, Cornell University, Ithaca, New York

J. M. BELLAMA (203), Department of Chemistry, University of Maryland, College Park, Maryland

JOHN W. EASTON (271), Department of Chemistry, Ryerson Polytechnical Institute, Toronto, Ontario, Canada

L. LUNAZZI (335), Institute of Organic Chemistry, University of Bologna, Bologna, Italy

F. W. McLAFFERTY (1, 91), Department of Chemistry, Cornell University, Ithaca, New York

GAIL M. PESYNA* (91), Department of Chemistry, Cornell University, Ithaca, New York

JOHN K. SAUNDERS (271), Département de Chimie, Université de Sherbrooke, Sherbrooke, Quebec, Canada

J. SCHRAML (203), Institute of Chemical Process Fundamentals, Czechoslovak Academy of Science, Prague, Czechoslovakia

ROBERT L. STRONG (157), Department of Chemistry, Rensselaer Polytechnic Institute, Troy, New York

* Present address: Committee on Science and Technology, U. S. House of Representatives, Washington, D. C.

Preface

No one could have foreseen the tremendous impact of physical methods upon organic chemistry when this treatise was started in the early 1950's. Not only have many new tools become available to the chemist interested in elucidating organic structure, but the tools themselves have achieved increasing sophistication. This is illustrated in this volume by the first two chapters dealing with exciting extensions of the well-known mass spectrometric technique. Fleetingly short-lived species are brought "to light" in the section on flash photolysis. In the concluding three chapters the theme of nuclear magnetic resonance is further explored for organosilicon chemistry, for solutes dissolved in liquid crystals, and finally for the role of the nuclear Overhauser effect in structure elucidation.

The Editors have again tried to combine accounts of new technology and techniques with descriptions of extensions of those methods that have proved so useful during the last two decades. As for the previous volumes, credit for the surveys goes to the individual authors; blame for errors of omission or commission must go to the Editors.

F. C. Nachod
J. J. Zuckerman
Edward W. Randall

Contents of Other Volumes

Amino Acid Sequencing of Oligopeptides by Mass Spectrometry

1

PATRICK J. ARPINO AND F. W. McLAFFERTY

I. INTRODUCTION

Mass spectrometry (MS) was recognized as a potentially valuable tool for oligopeptide sequencing a decade ago. The specific sequence information which can be obtained from fragmentation patterns has been described by

Biemann, Stenhagen, Lederer, Weygand, Shemyakin, to name only a few, and is now well documented.[1-20] However, few biochemists use MS as a routine method for peptide sequencing, and still favor the classical procedure involving chemical and enzymatic degradations and separations.[21-26a] However, a rapidly growing number of laboratories are now reporting successful applications of MS in sequencing.

In both the classical and the MS analysis of a long-chain polypeptide, such as a protein, the polymer must be broken into smaller units. This is generally done by a specific chemical reagent, such as cyanogen bromide, or by a peptidase; this is followed by fractionation and isolation of the individual oligopeptides by chromatography and/or electrophoresis. For the resulting oligopeptides which are too large, this procedure is repeated with a different reagent. In the classical method, the amino acid composition is obtained on each oligopeptide (subnanomolar level) by complete acid hydrolysis followed by chromatography. The N-terminal and C-terminal amino acids of each oligopeptide are determined by enzymatic or chemical methods (micro-dansylation sensitive to the picomolar range[21] and Edman degradation in the nanomolar range have been described).[22-24] The complete sequence is obtained by repetition of the Edman degradation, a procedure which is critically dependent on the purity of the starting oligopeptide. The previous analytical stages are repeated using a different chain cleavage to identify overlapping amino acid sequences, which allows the oligopeptide information to be combined to give the whole structure. The complex degradation, isolation, purification, and Edman procedures involved have become very sophisticated; many steps have been fully automated and require only a few nanomoles of peptide. Recent determinations of peptide chains with several hundred amino acid residues[25, 26a] serve as spectacular examples of the efficacy of the classical method, but it should be emphasized that these remarkable results necessitated many years of effort.

The most obvious drawback to the use of MS is its complexity and cost, especially with high resolution and automated data handling techniques. Mass spectrometry can only be applied directly to oligopeptides (<10 residues), and these usually must be derivatized to increase their vapor pressure. Thus the MS sequencing of a natural protein also requires an initial procedure of degradation and separation, and so has been used mainly as an alternative to the Edman degradation. Probably the most important uses to date of MS have been for samples which cannot be sequenced by Edman degradation (difficult or unusual[26a] amino acids) and in those cases in which a separation could not give oligopeptides in the prerequisite purity. However, there are promising new MS techniques which could become much more sensitive and much more highly automated than the present Edman procedure, so that with further development a degradation/partial separation/ MS procedure could be the method of choice for polypeptide sequencing.

Because of the numerous excellent reviews already available,[1-20] this chapter will emphasize the most accepted methods and promising recent developments. Topics will include the main types of mass spectral fragmentation under electron and chemical ionization; special derivatives for increasing oligopeptide volatility and optimizing the fragmentation pattern; analysis of samples of lower volatility by special techniques such as direct chemical ionization and field desorption; information from metastable ion and high-resolution data; data collection and interpretation, including computer techniques; examples of peptide sequencing by MS; direct coupling of separation methods such as fractional vaporization and gas and liquid chromatography with MS; and future possibilities for automated sequencing of subnanomole quantities of polypeptides.

Because MS methods have been used mainly as a substitute for the Edman degradation, some features of this should be emphasized. The length of the oligopeptide is not critical ($\sim$ 5–80 units), but it should be pure and contain a nonblocked N-terminal amino acid. The peptide is reacted with phenylisothiocyanate in dilute alkali followed by a cyclization step under acidic conditions which sets free the phenylthiohydantoin derivative of the N-terminal amino acid, and produces a shorter peptide (by one residue) with a new N-terminus (see Scheme 1). Distinction between glutaminyl (or aspara-

$$R'\!-\!N\!=\!C\!=\!S + H_2N\!-\!\underset{\underset{R_1}{|}}{CH}\!-\!\underset{\underset{O}{\|}}{C}\!-\!NH\!-\!\underset{\underset{R_2}{|}}{CH}\!-\!\underset{\underset{O}{\|}}{C} \ldots \longrightarrow$$

$$R'\!-\!NH\!-\!\underset{\underset{S}{\|}}{C}\!-\!NH\!-\!\underset{\underset{R_1}{|}}{CH}\!-\!\underset{\underset{O}{\|}}{C}\!-\!NH\!-\!\underset{\underset{R_2}{|}}{CH}\!-\!\underset{\underset{O}{\|}}{C} \ldots \longrightarrow R'N\!-\!\underset{\underset{S}{\|}}{C}\!-\!NH\!-\!\underset{\underset{R_1}{|}}{CH}\!-\!\underset{\underset{O}{\|}}{C} +$$

$$H_2N\!-\!\underset{\underset{R_2}{|}}{CH}\!-\!\underset{\underset{O}{\|}}{C} \ldots$$

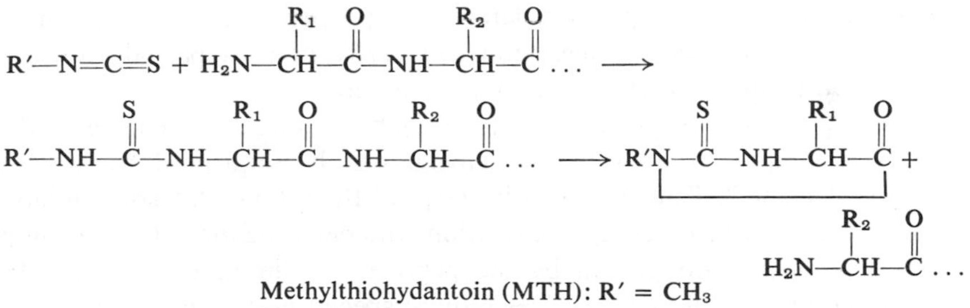

Methylthiohydantoin (MTH): R' = CH_3
Phenylthiohydantoin (PTH): R' = C_6H_5

SCHEME 1.

ginyl) from glutamic (or aspartic) residues from Edman degradation is often difficult, although it is straightforward by MS. In the automated version of the solid-phase Edman degradation about 20 cycles per day can be carried out. Though 60 cycle analyses have been reported,[23] incomplete reaction, nonspecific peptide chain cleavage, and possible minor contaminations apparently limit the routine analysis to 20 cycles, yielding identification of about 10–14 amino acid residues and information on the following 5 or 10. Sample requirements have been reduced dramatically in recent years; analyses of quantities as small as 35 nmoles of an octapeptide,[24] and 240 nmoles of a 21 amino acid peptide[23] have been reported.

On the other hand, present techniques of MS analysis usually require an equivalent amount of sample. Although sequencing has been accomplished on 10–20 nmoles of tetra- to hexapeptides converted to their O,N-permethyl-ated acetyl esters,[27] about 100–500 nmoles of the sample are generally required[28–34] for the mass spectrum of a middle length (6–9 residues) oligo-peptide, even under low resolution. Note, however, that this limitation apparently is due to the amount of sample required in the derivatization steps, as adequate mass spectra can be obtained on nanogram quantities of normal samples. Thus high-resolution studies[17, 28] have also employed 100–200 nmole peptide samples before acetylation and permethylation. Because of their zwitterionic structure and the polarity of their functional groups, in the past it has only been possible to run nonderivatized compounds as large as dipeptides and a few tripeptides with nonpolar residues by MS[1–20]; the restriction is even greater when a gas chromatographic inlet is used (GC/MS). However, preliminary results with two other techniques are highly promising. Field desorption ionization (FD) can supply molecular weight and limited sequence information on nanomole underivatized samples as large as nona-peptides.[35] *Direct* chemical ionization has given complete sequence informa-tion on 0.1 nmole of an underivatized tetrapeptide.[36]

However, at present the most useful sequence information is obtained from derivatized peptides utilizing electron ionization. Mass spectrometry in general can analyze O,N-permethylated N-acetyl peptides esters containing up to 8–10 amino acid residues, and can derive at least partial sequence information from mixtures of several of such peptides.

Note that an eicosapeptide, which is amenable to direct sequencing by the Edman technique, by appropriate cleavage can be sequenced by MS, as shown in Scheme 2. Thus the repetitive steps of the Edman degradation have been replaced by other cleavage, separation, and derivatization steps. Adding to the complexity brought on by the peptide size limitations of the MS analysis, certain amino acid residues require special pretreatments to give a suitable derivative. Thus as yet there appears to be no general approach to MS peptide analysis, but only a large set of different possible strategies. In other words, there appear to be many MS techniques promising for particular aspects of peptide sequencing, but it is still difficult to fit these into one "peptide analyzer"; there are too many parameters that must be chosen by an experienced scientist. Note that only a very limited number of MS analyses have been carried out on unknown peptides which are amenable to an Edman degradation.[34, 37]

Thus speed and sensitivity would be improved if the MS method did not require careful separation, purification, and individual derivatization of each degradation oligopeptide, and were independent of the nature of the different amino acid residues. New strategies to achieve this may take advantage of the fact that MS is able to deal with mixtures of three to four peptides,[38–42] and

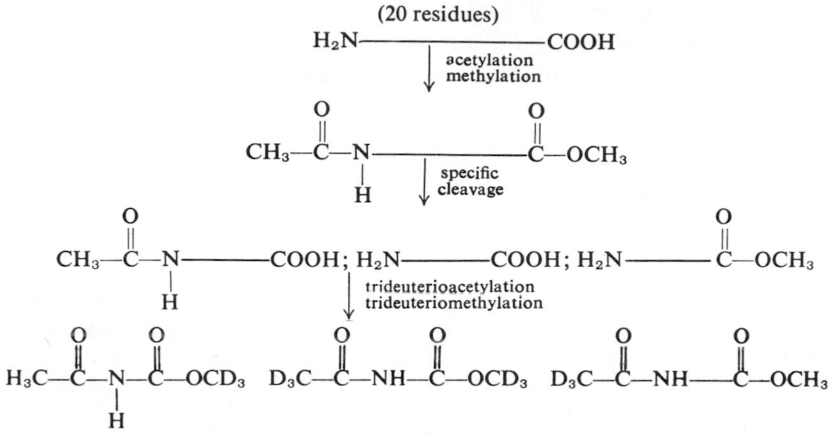

(20 residues)

$$H_2N\text{———————}COOH$$

$$\downarrow\ \begin{array}{l}\text{acetylation}\\\text{methylation}\end{array}$$

$$CH_3\overset{\displaystyle O}{\overset{\|}{-}}C-\underset{\underset{H}{|}}{N}\text{————}\overset{\displaystyle O}{\overset{\|}{-}}C-OCH_3$$

$$\downarrow\ \begin{array}{l}\text{specific}\\\text{cleavage}\end{array}$$

$$CH_3\overset{\displaystyle O}{\overset{\|}{-}}C-\underset{\underset{H}{|}}{N}\text{————}COOH;\ H_2N\text{————}COOH;\ H_2N\text{————}\overset{\displaystyle O}{\overset{\|}{-}}C-OCH_3$$

$$\downarrow\ \begin{array}{l}\text{trideuterioacetylation}\\\text{trideuteriomethylation}\end{array}$$

$$H_3C\overset{\displaystyle O}{\overset{\|}{-}}C-\underset{\underset{H}{|}}{N}\overset{\displaystyle O}{\overset{\|}{-}}C-OCD_3\quad D_3C\overset{\displaystyle O}{\overset{\|}{-}}C-NH\overset{\displaystyle O}{\overset{\|}{-}}C-OCD_3\quad D_3C\overset{\displaystyle O}{\overset{\|}{-}}C-NH\text{————}\overset{\displaystyle O}{\overset{\|}{-}}C-OCH_3$$

SCHEME 2.

that powerful separation methods such as gas and liquid chromatography can be coupled directly to MS.[17, 43–47a]

A number of recent papers,[48–61] mainly from Lovins and co-workers,[48–55] report the use of MS to identify the products of an Edman degradation. Other methods use MS to identify the N-terminal residues of peptides after an appropriate degradation[62–65]; the method of Gray and del Valle[64] appears especially promising as a complementary technique. For details of these more specialized MS applications the reader should consult the original literature.

II. ELECTRON IMPACT FRAGMENTATION OF DERIVATIZED OLIGOPEPTIDES

General fragmentation modes of oligopeptides under electron impact (EI) have now been elucidated in detail.[1–20] We will consider these in two parts: the cleavage of the peptide backbone, and the fragmentation modes characteristic of the different residues. Although it is the ions resulting from the former that provide the sequence information, ions from the latter also provide useful information, and both modes need to be understood to decipher a mass spectrum properly. Although few studies have been reported on photoionization[65a] and field ionization,[65b] these fragmentation pathways are generally similar to those under EI.

A. Fragmentation of the Backbone

1. N-Terminus Sequence Peaks

Oligopeptides possess a repetitive amide backbone which cleaves on either side of the carbonyl group to form preferentially the aldimine (A_i^+) and acylium (B_i^+) ions (Scheme 3), probably because the positive charge is better stabilized on that side. However, some C-terminal fragments, C_i^+ and D_i^+, are observed, especially if the ester group Y can stabilize the positive charge

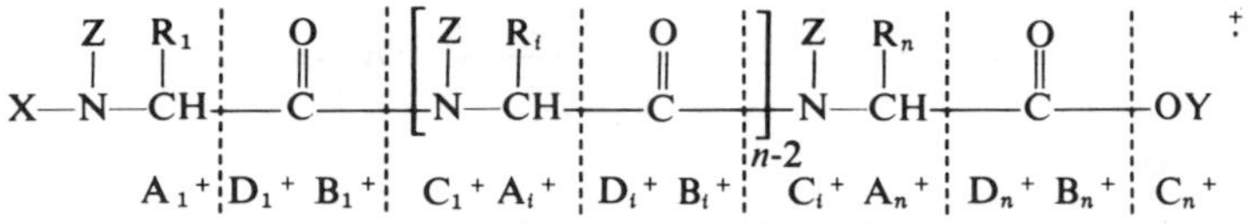

SCHEME 3. X = N-terminus protective group; Y = C-terminus protective group; Z = H, CH_3, or CD_3; R_i = side chain characteristic of amino acid in the ith position from the N-terminus. A_i^+ (aldimine): X—$(NZ$—CHR_j—$CO)_{i-1}N^+Z=CHR_i$; B_i^+, (acylium): X—$(NZ$—CHR_j—$CO)_{i-1}NZ$—CHR_i—$C{\equiv}O^+$; Note that A_i^+, B_i^+, C_i^+, and D_i^+ result from cleavages at the carbonyl group of the ith amino acid counting from the N-terminus. Thus the sum of the masses of A_i^+ plus D_i^+, and also the sum of the masses of B_i^+ plus C_i^+, equal the mass of the molecular ion.

(see Section II,A,4). Aldimine ions (A_i^+) originate partly by direct cleavage of the chain, and partly by decarbonylation of an amino acyl ion B_i^+ (see Section II,A,2) as evidenced by metastable transitions.[5, 6, 17, 66] Other metastables show the occurrence of sequence-to-sequence transitions such as $B_i^+ \rightarrow B_i^+{}_{-1}$ and $B_i^+ \rightarrow A_j^+$ (Scheme 4), the detection of which provides valuable data on the position of individual amino acids in the sequence, as discussed below (see Section II,A,5).

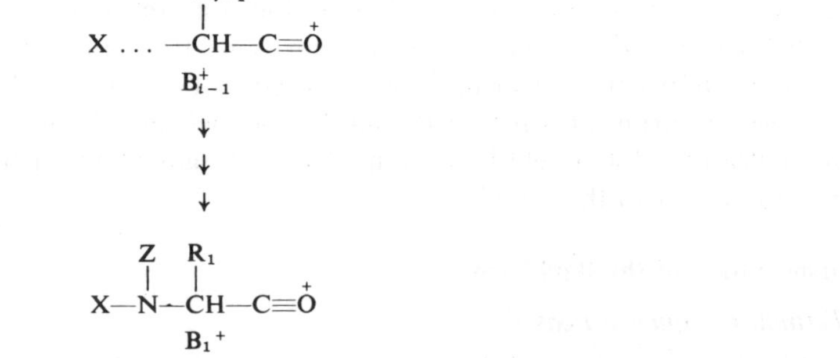

SCHEME 4.

The A_i^+ and B_i^+ ions are the so-called sequence peaks; $2n$ of such ions are expected from an oligopeptide with n residues. The relative abundance of A_i^+ and B_i^+ depends on many factors such as the nature of the N-terminus derivative, the length of the peptide, and the nature of the C-terminus derivative. Aromatic,[67] adamantoyl,[68] and trifluoroacetyl[66] N-blocking groups promote ionization at the aldimine peaks (A_i^+), while N-acyl peptides enhance the acylium ion (B_i^+) abundances. The latter situation is noteworthy for O,N,S-permethylated N-acetyl peptide esters,[29] such as that of Fig. 1.[69]

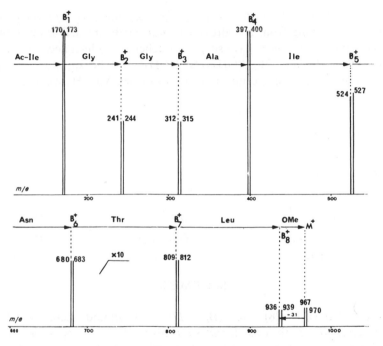

FIG. 1. Simplified mass spectrum of O,N-permethylated N-acetyl (1:1 $CH_3CO:CD_3CO$) octapeptide[69]; isotopic ions and ions of $< 1\%$ relative abundance have been omitted. Sequencing of the peptide is straightforward as the B_i^+ ions are of dominant abundance.

N-Trifluoroacetyl (TFA)[66] and N-trimethylsilyl (TMS)[69a] peptide esters sometimes give low abundance sequence ions (both A_i^+ and B_i^+), making sequencing of these derivatives more difficult.

Aldimine ions (A_i^+) are intense generally only in short peptides (di- and tri-, see Fig. 2); acylium peaks (B_i^+) are generally abundant in higher (Fig. 1) and often in lower peptides. However, even with O,N-permethylated N-acetyl peptide methyl esters, neither A_i^+ or B_i^+ are of significant intensities after the 8th or the 10th fragment (B_8^+, B_{10}^+ upward).[29, 70]

The deduction of the sequence from the mass spectrum is based mainly upon the identification of the sequence peaks, either from the N-terminal ion (A_1^+ or B_1^+) upward, or from the molecular ion downward. Special N-terminal derivatives facilitate the recognition of these peaks[28]; these include $CBrF_2CO-$ (characteristic isotopes)[71]; mixed $1:1$ $CH_3(CH_2)_n-CO-$ and $CH_3(CH_2)_{n+1}-CO-$[72]; $CH_3(CH_2)_n-CO$ and $CD_3(CH_2)_n-CO$ ($n = 0-20$,[42, 64, 69, 72] see Fig. 1); CF_3CO-, $CClF_2CO-$, etc.[73]

2. Aldimine Variants, $B_i^+ - 28$ and $B_i^+ - 27$

The aldimine ion (A_i^+) ($B_i - 28)^+$ is often accompanied by a satellite ion at $(A_i + 1)^+$ arising from a hydrogen migration. From deuterium labeling evidence, Weygand et al.[66] have suggested Scheme 5, which accounts for the formation of the ion $CF_3C(OH)=N^+HCHR_1$ in the spectra of N-TFA peptide esters. From deuterium labeling evidence, Van Heijenoort et al.[72]

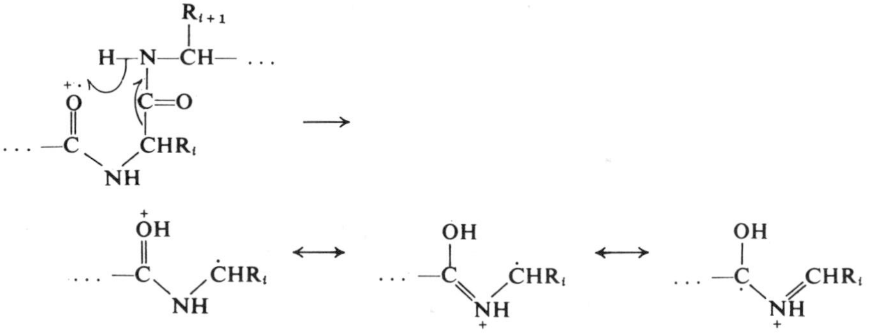

SCHEME 5.

suggested that $(B_i - 28)^+$ and $(B_i - 27)^+$ ions should occur only for α-amino acids, but both ions should be rare or absent if the amino acid is linked through another position, such as in ε-lysyl, β-aspartyl, α-glutamyl, β-alanyl,[3, 10, 72] or β-lysyl[74] peptides, although evidence of significant $(B_i - 28)^+$ peaks in β-alanine-containing peptides has been recently reported.[75] In the case of $(B_i - 28)^+$ apparently the elimination of CO from

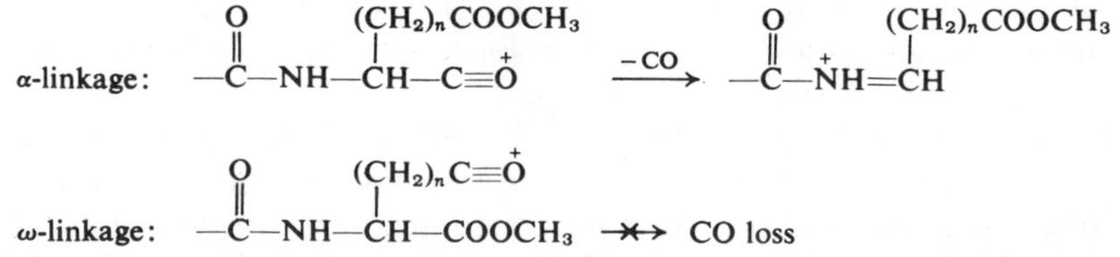

SCHEME 6.

B_i^+ is not aided by the participation of a neighboring amide group (Scheme 6).

3. Pseudo Sequence Ions

A different type of cleavage of the main peptide chain can occur at aromatic and heterocyclic residues (Phe, His, Tyr, and especially Trp),[66, 76–78] and at acidic residues (Asp, Asn).[69, 78, 78a] It has been also reported in the case of Cys derivatives.[11] In Scheme 7 the N-terminal portion of the peptide is

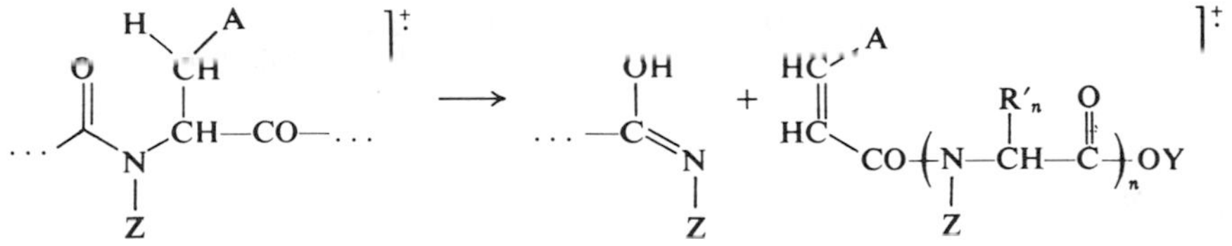

SCHEME 7.

A: C_6H_5, Phe; C_6H_4OZ, Tyr; 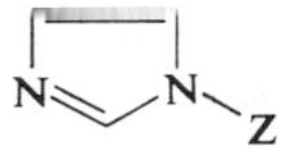 , His; 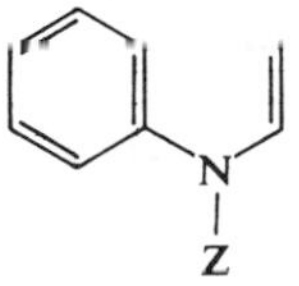 , Trp;

$COOY$, Asp; $CONZ_2$, Asn; $S—CH_2—CO_2CH_3$ in S-carboxymethylcysteine; X, Y, Z see Scheme 3.

lost. The resulting odd-electron ion, sometimes referred to as the pseudo molecular ion, will be particularly abundant if the aromatic or heterocyclic residue is in the C-terminal position. The pseudo molecular ion can also give rise to pseudo sequence peaks (A_i', B_i') with the aromatic or acidic side chain being the new N-terminal group. These even-electron ions are sometimes among the most prominent peaks in the spectrum (Scheme 8). Failure

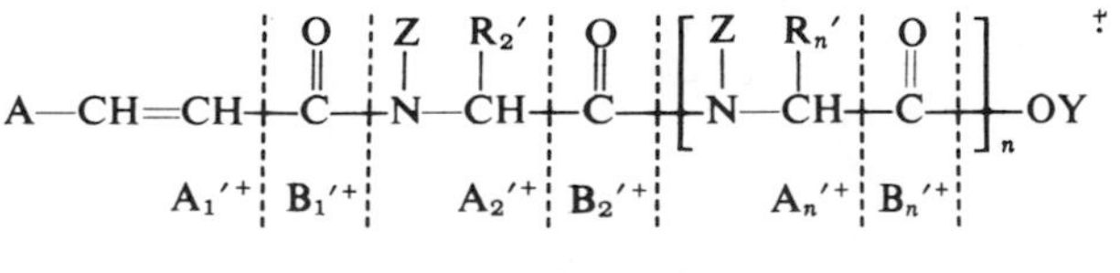

SCHEME 8.

to recognize the pseudo acylium type fragmentation is unfortunately a common mistake, and unusual fragmentation pathways have sometimes been suggested[13] instead of the mechanism proposed above. In the same manner

as the normal sequence ions, the pseudo acylium ions $B_i'^+$ are enhanced in
O,N-permethylated N-acetyl peptide methyl esters.

A second rearrangement is observed[78] when two of such residues are
adjacent in the chain (Tyr-Tyr, Trp-Trp, Trp-Tyr, Asn-Asn), and presence of
ions at 30 mass units above $B_i'^+$ for permethylated peptides ($Z = CH_3$) is
good evidence of such partial sequences (see Scheme 9). This accounts for the

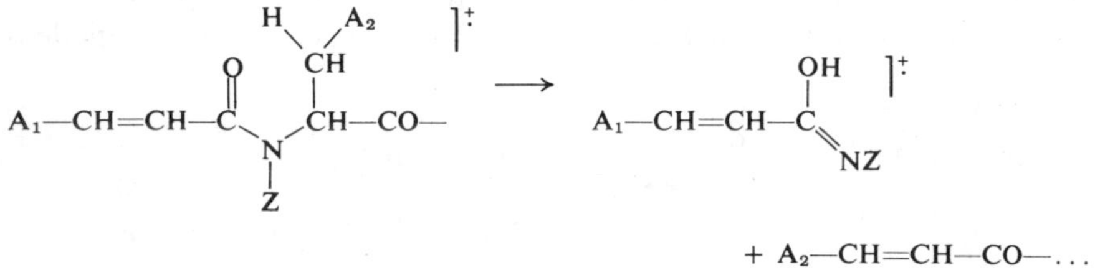

$$\left[A_1\!-\!CH\!=\!CH\!-\!\underset{\displaystyle \underset{Z}{\overset{\displaystyle N}{|}}}{\overset{\displaystyle \overset{O}{\|}}{C}}\underset{}{\overset{\displaystyle \overset{H}{\diagdown}\,\overset{A_2}{\diagup}}{\underset{}{\overset{CH}{\diagdown\diagup}}CH}\!-\!CO\!-\! \right]^{\ddot{+}} \longrightarrow \left[A_1\!-\!CH\!=\!CH\!-\!\underset{NZ}{\overset{OH}{\underset{\diagdown}{\overset{|}{C}}}} \right]^{\ddot{+}}$$

$$+ \; A_2\!-\!CH\!=\!CH\!-\!CO\!-\!\ldots$$

SCHEME 9.

presence of a strong m/e 191 peak in reported spectra of peptides containing
the partial sequence Tyr-Tyr[34] or Tyr-His[13] or the m/e 156 peak in Asn-
Asn,[34] though the authors gave no explanation for its occurrence.

4. C-Terminus Sequence Peaks

Retention of the positive charge on the C-terminal end accounts only for
the formation of minor ions, except for esters such as phenyl which stabilize
the positive charge.[79] Their detection is easier for trideuteriomethyl esters[28, 80]
or mixed $1:1$ CH_3/CD_3 esters.[80] The abundance of C-terminal ions generally

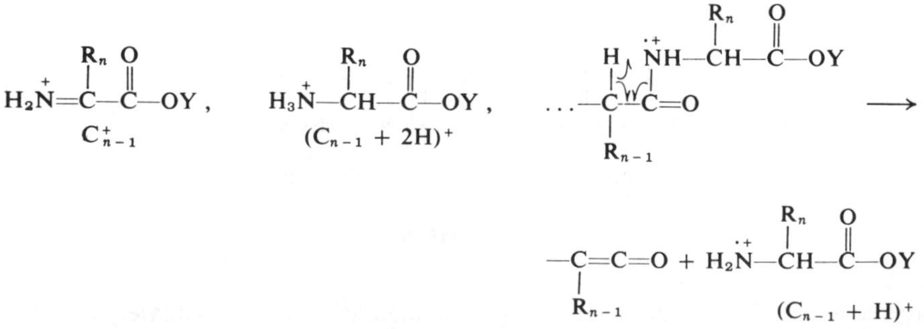

$$\underset{C_{n-1}^+}{\overset{+}{H_2N}\!=\!\underset{\overset{|}{R_n}}{C}\!-\!\overset{\overset{O}{\|}}{C}\!-\!OY} \, , \qquad \underset{(C_{n-1}+2H)^+}{\overset{+}{H_3N}\!-\!\underset{\overset{|}{R_n}}{CH}\!-\!\overset{\overset{O}{\|}}{C}\!-\!OY} \, , \qquad \ldots\!-\!\underset{\underset{R_{n-1}}{|}}{\overset{\overset{H}{\diagup}\;\overset{\centerdot+}{NH}\!-\!\underset{\overset{|}{R_n}}{CH}\!-\!\overset{\overset{O}{\|}}{C}\!-\!OY}{C\!\!=\!\!C\!=\!O}} \longrightarrow$$

$$-\!\underset{\underset{R_{n-1}}{|}}{C}\!=\!C\!=\!O \; + \; \underset{(C_{n-1}+H)^+}{\overset{\centerdot+}{H_2N}\!-\!\underset{\overset{|}{R_n}}{CH}\!-\!\overset{\overset{O}{\|}}{C}\!-\!OY}$$

SCHEME 10.

decreases rapidly with increasing size. The most common $C_i{}^+$-type ions containing only the C-terminal amino acid (cleavage at the carbonyl group of the $n-1$ residue) are $C_n^+{}_{-1}$, $(C_{n-1} + H)^+$, and $(C_{n-1} + 2H)^+$. The latter are formed by rearrangement of one and two hydrogen atoms, respectively; the $C_n^+{}_{-1}$ may also be formed with rearrangement to give the stable immonium structure (Scheme 10). The $(C_{n-1} + H)^+$ ions could arise by a reaction similar to that postulated for immonium ion formation (Section II,B,1). The $(C_{n-1} + 2H)^+$ ions are usually the most abundant of these C-terminal ions, and have been observed with the $C_n^+{}_{-1}$ ions in the spectra of many N-acetyl and N-TFA peptide methyl esters.[2, 28, 66, 71, 79, 81, 82]

Poor stabilization of the positive charge on the N-terminal ions of N-TFA and N-heptafluorobutyryl peptide esters, as evidenced by the low abundance of the $A_i{}^+$ and $B_i{}^+$ sequence ions, is also evidenced by the appreciable abundance of ions $D_n^+{}_{-2}$ and $D_n^+{}_{-1}$,[79, 82] which may result from the other possible cleavage of the bond adjacent to a carbonyl group (Scheme 11).

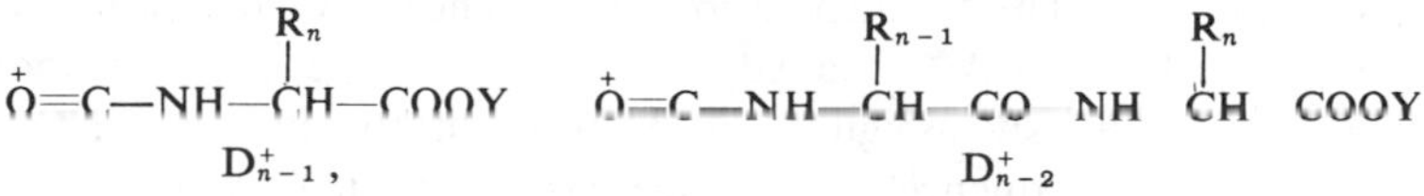

$$\overset{+}{O}=C-NH-\overset{\overset{\displaystyle R_n}{|}}{C}H-COOY \qquad \overset{+}{O}=C-NH-\overset{\overset{\displaystyle R_{n-1}}{|}}{C}H-CO \quad NH \quad \overset{\overset{\displaystyle R_n}{|}}{C}H \quad COOY$$
$$D_{n-1}^+ , \qquad\qquad\qquad\qquad D_{n-2}^+$$

SCHEME 11.

C-Terminal ions thus usually provide limited information on the sequence, although they are of real value if no molecular ion is observable. In a particular example[69] of an O,N-permethylated peptide, a complete set of C-terminal ions was found in addition to the N-terminal ions, allowing the reconstruction of the sequence shown in Scheme 12.

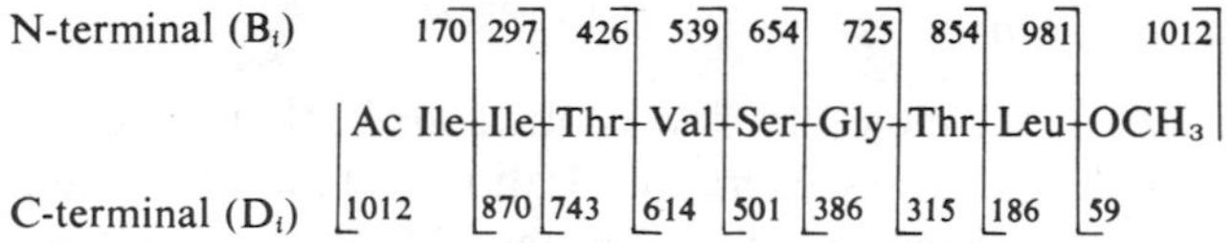

N-terminal (B_i)	170	297	426	539	654	725	854	981	1012
	Ac Ile	Ile	Thr	Val	Ser	Gly	Thr	Leu	OCH$_3$
C-terminal (D_i)	1012	870	743	614	501	386	315	186	59

SCHEME 12.

5. Metastable Transitions

Unimolecular metastable ion (MI) and collisional activation (CA) ion decompositions occurring in a field free region of the ion flight path in the mass spectrometer are of value because they can provide information on the identity of both the precursor ($m_1{}^+$) and daughter ($m_2{}^+$) ions of a particular decomposition, $m_1{}^+ \rightarrow m_2{}^+$. These ions generally appear as diffuse peaks at noninteger mass values in the normal spectrum and can be examined with higher sensitivity and specificity with special methods such as the Barber-Elliot-Major defocusing technique utilizing a double-focusing magnetic instrument. The transitions observed in MI are the unimolecular decompositions requiring the least energy; a much larger proportion of the possible reactions can be seen by CA because in this technique the internal energy of the precursor ion is increased substantially in the drift region by collision with a neutral molecule.

Metastable ion[34,37,83] and CA[84] spectra are helpful for the structure determination of peptides in several ways. They provide unambiguous distinction between Leu and Ile residues[84] (see Section II,B,2). They are especially valuable in the sequence determination of mixtures[38,39]; the precursor and the daughter ions arise from the same component in the mixture, identifying the peaks in the mixture spectrum which arise from the individual components. The following types of reactions can be particularly helpful in this regard. The loss of carbonyl ($B_i^+ \rightarrow A_i^+$) provides no sequence information, but it can provide useful confirmation of a postulated N-terminal amino acid. The "sequence ion $\rightarrow$ sequence ion" reactions such as $B_i^+ \rightarrow B_{i-1}^+$ and $B_i^+ \rightarrow A_j^+$ connect two ions from the same component, and thus unambiguously assign their position in the peptide. The most frequently observed sequence $\rightarrow$ sequence MI peaks in O,N-permethylated N-acetyl peptide esters correspond to the loss of one complete amino acid ($B_i^+ \rightarrow B_{i-1}^+$). They have been regularly found in this laboratory for $B_2^+ \rightarrow B_1^+$, and often for $B_3^+ \rightarrow B_2^+$ and $B_4^+ \rightarrow B_3^+$, in tri- and tetra-peptides.[38,39] "Sequence ion $\rightarrow$ immonium ion" (see Section II,B,1) metastable transitions are particularly abundant for the N-terminal residue (elimination of a neutral ketene from A_1^+), but the transitions to form immonium ions for residues in other positions are not always found.

6. The Molecular Ion

The intensity of the molecular ion is often weak, though some derivatives (see Section V,A,3) and low energy ionization techniques [chemical ionization (CI),[36] FD[35]] enhance it. Loss of water, $(M - 18)^+$, or loss of a methyl (particularly in the case of TMS derivatives)[43,69a] are sometimes useful as indirect evidence for the molecular ion. Dehydration is a general reaction affecting peptides, especially those containing Ser and Thr. Shemyakin et al.[11] have suggested that water loss in other peptides involves the oxygen of the peptide bond or intramolecular cyclization (see Scheme 13).

The protonated molecular ion, $(M + 1)^+$, can be observed under normal EI operation at higher sample pressure,[17,85] as well as under CI conditions. $(M + 14)^+$ and $(M + 28)^+$ ions are encountered as artifacts in N-acyl peptide esters (especially those containing His); bimolecular transprotonation and transmethylation account for these reactions which yield confusing ions.

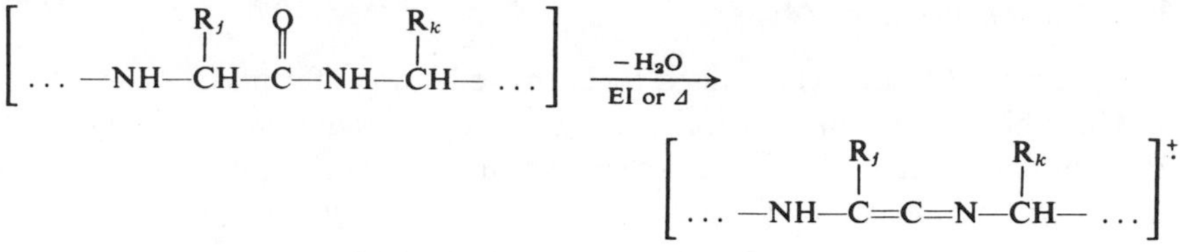

$$\left[\ldots -NH-\underset{\underset{R_j}{|}}{CH}-\underset{\underset{\parallel}{O}}{C}-NH-\underset{\underset{R_k}{|}}{CH}- \ldots \right] \xrightarrow[\text{EI or } \Delta]{-H_2O}$$

$$\left[\ldots -NH-\underset{\underset{R_j}{|}}{C}=C=N-\underset{\underset{R_k}{|}}{CH}- \ldots \right]^{\ddagger}$$

SCHEME 13.

An $(M - 4)^+$ ion is also encountered in histidine peptides (see Section II,B,6). O-TMS-polyamino alcohol derivatives show no molecular ion.

Note that observation of a molecular ion is not a necessity for complete sequencing of an oligopeptide. Even if sufficient overlap in the C- and N-terminal sequences is not found, the sequence can often be elucidated by using in addition the amino acid composition of the peptide.

7. N-Protective Group Fragmentations

Chain fragmentation can sometimes involve the N-terminal protective group. Rearrangement formation of ions by Scheme 14 is noteworthy in the mass spectra of synthetic N-acyl peptides or natural peptidolipids.[11, 72] N-Aryl[86] and alkyl[87] oxycarbonyl derivatives can lose the N-protective group either thermally or under electron impact, possibly as shown in Scheme 15.

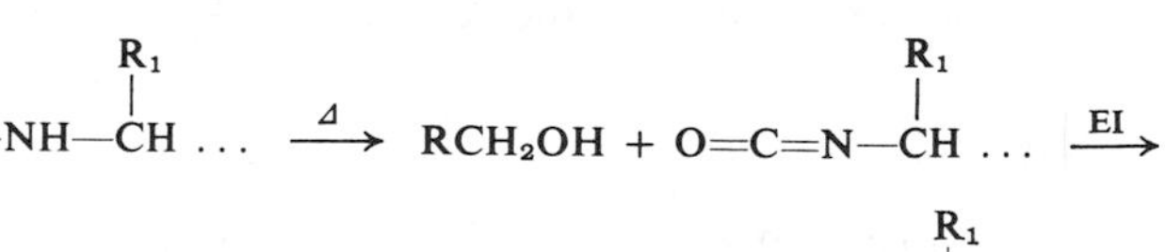

SCHEME 14.

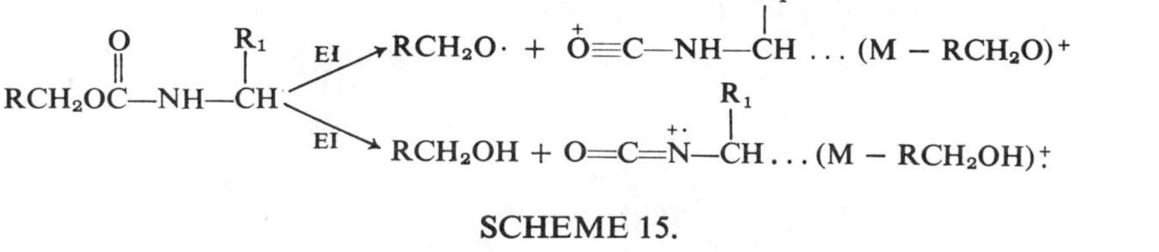

$$\underset{\underset{\parallel}{O}}{RCH_2OC}-NH-\underset{\underset{R_1}{|}}{CH} \ldots \xrightarrow{\Delta} RCH_2OH + O=C=N-\underset{\underset{R_1}{|}}{CH} \ldots \xrightarrow{EI}$$

$$O=\overset{+\cdot}{C}=N-\underset{\underset{R_1}{|}}{CH} \ldots (M - RCH_2OH)^{\ddagger}$$

$$\underset{\underset{\parallel}{O}}{RCH_2OC}-NH-\underset{\underset{R_1}{|}}{CH} \underset{EI}{\overset{EI}{<}} \begin{array}{l} RCH_2O\cdot + \overset{+}{O}{\equiv}C-NH-\underset{\underset{R_1}{|}}{CH} \ldots (M - RCH_2O)^+ \\ \\ RCH_2OH + O=\overset{+\cdot}{C}=N-\underset{\underset{R_1}{|}}{CH} \ldots (M - RCH_2OH)^{\ddagger} \end{array}$$

SCHEME 15.

8. Polyaminoalcohols

Polyaminoalcohols, as obtained from the reduction of peptides (see Section VI,C), exhibit simple mass spectra owing to the reduced number of fragmentation initiating sites.[2, 17, 88] The α-cleavage of the repetitive ethylenediamine group, with stabilization of the positive charge on either side of the cleaved bond, accounts for the presence in the spectra of both an N-terminal (RA_i^+) and C-terminal (RD_{i-1}^+) series of ions bearing the sequence information (see Scheme 16).

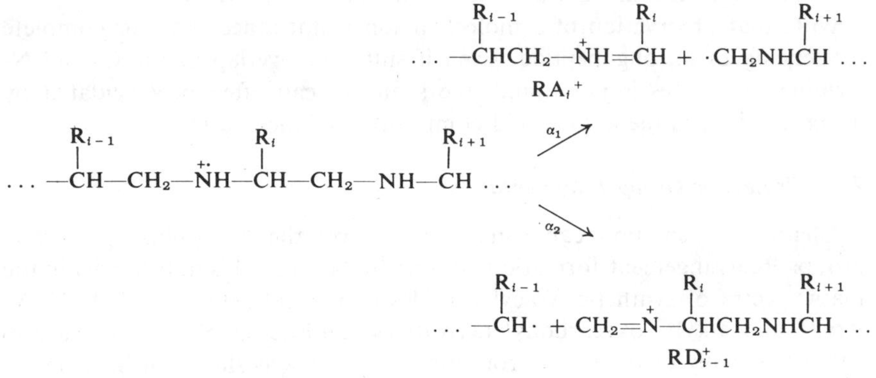

$$
\begin{array}{ccc}
R_{i-1} & R_i & R_{i+1} \\
| & | & | \\
\dots -CHCH_2-\overset{+}{N}H\!=\!CH & + & \cdot CH_2NHCH \dots \\
& RA_i^+ &
\end{array}
$$

$$
\begin{array}{ccccc}
R_{i-1} & & R_i & & R_{i+1} \\
| & & | & & | \\
\dots -CH-CH_2-\overset{+\cdot}{N}H-CH-CH_2-NH-CH \dots & & \nearrow^{\alpha_1} \\
& & & & \searrow_{\alpha_2}
\end{array}
$$

$$
\begin{array}{ccc}
R_{i-1} & R_i & R_{i+1} \\
| & | & | \\
\dots -CH\cdot + CH_2\!=\!\overset{+}{N}-CHCH_2NHCH \dots \\
& RD_{i-1}^+ &
\end{array}
$$

SCHEME 16.

B. Peaks Characteristic of Individual Amino Acids

Specific cleavages of R_i groups characteristic of individual amino acids include both simple C—C cleavages and hydrogen rearrangements. Though the resulting ions make the mass spectra more complex, their recognition is valuable for confirming the presence of a given residue and supplying information on the amino acid composition of the peptide.[39] Such fragmentations have been studied extensively by the Russian workers.[5, 11, 12]

1. Immonium Ions

The stable even-electron ion $Z\overset{+}{H}N\!=\!CHR_i$ characteristic of the ith amino acid can be formed from the corresponding aldimine ion (A_i^+) by rearrange-

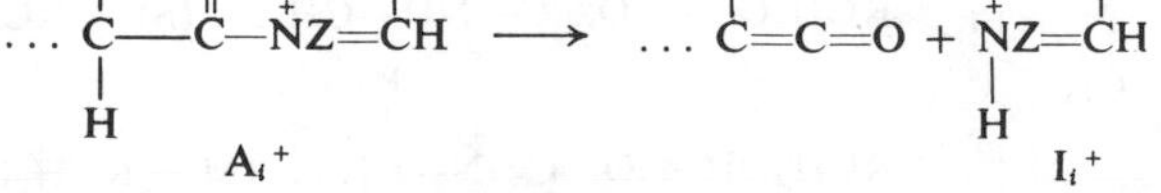

$$
\begin{array}{cccc}
R_{i-1}\ \ O & R_i & R_{i-1} & R_i \\
| \quad\ || & | & | & | \\
\dots C-C-\overset{+}{N}Z\!=\!CH & \longrightarrow & \dots C\!=\!C\!=\!O + \overset{+}{N}Z\!=\!CH \\
| & & | & \\
H & & H & \\
\quad A_i^+ & & & I_i^+
\end{array}
$$

SCHEME 17.

ment (Scheme 17). Obviously these give no direct sequence information. The occurrence of immonium ions has been reported for both nonpermethylated[11, 77, 85] (Z = H) and permethylated[39, 69, 84] (Z = CH_3) N-acetyl peptide esters.

The immonium ion of the N-terminal residue in N-acetyl peptides (permethylated or not) is usually more intense than the other immonium ions, probably because of the increased tendency to lose a neutral ketene from A_1^+.[39, 77] With CD_3CO as the N-protecting group, the corresponding shift of one amu clearly identifies the N-terminal residue. In addition to the α-H transfer, deuterium labeling has shown that in N-acetyl peptides a hydrogen atom from the —NH group can also be transferred during elimination of the immonium from the aldimine ion[77] (Scheme 18).

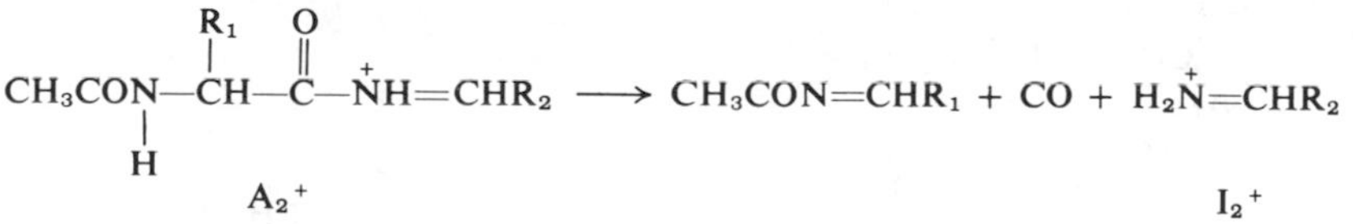

$$CH_3CON\overset{\overset{\displaystyle R_1}{\displaystyle |}}{\underset{\underset{\displaystyle H}{\displaystyle |}}{CH}}\overset{\overset{\displaystyle O}{\displaystyle ||}}{C}\overset{+}{N}H{=}CHR_2 \longrightarrow CH_3CON{=}CHR_1 + CO + H_2\overset{+}{N}{=}CHR_2$$

$$A_2^+ \qquad\qquad\qquad\qquad\qquad\qquad\qquad I_2^+$$

SCHEME 18.

Immonium ions from tryptophan, tyrosine, and histidine are sometimes not very intense; fortunately for these and other aromatic amino acid residues, characteristic cleavages yield prominent $ArCH_2^+$ ions (see Section II,B,6). On the other hand, the immonium ion from proline (and hydroxyproline) is usually intense irrespective of the position of the proline in the chain, but the residues between the amine end and Pro exhibit weak immonium ions.[77] Immonium ions from threonine and methionine are often rare or absent.[85] Immonium ions from lysine residues cyclize to yield further characteristic ions in the case of α-lysine and β-lysine.[74]

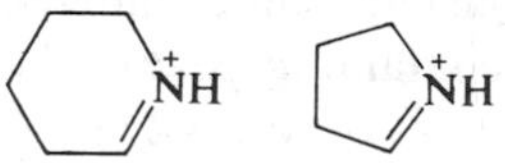

2. Amino Acids with Nonpolar R Groups

The amino acids glycine, alanine, valine, leucine, isoleucine, and proline are the least polar because of their saturated hydrocarbon side chains. It is especially important to recognize the rearrangement reactions possible for the longer side chains to avoid confusion of the products with isomeric ions which could arise from glycine or alanine analogs.

  a. *Pseudo Glycine Ions.* Glycyl residues seldom cause problems, except when they are in the C-terminal position. N-Acyl peptide esters of the latter type give a low abundance of the aldimine ion, $A_n{}^+$.[11] Also, *O,N*-permethylation of peptides has been reported to give *C*-methylation more often for C-terminal glycyl peptides.[29, 69] In peptides containing consecutive glycine residues, the $B_1{}^+$ ions in which the glycines have become the terminal residues are weak.

Identification of a glycyl residue in the peptide chain should be done with care, as many residues may lose their side chain with hydrogen rearrangement to yield an apparent glycine residue.[2, 3, 39, 66, 77, 89–91] This occurs at saturated side chains of Val, Leu, and Ile. The resulting enolic ion can also lose H_2O,

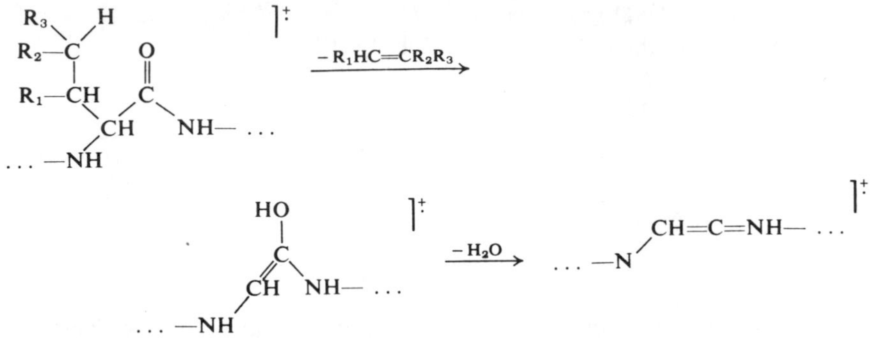

$$\left[\begin{array}{c} R_3 \quad H \\ R_2{-}C \qquad O \\ R_1{-}CH \quad C \\ CH \quad NH{-}\cdots \\ \cdots{-}NH \end{array}\right]^{+\cdot} \xrightarrow{-\,R_1HC{=}CR_2R_3}$$

$$\left[\begin{array}{c} HO \\ C \\ CH \quad NH{-}\cdots \\ \cdots{-}NH \end{array}\right]^{+\cdot} \xrightarrow{-\,H_2O} \left[\cdots{-}N \begin{array}{c} CH{=}C{=}NH{-}\cdots \end{array}\right]^{+\cdot}$$

	R_1	R_2	R_3	Loss of mass units	Loss of mass units
Val	CH_3	H	H	42	60
Leu	H	CH_3	CH_3	56	74
Ile	CH_3	CH_3	H	56	74

SCHEME 19.

but this is not a reliable fragmentation to differentiate such apparent glycine residues from true glycine-containing peaks. Thus peptides containing Val (or Leu, Ile) will contain a series of ions 42 and 60 (or 56 and 74) mass units below the expected $A_i{}^+$, $B_i{}^+$, and $M^{+\cdot}$ ions. Such pseudo glycyl residues can also arise by a similar rearrangement elimination at Ser and Thr, as described in Section II,B,3.

Note that the mechanism shown in Scheme 19 can take place at amino acids that are *not* at a terminus, and for which it may be difficult to write a formal localized charge or radical at the rearrangement site (such rearrangements have been termed "incipient radical site" reactions[92]). Thus if $R_i =$ Leu, not only will such a reaction at the C-terminus of the fragment ion[17] yield $(B_i - 56)^+$ and $(A_i - 56)^+$ ions, but other $(B_{i+n} - 56)^+$ and $(A_{i+n} - 56)^+$ ions should also be formed by this rearrangement in larger ions.[90, 91] For example, the tetrapeptide[91] *n*-Dec-Leu-Ala-Ala-Ala-OMe shows abundant ions at $(M - 56)^+$, $(B_3 - 56)^+$, $(B_2 - 56)^+$, but only a small $(B_1 - 56)^+$. Other similar examples on Val, Leu, Thr, Ser, and Met peptides show clear evidence that such side-chain elimination can occur for many $M^{\ddagger}$, $A_i{}^+$, and $B_i{}^+$ ions as long as at least one of these residues is in any position *in the ion*. For this rearrangement elimination of a stable neutral molecule, as well as for similar reactions cited elsewhere in this review, it is possible that some of the observed product ions have resulted from *thermal* eliminations.

When Ile or Val are in the N-terminal position the odd-electron ion $CH_3CO—NH—CH=C=O^{\ddagger}$ (one hydrogen less than the corresponding product shown in Scheme 19) has been observed by Heyns and Grütz-macher.[77, 89] The abundance of this ion does not appear to be a reliable way, however, to distinguish N-terminal Leu and Ile. It is also reported[5, 11] that branched alkyl radicals such as $C_3H_7 \cdot$ can be lost from valine-containing ions.

b. Distinguishing Leu and Ile. In general the mass spectra of peptides which differ only by a leucine residue replaced with isoleucine are very similar (Fig. 2). Differentiation of such oligopeptides by MS is a long-standing problem, which has been generally found to be difficult unless reference spectra of the isomers are available.

A possible differentiating reaction is the partial elimination of the side chain of Leu or Ile. Leu preferably eliminates a propyl radical or a neutral propylene (thus leaving an apparent alanyl residue), whereas Ile preferably eliminates an ethyl radical or a neutral ethylene (thus leaving an apparent norvalyl residue), as shown in Scheme 20. Such differences are often very

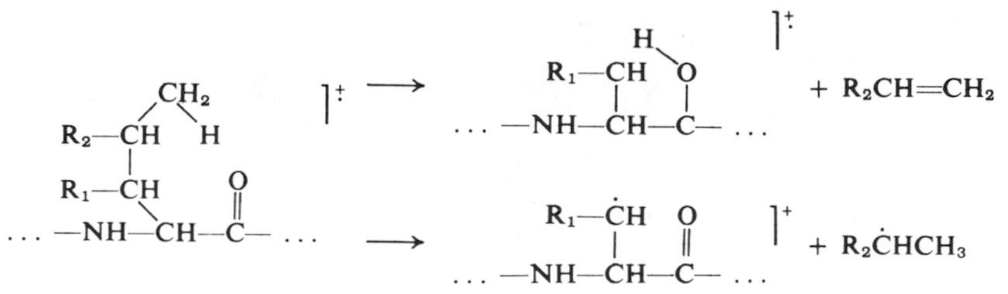

SCHEME 20.

small, and these product ions are less important than those of the reaction shown in Scheme 19, so that they do not provide unambiguous evidence for the presence of Leu or Ile. Previous studies by Weygand *et al.*[66, 83] had pointed out other fragment ions that could give evidence for the distinction of these isomers when they are in the N- or C-terminal position. The TFA N-terminal Ile ion formed by reaction shown in Scheme 5 should give[66] products of m/e 154 and 168 in forming a conjugated ion by the reaction shown in Scheme 21, while the Leu counterpart ion should give m/e 140 by

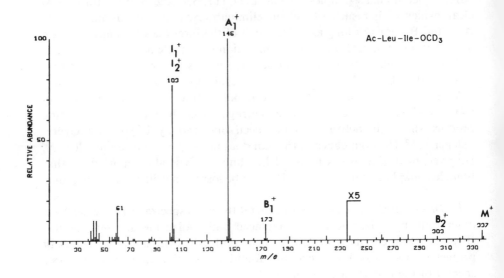

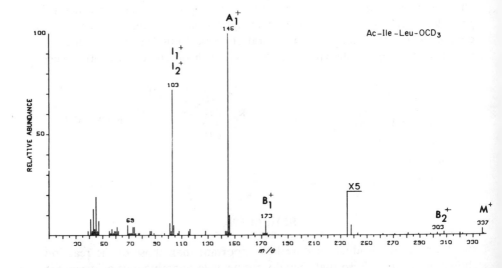

FIG. 2. Mass spectra of the dipeptides Ac-Leu-Ile-OCD$_3$ and Ac-Ile-Leu-OCD$_3$ illustrating the increased abundance of A$_i^+$ ions in small peptides. As discussed in Section II,B,2, the EI spectral behavior of Leu and Ile are often very similar, and special techniques, such as the MI or CA spectra of the immonium ion at m/e 103, are required to distinguish these.[84]

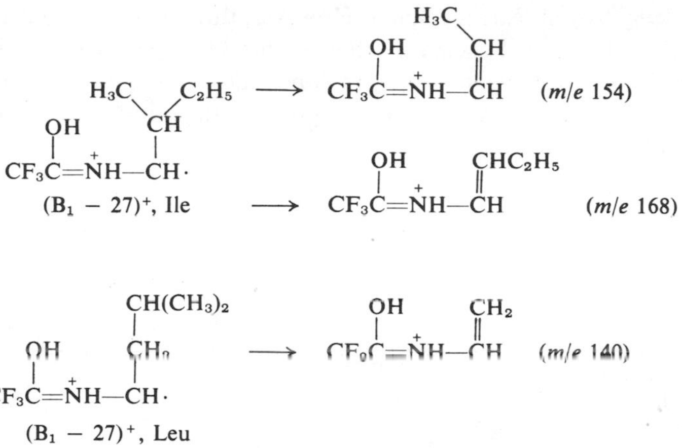

SCHEME 21.

the loss of $C_3H_7\cdot$. Also, the formation of $(M - 127)^+$ by double hydrogen rearrangement, Scheme 22, should be more favorable for C terminal Ile because a tertiary hydrogen is involved.[83]

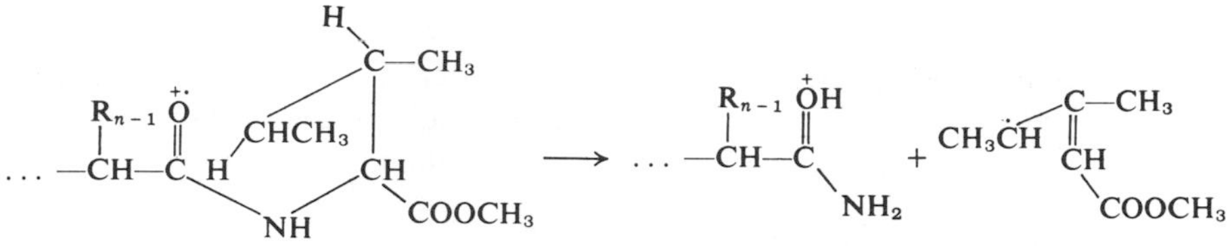

SCHEME 22.

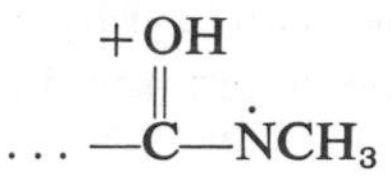

Recent studies have shown that Leu and Ile can be unambiguously distinguished in a peptide containing only one of the two isomers by use of the MI spectrum, or even better from the CA spectrum of the immonium ion.[84] As shown in Scheme 23, isomeric immonium ions I^+ from a Leu or a Ile peptide undergo metastable decomposition via different routes. The presence of a "metastable peak" at m/e 36.1 ($103 \rightarrow 61$) indicates a Leu residue, whereas m/e 64.3 ($103 \rightarrow 69$) is much more intense in Ile than in Leu. Other transitions in the MI and especially in the CA spectra can provide additional data for distinction of such isomers. An MI and CA reaction favored for C-terminal Ile apparently involves the formation of an $(M - 128)^+$ ion, by a single hydrogen rearrangement. However, this reaction closely resembles the reaction shown in Scheme 22, the double hydrogen rearrangement observed by Prox and Weygand for N-TFA peptide esters[83]; because of the poor resolution of MI and CA spectra it is possible that this[84] is actually $(M - 127)^+$ ion formation.

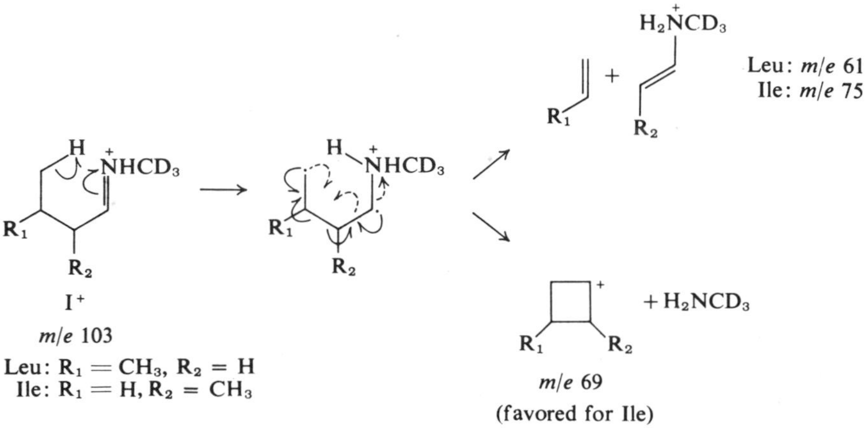

SCHEME 23.

c. *Proline.* N-Acyl peptide esters with a C-terminal Pro sometimes give small ions corresponding to the opening and the contraction of the heterocyclic ring (Scheme 24).[5, 11, 91, 93] The unusual abundance of the immonium

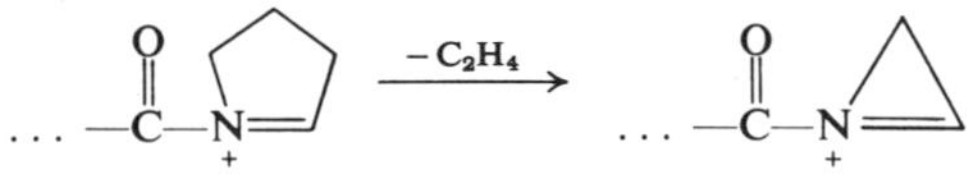

SCHEME 24.

ion from Pro has been already noted. Normal sequence ions (B_i^+ and A_i^+) are sometimes small in Pro peptides.[77]

3. Hydroxyamino Acids

The hydroxy group and, in permethylated peptides, the methoxy group are characteristic of serine, threonine, hydroxyproline, and homoserine. For Ser and Thr, rearrangement of the hydroxyl hydrogen gives side-chain elimination (Scheme 25) of the type noted for Val, Leu, and Ile (see Scheme 19).[1, 15, 17, 93, 95]

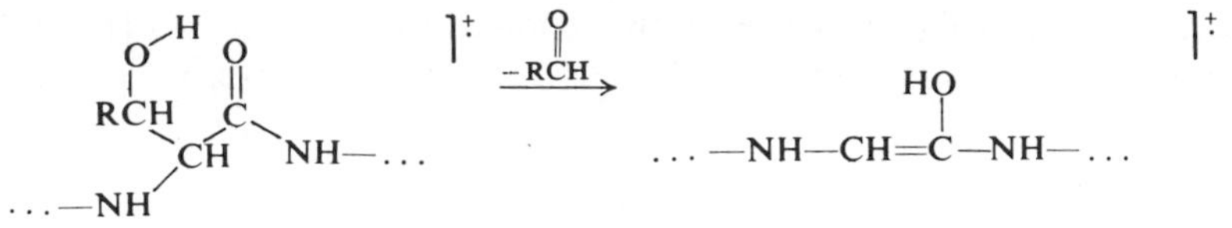

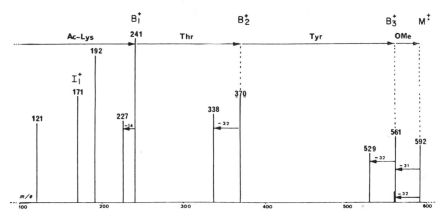

Ser: R=H; Thr: R=CH₃

SCHEME 25.

Loss of water[5, 6, 11, 77, 93] and, in permethylated peptides, methanol[6, 29, 69] (Fig. 3) is common for these residues, but is not a highly specific characteristic. The loss of water in *N*-acetyl peptides which do not contain hydroxyls was described in Section II,A,6, and loss of water in Asp peptides is

FIG. 3. Simplified mass spectrum of an *O,N*-permethylated *N*-acetyl tripeptide, Ac-Lys-Thr-Tyr-OCH₃[69]; isotopic ions and ions of < 1% relative abundance have been omitted. Note the loss of CH₃OH from the M⁺·, B₂⁺, and B₃⁺ ions which contain Thr. Also note the Tyr pseudo sequence C-terminal peak at *m/e* 192 (Section II,B,6) and the characteristic ion of Tyr at *m/e* 121.

described in Section II,B,5. These ions are more intense for higher source block temperatures,[3, 34] thus suggesting that the above eliminations can be thermally induced.

An aldimine ion minus 17 found in *N*-acetyl peptides containing Ser and Thr has been explained as shown in Scheme 26.[77]

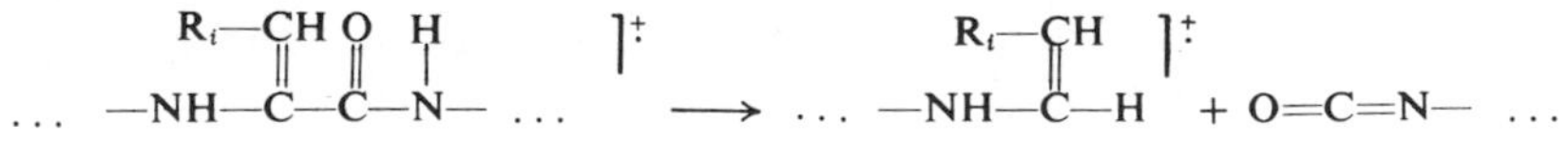

(M − H₂O)⁺; Ser: R = H; Thr: R = CH₃

SCHEME 26.

In a manner similar to the reactions reported for Val (Section II,B,2),[5,11] Thr can also lose its side chain, $CH_3CH(OH)\cdot$, by α-cleavage.[93]

4. Sulfur-Containing Amino Acids

Methionine-containing peptides may undergo a partial or total loss of the side chain, as a radical or as a neutral molecule. The loss of 48 and/or 74 amu observed in normal[5,11,17,85,91,93] and permethylated[94-96] peptides, and the simultaneous presence of *m/e* 61, is a good indication of the presence of a methionine residue in the chain (Scheme 27). These peaks are particularly valuable as Met produces only a weak immonium ion (*m/e* 104).[85]

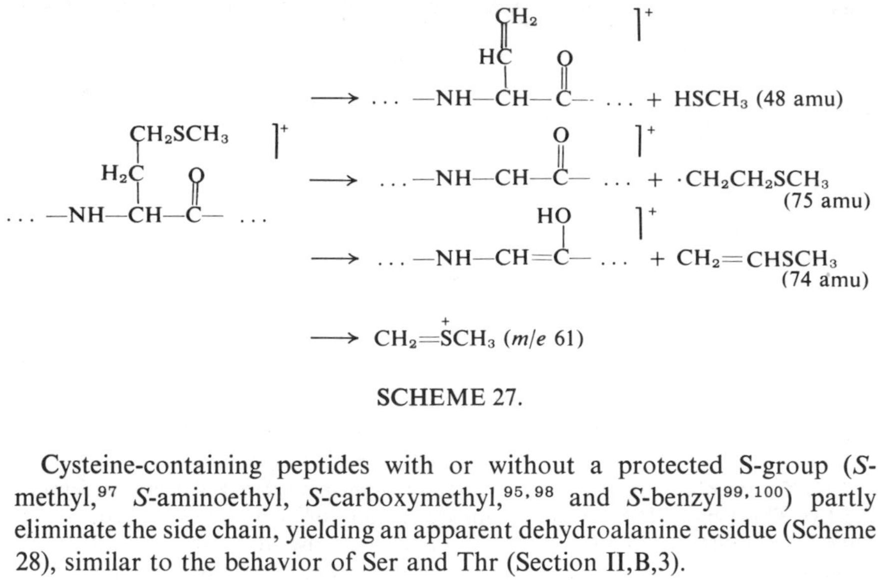
CH2SCH3
H2C
O
... —NH—CH—C— ...
CH2
HC
O
... —NH—CH—C— ... + HSCH3 (48 amu)
O
... —NH—CH—C— ... + ·CH2CH2SCH3
(75 amu)
HO
... —NH—CH=C— ... + CH2=CHSCH3
(74 amu)
CH2=S+CH3 (m/e 61)

SCHEME 27.

Cysteine-containing peptides with or without a protected S-group (*S*-methyl,[97] *S*-aminoethyl, *S*-carboxymethyl,[95,98] and *S*-benzyl[99,100]) partly eliminate the side chain, yielding an apparent dehydroalanine residue (Scheme 28), similar to the behavior of Ser and Thr (Section II,B,3).

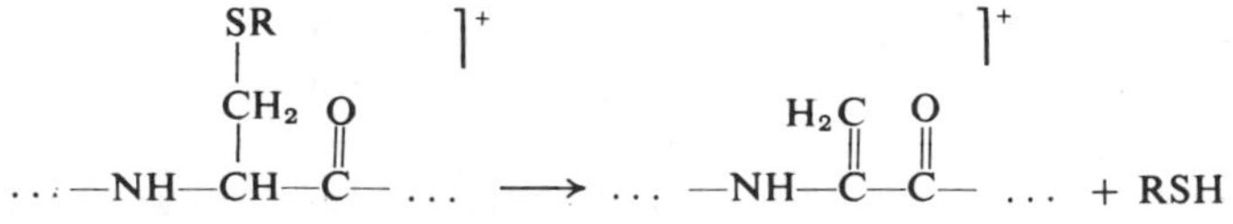
SR
CH2
O
... —NH—CH—C— ...
H2C
O
... —NH—C—C— ... + RSH

SCHEME 28.

5. Dicarboxylic Acids and Amides

The sequence peaks can provide differentiation between α- and ω-linkages in Asx and Glx peptides (see Section II,A,2). The ω-acid group is usually derivatized during methyl esterification; the resulting group can lose under

EI a small neutral molecule (H_2O, CH_3OH) or a small radical ($CH_3O\cdot$, $CH_3OOC\cdot$).[72, 101–103]

a. Asp, Asn. The β-terminal group of β-methylaspartyl and β-methylasparaginyl residues may be converted to a ketene; the loss of 32 is often more favored than the simple loss of 31 ($\cdot O—CH_3$) (see Scheme 29).[101–103]

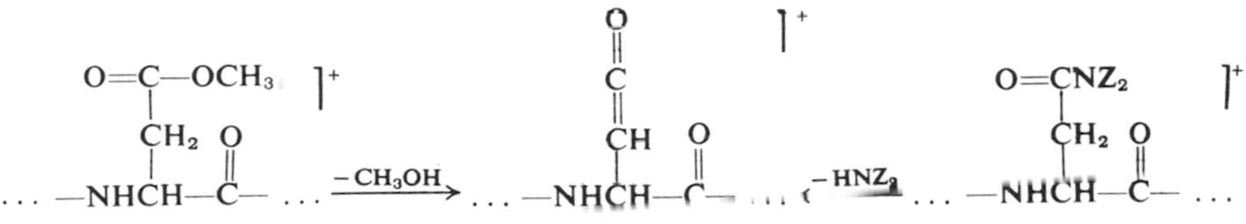

$$O=C-OCH_3 \;]^+$$

$$\underset{\displaystyle \cdots -NHCH-C-}{\overset{\displaystyle CH_2 \; O}{}} \cdots \xrightarrow{-CH_3OH} \cdots \underset{\displaystyle -NHCH-C-}{\overset{\displaystyle CH \;\; O}{\overset{\displaystyle C}{\overset{\displaystyle \|}{\overset{\displaystyle O}{}}}}} \cdots \xrightarrow{-HNZ_2} \cdots \underset{\displaystyle -NHCH-C-}{\overset{\displaystyle CH_2 \; O}{\overset{\displaystyle O=CNZ_2}{}}} \cdots \;]^+$$

SCHEME 29.

Pseudo amino acid fragmentation (see Scheme 8) in Asp and Asn peptides yields characteristic ions (Scheme 30), and sequence ions starting from them.[69, 78]

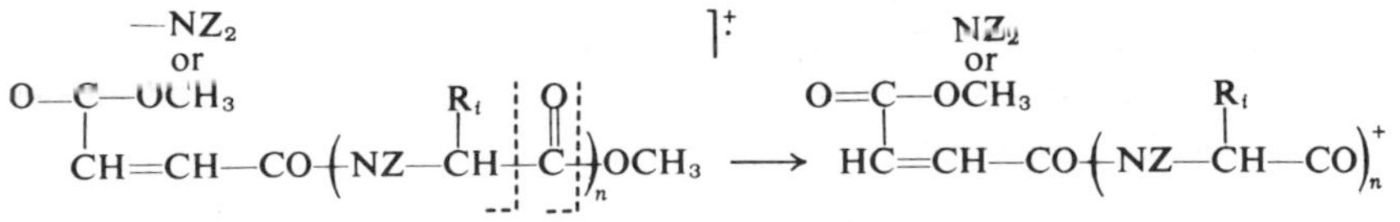

$$\cdots \xrightarrow{\hspace{3cm}} \cdots$$

Asp: *m/e* (144 + residue); B_i^+ = *m/e* (113 + residue)
Asn, Z = H: *m/e* (129 + residue); B_i^+ = *m/e* (98 + residue)
Z = CH_3: *m/e* (157 + residue); B_i^+ = *m/e* (126 + residue)

SCHEME 30.

b. Glu, Gln. The γ-group of Glx may also be converted to a ketene[101]; however, in this case the subsequent elimination of ketene can also occur[5, 11] to yield an apparent dehydroalanyl residue. This reaction is useful for characterization, as it does not occur for γ-linked Glx peptides (Scheme 31).

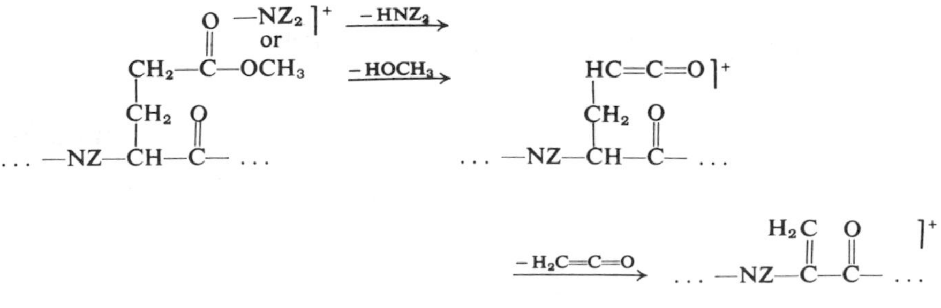

SCHEME 31.

Thermal cyclization of α-(γ-methylglutamyl)-containing N-acyl peptides i· the mass spectrometer apparently is the origin of the (glutamyl − 18)⁺ peak observed in their spectra. This elimination does not occur in γ-(α-methylglutamyl) residues, providing an additional method for distinguishing the α- and γ-linkage at Glu (Scheme 32).[103]

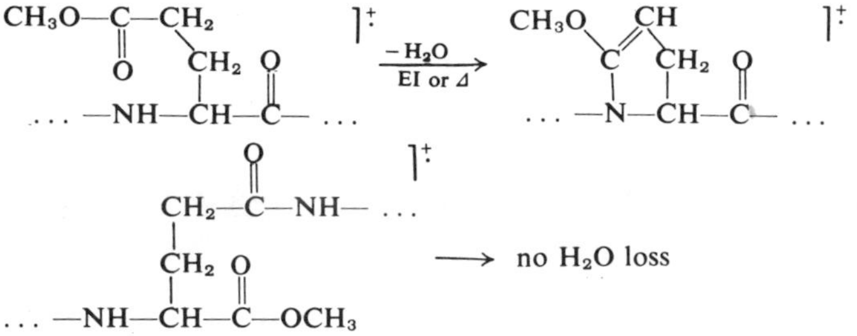

$$
\begin{array}{c}
\text{CH}_3\text{O}-\overset{\displaystyle\parallel}{\underset{\displaystyle O}{C}}-\text{CH}_2 \\
\underset{\displaystyle\ldots -\text{NH}-\text{CH}-\overset{\parallel}{\underset{O}{C}}- \ldots}{\overset{\displaystyle\text{CH}_2}{} \;\; O}
\end{array}
\Bigg]^{\ddagger}
\xrightarrow[\text{EI or }\Delta]{-\text{H}_2\text{O}}
\begin{array}{c}
\text{CH}_3\text{O} \quad \text{CH} \\
\text{C} \quad \text{CH}_2 \quad O \\
\ldots -\text{N}-\text{CH}-\text{C}- \ldots
\end{array}
\Bigg]^{\ddagger}
$$

$$
\begin{array}{c}
\text{CH}_2-\overset{O}{\overset{\parallel}{C}}-\text{NH}- \ldots \\
\text{CH}_2 \quad O \\
\ldots -\text{NH}-\text{CH}-\overset{\parallel}{C}-\text{OCH}_3
\end{array}
\Bigg]^{\ddagger}
\longrightarrow \text{ no H}_2\text{O loss}
$$

SCHEME 32.

N-Terminal γ-(α-methylglutamyl) and α-(γ-methylglutamyl) can undergo a hydrogen rearrangement whose final products clearly distinguish the two peptides, as shown in the mass spectra of N-ethoxycarbonyl dipeptide esters[104]; mechanisms other than Scheme 33 are also possible for the formation of the (M − 148)⁺ ion.

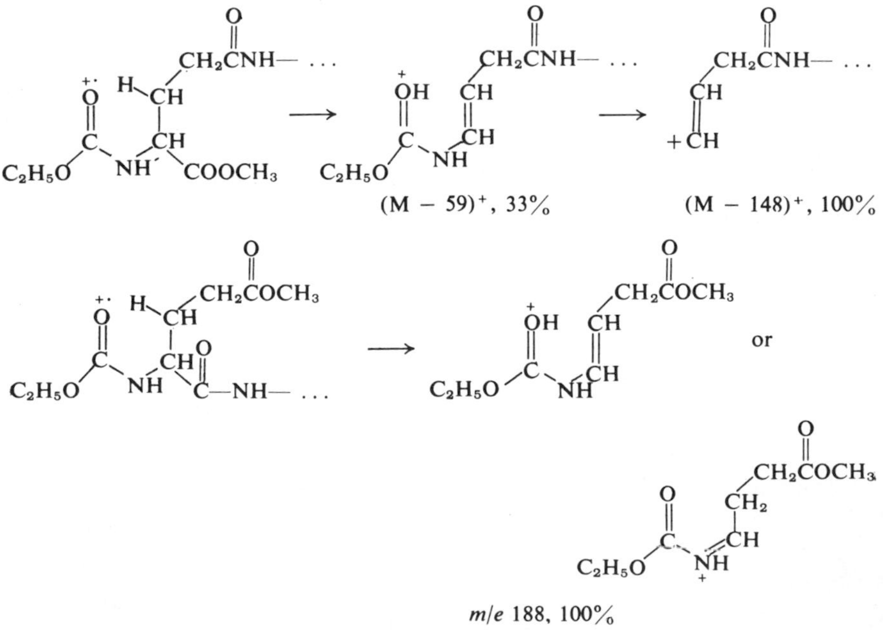

$(M − 59)^{+}, 33\%$ $(M − 148)^{+}, 100\%$

or

m/e 188, 100%

SCHEME 33.

N-Terminal Glu and Gln in *N*-acetyl peptides and permethylated peptides also show a series of sequence ions starting at a N-terminal pyroglutamyl group (Scheme 34; compare with Scheme 32). The initial ion of the series at

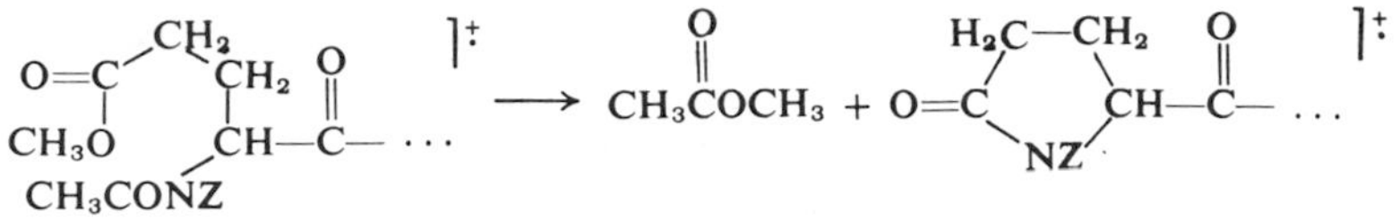

Glu; Gln, loss of CH_3CNZ_2

SCHEME 34.

m/e 112 (Z = H; if Z = CH_3, m/e 126) is weak or is not observed, although the corresponding ion at m/e 84 (98 if Z = CH_3) from loss of CO is characteristic for N-terminal Glx and pyro-Glu residues.[13, 30, 38, 39, 42, 78a, 85, 105–115] The relative intensities of normal sequence ions and sequence ions with a *N*-pyrrolidone carboxylic end are difficult to predict as the cyclization (easier with Gln) may occur in the mass spectrometer or during a derivatization step. Natural pyroglutamyl peptides do not pose any unusual problem and show sequence ions starting at m/e 112 (126). Because Edman degradation is not applicable to blocked amine ends, mass spectrometry in this specific case is uniquely useful as a sequencing tool.

6. *Aromatic and Heterocyclic Amino Acids*

Peptides containing Phe, Tyr, His, or Trp undergo a series of specific cleavages. The cleavage of the benzylic C_α—C_β bond with charge retention on the aromatic residue accounts for the very characteristic and abundant ions found in the spectra of both permethylated[34, 39, 78] and nonpermethylated peptides.[66, 76, 77, 89, 93, 116] Alternatively, it has also been reported for acyl peptide esters that the elimination of the side chain as a radical may leave ions

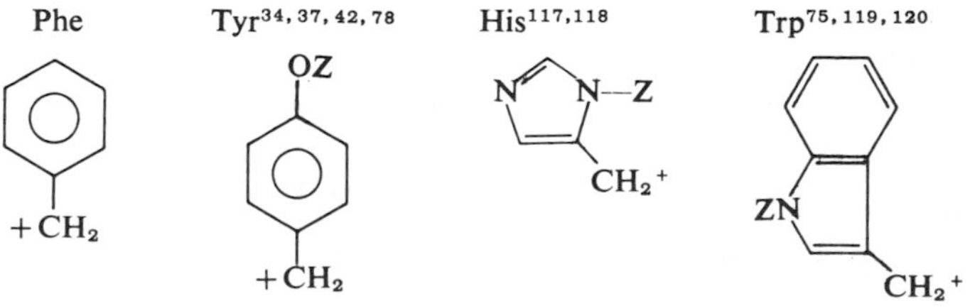

	Phe	Tyr[34, 37, 42, 78]	His[117,118]	Trp[75, 119, 120]
				m/e
Z = H:	91	107	81	130
Z = CH_3:	91	121	95	144

at $(M - 91)^+$ and $(M - 81)^+$ in N-acyl peptide esters containing Phe or His, respectively. However in Tyr and Trp the cleavage of the C_α—C_β bond is more often accompanied by a hydrogen rearrangement, thus leaving $(M - 106)^{\pm}$ and $(M - 129)^{\pm}$ ions.[11] In the case of Tyr, deuterium labeling[66] has shown that this involves the migration of the phenolic hydrogen.

Pseudo amino acid type fragmentations (see Scheme 7) also produce characteristic abundant ions reflecting the partial sequence between the aromatic or heterocyclic residue and the C-terminal end. This is most marked at Trp, then at Tyr (see Fig. 3).

The simultaneous presence in the spectrum of one of these pseudo N-terminal ions and of a satellite at (NH—Z) amu higher indicates the occurrence in the peptide chain of two adjacent aromatic or heterocyclic residues,[78] as explained by the reaction shown in Scheme 9.

Another reaction (Scheme 35) reported for N-acyl peptides containing Tyr involves a rearrangement of two hydrogen atoms.[11] The further loss of NH_3

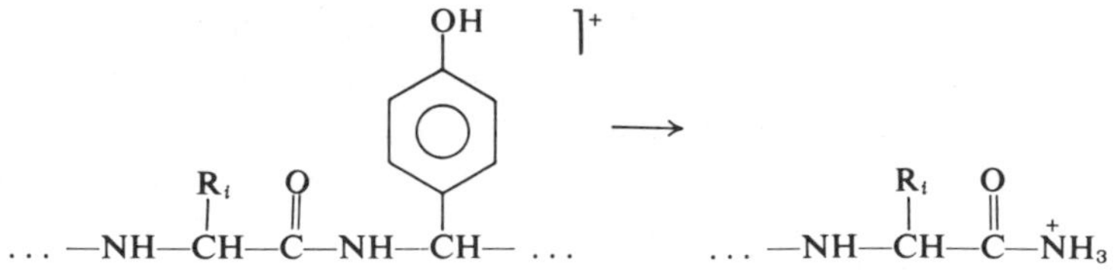

SCHEME 35.

to give the $B_i{}^+$ ion provides evidence for this mechanism, although the reaction may instead involve the sequence shown in Scheme 22.

Histidine (and to a lesser extent tryptophan and N_δ-pyrimidyl-Orn) N-acyl peptides (nonpermethylated) have been reported to show satellite peaks at $+14$ amu (and even $+28$) for all ions containing that residue, including the molecular ion.[5, 28, 93, 121] A mechanism suggesting a bimolecular alkylation from a methyl of the esterifying C-terminal acid group is supported by deuterium labeling.[28] The simultaneous occurrence of intermolecular N-methylation and peptide dehydration in His peptides accounts for the vanishingly small molecular ion and for the misleading $(M - 4)^{\pm}$, probably $(M + 14 - 18)$, ion observed in their spectra.[121] O,N-Permethylated histidine (and also Trp) N-acetyl peptide esters exhibit satellite peaks at $+14$ or -14 amu owing to quaternary ammonium formation[6, 39] or incomplete N-methylation during the permethylation reaction (see Section V,D,2).

7. Basic Amino Acids

Derivatized Lys and Orn do not cause any unusual problems, except when they are in the C-terminal position. In such a case cyclization may occur, as reported for a permethylated peptide with a C-terminal Orn.[122] The resulting ion subsequently undergoes the main amino acid type fragmentations (Scheme 36).

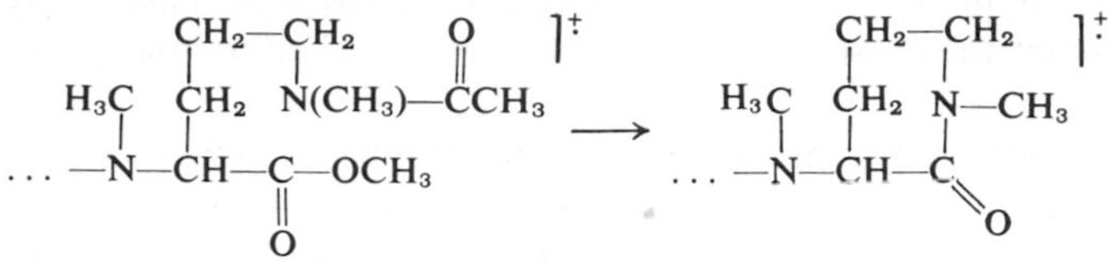

$$
\left[
\begin{array}{c}
CH_2{-}CH_2 \quad\quad O \\
| \quad\quad | \quad\quad\; \| \\
H_3C \quad CH_2 \quad N(CH_3){-}CCH_3 \\
\;\;\; | \quad | \\
\cdots {-}N{-}CH{-}C{-}OCH_3 \\
\quad\quad\quad\quad \| \\
\quad\quad\quad\quad O
\end{array}
\right]^{\ddagger}
\longrightarrow
\left[
\begin{array}{c}
CH_2{-}CH_2 \\
| \quad\quad | \\
H_3C \quad CH_2 \quad N{-}CH_3 \\
\;\;\; | \quad | \\
\cdots {-}N{-}CH{-}C \\
\quad\quad\quad\quad\;\; \backslash\!\!=\!\!O
\end{array}
\right]^{\ddagger}
$$

SCHEME 36.

$$
\left[
\begin{array}{c}
O \\
\| \\
CH_3C{-}NH \quad (CH_2)_{1\ or\ 2} \\
\backslash \quad\quad\quad / \\
O \quad CH_2 \quad CH_2 \quad O \\
\| \quad\quad | \quad\quad\; \| \\
CH_3C{-}N{-}CH{-}C{-} \cdots \\
\quad\quad | \\
\quad\quad H
\end{array}
\right]^{\ddagger}
\longrightarrow \longrightarrow
\begin{array}{c}
(CH_2)_{1\ or\ 2} \\
/ \quad\quad\quad \backslash \\
O \quad CH_2 \quad CH_2 \\
\| \quad\quad | \quad\quad / \\
CH_3C{-}N{=}CH \\
\quad\quad +
\end{array}
$$

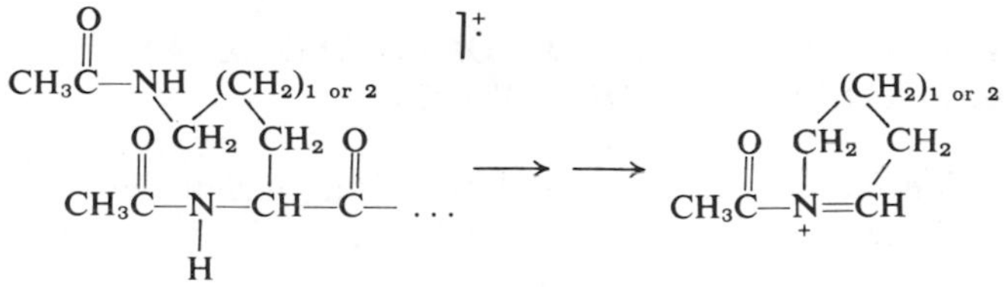

SCHEME 37.

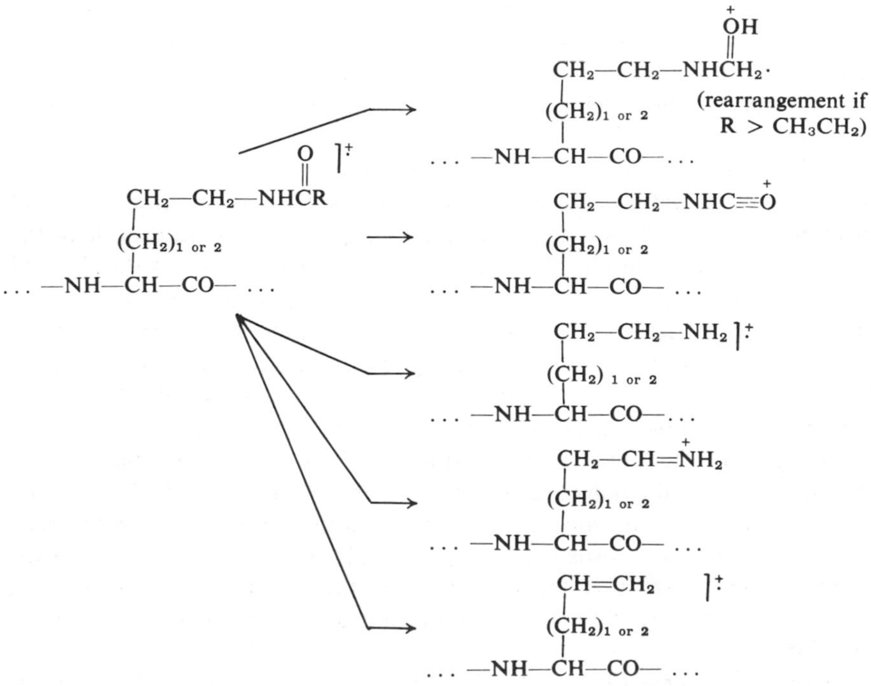

SCHEME 38.

Acetyl peptides with an N-terminal Lys or Orn exhibit an abundant ion[93, 123] which is the cyclization product of A_1^+ (Scheme 37).

In addition to the normal sequence peaks, minor ions due to the elimination of neutral molecules, or small radicals from the ω-acyl amide group have been reported in the spectra of N-acyl peptides[123] containing Lys and Orn (Scheme 38).

Underivatized arginine adversely directs the fragmentation; at best, sequence peaks up to but not including the arginine residue may be observed. Derivatization or removal of the Arg residue, whenever possible, is advisable (see Section V,A).

III. CHEMICAL IONIZATION OF DERIVATIZED AND UNDERIVATIZED OLIGOPEPTIDES

Chemical ionization (CI) is a recent promising technique which has attracted considerable interest over the past few years.[124, 124a] Ionization of the sample molecules is effected in an ion plasma which is created in the ion source by electron bombardment of a reagent gas at a relatively high pressure (~ 1 Torr). For example, CH_4 as a reagent gas produces CH_5^+ ions, which protonate the sample molecules to give $(M + H)^+$ ions. Some reagent gases such as n-pentane cause hydride abstraction, giving $(M - H)^+$ ions.[124] The technical modifications to convert a conventional EI mass spectrometer into CI mode are now well documented,[124] and most commercial instruments now offer that option.

"Direct chemical ionization" is a promising method for obtaining mass spectral data on less volatile peptides, including some underivatized tetra- and pentapeptides.[36, 47] This is accomplished by exposing the solid peptide to the ion plasma directly in the ion source, which apparently lowers the thermal energy requirements of the vaporization process. The solid can be introduced into the ion source on the outside of the probe; alternatively, the sample is dissolved in a solvent used as the ionizing reagent, and the solution is sprayed into the ion source. Friedman and co-workers have shown that the surface on which the sample is deposited for ion source evaporation is critical; using Teflon lowers the sample temperature required for vaporization.[125, 125a] The solution introduction of the sample for direct CI thus avoids the necessity of vaporization of the solid sample from another surface; the very dilute solutions (0.01–0.1%) usually employed should give microcrystallites of the sample immediately after the flash vaporization of the solvent in the ion source. Sample sizes of 10^{-8}–10^{-9} g are sufficient to obtain a mass spectrum by direct CI.

A limited number of derivatized [31, 39, 126] and underivatized[36, 85, 125, 125a, 127–131] peptides have been examined by CI. The most obvious advantage of

CI in comparison to EI is the much larger abundance of the even-electron $(M + 1)^+$ or $(M - 1)^+$ ions in CI versus the odd-electron $M^{\ddagger}$ ion from EI; a separate CI run is often made to provide more reliable molecular size information. Most reported CI studies have utilized methane as the reagent gas; the CH_5^+ and $C_2H_5^+$ produced protonate the sample molecules to give the $(M + 1)^+$ peak and low abundance satellites at $(M + 15)^+$ and $(M + 29)^+$. The fragmentation pathways under CI are far less well documented than the fragmentations under EI; however, some general types of cleavages have been discerned.[124a] Dehydration is pronounced in underivatized and N-acyl peptide esters to yield $(M + H - H_2O)^+$; the loss of two H_2O molecules can also be observed if Thr or Ser is present. The $(M + H - H_2O)$ ion is of low abundance or absent in O,N-permethylated peptides, but the loss of CH_3OH is then common. Sometimes other neutral molecules ($HCOOH$, NH_3, H_2S, etc.) are also easily lost from the $(M + H)^+$ ion of an underivatized peptide; such even-electron fragment ions are usually the most abundant in the CI spectrum.

Because of the relatively large number of collisions occurring in the higher pressure CI source, the sample ions are largely thermalized and their internal energies are substantially influenced by the source temperature. The choice of reagent gas is also critical; a strong acid (H_3^+ from hydrogen or CH_5^+ from methane) or a weak acid ($C_4H_9^+$, $CH_3OH_2^+$, or CH_3CNH^+ from isobutane, methanol, or acetonitrile) gives $(M + H)^+$ ions of high or low internal energies, respectively, so that the amount of fragmentation can be varied.[132] Although many substances exhibit a simpler CI spectrum than EI spectrum, this is not necessarily true for peptides (compare Figs. 4a and 4b). The fragmentation of the peptide backbone is the main CI cleavage, paralleling the behavior under EI (Scheme 39). However, although this seldom results

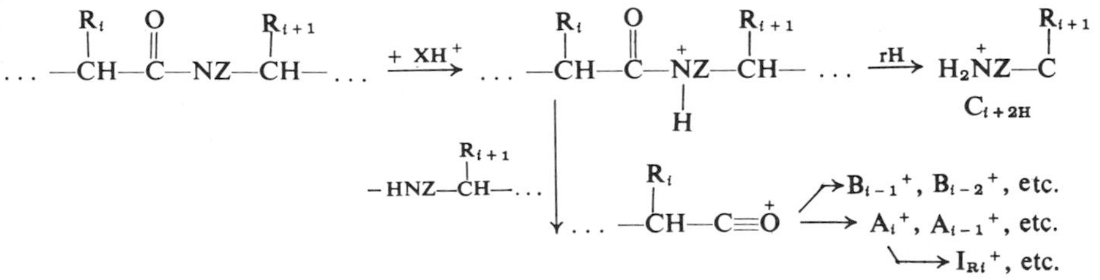

SCHEME 39.

in abundant C-terminal ions under EI, with CI the C-terminal fragments (Fig. 4b, also Figs. 5 and 6a) produce common ions with the rearrangement of two hydrogen atoms; a complete set of C_{i+2H}^+ ions are often observed under CI for both derivatized and underivatized peptides.

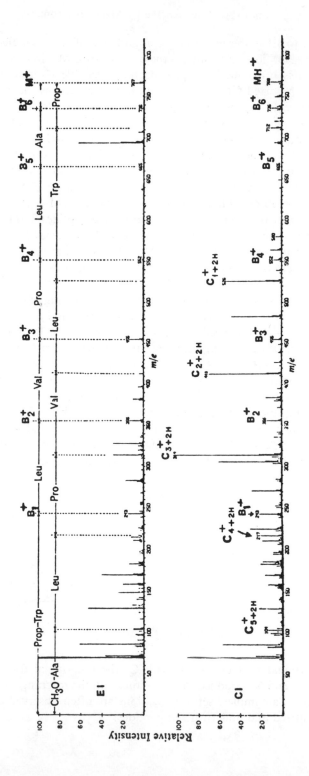

FIG. 4. The EI and CI mass spectra of C_2H_5CO-Trp-Leu-Val-Pro-Leu-Ala-OCH_3 (Milne *et al.*[126]; ion source temperature $250°–300°$; methane as the ionizing gas).

Kiryushkin et al.[126] have suggested that the formation of C_{i+2H}^{+} involves the migration of the hydrogen atom from the neighboring peptide bond (Scheme 40), based on the relatively low abundance of C_{0+2H} (loss of the N-acyl group with rearrangement of 2H), and the absence of C_{i+2H} from the unique O,N-permethylated peptide they had prepared.

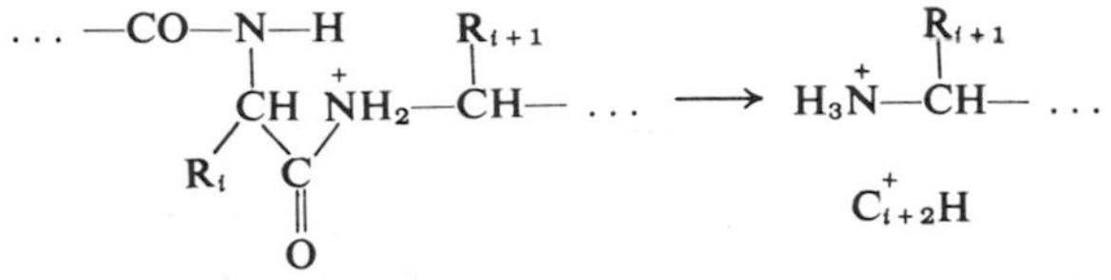

SCHEME 40.

However, C_{i+2H} ions (though of relatively lower abundance) have been observed by others in a series of O,N-permethylated N-acetyl peptide esters,[31, 39] which suggests that the hydrogen atom may migrate from a more remote source, or from the next α-carbon atom, in a fashion similar to the expulsion of an immonium ion from the aldimine ion (Scheme 41; see Section II,B,1). These two complementary sets of sequence ions are an attractive advantage of the CI spectra, although this increases the possibility of isobaric C_{i+2H}^{+} and B_i^{+} ions (for example in H-Met-Met-Leu-OH). A_i^{+} ions are in general small or absent.

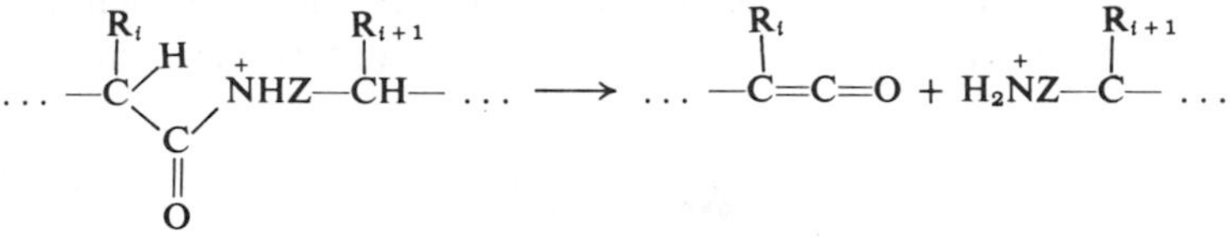

SCHEME 41.

The B_1^{+} ion is of low abundance or absent in the CI spectra of underivatized peptides. However, in the CI spectra of N-acetyl peptide esters[126] and O,N-permethylated N-acetyl peptide esters[31, 39] (Figs. 4b, 5 and 6a), in general B_i^{+} ions are observable, and are often among the most abundant ions. The B_2^{+} ion is the base peak in most of the underivatized peptides,[36, 125a, 128, 133] and often the most abundant ion in N-acetyl peptides (permethylated[31] or not[126]), but its abundance relative to that of B_3^{+} and higher B_i^{+} ions decreases from N-propionyl to N-decyl peptides.[126] This suggests to us that peaks corresponding to B_2^{+} ions arise (Scheme 42) in part by cyclization of $(M + H)^{+}$ to the protonated diketopiperazine ion $DKP^{+}_{i \cdot i+1}$. The formation of a $DKP^{+}_{1 \cdot 2}$ from dipeptides is reported[128] to be independent of the source temperature.

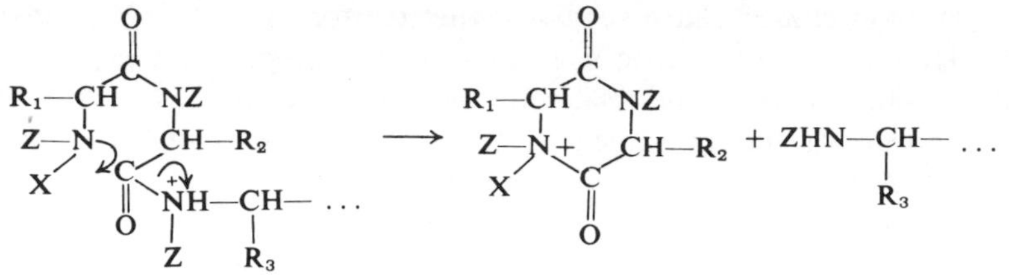

SCHEME 42.

It appears that underivatized tri-, tetra-, and pentapeptides also give ions corresponding to all of the dipeptide fragments made with peptide pairs adjacent in the main chain,[36, 133] possibly in the form of their protonated DKP. B_2^+ is isobaric with the DKP of the first two N-terminal residues, but other DKP's may be found at masses other than B_i^+ or C_i^+. These DKP ions may in turn lose water (see, for example, Fig. 6a). The simultaneous presence of all of these ions can seriously complicate the interpretation of the spectrum of an unknown sample. Although it seems that the formation of a protonated DKP is induced by the reagent gas, the resulting situation is similar to the original studies on underivatized peptides by EI where thermal cyclization would yield DKP ions (nonprotonated) and complex mass spectra.[2, 134, 135, 137a]

Because of lack of data, CI fragmentations characteristic of the different residues are not fully elucidated. The elimination of a neutral molecule (H_2O, $HSCH_3$, CH_3OH, $HCOOH$, NH_3) is easier under CI than under EI, especially for underivatized peptides, but molecular rearrangements should be

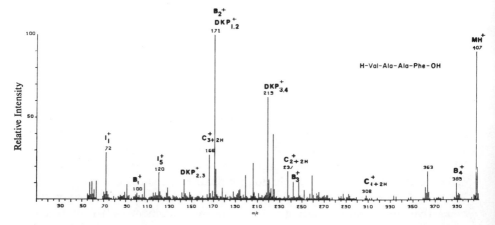

FIG. 5. The direct chemical ionization spectrum of underivatized Val-Ala-Ala-Phe (Baldwin and McLafferty[36]; source temperature 160°; methane as the ionizing reagent gas).

less favored. Immonium and other ions characteristic of particular residues have also been observed[36,128]; they provide information similar to that derived from their presence in EI spectra.

Figure 4 compares the CI (methane) and EI spectra of an *N*-acyl peptide ester.[126] Figure 5 shows the direct CI spectra of an underivatized peptide,[36] and Fig. 6 shows a comparison of the field ionization (FI), FD,[35,133] and CI spectra of an underivatized pentapeptide. Note in Fig. 6 that neither the CI nor FI spectrum gives an $(M + H)^+$ peak; and, as discussed in the next section, although FD does give this valuable peak, in this case it shows only partial sequence information.

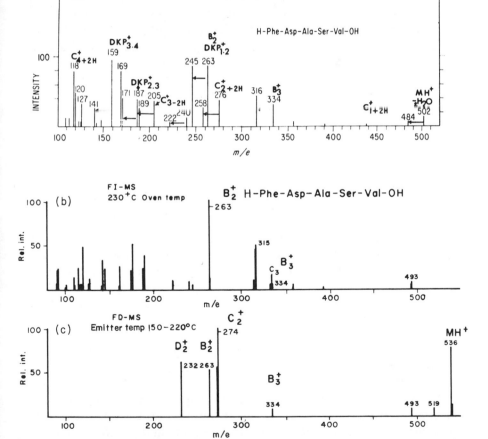

FIG. 6. Mass spectra of underivatized Phe-Asp-Ala-Ser-Val: (a) proton transfer mass spectrum, bombardment with NH_4^+ in a tandem mass spectrometer[133]; (b) field ionization mass spectrum, inlet temperature 230°C[35]; (c) field desorption mass spectrum, emitter temperature 150°–220°C.[35]

The preliminary reports by different groups concerning the possible sequencing of underivatized peptides, thus avoiding the disadvantages of the derivatization procedures, were very hopeful. It appears, however, that a much simpler and more meaningful spectrum is generally obtained from a derivatized sample.

IV. FIELD DESORPTION IONIZATION OF UNDERIVATIZED OLIGOPEPTIDES

Field ionization is a process in which an electron is removed from a gaseous molecule electrostatically in a high field gradient ($\sim 1\,V/\text{Å}$, overall field of ~ 10 kV) produced at a sharp point or edge.[35, 124a] Quantum tunneling allows the electron to be removed at minimum energy, so that the molecular ions produced are at near thermal energies and exhibit a very low degree of fragmentation; fragment information can be obtained from collisional activation spectra, however.[136] FI usually does not supply complete sequence information on derivatized oligopeptides, and thus has been applied mainly to supplement information from EI.[65b]

However, the recent development of field desorption (FD)[35] may be a highly promising breakthrough, as in particular cases it has provided sequence information on large underivatized oligopeptides.[35, 137, 137a] In FD the sharp FI points are semiconducting organic microneedles grown by vacuum pyrolysis; the peptide is deposited from solution directly on these emitter needles. These are given a high (~ 10 kV) positive potential relative to the ion source exit slit, and the sample apparently is ionized by direct tunneling of the electron to the needle surface, the ions being repelled from the positive emitter toward the exit field. Winkler and Beckey[35] obtained an intense molecular ion from the underivatized nonapeptide Arg-Gly-Gly-Gly-Pro-Gly-Gly-Gly-Arg, with partial sequence information. However, heating of the emitter increases fragmentation,[35, 137, 137a] so that in particular recent cases complete sequence information is reported to be obtainable by FD. A dramatic example[137a] is the underivatized nonapeptide bradykinin, Arg-Pro-Pro-Gly-Phe-Ser-Pro-Phe-Arg. About 10^{-6} g of sample was required, but freeze drying of the sample solution on the emitter apparently reduces this to $\sim 10^{-7}$ g or possibly less.[137]

The real utility of FD for peptide sequencing is rapidly being clarified by vigorous research programs in several laboratories. Progress has been made on a number of serious problems. Reactions at the emitter surface producing bimolecular products are found to be dependent on surface conditions, temperature, and impurities. Complete removal of inorganic cations such as Na^+ and K^+ is critical. The fragmentation pattern is a sensitive function of emitter temperature, and ion production fluctuates, so that sophisticated data collection techniques are helpful; photoplate recording of high resolution data appears to be especially desirable.[137] $M^{\ddot{+}}$ and/or $(M + H)^+$ ions are formed; at present it is not possible to predict the number of hydrogen atoms (0–3) transferred to or from the ionized fragment produced in backbone or side-chain cleavage at a particular emitter temperature. Continuous sample introduction, such as utilized in GC/MS and LC/MS (Section VI) is not possible for FD. A variety of problems are apparent in data interpretation at present,[138, 139] but it is obvious that FD has a very exciting potential for obtaining sequence information from peptides whose volatility is much too low for other techniques.

V. DERIVATIZATION METHODS

Except for special techniques such as FD and direct CI (Sections III and IV), peptides are not directly amenable to mass spectrometry. Free peptides are of low volatility[2, 134, 134a, 137a] due to their zwitterionic structure[140] and intermolecular hydrogen bonding.[72] To increase sample volatility, a variety of pretreatments have been proposed for cyclic peptides, peptide antibiotics,[141] peptide alkaloids,[142–145] glyco-[146] or lipopeptides, and oligopeptides from the cleavage of long-chain natural proteins. For the latter, as mentioned earlier, long chains must be broken into smaller units. For direct probe sample

Zwitterionic structure of N,C-terminus unsubstituted peptides

Hydrogen bonding between amide bonds

introduction into the mass spectrometer, derivatized oligopeptides of 8–10 amino acid residues often yield complete sequence information, while partial information can be obtained from peptides with as many as 20 residues (the number depends on the particular residues present). For GC separation prior to MS introduction, special derivatives make it possible to examine further di-, tri-, and some tetrapeptides.

Derivatization always involves protection of the N-terminus amino group (unless it is naturally blocked), and usually esterification of the free carboxylic group. Further derivatization procedures (permethylation, reduction, etc.) are less general, with the recommended procedure dependent on the laboratory performing the sequencing. There have been exhaustive studies of different protective groups and their MS behavior. Many sophisticated derivatives, which have been proved to be useful in peptide chemistry, have been tried: some may find use in specific cases, where enhancement of the molecular ion signal or of specific fragment ions are desired; the comparison between mass spectra of several derivatives of the same peptide (when enough material is at hand) may give complementary information. However, for a generally applicable strategy for the sequencing of unknown peptides by MS, particular attributes are desirable. A good derivative (1) should be thermally stable and increase the volatility of the peptide; (2) should give abundant and recognizable sequences ions (series of ions starting with the C- and/or N-terminus amino acid residue); (3) should give a recognizable molecular ion; and (4) should be produced from reactions giving minimum side products which are applicable at the submicromolar level with good yields (for example, reactions which involve addition of volatile reagents to the peptide sample followed by evaporation of coproducts and excess reagents).

Historically, the first general strategy, proposed by Biemann and co-workers in the late 1950's was the complete reduction to polyamines.[2, 17, 88] Although for general use this has been superseded by other methods, it has recently found use in an approach involving GC separation, and will be discussed in Section VI,C.

Because of the large amount of work reported, it is impossible to review in detail here the preparative methods, the difficulties encountered with particular structures, and the MS fragmentation characteristics for all of the derivatives tried for peptides. Comparative studies have been made,[79, 86] taking different peptides as models for a variety of different amine or acid derivatives. These studies indicate that N-acetyl peptide methyl esters are the best compromise in terms of ease of preparation, volatility, and MS fragmentations. An important modification is O,N-permethylation (which makes esterification unnecessary), whose utility has been demonstrated by the classic work of Lederer's group and more recent results from other laboratories.

A. Modification of Problem Amino Acids

The presence in the peptide of certain trifunctional amino acids such as the sulfur-containing (Cys, Met) and basic amino acids (His, Arg) often cause reduced volatility and more complex mass spectra. It has been proposed that these be removed when they are in terminal positions,[6, 11, 16] providing that the amino acid composition or the identity of the amine and acid ends are known, or that pure tryptic peptides are investigated (Arg would be at the C-terminus). Thus removal of the C-terminal acid by carboxypeptidase B, or of part of the C-terminal end by carboxypeptidase A + B, or elimination of an N-residue by one step of Edman degradation are suitable methods.[11, 69]

It is more common that the problem residue is in a nonterminal position, in which case its chemical modification is an alternative solution. Some derivatization schemes proposed earlier have since been superseded by more general methods, especially permethylation (Section V,D). Thus with this the complex method[147] suggested for opening the ring of histidine is unnecessary. Sulfur-containing peptides are smoothly desulfurized through reduction with Raney nickel in dimethylformamide at 20°.[6, 122, 148, 149] Methionine gives aminobutyric acid residues,[77] but cysteine gives alanine,[150] leading to an ambiguity with naturally occurring alanine unless deuterated Raney nickel is used[6, 11]; again, most laboratories no longer find these special schemes necessary for these residues.

The one remaining exception is arginine, whose derivatization has been studied in a number of laboratories. Although diacylation of the guanidine group of Arg has been unsuccessful,[151] two ways appear satisfactory for converting arginine residues into compounds amenable to the mass spectrometer.

Hydrazinolysis (Scheme 43) of the guanidino group is rapid (15–60 minutes)[95, 151]; this yields an ornithine residue which can be acylated, and if necessary permethylated. If acidic residues (Asp, Glu) are present in the peptide, care should be taken to avoid peptide bond cleavage[95, 122, 123, 151]; 10–60% peptide bond cleavage has been reported.[17] A C-terminal Arg gives upon hydrazinolysis a C-terminal Orn, which can undergo cyclization (Scheme 43).[122]

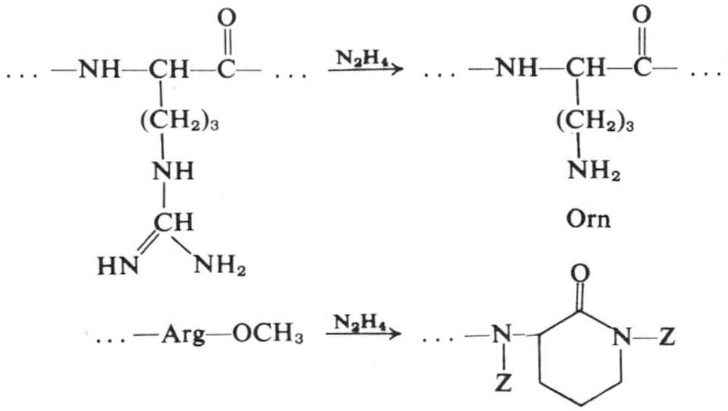

SCHEME 43.

Condensation of the guanidine group with dicarbonyl compounds to a N_δ-2-pyrimidinyl ornithine derivative is the alternative method.[5, 6, 95, 151–154] 1,1,3,3-Tetramethoxypropane is convenient, except that under acidic conditions it is not suitable for Trp-containing peptides (Scheme 44A).[123, 151] Condensation with 1,2-cyclohexanedione (Scheme 44B)[154, 155] or acetylacetone (Scheme 44C)[33, 42, 46, 95, 123, 151, 153] under mildly alkaline conditions is also used. With the latter reagent, which yields a N_δ-2-(4,6-dimethylpyrimidyl) ornithine[11, 33, 42, 60, 95, 152, 153] residue, the reaction requires several hours.[95]

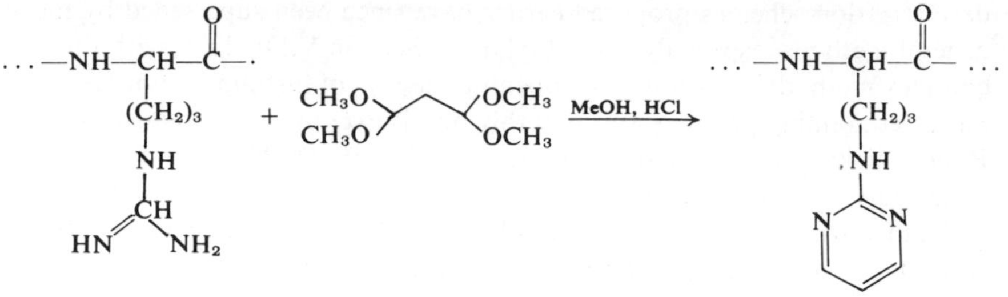

N_δ-2-pyrimidinyl ornithine

SCHEME 44A.

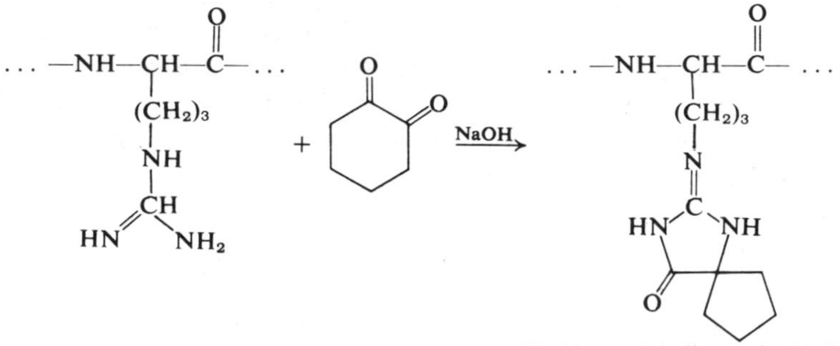

N_δ-(4-oxo-1,3-diazaspiro[4.4]non-2-ylidene) ornithine

SCHEME 44B.

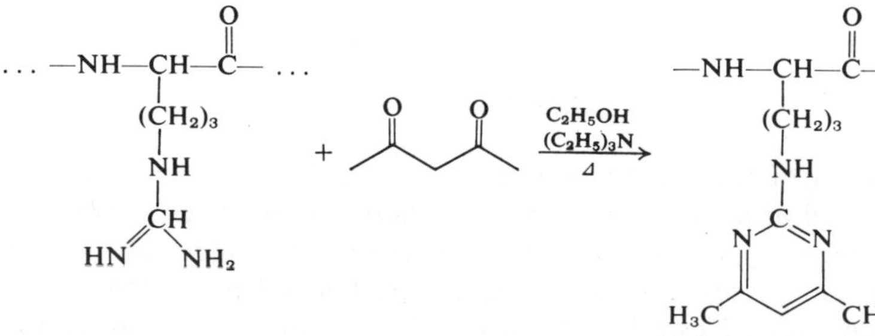

N_δ-2-(4,6-dimethylpyrimidinyl)ornithine

SCHEME 44C.

B. N-Terminus Blocking Groups

Compromises between fragmentation and volatility have always been a problem. Trifluoroacetyl esters were among the first derivatives tried, as they were known to increase greatly the volatility of polar groups. They were first successfully tried on di- to tetrapeptides by Andersson[140] and Stenhagen,[81] then carefully studied by Prox and Weygand in 1965.[79,83] N-Acetyl peptides were introduced by Heyns and Grützmacher in 1963.[89] Probably the most dramatic breakthrough came in 1965 when the Russian and French workers[90,156] found useful sequence information from the spectra of peptides containing long-chain (8–20 carbon atoms) N-acyl groups. The large N-acyl group moves all of the N terminal sequence peaks to higher masses, minimizing their confusion with small fragment ions bearing no sequence information (immonium ions, side-chain fragments). Mixed 1:1 N-acyl derivatives[72] such as $n\text{-}CH_3(CH_2)_mCO$ plus $n\text{-}CH_3(CH_2)_{m+1}CO$, or $CH_3(CH_2)_pCO$ plus $CD_3(CH_2)_p\,CO$ also can provide labeling of the N-terminal end. The advantages derivable from such long-chain N-acyl, groups were amply demonstrated by elucidating the structure of several natural peptidolipids (see Section V,B,2).

1. N-Trifluoroacetylation

N-Trifluoroacetyl derivatives are generally more difficult to prepare than their N-acetyl counterparts. They are obtained by reaction with $(CF_3CO)_2O$ in CF_3COOH, or CF_3COOCH_3 in CH_3OH under mildly alkaline (pH 8) conditions; methyl esterification is usually also performed on the carboxyl end.[71,79,81,83,99,100,158,165] The reaction requires 15 minutes under reflux or 7–8 hours at room temperature.

From the point of view of volatility, a TFA peptide gives a spectrum at 30–50° below the source block temperature necessary for a N-acetyl peptide, and even below that for an N-acetyl permethylated peptide. The unique elemental composition of this N-terminal blocking group also serves to label the amine end ($CF_2Br\,CO$ would be even more advantageous).[71] However, the interpretation of their mass spectra, especially for larger peptides, is less straightforward than with the other N-blocking groups, in part due to the high electron-withdrawing power of the TFA group. Only a few practical applications of TFA derivatives to biological problems have been reported. The structures of a trifunctional[165] and a tetrafunctional collagen cross-link have been elucidated from the mass spectra of their 1:1 trifluoro- and difluoroderivatives.[73,79] The TFA derivative of the pentadecapeptide scotophobin, thought to be a brain-coded compound causing a specific behavior in rats, gave a series of small di-, tri-, and tetrapeptide fragments under mass spectrometric conditions, probably from pyrolytic cleavage in the source. The recombination of these fragments led to a proposed structure [13,160,161] which has been severely criticized.[166]

2. Long-Chain N-Acyl Groups

Long-chain *N*-acyl derivatives have played a key role in the history of peptide sequencing by mass spectrometry. The elucidation of the structure of fortuitine[157] was the first real sequence problem solved by mass spectrometry.

Fortuitine: $CH_3(CH_2)_n$—CO-Val-(Me)Leu-Val-Val-(Me)Leu-Thr(Ac)-Thr(Ac)-Ala-Pro-
OMe   ($n = 18, 20$)   (157)

From M. Johnei: CH_3—$(CH_2)_n$—CO-Phe-Ile-Ile-Phe-Ala-OCH$_3$ ($n = 14, 16, 18, 20$)
(172)

Peptidolipin NA: CH_3—$(CH_2)_n$—

CH—CH$_2$—CO-X-Val-Ala-Pro-alloIle-Y-NH—CH—CO—O
                                                    CHOH
                                                    CH$_3$

	X	Y	n	
NA	Thr	Ala	16	(170)
Val⁶	Thr	Val	16, 17, 18	(171)
α-AB¹	α-AB	Ala	16, 17, 18	(α-AB = α-Aminobutyryl) (171)

Isariin: $CH_3(CH_2)_8$—CH—CH$_2$—CO-Gly-Val-Leu-Ala-Val-O   (102)

Mycoside C$_b$ (partial structure): $CH_3(CH_2)_n$—   O—
                        CH=CH—CH(OCH$_3$)—CH$_2$—CO-Phe-Thr-Ala   (173)

Surfactin: $(CH_3)_2$—CH—$(CH_2)_9$—

CH—CH$_2$—CO-Glu-Leu-Leu-Val-Asp-Leu-Leu-O   (168)

Viscosin: $CH_3(CH_2)_6CH(OH)CH_2CO$-
                    Leu-Glu-Thr-Val-Leu-Ser-Leu-Ser-Ile   (169)

                                            O
Esperin: $C_{10}H_{21}$—CH—CH$_2$—CO-Glu-Asp-Val-Leu-Leu-OH   (174)

Stendomycin: $(CH_3)_2CH$—$(CH_2)_{10}$—
        CO-Pro-(Me)Thr-Gly-Val-Ile-Ala-ΔAbu-Thr-Val-Val-Thr-Ser-Ile-B   (177)

B =
HN
N   N   NH$_2$ COOH
CH$_3$ CH$_3$

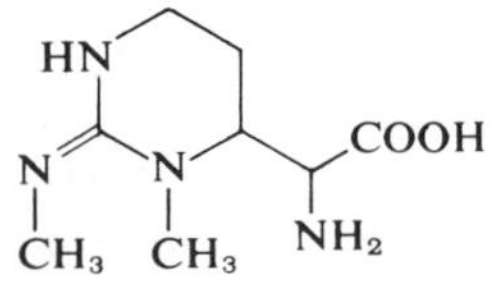

SCHEME 45.

In a concurrent significant development, Shemyakin and co-workers recognized most of the fragmentation pathways of the side chains of the commonly found amino acids by the examination of a large collection of synthetic N-acyl peptide esters.[5, 11, 12, 93, 101, 123, 156, 167]

N-Blocking with large acyl groups is best obtained by reacting the peptide with a succinimide ester.[72, 156] The most serious drawback of such derivatives is the increase in the weight of the molecule; in many cases the high mass peaks, including the molecular ion, cannot be observed. The main practical application of such derivatives has been as models for mass spectrometry. As a consequence the structures of a number of natural peptidolipids containing large N acyl groups have been elucidated,[4, 6, 135a, 157, 168–179] but for natural peptides with a free amine end it was not necessary to add such a large increment to an already complex molecule. The sequences of the peptides Ile-Leu-Gly-Asp-Val-Phe from pig pepsin, and Leu-Glu-Ala-Leu-Lys and Val-Glu-Glu-Arg from pig heart amino transferase have been elucidated after such N-acylation and methylation.[167] With the development of a detailed understanding of peptide fragmentation mechanisms, the effect of the large acyl groups was no longer needed to decipher the mass spectrum.

Structures of a number of natural peptidolipids have been elucidated or confirmed (totally or in part) from their mass spectra (see Scheme 45) (references given in parentheses to the right of the reaction).

3. Varieties of Derivatives

Acetylation, to be discussed in the next section, is by far the most widely used N-acyl derivative. However, the mass spectra of a surprising variety of derivatives of synthetic peptides have been reported: t-butyloxycarbonyl,[65b] methoxycarbonyl,[86] ethoxycarbonyl, [86, 180] benzyloxycarbonyl,[65b, 67, 79, 86, 87, 181] methylaminocarbonyl,[86] phenylaminocarbonyl,[86] phthaloyl,[67, 79, 86, 182] benzoyl,[67, 79, 182] pentadeuteriobenzoyl,[183] thiobenzoyl,[62] cyclohexanecarbonyl,[67] pentafluorobenzoyl,[67, 79] p-chlorobenzoyl,[67, 79] 2,4-dinitrophenyl,[67, 79, 184] 1-naphthoyl,[67] 2-naphthoyl,[67] adamantoyl,[68] naphthalene-1,8-dicarboxyl,[67] naphthalene-2,3-dimethylene,[67] benzenesulfonyl,[67, 79] tosyl,[67] dansyl (DNS),[67] 1-naphthalenesulfonyl,[67] benzylidene,[67] salicylidene,[67] β-indolylmethylidene,[67] p-nitrobenzylidene,[67] p-cyanobenzylidene,[67] α-phenyl-p-dimethylaminobenzylidene,[67] p-dimethylaminocinnamilidene,[67] p-methoxybenzylidene,[67] 2-, 3-, and 4-pyridylmethylidene,[67, 185] acetylacetonyl,[63, 67] p-dimethylaminobenzylidene (DMB)[67, 185] 4-dimethylamino-1-naphthylidene (DMN),[67, 185] 2-hydroxy-1-naphthylidene,[67] guazulylacetyl,[186] trimethylsilyl,[43, 69a, 187] heptafluorobutyryl,[82] pentafluoropropionyl,[46] N-[2-(3-alkylpyragen)-2-on-1-yl]acyl,[188] and chlorodifluoroacetyl.[73]

Many of the derivatives are those used in peptide analysis and synthesis. Aromatic N-derivatives were tried in an attempt to stabilize the positive charge at the amine end, and thus enhance the abundances of the sequence-determining peaks and the molecular ion; the best results have been obtained with dansyl (DNS), p-dimethylaminobenzylidene (DMB), and 4-dimethyl-amino-1-naphthylidene (DMN) derivatives. Instability or difficulties of synthesis have discouraged general use for most. The Schiff base derivatives recently suggested by Day[67, 185, 189] and co-workers have not yet been tried by others.

Very few of this long list of possible derivatives have actually been used in real problems. Mass spectral sequence information has been obtained from the N,S-ethoxycarbonyl derivative of an intracellular peptide from the myce-lium of cephalosporium sp2.[104] A tetrapeptide (Ala-Ser-Asp-Orn) from the peptide chain of a glycopeptide from the posterior lobe of pig pituitaries was fully sequenced from the mass spectra of its O,N-permethylated N-ethoxy-carbonyl derivative.[78a] The sequence of S-carboxymethylated porcine insulin A chain was obtained after enzymatic cleavage and subsequent GC/MS of the pentafluoropropionyl dipeptide esters.[46] Trimethylsilyl (TMS) derivatives of peptides appear to give poor results in sequencing,[69a, 187] and incomplete reaction[69a] for O,N-pertrimethylsilylation; however Biemann[43] has reported interesting results from the selective O-TMS of polyaminoalcohols obtained by reduction of small peptides (see Section VI,C).

4. N-Acetylation

The acetyl derivative is by far the most common means used for N-blocking of peptides.[89] Diverse acetylation conditions have been described, using acetic anhydride in glacial acetic acid,[37, 85, 89] methanol,[122] or water.[42, 190] Simultaneous N-acetylation and methylation of the C-terminus carboxyl group by $(CH_3CO)_2O/MeOH$ is possible but gives unsatisfactory results,[86, 190, 191] and the two reactions should be done in two steps. Lederer's group suggests[19, 192] initial acetylation of the peptide by a methanolic solution (others[97] have suggested pH adjustment by triethylamine) of acetic anhydride (Ac_2O:MeOH, $1:3$[19] or $1:4$[122]) at room temperature for a few hours (varying the time from 15 minutes to 18 hours; others[32] have found 3 hours to be optimum), followed by evaporation of the solvent and reagent under vacuum. A trace of water is added when the peptide is insoluble in methanol alone. Very insoluble peptides and proteins are acetylated in pyridine[64, 69] in 15–30 minutes. Acetylation following methylation of the C-terminus carboxyl group has been described.[28, 37] To distinguish natural O-acetyl peptides,[157] or to label the N-terminus residue for easier interpretation of the mass spectrum,[42, 64, 72] pure trideuterioacetic anhydride or an equimolecular mixture of acetic and trideuterioacetic anhydrides can be used.

In methanol, acetylation of the N-terminal amine group is accompanied by acetylation of the primary amino groups in the side chains of lysine ornithine, and S-β-aminoethylcysteine. The guanidine group of arginine is unaffected, but acetylation on the N_δ-position in N_δ-2-(4,6-dimethylpyrimidyl) ornithine, a derivative of arginine, may occur, and a method which avoids this has been reported.[42] Amide groups of asparagine and glutamine residues, and heterocyclic rings in histidine and tryptophan residues are unaffected. N-Acetylation of N-terminal glutamic[34, 192-194] and glutamine residues and their corresponding mass spectra have been reported. However, cyclization occurs very easily,[42, 64] although it is difficult to know if this happens during derivatization of the oligopeptide, in the cleavage of the original protein (the reaction is rapid under acidic conditions), thermally, or under electron impact in the mass spectrometer. In practice the mass spectrum shows a series of ions starting at a pyrrolidone-carboxylic acid residue (see Scheme 46).

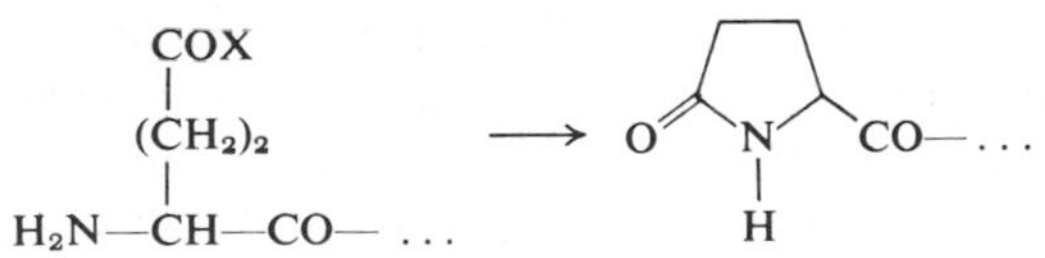

SCHEME 46.

Hydroxyl groups usually are not affected, but O-acetylation of threonine and serine residues may occur, especially during acetylation in pyridine (this has also been reported with methanol[32]). This rarely happens on tyrosine, but natural O-acetyltyrosine-containing peptides are known. If permethylation follows, any O-acetyl groups will be converted to O-methyl groups.

Incomplete reactions[190] and artifacts due to chain cleavages during acetylation in methanol have occasionally been reported.[11, 86, 167, 190, 195] The latter appear to occur when excessive reaction times (more than 10 hours) are allowed, and have not been encountered under Thomas conditions.[122]

C. C-Terminus Blocking Groups

Methylation, as first described by Stenhagen,[81] is the most common reaction for derivatization of the C-terminal carboxylic acid group. Comparative studies[79] have shown that other derivatives such as ethyl,[67] t-butyl,[93] phenyl, and thiophenyl esters,[181] and anilides[159, 196] do not give major improvements in the volatility of the peptide ester, although aromatic groups can better stabilize the positive charge and thus produce more abundant C-terminus ions in the mass spectrum.

When permethylation of the peptide is planned, no esterification step is necessary, as both will be accomplished by this reaction. Any other carboxyl-esterifying group will be also displaced by methyl during permethylation. As a larger increase in volatility is to be expected from the permethylation than from the choice of an ester group other than a methyl, use of variety of peptide esters is rare, with the notable exception of trideuteriomethyl esters.[77] When permethylation is not used, methylation of the carboxyl group is more often done after acylation, but the reverse has also been suggested,[28, 37] especially in cases of very methanol-soluble peptides, such as those containing S-benzyl cysteine.[99] A number of reagents and conditions have been proposed such as reaction in methanol with catalytic amounts of $SOCl_2$,[5, 67] reaction with diazomethane in methanol,[13, 99, 156] reaction in methanolic HCl,[28, 37] or pyrolytic conversion of the trimethylanilinium salt in the solid probe of the mass spectrometer.[159, 196]

Amide bond cleavage during methylation under acid conditions such as MeOH:HCl has been reported,[13, 86, 191] probably due to methanolysis (note that acid solvolysis of an Asp- or Asn-containing peptide, on either side of these residues, is at least 100 times more rapid than for other residues).[197] Modification of carboxyl groups in proteins under very mild conditions has been described,[198] though the procedure (activation of the carboxyl groups by a water soluble carbodiimide, and subsequent displacement by glycine methyl ester) is too complex for routine analysis of trace amounts of peptides.

Carboxyl groups in side chains (Glu, Asp) are esterified simultaneously with the C-terminus. During methylation with diazomethane, N-acetyl-cysteinyl peptides are also S-methylated.[97] N-Methylation of His and L-2-pyrrolidone-5-carboxylyl (PCA) has been noted.[129]

D. O,N,S-Permethylation

Among the "fortunate" factors which contributed to the original success of the structure elucidation of the peptidolipid fortuitine[157] was the fact that three of the nine residues had substituted amido groups (two natural N-methyl leucines and one proline); it was also noted[72] that the tetrapeptide H₂N-allo-Ile-Pro-Sar-MeVal-OMe gives a good mass spectrum, even with a free N-terminal residue. The increase in volatility was attributed[72] to the lack of hydrogen on amide groups; this was confirmed by experimental studies,[199, 199a] and explained in terms[29] of elimination of intermolecular hydrogen bonding between —NH— and $\diagup$C=O groups. It was shown also that O,N,S-permethylation of N-acetyl peptide esters gave simpler mass spectra in many cases, with more abundant sequence-determining ions. These advantages have led to careful studies of the permethylation reaction, and although it still has particular drawbacks, it is at present used in many laboratories. O,N,S-Pertrideuteriomethylation has been recommended for the distinction of natural O- or N-methyl residues[41, 117, 170, 177] and extramethylation artifacts (see Section V,D,2). Permethylation of N-TFA derivatives has been reported to cause cleavage of the TFA group under the strongly basic conditions required for permethylation[44], although the mass spectrum of O,N-permethylated N-TFA bradykinin has been reported.[33] The mass spectrum of an O,N-permethylated N-ethoxycarbonyl peptide has also been used in the sequencing of the peptide part of a glycopeptide,[146] but fails to show clearly superior peaks.

The permethylation reaction involves the action of a strong base on the peptide in an anhydrous medium, with subsequent attack of the N anion on a methyl derivative with a good leaving group. Iodide has been used in all reported cases except one in which dimethyl sulfate was suggested (see Scheme 47).[200]

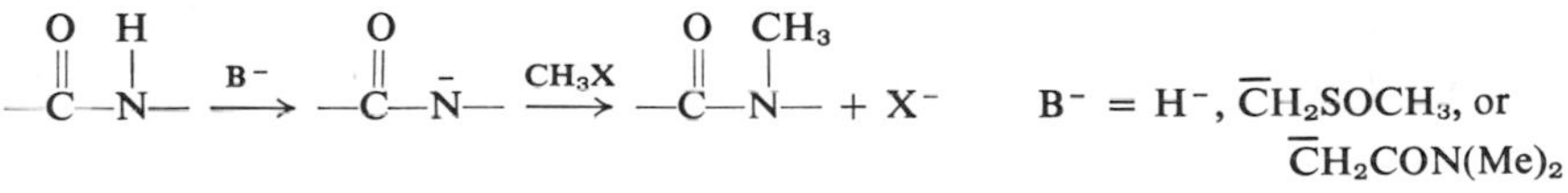

$$\underset{\underset{\text{C}}{\overset{\overset{\text{O}}{\|}}{\,}}-\underset{\underset{\text{N}}{\overset{\overset{\text{H}}{|}}{\,}}- \xrightarrow{\text{B}^-} \underset{\underset{\text{C}}{\overset{\overset{\text{O}}{\|}}{\,}}-\overset{-}{\underset{\text{N}}{}}- \xrightarrow{\text{CH}_3\text{X}} \underset{\underset{\text{C}}{\overset{\overset{\text{O}}{\|}}{\,}}-\underset{\underset{\text{N}}{\overset{\overset{\text{CH}_3}{|}}{\,}}- + \text{X}^-$$

$\text{B}^- = \text{H}^-,\ \overline{\text{C}}\text{H}_2\text{SOCH}_3,\ \text{or}$
$\overline{\text{C}}\text{H}_2\text{CON(Me)}_2$

SCHEME 47.

1. Results

As summarized in Table I, all exchangeable hydrogens are replaced by methyls, yielding the following:

1. N-Methylation of all amide bonds, including acylamide groups in N-terminal residues and the side chains of N-acylated lysine and ornithine (N-acylation must precede permethylation—see below), and formation of tertiary N,N-dimethylamide groups on the side chains of Gln and Asn.

2. N-Methylation of the heterocyclic rings of Trp, His, and modified Arg residues.

3. Methyl esterification of the free carboxyl of the C-terminus residue, and of the carboxyl groups in side chains of Asp and Glu, or displacement by a methyl of other esterifying groups at these positions.

4. O-Methylation of hydroxyl groups in side chains of Thr, Ser, Tyr, or displacement of possible O-acetyl groups (natural O-acetyl residues or artifacts from the N-acetylation reaction).

5. S-Methylation of cysteine residues.

2. Side Reactions

Some unwanted side effects can accompany this reaction, and it is important to be aware of their possible occurrence. None of the experimental conditions described below (Section V,D,3) completely avoids these artifacts for all types of peptides.

TABLE I

Permethylation Reactions

Functional group	Location	Normal products	Artifacts
H_2N-	N-Terminal amines N_δ-lysine N_γ-ornithine	$-N(CH_3)CH_3$	$CH_3-N^+(CH_3)CH_3$ I^- formed easily; prior N-acylation necessary[64]
$CH_3CO-NH-$	N-terminal acylamide N_δ-lysine N_γ-ornithine	$CH_3C(=O)-N(CH_3)-$	Slow reaction with Lys and Orn; possible undermethylation
R_f, $-CH-C(=O)-NH-$	Peptide bond	R_f, $-CH-C(=O)-N(CH_3)-$	
$NH_2-C(=O)-$	Asparagine Glutamine	$(CH_3)_2N-C(=O)-$	
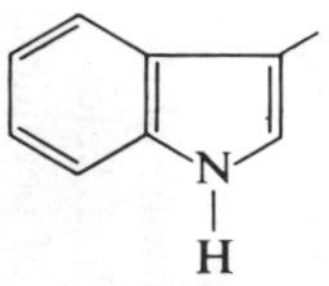 (indole N—H, Tryptophan ring)	Tryptophan	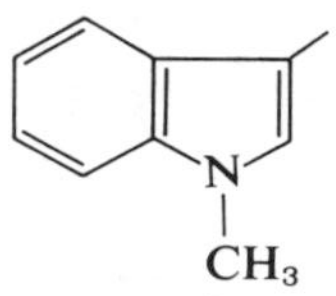 (indole N—CH₃ ring)	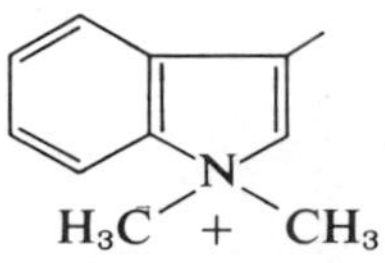 (indole $H_3C-N^+-CH_3$ quaternary, I^-) or incomplete N-methylation

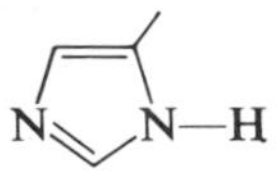	Histidine	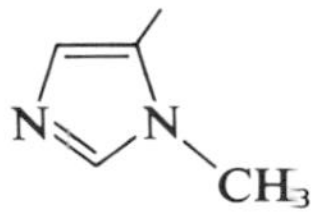	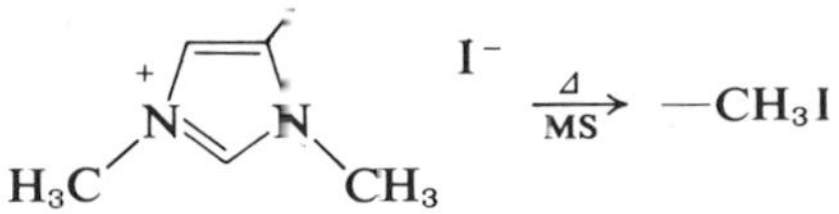 *N*,*N*-Dimethylimidazolium ion Incomplete *N*-methylation is also possible
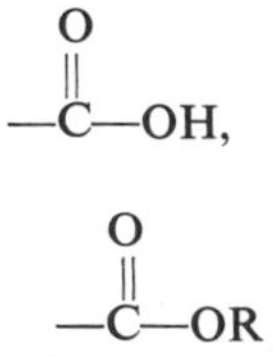	C-terminal carboxyl Asp Glu	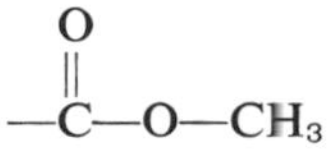	*C*-Methylation of Asp 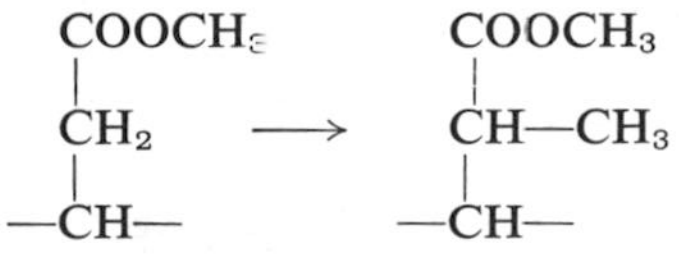
$—CH_2—$	Gly Ala Val	Unchanged	*C*-Methylation
$CH_3—S—$	Methionine	Unchanged	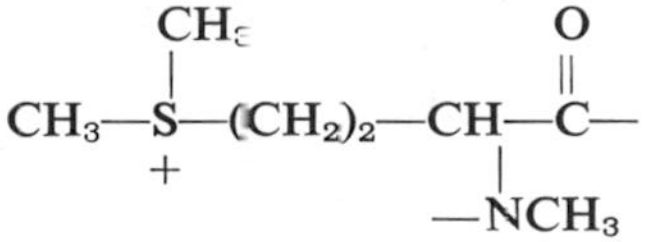 sulfonium 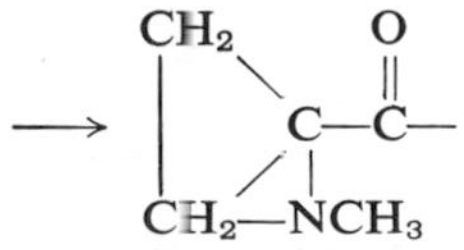 or

TABLE I (*Continued*)

Functional group	Location	Normal products	Artifacts
$HS{-}CH_2{-}CH{-}$, $R{-}S{-}CH_2{-}CH{-}\mid$	Cysteine	$CH_3{-}S{-}$	$CH_2{=}CH{-}CH{-}C({=}O){-}$, also[39] $-NCH_3$
	$R = -CH_2{-}COOMe$		$CH_3{-}S{-}(CH_2)_2{-}C(CH_3){-}C({=}O){-}$ *C*-Methylation $-NCH_3$
	$R = -CH_2CONH_2$		$CH_3{-}S(CH_3){+}{-}CH_2{-}CH{-}C({=}O){-}\rightarrow$ $-NCH_3$
	$R = CH_2{-}NH{-}COCH_3$		$CH_2{=}C{-}C({=}O){-} + (CH_3)_2S$ $-NCH_3$
$-OH$, $-O{-}C({=}O){-}CH_3$	Serine Threonine Tyrosine	$-O{-}CH_3$	Dehydroalanine Displacement of *O*-acylated groups

a. Undermethylation. Heterocyclic rings of histidine, tryptophan,[167] *N*-acetyllysine, and ornithine may undergo incomplete N_ω-methylation,[29,167,193] especially in reactions using an "equimolar" (see Section V,D,3) amount of methyl iodide.[95] This causes lower homologs of the sequence peaks to appear in the mass spectrum. However, *overmethylation* is more frequent and troublesome.

b. C-Methylation. This is the most unfavorable and irreversible artifact. Denaturation of glycine[39,193] (especially facile when Gly is in the C-terminal position[97,194] or next to a proline residue[190]) to produce alanine, of aspartic[5,111] acid to give an isomer of glutamic acid, of valine to produce an isomer of leucine/isoleucine, and of serine to give threonine are common side reactions, as are *C*-methylation of histidine, tryptophan, and glutamic acid.[6,193] Complete methylation on the C_α of methionine in the C-terminal position has also been noted.[39] Pertrideuteriomethylation makes it possible to detect such artifacts.

c. Ammonium and Sulfonium Salts. These salts are water soluble and may be lost during the work-up of the permethylation reaction. They are fairly involatile, yielding poor mass spectra, and thermal elimination in the source of the mass spectrometer may denaturate some residues.

d. Ammonium Salts. Primary amine groups in N-terminal residues and in side chains of lysine and ornithine on permethylation yield *N*-trimethyl quaternary ammonium salts, making preliminary acetylation mandatory.[122] Tryptophan and especially histidine[147] may give quaternary ammonium iodide salts; however, in the latter case this salt loses methyl iodide thermally in the mass spectrometer so that the spectrum of ordinary permethylated histidine is recorded.[6,39] Quaternization of the N-terminal amine is actually used to advantage in rapid determination of N-terminal residues by the method of Gray and del Valle.[64]

e. Sulfonium Salts. Formation of sulfonium iodide salts in sulfur-containing residues is one of the reasons given for desulfurization of peptides prior to permethylation.[6,122,148] Recent permethylation techniques not using Ag_2O are not particularly subject to this constraint. Formation of the sulfonium iodide salt of methionine leads to poor peptide spectra, with sequence peaks only up to, and not including, methionine. A peak at m/e 142 is usually observed for the methyl iodide lost from the salt in the mass spectrometer.[122] Pyrolytic decomposition to cyclopropane-containing residues has been reported in one case.[194] Temporary protection by oxidation of methionine to the sulfoxide, followed by subsequent *N*-methylation and reduction to the *N*-methylmethionine residue has been suggested,[94] though these additional experimental steps are obviously inconvenient.

Sulfonium iodide salts of cysteine easily lose methyl mercaptan to yield a dehydroalanine residue, but attempts to achieve a quantitative elimination during the permethylation have been unsuccessful.[97] The presence of dehydroalanine does not make the mass spectrum uninterpretable; sequence peaks and molecular ions are observed, but difficulties arise when partial elimination gives a mixture of dehydroalanine and methylcysteine peptides.

Cysteine peptides are normally isolated from proteins as derivatives (carboxy methylated, or S_β-ethylamino) whose mass spectral behavior is generally satisfactory; however overmethylation and elimination to dehydroalanine are also possible.

f. Arginine. None of the permethylation methods is effective for arginine-containing peptides.[192] Derivatization prior to permethylation is mandatory. Many possibilities are offered (see Section V,A), and the resulting ornithine or heterocyclic derivative can be permethylated to give a satisfactory mass spectrum if some care is taken to avoid quaternization of the N_δ in pyrimidylornithine.

3. Permethylation Methods

The original Kuhn method (Ag_2O, CH_3I, dimethylformamide), adapted for peptides by Das et al.,[199] and used by others,[29, 155] was soon abandoned because of difficulties of repeatability, side reactions, and cleavage of peptide bonds, especially with Asp, Glu, Met, and Trp residues.[6, 192, 194]

The less drastic Hakomori reaction,[201] using the sodium derivative of dimethyl sulfoxide (dimsyl sodium, Na^+ $^-CH_2SOCH_3$) as the base, was tried first for peptides by Vilkas and Lederer,[202] and improved by Thomas.[192] This reaction (Scheme 48) proceeds smoothly, in 30–40 minutes at room temperature in an excess of base (approximately tenfold) and methyl iodide

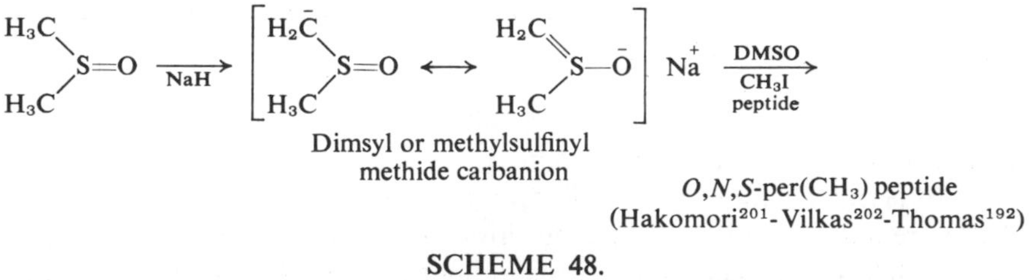

Dimsyl or methylsulfinyl
methide carbanion

O,N,S-per(CH_3) peptide
(Hakomori[201]-Vilkas[202]-Thomas[192])

SCHEME 48.

(> tenfold excess) over the peptide, if precautions are taken to eliminate all traces of water. Clean permethylation of as low as 20 nmoles of peptides has been reported.[27] This method is still widely used.[39, 40, 42, 64, 78, 146]

A variation of the method,[30] using the sodium derivative of dimethylacetamide as the base yields similar results. A milder base, quaternary ammonium hydroxide, has been suggested[185, 203, 204] for clean methylation of methionine and histidine. The use of sodium hydride as the base in dimethyl-formamide, proposed by Coggins and Benoiton[205] has proved to be effective for certain peptides[70, 105]; this reaction is slower in acetonitrile unless Me_2SO_4 is used instead of CH_3I.[200] In a comparative study Thomas[193] showed that sodium hydride produces more C-methylation of Asp, Glu, and Gly than does dimsyl anion; gas–liquid chromatography (GLC) studies have confirmed this observation.[44] However, others[95, 97, 190, 200] found substantial artifact production (C-methylation, sulfonium salt formation for cysteine and methionine) under Thomas conditions. It was shown later that, irrespective of the base used (hydride or dimsyl anion)[32, 33, 96, 97, 200, 206] C-methylation and sulfonium ion formation could be dramatically reduced by adding MeI only in an amount equivalent to the amount of the base, and using these in a 5- to 10-fold excess of the number of sites in the peptide; this is referred to as the "equimolar procedure." Reaction times from 5 minutes to 3 hours have been tried, with 1 hour found to be the optimum.[32] Note that because the amount of peptide and the number of sites to be permethylated are not always known for the sample, and because the strength of the base may be variable, undermethylation is possible.[95]

Recently Morris[95, 117] has suggested that under excess methyl iodide (Thomas) conditions, clean and complete O,N,S-methylation at the nanomolar level can be done in times as short as 60 seconds (the first permethylations required 3 days![199]); this method apparently is applicable to all known amino acids, with the exception of arginine and probably of cysteine.[95, 207, 208] If these promising results are confirmed, this should provide the most convenient permethylation method yet proposed.

For the permethylation reaction a wide variety of experimental conditions have been proposed; until the best of these have been thoroughly evaluated, the possible artifacts should be always considered. However, very impressive results in peptide sequencing applied to real biochemical problems have been obtained using O,N,S-permethylated peptide esters prepared under one of the above reactions.

4. Natural Peptides Sequenced as O,N-Permethylated N-Acetyl Derivatives

Many recent examples of MS applications have involved N-terminal pyroglutamyl peptides (Table II) because classical methods fail to sequence them: the structures of feline gastrin,[30] a heptapeptide from the zymogen of phospholipase A,[105] and a docosapeptide from a pig immunoglobulin λ chain[105, 105a] have been partly elucidated from their mass spectra. Two hypothalamic hormones (the thyrotropin releasing factor, TRH,[13, 85, 106–110, 129] and the luteinizing releasing hormone LRH)[42, 111, 112] have been identified. A crustacean color change hormone[113] gave poor spectra, but permethylation was not attempted. A series of bradykinin-potentiating peptides from snake

TABLE II
Oligopeptides with a Blocked N-Terminus

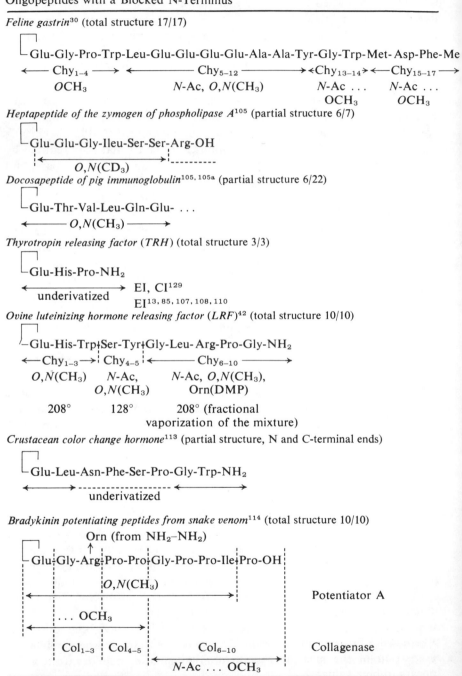

Feline gastrin[30] (total structure 17/17)

Glu-Gly-Pro-Trp-Leu-Glu-Glu-Glu-Glu-Ala-Ala-Tyr-Gly-Trp-Met-Asp-Phe-Me

←— Chy_{1-4} —→ ←————— Chy_{5-12} —————→ ‹Chy_{13-14}› ←—Chy_{15-17}—→

OCH_3 N-Ac, $O,N(CH_3)$ N-Ac ... N-Ac ...
                                                           OCH_3 OCH_3

Heptapeptide of the zymogen of phospholipase A[105] (partial structure 6/7)

Glu-Glu-Gly-Ileu-Ser-Ser-Arg-OH

←——— $O,N(CD_3)$ ———→ - - - - - - - - -

Docosapeptide of pig immunoglobulin[105, 105a] (partial structure 6/22)

Glu-Thr-Val-Leu-Gln-Glu- ...

←——— $O,N(CH_3)$ ———→

Thyrotropin releasing factor (TRH) (total structure 3/3)

Glu-His-Pro-NH₂

←——————→ EI, CI[129]
   underivatized   EI[13, 85, 107, 108, 110]

Ovine luteinizing hormone releasing factor (LRF)[42] (total structure 10/10)

Glu-His-Trp⊢Ser-Tyr⊢Gly-Leu-Arg-Pro-Gly-NH₂

←—Chy_{1-3}—→ Chy_{4-5} ←——— Chy_{6-10} ———→

$O,\bar{N}(CH_3)$ N-Ac, N-Ac, $O,N(CH_3)$,
                   $O,N(CH_3)$ Orn(DMP)

   208°   128°   208° (fractional
                                  vaporization of the mixture)

Crustacean color change hormone[113] (partial structure, N and C-terminal ends)

Glu-Leu-Asn-Phe-Ser-Pro-Gly-Trp-NH₂

←——————→ - - - - - - - - - - - - - - - ←——————→
           underivatized

Bradykinin potentiating peptides from snake venom[114] (total structure 10/10)

                     Orn (from NH₂–NH₂)
                          ↑
Glu⊦Gly-Arg⊦Pro-Pro⊦Gly-Pro-Pro-Ile⊦Pro-OH

                   $O,N(CH_3)$

←——————————————————————————————→   Potentiator A

... OCH_3

←——————————→

Col$_{1-3}$ ¦ Col$_{4-5}$ ¦   Col$_{6-10}$ Collagenase

                   N-Ac ... OCH_3

TABLE II (*Continued*)

Two peptides from the bovine paracasein K^{78a}

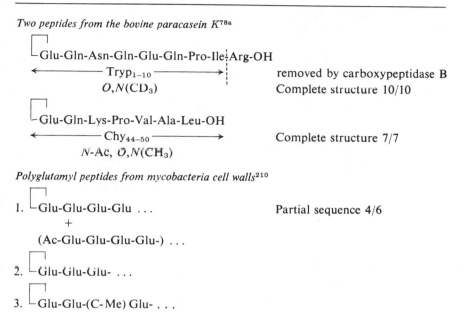

⌐Glu-Gln-Asn-Gln-Glu-Gln-Pro-Ile¦Arg-OH	
←——————— $Tryp_{1-10}$ ———————→¦	removed by carboxypeptidase B
$O,N(CD_3)$	Complete structure 10/10
⌐Glu-Gln-Lys-Pro-Val-Ala-Leu-OH	
←——————— Chy_{44-50} ———————→	Complete structure 7/7
N-Ac, $O,N(CH_3)$	

Polyglutamyl peptides from mycobacteria cell walls[210]

1. ⌐Glu-Glu-Glu-Glu ...   Partial sequence 4/6
       +
   (Ac-Glu-Glu-Glu-Glu-) ...

2. ⌐Glu-Glu-Glu- ...

3. ⌐Glu-Glu-(C-Me) Glu- ...

venom[114, 209] and two enzymatic peptides from bovine paracaseine K^{116} have been sequenced. Three polyglutamyl peptides from mycobacteria cell walls[210] gave partial N-terminal sequences after permethylation.

A number of oligopeptides with a nonblocked N-terminus (which could thus be sequenced by classical methods) obtained from long-chain natural proteins have been sequenced by MS (Table III): some chymotryptic peptides from the silk fibroin of *Bombyx mori*,[37] seven peptides from the enzymatic hydrolysis of *Staphylococcus aureus* penicillinase, one from a cytochrome, two adjacent chromatographic fractions of a tryptic digest of a denatured pepsin,[34] and a mutant peptide from an abnormal hemoglobin and a contaminating peptide.[120] The examples of the chromatographic fractions show the advantage of MS for sequencing mixtures of two or three peptides (see also Franêk *et al.*[70] and Hodges *et al.*[190]). Twenty-nine small di- to hexapeptides resulting from the elastase digest of a dihydrofolate reductase have been separated on a cation exchange resin and sequenced from their mass spectra.[207]

Fourteen chymotryptic and two tryptic peptides from the protein of the yellow turnip mosaic virus (189 residues) have been sequenced by MS.[69] Although MS failed to sequence some of the chymotryptic peptides, other oligopeptides proved to have been wrongly sequenced by classical methods; comparison of the mass spectrometric and classical sequencing methods appears to be a valuable way to find the correct structure.[69]

TABLE III

Oligopeptides with a Nonblocked N-Terminus

Chymotryptic peptide from silk fibroin of Bombyx mori[37]

H_2N-Gly-Ala-Gly-Val-Gly-Ala-Gly-Tyr-OH	Full structure, 8/8
H_2N-Gly-Ala-Gly-Ala-Gly-Ala-Gly-Tyr-OH	Full structure, 8/8
H_2N-Gly-Val-Gly-Tyr-OH	Full structure, 4/4

as *N*-Ac, *O,N*(CH₃); some peptides contaminated by minor peptides

Enzymatic hydrolysis peptides from Staphylococcus aureus penicillinase[34]

H_2N-Leu-Glu-Gln-Val-Pro-Tyr-OH   Full structure, 6/6
H_2N-Val-Gly-Lys-Asp-Leu-X-X-OH   XX = Leu-Thr, Thr-Leu, Val-
←─────────────────────────→   Asp, Asp-Val, partial
                                                      structure, 5/7

H_2N-Ser-Lys-Glu-Asn-Lys-Lys-Phe-OH···   Partial structure, 6/7
       ←──────────────→

H_2N-Gly-Lys-Thr-Leu-OH   Full structure, 4/4
H_2N-Glu-Leu-Glu-Leu-Asn-Tyr-Tyr-OH   Partial structure, 6/7
←─────────────────→

H_2N-Glu-Glu-Val-Pro-Tyr-OH   Full structure, 5/5
H_2N-Ser-Pro-Leu-Leu-Glu-Lys-Tyr-OH   Full structure, 7/7

Enzymatic hydrolysis peptide from a cytochrome[34]

H_2N-Ala-Lys-Trp-OH   Full structure, 3/3

Two adjacent fractions from the LC separation of a tryptic digest of a denatured pepsin[34]

Fraction A: H_2N-Val-Gly-Leu-Ala-Pro-Val-Ala-OH   Full structure, 7/7
Fraction B: mixture of 3 peptides separated by
    fractional vaporization
    200°: H_2N-Val-Gly-Leu-Ala-Pro-Val-Ala-OH   Full structure, 7/7
    220°: H_2N-Ala-Asn-Asn-Lys-OH   Full structure, 4/4
                                                          Partial structure, 5/7

    260°: H_2N-Glu-Tyr-Tyr-Thr-Val-Phe⸽Asp-Arg-OH   partial structure, 5/7
          ←─────────────────────────→⸽

          ⌐┐
          └Glu-Tyr-Tyr-Thr-Val-Phe⸽Asp-Arg-OH
          ←──────────────────────→⸽

all as *N*-Ac, *O,N*(CH₃), but unmodified Arg

Mutant peptide of an abnormal hemoglobin[120]

──────→ { Leu-Leu-Gly-Asn-Val-Leu-Phe
                                    ↑
           Leu-Leu-Val-Val-Tyr-Pro-Trp (contaminating)

The sequence of the "normal" peptide was known as Leu-Leu-Gly-Asn-Val-Leu-Val-Cys. The "mutant" was known to have one Val replaced by Phe; the mass spectrum of the mutant peptide unambiguously determined which Val had been substituted.

TABLE III (*Continued*)

Sequences of the 29 major components of the elastase digest of dihydrofolate reductase obtained by mass spectrometry[207]

Asn-Val-Val-Leu	Lys-Glu-His-Leu
Gly-Gly-Ala-Gln-Leu	Tyr-Ala-Lys-Glu-His-Leu
Ala-Gly-Ser-Phe-Gly-Gly	Phe-Leu-Trp
Asp-Thr-Lys-Met-Leu	Asp-Val
Met-Val-Val	Met-Leu-Pro
Val-Val-Val	Thr-Ala-Phe-Leu-Trp-Ala-Glu
Phe-Ala Tyr-Ala	Leu-Met-Val-Val
Trp-Ala-Glu	Phe-Thr
Ala-Ala	His-Thr-Thr
Trp-Glu	Glu-Asp-Thr
Phe-Lys	Asp-Thr-Asn
Gly-Lys-Leu	Val-Val-Val-His-Asp-Val-Ala
Arg-Leu-Ala	Glu-Val-Trp
Glu-Lys-Lys	
Leu-Lys-Lys	
Leu-Pro Leu-Asn-Trp	
Thr-Tyr	

Sequence of the oligopeptides from the chymotryptic digest of the protein of the yellow turnip mosaic virus[69]

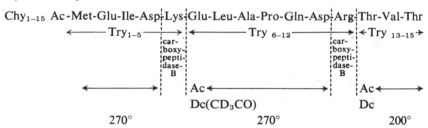

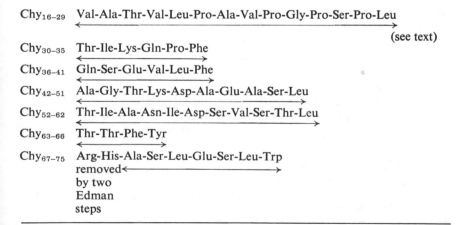

TABLE III (*Continued*)

Chy$_{96-111}$ Val-Pro-Ala⋮Asn⋮Ser-Pro-Val-Thr-Pro-Ala-Gln⋮Ile-Thr-Lys-Thr-Tyr

⟵————⟶   (1) selective acidic
cleavage at Asn

(2) papain digestion

⟵————————Pap$_{100-106}$————————⟶ ⟵ Pap$_{107-111}$ ⟶

from the mixture

Chy$_{112-116}$ Gly-Gly-Gln-Ile-Phe
⟵————————⟶

Chy$_{117-125}$ AeCys⋮Ile-Gly-Gly-Ala-Ile-Asn-Thr-Leu⋮

removed   ⟵————————————————⟶
by one   (Ae = aminoethyl)
Edman step

Chy$_{124-128}$ Thr-Leu-Ser-Pro-Leu
⟵————————⟶

Chy$_{170-177}$ Ile-Ile-Thr-Val-Ser-Gly-Thr-Leu
⟵————————————⟶

Chy$_{184-189}$ Ile-Thr-Asp-Thr-Ser-Thr
⟵————————⟶

Somatostatin[119]

H-Ala-Gly-Cys-Lys⋮Asn-Phe-Phe-Trp-Lys⋮Thr- Phe-Thr-Ser-Cys-OH

⟵————Try$_{5-9}$————⟶

N-Ac, $O,N(CH_3)$
and N-Ac, $O,N(CD_3)$

Peptides from cheese[190]

155° H-Glu-Val-Leu-Asn-OH
⟵————————⟶

200° H-Asn-Glu-Asn-Leu-Leu-OH
⟵————————⟶

260° H-Ala-Pro-Phe-Pro-Glu-Val-Phe-OH
⟵————————————⟶

N-Ac, $O,N(CH_3)$, fractional vaporization of the mixture

N-terminal sequence of a tryptic octadecapeptide from pig immunoglobulin λ-chain[70]

Classical methods   Ala-Thr-Leu-Thr-Ile-Thr-Gly-Ala-Glx . . .
Ala   Leu   Ala
Gly   Gly

Mass spectrometry   Ala⋮Ala⋮Leu-Thr-Leu-Thr-Gly-Ala-Gln-Ala- . . .
mixture of 2 peptides   ⎱   Ala⋮Thr⋮Leu-Thr-Leu-Thr-Gly-Ala-Gln-Ala . . .
(partial sequence, 10/18) ⎰ ⟵————————————————⟶
N-Ac, $O,N(CH_3)$—

Peptic peptide from merino wool[212]

Ac-Ala-(Cm)Cys-(Cm)Cys-OH
↓   (Cm = carboxymethyl)
Ac-Ala-Ala-Ala-OH

TABLE III (*Continued*)

DAP Peptides from mycobacterial cells[213]

H-Ala-κ-Gln-DAP-Ala-OH
H-Ala-κ-Glu-DAP-Ala-OH
H-Ala-κ-Gln-DAP-OH

α-Melanocyte stimulating hormone[117]

Ac-Ser-Tyr-Ser-Met-Glu-His-Phe-Arg-Trp-Gly-Lys-Pro-Val-NH$_2$
  ↓ CNBr
Ser-Tyr-Ser-Hse – – – – – ↓ chymotrypsin (Hse = homoserine)
←————————→
N-Ac, *O*,*N*(CH$_3$), 4/4

   Chy$_{5-7}$ Chy$_{10-13}$
←————→ – – – ←————————→
N-Ac, *O*,*N*(CH$_3$), 2/3 *N*-Ac, *O*,*N*(CH$_3$), 4/4

Bradykinin potentiating peptides from tryptic digest of rabbit and bovine albumin[214]

H-Leu-Val-Glu-Ser-Ser-Lys-OH 6/6
H-Thr-Pro-Val-Ser-Gly-Lys-OH 6/6
 N-Ac, *O*,*N*(CH$_3$)

Peptide from bovine α$_{s1}$-casein[215]

¦Gly-Leu-Pro-Glu-Pro-Phe¦Pro-Gln 6/8
←————————→
 N-Ac, *O*,*N*(CH$_3$)

Tryptic peptide from tropoelastin[216]
H-Ala-Ala-Ala-Lys-OH
H-Ala-Ala-Lys-OH
 N-(Ac), *O*,*N*(CH$_3$), *N*-(Dc), *O*,*N*(CH$_3$)

Other examples of nonblocked oligopeptides include another hypothalamic hormone, somatostatin,[119] a chymotryptic digest of α-lactalbumin, a milk protein,[211] a mixture of three peptides from cheese by their fractional vaporization in the mass spectrometer,[190] and two octadecapeptides from pig immunoglobulin λ-chain differing only by the nature of the residues in position 2.[70] A peptic tripeptide from reduced merino wool[212] was sequenced after reductive desulfurization and permethylation.[212] Three DAP- (2,6-diaminopimelic acid) containing peptides resulting from the enzymatic digestion of mycobacterial cell walls were sequenced by MS after acetylation and permethylation[213]; the mass spectra show clearly how the DAP residue is linked to the peptide. A tridecapeptide containing 12 different residues from α-melanocyte-stimulating hormone[118] was sequenced after cleavage by CNBr, trypsin, or chymotrypsin; although other data were incorporated in the analysis, MS was the only sequencing tool, and the complete structure was obtained.

Two bradykinin-potentiating peptides among those resulting from the tryptic digest of rabbit and bovine albumin were sequenced from their mass spectra,[214] as were a peptide from the A variant of bovine α_{S1}-casein,[215] and two tryptic peptides from tropoelastin.[216] Recently the enzyme ribitol dehydrogenase was investigated by MS and some sequences were obtained.[208] These extensive examples give clear evidence that the method has left the experimental stage, and that it has become accepted by many laboratories. However, there are promising alternative methods for the sequencing of oligopeptides by MS under development, as discussed separately.

VI. ANALYTICAL TECHNIQUES

A variety of special methods for sample handling, data acquisition, and data interpretation have been proposed. The more important techniques will be discussed in this section.

A. Vaporization of Low-Volatility Samples

Direct chemical ionization (CI, Section III) and field desorption (FD, Section IV) make it possible to obtain mass spectral data at lower sample temperatures than those necessary for the normal direct probe operation, and thus to examine higher molecular weight peptides. Although FD is by far the most promising method for analyzing such large peptides, Friedman and his co-workers have shown[125, 125a] that it is possible to minimize thermal decomposition accompanying direct probe sample introduction by optimization of the heating rate and proper selection of the surface material from which the sample vaporizes. It was shown that rapid sample heating rates of up to 12°/second (complete sample evaporation in < 30 seconds) substantially reduces pyrolysis, as does use of a Teflon sample probe in a Teflon-lined ion source.[125, 125a] Proper desorption is critically dependent on the absence of trace nonvolatile impurities (note the similar sensitivity of FD to inorganic impurities, Section IV). Underivatized Arg-containing peptides gave useful mass spectral information by these techniques using ion bombardment ionization.[125a]

1. Underivatized Peptides

Until recently, gross thermal decomposition accompanied any attempt to obtain mass spectral information from molecules larger than mono- or dipeptides whose terminal amine and carboxyl groups were free.[1, 2, 134, 134a] An attempt to derive useful sequence information by flash pyrolysis of peptides was unsuccessful, with the main product being the diketopiperazine formed from the N-terminal residue pair.[135] Cyclic peptides and depsipeptides[217-225] can often be analyzed without derivatization because they provide their own N- and C-blocking. Their mass spectra may be either simpler or more complex than the isomeric linear peptide, depending on how many of the possible ring openings are induced by the electron impact, but they generally exhibit a fairly abundant molecular ion. As they resist sequencing by classical methods, their mass spectrometric behavior is well documented and the reader is referred to general reviews[4, 77, 141] and articles[217-225] on the subject.

The rapid heating technique has[125a] been used to obtain mass spectral data on Arg-Pro, Arg-Pro-Pro, Arg-Pro-Pro-Gly, Arg-Pro-Pro-Pro-Gly-Phe, Ser-Pro-Phe-Arg, Pro-Phe-Arg, Phe-Arg, and bradykinin itself. No molecular ion species are observed; fragmentation at peptide linkages and elimination of small molecules (H_2O, HCN, NH_3, H_2NCONH_2, $HCOOH$, etc.) is extensive. Applications of direct CI and FD are discussed in Sections III and IV.

CI-MS has proved to be useful in the elucidation of the hypothalamic hormone TRF,[129] for which a very small amount had been obtained in a relatively pure form from the extraction of more than 300,000 ovine hypothalami. Although the structure of the tripeptide and the synthetic reference material had been obtained, a CI spectrum from a few nanomoles of the free tripeptide gave clear evidence of its sequence.

B. Sequences of Mixture Components, Directly and by Fractional Vaporization

It has been pointed out[38] that the high amount of specific information available in a mass spectrum can provide sequence information on several oligopeptides present in a mixture. Identification of impurities in mixtures is a well-known attribute of MS.[92] Oligopeptide impurities had been identified in isolated cases[37, 150] before this was proposed as a general technique for sequencing and applied to more complex mixtures.[38, 39]

In addition, the mass spectral peaks belonging to individual oligopeptides can often be identified by fractional vaporization of the sample from the direct probe in the MS ion source. As illustrated by Fig. 7,[39] all of the peaks arising from a common component should reach their maximum abundance in the spectrum simultaneously, and parallel behavior is shown by plotting the logarithm of the absolute intensities of the peaks vs the probe temperature. Fractional vaporization plots with more than one maximum indicate that more than one component contributes to a particular peak.[39, 226] Separation and identification of mixtures of synthetic peptides and mixtures of peptides derived from natural proteins have been reported.[39-42, 69, 120] For this purpose high-resolution data are of course more specific,[38, 39] but identification of several mixture components from low-resolution mass spectra have also been reported.[34, 40, 41, 120]

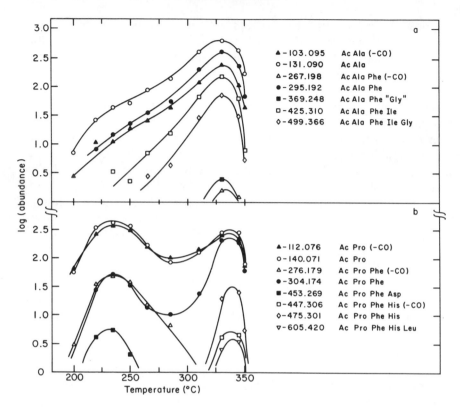

FIG. 7. Variation of ion abundance with sample temperature for the mass spectra of a mixture of Ala-Phe-Ile-Gly-Leu-Met, Pro-Phe-His-Leu-Leu, and Pro-Phe-Asp.[39]

C. Gas Chromatography/Mass Spectrometry (GC/MS)

A scheme for oligopeptide sequencing involving prior GC separation was suggested as early as 1960 by Biemann.[2, 17, 88] In 1967 Prox and Weygand[83] successfully subjected a cyclopeptide to GC/MS, but it is only recently that GC/MS methods have been examined seriously; their potential for sequencing may rival the permethylation route. The principle is simple: the natural protein is cleaved into small fragments (di- or tripeptides), which are separated by GC and identified on-line by MS; repeating this procedure with a different cleavage method yields overlapping sequence information from which the structure of the original protein is reconstructed. (GC/MS of the fragment released from an Edman degradation has also been suggested.)[227, 228]

In order for GC/MS to be a general method, it is mandatory that at least the distinction and identification of all of the possible dipeptides resulting from the combination of the natural amino acids can be done unambiguously;

lack of information on only one dipeptide would hamper the entire procedure. If two particular amino acids occur more than once as an adjacent pair in the polypeptide, it will be necessary to identify the corresponding possible tripeptide fragments containing this pair to identify the pair positions.

The problems involved in this method are rather different from those previously encountered. In GC/MS the gas chromatograph limits the method more than the mass spectrometer does; in general any substance which elutes from the GC column can be analyzed by MS. Gas chromatographic separation of even an underivatized dipeptide is impossible. Partial data on pyrolysis GC/MS of underivatized dipeptides have been reported, but it does not appear that such an approach can provide routinely useful identification and sequence information.[229] Thus it is necessary to prepare derivatized peptides which behave well for GC separation as well as for the previously stated requirements (see Section V) of derivatives. Thermal stability is a more severe criterion for GC/MS than for solid probe introduction, and decomposition may take place in the GC or in the interface. To make those parts of the system more inert, silanized packing and glass columns are used in the GC and a silanized glass frit or jet in the interface.[46] The choice of the derivative also requires special consideration. Biemann originally suggested a GC/MS method in which the mixture of N-acetyl peptide esters is reduced to the polyamino alcohols with $LiAID_4$ (the deuteride is preferred over the hydride for better distinction of the N- and C-terminal fragments), because polyamino alcohols are less polar and give simpler mass spectra[88] than the corresponding peptide. The current proposal by Biemann also utilizes the reduction of the peptides to polyamino alcohols followed by silylation of the latter; the resulting derivative behaves well with both GC and MS (see Scheme 49). An example in which the method provided nearly complete sequence information on an eicosapeptide is shown in Fig. 8.

Previous attempts to use TMS derivatives for peptide[69a,187] sequencing had shown that O,N-pertrimethylsilylation was often incomplete, and that the fragment ions fail to exhibit clear sequence information. Also, the increase in the molecular weight of the derivative is severe (a simple dipeptide with bifunctional amino acids takes 3 TMS groups, thus increasing its MW by 216). (This survey found only one example of the application of GC/MS to TMS dipeptides—the case of some natural pyroglutamyl dipeptides from a mushroom.[187]) O,N-Pertrimethylsilylated polyaminoalcohols appear to suffer part of the same failing, and selective O-silylation has been achieved instead.[43,230,231] Mass spectra of O-TMS derivatives of polyamino alcohols exhibit enhancement of the fragment derived from the C-terminus, so that their spectra generally provide clear sequence information.[43,88,230] However, these spectra fail to exhibit a molecular ion, and even the $(M - 15)^+$ ion may be of low abundance.

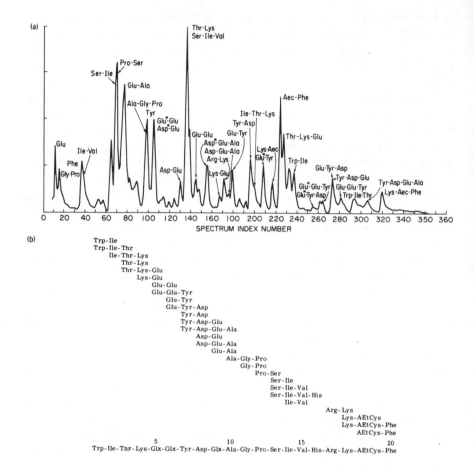

FIG. 8. (a) Total ionization plot of *O*-trimethylsilylated polyamino alcohols obtained by derivatization of an acid hydrolysate (6 *N* HCl, 110°, 20 minutes) of the C-terminal cyanogen bromide fragment (an eicosapeptide) of actin. Derivatives formed by elimination of trimethylsilanol from the corresponding trimethylsilylated polyamino alcohols are marked with an asterisk on the affected amino acid. (b) Reassembled sequence using the identified oligopeptide fragments; Ser-Ile-Val-His was identified by fractional vaporization of an aliquot of the derivatized sample mixture introduced directly into the mass spectrometer ion source.[230]

From the compilation of preliminary data, it appears that most of the *O*-TMS polyaminoalcohols derived from dipeptides are amenable to GC/MS[43, 230] with, however, some notable exceptions: amide and acid residues cannot be distinguished (Asp and Asn, Glu and Gln) because both functional groups are reduced to a common primary alcohol upon reaction

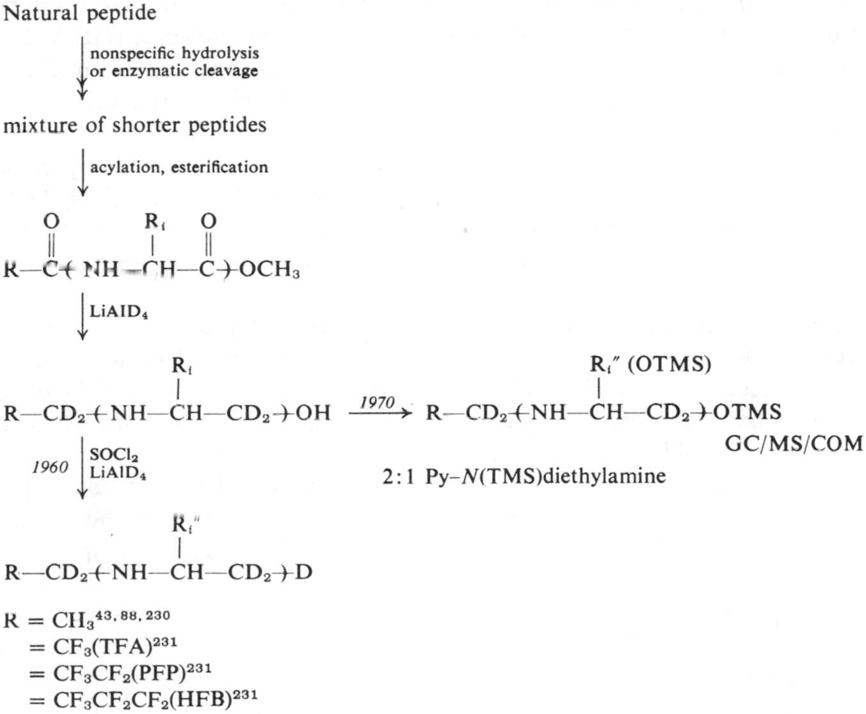

Natural peptide

$\downarrow$ nonspecific hydrolysis
or enzymatic cleavage

mixture of shorter peptides

$\downarrow$ acylation, esterification

$$R-\overset{O}{\overset{\|}{C}}+NH-\overset{R_i}{\overset{|}{C}}H-\overset{O}{\overset{\|}{C}})OCH_3$$

$\downarrow$ LiAID$_4$

$$R-CD_2+NH-\overset{R_i}{\overset{|}{C}}H-CD_2)OH \xrightarrow{\textit{1970}} R-CD_2+NH-\overset{R_i''(OTMS)}{\overset{|}{C}}H-CD_2)OTMS$$

GC/MS/COM

$\textit{1960}$ $\begin{vmatrix} SOCl_2 \\ LiAID_4 \end{vmatrix}$ 2:1 Py–N(TMS)diethylamine

$$R-CD_2+NH-\overset{R_i''}{\overset{|}{C}}H-CD_2)D$$

R = CH$_3$[43, 88, 230]
  = CF$_3$(TFA)[231]
  = CF$_3$CF$_2$(PFP)[231]
  = CF$_3$CF$_2$CF$_2$(HFB)[231]

SCHEME 49.

with LiAIH$_4$. Arginine is transformed into N-methyl Orn during the derivatization so that it does not need any special treatment. O-TMS derivatives may partly decompose in the GC by eliminating TMS-OH, especially in aminoalcohols from Glx and Asx. Cysteine must be protected (e.g., aminoethyl Cys). With the higher perfluoroalkanoyl derivatives apparently even His-containing dipeptides can now be eluted from the GC[231]; previously a complementary analysis by solid probe introduction was needed for these peptides.[230] Trp-containing di- or tripeptides can be analyzed by GC/MS, but some such derivatives of "difficult" peptides such as Trp-Trp or Trp-Glx can be difficult to elute from the GC.

Others have proposed GC/MS schemes not including the reduction step, involving only short derivatized peptides. For this, N-TFA[83, 99, 100, 162–164, 211] or N-PFP (pentafluoropropionyl)[46] have been shown to be effective: they are easier to prepare and more volatile than the O,N-permethylated N-acetyl peptide esters (the difference in the Kovats index is between 100 and 300 for an apolar stationary phase[44]). Permethylation has not been used[44] to increase the volatility further. However, the problems with this scheme appear to be more severe than those for the GC/MS of polyaminoalcohols. Not all of the possible dipeptides can be eluted from the GC column; Arg must be transformed into either Orn or Pyr(Orn). Cys must be protected (GC/MS of benzylcysteine peptides has been reported[99]); but there seems to be no way of dealing with a His peptide. N-TFA derivatives may also partly decompose in the GC.[46]

No direct comparison between the GC/MS behavior of N-TFA peptide esters and the corresponding O-TMS polyaminoalcohols has been done. Data are still scarce, and the reduction route has been advocated by only one group. The other aspects involved in GC/MS sequencing are more or less common to the analysis of both derivatives.

Previous studies have shown that base-line separation by GC can even be achieved for closely related isomers such as Leu-Ala and Ala-Leu,[99] Leu-Ile and Ile-Leu,[83] or α-and ω-linkages at acidic residues.[44] However, in these cases conditions were developed to achieve a specific separation and these conditions would be unrealistic for a wide variety of peptides.[230] It is probable that even with a column of high resolving power, the separation of the hydrolysate of a large natural protein is likely to require an hour with temperature programing over a wide interval, and will give a complex GC trace with many overlapping and unresolved peaks. (For technological reasons, and despite encouraging preliminary results,[232, 233] high-resolution GC/MS is not yet feasible, and one has to rely on low-resolution data.) Automatic recording of spectra from the GC/MS run appears to be highly desirable for the complex mixture of single amino acids, di-, tri-, and even some tetrapeptides. Biemann has demonstrated[230, 231, 234–236] that the high specificity of peptide mass spectra allows identification of components of incompletely resolved GC peaks (identification of peptides in mixtures[38, 39] and from fractional vaporization[38–42] was described in Section VI,B); a capability for computer data acquisition during cyclic scanning (a complete mass spectrum of the GC effluent every few seconds) is very helpful for such mixture identification by the use of "mass chromatograms." Biemann[235, 236] has also shown that identifications are much more specific if the Kovats GC retention indices are also used; these can be predicted with good accuracy for the dipeptides.

For such GC/MS methods the cleavage of the natural protein was originally done by acid hydrolysis or nonspecific enzymatic degradation. The course of these reactions cannot be accurately controlled, and the resulting mixture usually contains a variety of fragments of different lengths. Some of these products are useless and thus wasteful; the simple amino acids give no sequence information and the higher peptides do not elute from the GC column. The advent of a selective diaminoacidyl peptidase[237] (DAP I or cathepsin C) was greeted by several groups as a promising solution to this problem. In theory, two GC/MS runs, one after degradation by DAP I and the other after one Edman degradation step and DAP I degradation, generate two sets of overlapping dipeptides. However, the reconstitution of the sequence from this information, although straightforward,[43,45,46] is not necessarily unambiguous (for example, a synthetic tetradecapeptide containing 5 Gly[45]). Unfortunately the enzymatic cleavage stops at Pro and is ineffective for N-terminal Arg or Lys peptides. Application of this technique to insulin A is shown in Fig. 9.[46]

So far a limited number of oligopeptides have been sequenced by GC/MS. Examples utilizing acidic hydrolysis and N-TFA derivatives include the structure elucidation of a cyclic nonapeptide[83]; the control of synthetic peptides from the Merrifield solid phase method[99,162]; and the structure of antanamide, a cyclic decapeptide,[100,104] and of an oxytocin analogue.[100] Diaminoacidyl peptidase and N-TFA peptides have been used for the

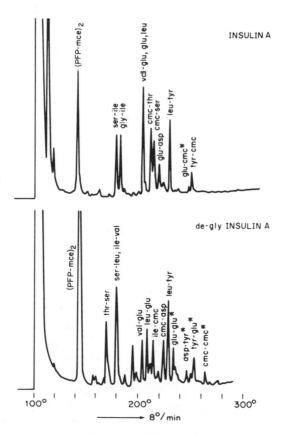

FIG. 9. The GC/MS total ion monitor recordings of the digests of carboxymethylated insulin A chain and de-Gly insulin A chain. For those dipeptides marked with an asterisk, partial decomposition was observed.[46]

sequencing of a synthetic tetradecapeptide from the tryptic hydrolysate of cytoplasmic aspartate aminotransferase.[45] Diaminoacidyl peptidase I and N-PFP have been used for a porcine carboxymethylated insulin A chain containing 21 residues.[46] Acidic and enzymatic cleavages (including DAP I) followed by reduction and derivatization to O-TMS polyamino alcohols have been used to sequence a C-terminal cyanogen bromide fragment of actin[230] and a ribonuclease S peptide.[43]

Peptides Sequenced by GC/MS:

1. *Tryptic peptide from cytoplasmic aspartate aminotransferase*[45]
   H-Val-Gly-Gly-Val-Glu-Ser-Leu-Gly-Gly-Thr-Gly-Ala-Leu-Arg-OH

2. *Ribonuclease S peptide*[43]
   H-Lys-Gly-Thr-Ala-Ala-Ala-Lys-Phe-Glu-Arg-Glu-His-Met-Asp-Ser-Ser-
   Thr-Ser-Ala-Ala-OH

3. *Cyanogen bromide fragment of actin*[230]
   H-Trp-Ile-Thr-Cys-Glx-Glx-Tyr-Asp-Glx-Ala-Gly-Pro-Ser-Ile-Val-His-
   Arg-Lys-(Ae)Cys-Phe-OH   (Ae = aminoethyl)

4. *Porcine insulin A chain*[46]
   H-Gly-Ile-Val-Glx-Glx-(Cm)Cys-(Cm)Cys-Thr-Ser-Ile-(Cm)Cys-Ser-Leu-
   Tyr-Glx-Leu-Glc-Asx-Tyr-(Cm)Cys-Asx-OH   (Cm = carboxymethyl)

D. Liquid Chromatography/Mass Spectrometry (LC/MS)

Polypeptide sequencing methods that utilize information from the sequences of fragments must utilize information of overlapping fragments to arrange the pieces in the proper order. The overlap information will be unique only if it is sufficiently long so that it is not duplicated elsewhere in the polypeptide; di- and tripeptides often do not provide unambiguous overlap information. High-pressure liquid chromatography LC has no vapor pressure limitations, and recently the direct on-line coupling of LC and MS has been achieved by continuous introduction of the LC effluent solution into a chemical ionization (CI) source, using the LC solvent as the ionizing reagent gas.[47, 47a, 238, 239] Relatively complete sequence information has been found in the resulting CI spectra for a variety of oligopeptides using $< 10^{-7}$ g samples. Such introduction gives "direct CI" (Section VI,A), so that somewhat larger oligopeptides can be studied. Development of a system is in progress in which the polypeptide would be cleaved in a relatively nonspecific fashion (such as digestion by elastase[207] or thermolysin[208]) to larger oligopeptides (5–10 residues), derivatized, the resulting mixture separated by high-resolution LC, and the eluted fractions sequenced on-line from their CI mass spectra. For very large polypeptides simpler oligopeptide mixtures could be obtained by less complete hydrolysis or digestion; the larger peptide fragments would then be separated by exclusion chromatography and degraded again with a different reagent to give additional oligopeptides for LC/MS sequencing. Detection of $< 10^{-14}$ mole of solute in solution introduced into the MS in this manner has been achieved[239]; thus, this LC/MS method has the potential for sequencing of much smaller amounts of polypeptides than possible by present techniques.

E. Types of Analysis of Data

Sequencing of oligopeptides by MS is possible because the linear structure is determined unequivocally from the identity of the possible fragments which contain one end of the chain. The sequence peaks A_i^+ and B_i^+ carry this information for the N-terminus and the C_i^+ and D_i^+ for the C-terminus, as discussed in Section II,A.

1. Manual Sequencing

For low-resolution spectra, in which peaks are measured to the nearest integral mass number, sequencing can often be accomplished by judicious assignment of peaks in the mass spectrum; m/e values arising from the possible residues are given in Table IV. It is also possible to match a particular fragment to a mass interval between two prominent peaks, although the importance of peak abundance is an inherently subjective criterion. The procedure is easier if the molecular ion can be identified, which is difficult to do unambiguously, or is known from CI or FI, or a separate amino acid analysis; then peaks corresponding to the loss of a possible amino acid from $(M - OCH_3)^+$ or $(M - COOCH_3)^+$ can be sought, and so on. Weygand et al.[66] have suggested listing the ions in a tabular form (Differenzchema) for identification of the logical sequences starting from the molecular ion (see Fig. 10). The more general procedure is, however, to start from the low mass end, identify the N-terminal residue from the B_i^+ (and A_i^+) peaks, add on the masses of the possible amino acid units, search for a match between these possible B_2^+ ions and the important peaks in the spectrum, and so on. Each new fragment found should be checked with some of its characteristic ions (immonium ion, side-chain elimination), and the pseudo amino acid type fragmentation investigated if Trp, His, Phe, Tyr, or Asx are detected. Further confirmation is provided by complementary information such as amino acid analysis, known end residues, other classical tests, comparison of the mass spectra of different derivatives of the same peptide, fractional vaporization chemical or field ionization, and metastable transitions. Low-resolution data have been used in this manner in more than 90% of the reported cases in which sequencing has been achieved. Typical examples using this strategy with the low-resolution mass spectra of O,N-permethylated peptide esters are presented.

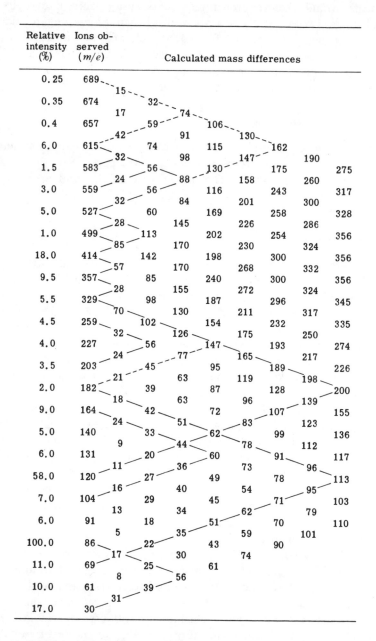

Relative intensity (%)	Ions observed (m/e)	Calculated mass differences

FIG. 10. "Differential scheme" of Weygand et al.[66] applied to the mass spectrum of N-TFA-Leu-Phe-Gly-Leu-Met-OCH₃.

Stehelin[69] has studied the chymotryptic digest of the oxidized protein of turnip yellow mosaic virus by MS. The 20 resulting oligopeptides were purified and isolated in pure form by a combination of chromatographic methods. The spectrum of one of these is presented in Fig. 11. Its N-terminal amine has been blocked with a 1:1 mixture of the acetyl and trideuterioacetyl groups which yield characteristic doublets separated by 3 mass units. From the 270° spectrum the sequence of 10 residues can be derived unambiguously,

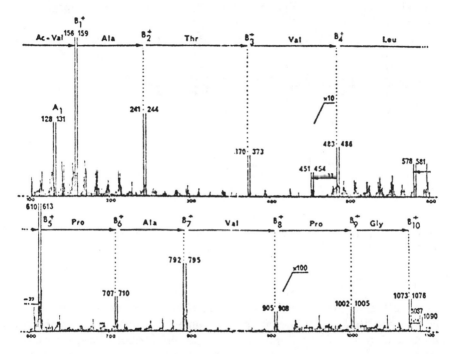

FIG. 11. Mass spectrum of an oligopeptide from turnip yellow mosaic virus after N-acetylation (1:1 $CH_3CO:CD_3CO$) and O,N-permethylation.[69]

but further sequence ions and the molecular ion are not seen. C-Methylation of Gly accounts for the m/e 1087, amino acyl ions are prominent, and the loss of MeOH is consistent with the presence of Thr. Thus the structure Val-Ala-Thr-Val-Leu-Pro-Ala-Val-Pro-Gly is indicated.

Another sample from the same protein was N-acetylated, digested with the nonspecific enzyme papaine, and the resulting mixture trideuterioacetyl-ated (Dc), thus labeling the newly formed N-termini. The mixture was directly submitted to the mass spectrometer. During fractional vaporization the following sequences were obtained:

160°: Ac-Val-Ala-OMe, Dc-Leu-Pro-Ala-OMe, Dc-Thr-Val-Leu-Pro-Ala-OMe
250°: Ac-Val-Ala-Thr-Val-Leu-Pro-Ala-OMe, Dc-Val-Pro-Gly-Pro-Ser-Pro-Leu-OMe

Thus, the reconstitution of the sequence of the tetradecapeptide is straight-forward:

Val-Ala-Thr-Val-Leu-Pro-Ala-Val-Pro-Gly-Pro-Ser-Pro-Leu

The work of Stehelin includes the identification of 19 other chymotryptic and tryptic peptides of this particular natural protein. For each case a strategy was chosen which fit best the oligopeptide, such as specific acid hydrolysis at Asp and Asn, selective or nonselective enzymatic cleavage, or sometimes removal of one or two residues by Edman degradation of carboxypeptidase. Similar work, but on a smaller scale, done in other laboratories [30, 34, 37, 42, 118] is listed in Section V,D,4; an example by Roepstorff[40] is shown in Fig. 12.

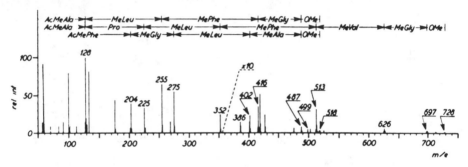

FIG. 12. Interpretation of the mass spectrum of a mixture of three peptides (vaporization at $150°$[40]) using manual processing.

2. Automated Sequencing

Soon after the first positive results were obtained on peptides by MS, it was realized that the nature of the data interpretation (finite number of amino acid residues in natural proteins and the logical procedure of the sequencing method) made it amenable to computerization.[9, 28, 240–242] Algorithms were developed to examine the mass spectrum of a totally unknown peptide and generate its sequence from the systematic investigation of all of the logical fragments. Computerized mass spectrometers are now more frequently found, making it possible to combine at least part of the interpretation with the data acquisition step.

The main objectives of such a computer program are to save the chemist's time and to force him to consider all possible combinations of the spectral data. The computer programs do *not* include everything that the chemist knows about interpretation of the data, especially concerning incorporation of *other* non-mass-spectral information on the sample. It is important to emphasize that the computer does not bring any "magic solution" for the sequencing of a peptide which would resist a thoroughly manual processing of the data. The computer is mainly a convenience when the sequence peaks

TABLE IV

Masses of Sequence Peaks of O,N-Permethyl-N-acetyloligopeptides

Amino acid residue	Residue unit mass		N-Terminal B$_1$ ions^a		C-Terminal C$_n$ ions^a		Immonium ions	Pseudo N-terminal	Pseuco C-terminal	Other characteristic ions
	CH$_3$	CD$_3$	CH$_3$	CD$_3$	CH$_3$	CD$_3$				
Gly	71	+3	114	+3	102	+6	44			+14 (C-methylation)
Ala	85	+3	128	+3	116	+6	58			
Pro	97	+0	140	+3	128	+3	70			
Val	113	+3	156	+3	144	+6	86			
Ser	115	+6	158	+3	146	+9	88			Loss of 32 (CH$_3$OH)
Leu, Ile	127	+3	170	+3	158	+6	100			Distinguish by m*
Thr	129	+6	172	+3	160	+9	102			Loss of 32 (CH$_3$OH)
Met	145	+3	188	+3	176	+6	Weak			Loss of 48 (CH$_3$SH)
Asp	143	+6	186	+3	174	+9	116	113	144	
Asn	156	+9	199	+3	187	+12	129	126	157	
Glu	157	+6	200	+3	188	+9	130			N-Terminal Glu, m/e 126, 98
Phe	161	+3	204	+3	192	+6	134	131	162	m/e 91
His	165	+6	208	+3	196	+9	Weak	135	166	m/e 95
Gln	170	+9	213	+3	201	+12	143			See Glu
Orn	184	+6	227	+6	215	+9	157			
Tyr	191	+6	234	+3	222	+9	164	161	192	m/e 121
Lys	198	+6	241	+6	229	+6	171			−14 (incomplete methylation)
Trp	214	+6	257	+3	245	+9	187	184	215	m/e 144
(Dmp)Orn	248		291	+6	279	+9	221			
Cys	131	+6	174	+6	162	+9	104			

^{a} In general, the A$_1$$^+$ and D$_n$$^+$ ions are less intense than the corresponding B$_1$$^+$ and C$_n$$^+$ ions, respectively; the masses of the former can be obtained by subtracting 28 from the masses of the latter.

can be unambiguously distinguished, such as for the high intensity peaks found in the spectra of simple peptides containing nonpolar residues (Gly, Val, Pro), or when previous knowledge of the amino acid composition limits the search for a match to a few possibilities. It is sometimes reported that a small peak is "significant for the proposed sequence" mainly because it justifies what the author supposes based on knowledge from other sources of data. In real problems with polar residues, and especially with unknown mixtures of peptides, intensities are not a reliable indication of the source of a peak, and other criteria have to be found: the elemental composition of the fragments,[28, 38, 39, 71, 240, 241, 243] or comparison with the fragments obtained under another ionization method (CI)[39, 126] or from different derivatives.

A recent study by Desiderio and co-workers[244] of a series of derivatized oligopeptides which are fragments of synthetic "scotophobin" illustrates the value of using a variety of isotopically labeled derivatives and computer analysis of the sequence peaks. For particular peptides the computer found more than one possible sequence, and the most probable was not always the correct sequence. A very extensive recent study by Morris and co-workers[208] describes sequencing of the enzyme ribitol dehydrogenase by degradation and mass spectrometric analysis of the partially separated oligopeptide mixtures. The success of this study certainly indicates that the method has come of age. The authors conclude that computer analysis of the mass spectral data is useful, but feel that there is no specific advantage of utilizing high-resolution data because ambiguities in the resulting sequences are so unimportant, despite the added complexity of the mixture spectra.

3. High-Resolution (Exact Mass) Data

The use of a wide variety of auxilary data, such as CI, FI, and CA spectra, metastable ions, and amino acid analyses, has been suggested to augment the mass spectral information in more complex cases. Surprisingly, the use of high-resolution data is uniquely controversial, attracting vociferous advocates both for and against. This method is based on the fact that measuring the mass of an ion with sufficient accuracy establishes its elemental composition unequivocally. Such measurements involve equipment of greatly increased cost and complexity; an on-line computer is desirable for exact mass determination of the whole spectrum. Achievable sensitivity is approximately tenfold lower unless special techniques are used. However, if proper equipment is available, which is the case in many modern research laboratories, it is difficult to see why this valuable additional information should not be utilized for more complex samples.

A sequence peak is identified if its exact mass, within the error of the experimental data, fits the predicted elemental composition. For example, B_1^+ in Ac-Pro-Gly-Phe-Gly-OCH$_3$ has the composition $C_7H_{11}O_2N$, or m/e

140.0712; a peak with an identical mass value indicates the presence of an N-terminal Pro. Of course, to check for all possible compositions requires the high-resolution recording of as many peaks as possible. Manual peak matching cannot be used because it necessitates 2–5 minutes per peak.[34,37] High-resolution mass spectra can be recorded on photoplates, with the masses measured by an automatic comparator–microdensitometer and the resulting data processed by the computer.[28,71] Real-time data collection and reduction during magnetic scanning is now commonly done by a dedicated on-line minicomputer.[245] Data collection giving the exact masses of as many as 600 peaks can be done every 2 minutes during fractional vaporization.[39] Sensitivities approaching those achievable by low-resolution recording of the mass spectrum appear possible for high-resolution spectra using sensitive photoplate recording or peak rescan in real time by computer feedback control.[232]

The algorithm used for finding the sequence from high-resolution spectra utilizes the logical steps involved in manual processing. Barber *et al.*[240] suggested starting from the molecular ion and subtracting the exact mass of —NH—CHR$_n$—COOCH$_2$ and —CO—NH—CHR$_n$—COOCH$_3$ for each of the possible amino acids until a matching peak (or peaks) is found in the spectrum. The procedure is then repeated to identify R$_{n-1}$, R$_{n-2}$, and so forth. It is therefore imperative that the molecular ion be identified unambiguously.

A more general method, not requiring knowledge of the amino acid composition, was suggested independently by Biemann[241,243] and McLafferty.[28,71] Basically, in this method the recognition of the sequence ions starts from the N-terminal end. An improved algorithm[38,39] allows the identification of unknown mixtures of oligopeptides. An example of its application to pure peptides is shown in Fig. 13.

High-resolution data are obtained from fractional vaporization (100°–350°) of 0.1 μmole of *O,N*-permethylated *N*-acetyl peptide esters. EI or CI spectra are recorded every 2 minutes on-line using a PDP-9 computer; low-resolution and metastable ion data are obtained in a separate scan.[29] A schematic representation of the sequencing algorithm is shown in Fig. 14. The sequence determination step is carried out off-line. The data are combined into one list and loaded into the core of the computer together with the list of the possible amino acid residues (including probable artifacts like *C*-MeGly or *C*-MeMet and dehydroalanine) and the list of the N-terminal, C-terminal, and side-chain derivatives. The accurate mass of the first possible amino acid is added to that of the N-terminal derivative (for example, permethylated Gly + acetyl = 71.0371 + 43.0184 = 114.0555), and a match for a possible B$_1{}^+$ ion is sought in the data list within the allowable error range (± 0.003 is a typical value). The accurate mass of CO (27.9449) is then subtracted, and a search is made for a possible A$_1{}^+$ peak. Then the next possible starting amino acid is tried, and so on. An N-terminus residue of a peptide

1 AMINO ACID SEQUENCE ANALYSIS

-SAMPLE NUMBER 66-2

N-TRIDEUTEROACETYL METHYL ESTER

- THE FOLLOWING ARE THE SEQUENCE PEAKS FOUND IN THIS PEPTIDE

SEQUENCE IDENTITY	FOUND	CALC	ERROR
A-1 ALA	89.079700	89.079410	-0.29
B-1 ALA	117.073170	117.074321	1.15
A-2 LEU	202.164000	202.163470	-0.53
B-2 LEU	230.158401	230.158380	-0.02
A-3 ALA	273.201201	273.200500	-0.70
B-3 ALA	301.196100	301.195491	-0.61
A-4 VAL	372.267500	372.268990	1.49
B-4 VAL	400.264310	400.263900	-0.41
A-5 VAL	471.336600	471.337400	0.80
B-5 VAL	499.333301	499.332310	-0.99
B-6 VAL	598.402000	598.400721	-1.28
MOL-ION	629.421000	629.419110	-1.89

- AMINO ACID SEQUENCE IN THIS PEPTIDE IS

*(D)AC-ALA *LEU *ALA *VAL *VAL *VAL*

1 AMINO ACID SEQUENCE ANALYSIS

-SAMPLE NUMBER 66-1

N-ACETYL METHYL ESTER

- THE FOLLOWING ARE THE SEQUENCE PEAKS FOUND IN THIS PEPTIDE

SEQUENCE IDENTITY	FOUND	CALC	ERROR
A-1 PRO	112.073300	112.076240	1.94
B-1 PRO	140.071199	140.071150	-0.05
A-2 GLY	169.095800	169.097748	1.95
B-2 GLY	197.090799	197.092659	1.86
A-3 PHE	316.164799	316.164299	-0.50
B-3 PHE	344.160500	344.159210	-1.29
A-4 GLY	373.189301	373.187046	-2.25
B-4 GLY	401.182598	401.181957	-0.64
MOL-ION	432.200401	432.200985	0.58

- AMINO ACID SEQUENCE IN THIS PEPTIDE IS

* AC-PRO *GLY *PHE *GLY *

FIG. 13. Computer-determined sequences using the high-resolution spectra of Ala-Leu-Ala-Val-Val-Val and Pro-Gly-Phe-Gly.[28]

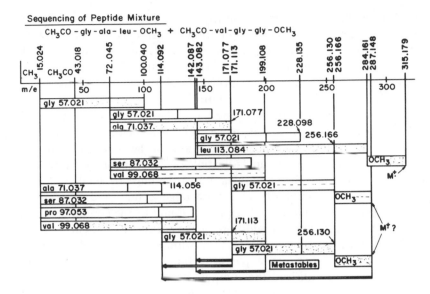

FIG. 14. Sequencing utilizing exact mass information from the mass spectrum of an oligopeptide mixture.[39]

component of the mixture is identified when both A_1^+ and B_1^+ are found. Whenever positive results are obtained, these exact masses are taken as starting data for the search for a dipeptide fragment. A search is attempted, again trying the different possible residues, but only the presence of A_2^+ or B_2^+ is required for identification. The process is repeated in search of possible tri-, tetra-, and higher peptides, until the search for an additional residue finds no suitable match for either A_i^+ of B_i^+ ions (see Fig. 9). For every partial sequence a possible molecular ion M^+ is sought by adding the accurate mass of the C-terminal ester group (OCH_3) to the mass of the last identified B_i^+ ion. When no molecular ion can be found, a search for indirect evidence of the molecular ion ($M^+ - CH_3$ or $M^+ - H_2O$) is attempted. When the two last attempts fail, a search for a possible missing residue is attempted by searching for a match for B_i^+ (last found) + each of the possible residues + OCH_3 (or + $OCH_3 - H_2O$).

Because peaks corresponding to tri- and higher peptide fragments may be so weak that they have not been recorded in the high-resolution spectrum, the low-resolution data of higher masses are also incorporated in the list. A similar sequence search is made for the C-terminal ions, until no further sequence matches are found. The amino acid composition indicated in these identification steps is confirmed by the search for immonium ions and characteristic aromatic fragments. Pseudo sequence peaks are investigated similarly, starting the search at possible "cinnamoyl residues" from Trp,

His, Tyr, or Phe, or the similar ion from Asx. Sequence → sequence meta-stable transitions are inspected to confirm the postulated dipeptide fragments (using the $B_1^+ \rightarrow A_1^+$ or the sequence → immonium transitions) and the N-terminal amino acid ($A_1^+ + \rightarrow I^+$). The identities of the postulated oligo-peptide components are also confirmed from the change of their relative abundances during fractional vaporization.

An added feature of Biemann's program[241] for pure peptides is that it stores and sums the intensities of the sequence peaks. When more than one sequence is found possible for a pure peptide, the value of the summed intensities for each sequence are taken as an indication of its probability. However, the highest sum does not always correspond to the most probable sequence.[244]

With increasing complexity of a pure oligopeptide, and especially with mixtures (Fig. 9), the number of sequences found by the computer increases, with sequences indicated in addition to those actually present in the mix-ture.[38, 39] Other possible combinations within the assigned error range, peaks from impurities, or apparent residues resulting from molecular rearrangements can mislead the computer search.[241] [Peaks indicating Gly or Ala can arise from side-chain rearrangement of other residues (see Section II,B,2).] The computer may also identify isomeric fragments to give a reverse sequence; for instance if both Ac-Pro-Gly... and Ac-Val-Pro... peptides are present in a mixture, the computer will also find (erroneously) Ac-Pro-Val. . . . Another source of confusion is the occurrence of combinations of residues which have identical elemental compositions (and thus the same exact mass), such as Lys and Val-Ala, Gln and Ala-Gly, Gly-Leu and Ala-Val. The possibility of these erroneous sequences emphasized the importance of including all available auxiliary information, such as metastable transitions, immonium ions, and residue characteristic ions.

Several bi-, tri-, and quaternary mixtures, including combinations of all of the known amino acids except Arg, were analyzed by this method. For example, for one mixture the computer identified five possible N-terminal residues: Ser($- H_2O$), Ala, Pro, Val, Leu (no distinction is made between Leu and Ile), and Phe, but the lack of immonium ions for Ser and Val, and the absence of phenyl ions eliminated Ser, Val, and Phe. Metastable ions con-firmed only Ala and Pro, so these were considered to be the only N-terminal residues in the mixture. Then the computer found peaks compatible with the following dipeptides: Ac-Ala-Gly..., Ac-Pro-Gly..., Ac-Ala-Ala..., Ac-Pro-Ser..., Ac-Ala-Phe..., Ac-Pro-Met..., Ac-Ala-Met..., Ac-Pro-Asp..., Ac-Ala-Tyr..., Ac-Pro-Phe..., and Ac-Pro-Asn.... The lack of immonium ions eliminated Ser and Asn, and metastable transitions confirmed only Ala-Phe and Pro-Phe. The following higher peptides were then indicated by the computer (verification by metastable transitions is not required due to sensitivity considerations; sequences confirmed by such transitions are shown in italics): Ac-*Ala-Phe-Gly*-Trp-Ser..., Ac-*Pro-Phe-Asp*-OCD₃, Ac-*Ala-Phe-*

Gly-Leu..., Ac-*Pro-Phe-His-Leu*-Ser..., Ac-*Ala-Phe-Val*-Ala..., Ac-*Pro-Phe*-Thr-Ala..., Ac-*Ala-Phe-Leu-Gly*..., Ac-*Ala-Phe*-Cys..., and Ac-*Ala-Phe*-Lys.... As no characteristic ions for Cys, Lys, Ser, Thr, Trp, and Val were found in the spectra, the only compatible sequences are Ac-Ala-Phe-Gly-Leu..., Ac-Pro-Phe-Asp-OCD₃..., Ac-Ala-Phe-Leu-Gly..., and Ac-Pro-Phe-His-Leu.... The following pseudo sequences were also identified, but only confirmed the previous findings (asterisks indicate pseudo terminal residues): *Phe-Asp..., *Phe-His-Leu.:., *His-Leu..., *Phe-Leu-Gly..., and *Phe-Gly-Leu.... The computer also identified the following C-terminal sequences: ...Phe-Asp-OCD₃, ...Leu-Leu-OCD₃, and ...Met-OCD₃. No higher fragments and only one molecular ion were identified by the program. The fractional vaporization plot (see Fig. 7) shows two maxima at 235° and 335° for Ac-Pro-Phe⁺, so that two peptides start with this partial sequence. Since Ac-Pro-Phe-His-Leu also reaches a maximum at 335°, this sequence is confirmed, but distinction of the sequences Ac-Ala-Phe-Leu-Gly and Ac-Ala-Phe-Gly-Leu is not possible. (Chemical ionization data had not been taken for this sample, but subsequent work shows that CI information giving molecular sizes would have been valuable at this stage.) Therefore, the sequences suggested by the computer are Ac-Pro-Phe-Asp-OCD₃, Ac-Ala-Phe-Gly-Leu..., Ac-Ala-Phe-Leu-Gly..., Ac-Pro-Phe-His-Leu..., ...Leu-Leu-OCD₃, and ...Met-CD₃; the composition of the mixture was in fact an equimolar amount of Ala-Phe-Ile-Gly-Leu-Met, Pro-Phe-His-Leu-Leu, and Pro-Phe-Asp. Thus the tripeptide and the parts of the larger peptides were correctly identified.

The apparent effort required to achieve unambiguous sequence information from these light-resolution data is somewhat in contrast to a recent report[208] of very extensive sequencing of mixtures from the degradation of real samples by Morris and co-workers utilizing only low-resolution data. This is most encouraging, suggesting that experience in mass spectral behavior and overlap of information from varied degradation products will make further automation possible without sacrificing accuracy.

Additional studies include mathematical models and theoretical predictions of sequencing results from low- and high-resolution data.[9, 211, 246–250] A program has also been written for low-resolution CI spectra.[126]

VII. THE FUTURE OF MASS SPECTROMETRY FOR PEPTIDE SEQUENCING

As reported above, there are a number of problem areas in classical polypeptide sequencing in which MS can be of value: (1) confirmation of oligopeptide sequences obtained by chemical means; (2) sequencing of N-blocked and other oligopeptides for which chemical techniques are not applicable;

(3) sequencing of oligopeptide samples which cannot be purified to the extent required by chemical methods; and (4) identification of abnormal amino acids[26a] or those containing synthetic chemical or isotopic labels. Note, however, that in these cases MS is supplementing, not displacing, the classical methods. For any general strategy utilizing MS which we can conceive, the polypeptide must be degraded by a reaction preferably attacking a few specific backbone bonds, the resulting oligopeptides separated, those that are still too large degraded by another specific cleavage reaction, the new oligopeptides separated, and so forth, just as now required by the classical approach. For a strategy in which MS is the sole means of sequencing the final oligopeptides, its lower molecular weight maximum will mean that many oligopeptides normally suitable for Edman sequencing would have to be further degraded and separated. However, there are only a limited number of reagents which give specific cleavages (cyanogen bromide, trypsin, chymotrypsin), so that in many cases the additional degradation of the pure oligopeptide could lead to a fairly complex mixture of smaller oligopeptides to be analyzed by the mass spectrometer. Such a mixture would be too complex for direct mixture analysis (see Section VI,B), even with fractional vaporization; a chromatographic technique would appear to be best for its separation. Thus we conclude that there is little incentive to develop a strategy based on MS sequencing of the oligopeptide fragments for those samples amenable to the classical technique with Edman sequencing.

Discouraging as this analysis may appear to be, development of such a strategy could be worthwhile if it could take advantage of the unusual sensitivity of MS. Present instruments are able to obtain a useable spectrum (and thus the sequence information) on a few *picomoles* of sample by EI and CI (FD apparently requires a somewhat higher amount); note also that $< 10^{-14}$ moles can be detected by single ion monitoring. Each step involving a chemical reaction and separation *in solution* demands a much larger amount of sample for handling and losses, so that if MS can replace steps of this kind the sensitivity should be improved. As stated above, we see no method at present to replace the initial polypeptide degradation and separation steps, but it should be possible to lower the initial sample requirement by replacing the final chemical sequencing of the oligopeptide which involves separate chemical reaction and separation steps for each residue.

For a method based on sequencing solely by MS, separation of the final mixture of oligopeptides suitable for sequencing would appear to be best done by a chromatographic method. Gas chromatography, particularly the scheme of Biemann[230, 231, 234–236] discussed in Section VI,C, is promising for this; with reduction and derivatization it appears that all of the possible dipeptides and many tripeptides can be simultaneously separated and sequenced by an automated GC/MS/computer system. This does pose a rather difficult problem in the degradation step, however; hydrolysis conditions under which a 30-residue oligopeptide gives fragments no larger than tripeptides would surely give a high proportion of mono- and dipeptides, raising the quantity of oligopeptide which must be hydrolyzed. Note that even if all possible dipeptides (and, in unfavorable cases, tripeptides) of a 30-residue oligopeptide are prepared and sequenced these will not provide unambiguous sequence information for the oligopeptide. Other advantages and drawbacks of the present GC/MS approach have been discussed in Section VI,C.

Less of the oligopeptide will be required if larger fragments can be used to provide the overlapping sequence information; separation of these is probably most feasible with some form of liquid chromatography Although FD/MS may be able to sequence directly oligopeptides of ~ 15 residues, it is not amenable to continuous chromatographic sampling, and so some other detector sensitive to 10^{-11} mole of peptide will be necessary to identify the chromatographic fractions for FD analysis. It would appear that a technique other than FD will also be necessary for quantitative analysis of the component amino acids at $\ll 10^{-12}$ mole sensitivity. For this problem we feel that the LC/MS/computer system (Section VI,D) could yield an even higher sensitivity, combined with relatively unambiguous results and automated operation. An accurate amino acid analysis on $< 10^{-12}$ mole of the pure oligopeptide should be possible by complete hydrolysis and on-line multiple ion monitoring of an LC separation. To obtain the sequence information on a minimum amount of the peptide, the degree of degradation is critical; this must proceed far enough to provide sufficient quantities of overlapping residues of oligopeptides of sufficient volatility, but the wasteful further degradation of these to mono- and dipeptides should be minimized. From the recent reports by Morris and co-workers[207, 208] nonspecific enzymes such as elastase and thermolysin appear to be very promising for this. Possibly the products of sufficiently low molecular weight could be removed chromatographically from the reaction at intervals, or the reaction progress monitored by running aliquots of the reaction mixture by FD or direct CI. The mixture of fragment peptides would then be acylated, esterified, separated by LC on a microcolumn[47] (flow rate sufficiently low so that no stream splitting is necessary), and the total effluent continuously introduced into the CI/MS for on-line sequencing of the eluted fractions. It is conceivable that 10^{-11} mole of a 30-residue oligopeptide could be sequenced in this way, making possible the total sequencing (including overlap information) of 10^{-10} mole or less of the original polypeptide sample. Initial efforts on the implementation of such a scheme in this laboratory appear to be promising.

A future alternative method which also is under consideration in this laboratory would utilize polypeptide degradation in the MS ion source, separation of the resulting oligopeptides in the high-resolution MS according to their mass, and sequencing by collisional activation of the separated oligopeptide ions. Although RF-heating pyrolysis of peptides gave discouraging results,[135] very recent work[251] indicates that mass spectral information can be obtained from compounds of very low volatility by alpha particle bombardment; this presumably produces extremely rapid local heating in the solid sample to effect vaporization and ionization. It might be possible that heating with high energy particles or laser irradiation in a CI source during direct solution introduction of the polypeptide would achieve pyrolysis, ionization, and thermalization producing stable oligopeptide ions which could then be separated mass spectrometrically. The individual mass spectral peaks would then be subjected to on-line collisional activation and energy analysis to provide a "CA spectrum" indicative of the amino acid sequence of each oligopeptide.[84] The whole process would be amenable to on-line computer data acquisition and feedback control,[232] providing a further potential method for automated polypeptide sequencing at the subnanomole level.

Acknowledgments

We are indebted to Drs. R. Venkataraghavan, H. K. Wipf, Philip Irving, and Malcolm McCamish, and Mr. B. G. Dawkins who have been active in the Cornell research program in peptide sequencing in the last few years. This research has been generously funded by the National Institutes of Health under grant GM16609. Completion of a manuscript of this size would not have been possible without the patience, perseverance, and secretarial skills of Miss Patricia L. Odell.

References

[1] H. Budzikiewicz, C. Djerassi, and D. H. Williams, in "Structure Elucidation of Natural Products by Mass Spectrometry," Vol. 2, p. 183. Holden-Day, San Francisco, California, 1964.

[2] K. Biemann, "Mass Spectrometry, Organic Chemical Applications," p. 295. McGraw-Hill, New York, 1962.

[3] J. H. Jones, Quart. Rev., Chem. Soc. 22, 302 (1968).

[4] E. Lederer and B. C. Das, in "Peptides" (Proceedings of the VIIIth European Peptide Symposium) (H. C. Beyerman, A. Van de Linde, and W. Maassen Van den Brink, eds.), p. 131. North-Holland Publ., Amsterdam, 1967.

[5] M. M. Shemyakin, Pure Appl. Chem. 17, 313 (1968).

[6] E. Lederer, Pure Appl. Chem. 17, 489 (1968).

[7] G. E. Van Lear and F. W. McLafferty, Annu. Rev. Biochem. 38, 289 (1969).

[8] B. C. Das and E. Lederer, Advan. Anal. Chem. Instrum. 8, 255 (1970).

[9] B. Sheldrick, Quatr. Rev., Chem. Soc. 24, 454 (1970).

[10] B. C. Das and E. Lederer, in "Topics in Organic Mass Spectrometry" (A. L. Burlingame, ed.), p. 255. Wiley (Interscience), New York, 1970.

[11] M. M. Shemyakin, Yu. A. Ovchinnikov, and A. A. Kiryushkin, in "Mass Spectrometry: Techniques and Applications" (G. W. A. Milne, ed.), p. 289. Wiley, New York, 1971.

[12] V. M. Lipkin, Yu. B. Alakhov, N. A. Aldanova, M. Yu. Feigina, A. A. Kiryushkin, A. I. Miroshnikov, Yu. A. Ovchinnikov, B. V. Rosinov, M. M. Shemyakin, and A. Vinogradov, in "Peptides" (Proceedings of the Xth

European Peptide Symposium) p. 232. (E. Scoffone, ed.), North-Holland Publ., Amsterdam, 1971.

[13] D. M. Desiderio, Jr., *Intra-Sci. Chem. Rep.* **5**, 391 (1971).

[14] R. G. Cooks and G. S. Johnson, *in* "Mass Spectrometry" (D. H. Williams, ed.), p. 139. Chemical Society, London, 1971.

[15] H. Budzikiewicz, *in* "Mass Spectrometry Proceedings of International School on Mass Spectrometry" (J. Marsal, ed.), p. 62. J. Stefan Inst., Lubljana, Yugoslavia, 1971.

[16] B. C. Das and E. Lederer, *in* "New Techniques in Amino Acid, Peptide, and Protein Analysis" (A. Niederwiesser and G. Pataki, eds.), p. 175. Ann Arbor Sci. Publ., Ann Arbor, Michigan, 1971.

[17] K. Biemann, *in* "Biomedical Applications of Mass Spectrometry" (G. R. Waller, ed.), p. 405. Wiley (Interscience), New York, 1972.

[18] P. A. Leclercq, P. A. White, K. Hägele, and D. M. Desiderio, *in* "Chemistry and Biology of Peptides," (J. Meienhofer, ed.) p. 687. Ann Arbor Sci. Publ., Ann Arbor, Michigan, 1972.

[19] B. C. Das and E. Lederer, *in* "Peptides" (C. H. Nesvadba, ed.), p. 253. North-Holland Publ., Amsterdam, 1973.

[20] H. R. Morris, *in* "New Techniques in Biophysics and Cell Biology" (R. H. Pain and B. J. Smith, eds.), p. 149. Wiley (Interscience), New York, 1973.

[21] V. V. Neuhoff, F. von der Haar, E. Schlimme, and M. Weise, *Hoppe-Seyler's Z. Physiol. Chem.* **350**, 121 (1969).

[22] P. Edman and A. Begg, *Eur. J. Biochem.* **1**, 80 (1967).

[23] R. A. Laursen, *Eur. J. Biochem.* **20**, 89 (1971).

[24] R. A. Laursen, M. J. Horn, and A. G. Bonner, *FEBS (Fed. Eur. Biochem. Soc.) Lett.* **21**, 67 (1972).

[25] G. M. Edelman, B. A. Cunningham, W. E. Gall, P. D. Gottlieb, U. Rutishauser, and J. Waxdal, *Biochemistry* **63**, 78 (1969).

[26] J. F. Thompson, C. J. Morris, and I. K. Smith, *Annu. Rev. Biochem.* **38**, 137 (1969).

[26a] T. H. Jukes and R. Holmquist, *J. Mol. Evol.* **2**, 343, (1973).

[27] J. Lenard and P. M. Gallop, *Anal. Biochem.* **34**, 286 (1970).

[28] M. Senn, R. Venkataraghavan, and F. W. McLafferty, *J. Amer. Chem. Soc.* **88**, 5593 (1966).

[29] D. W. Thomas, B. C. Das, S. D. Géro, and E. Lederer, *Biochem. Biophys. Res. Commun.* **32**, 199 (1968).

[30] K. L. Agarwal, G. W. Kenner, and R. C. Sheppard, *J. Amer. Chem. Soc.* **91**, 3096 (1969).

[31] W. R. Gray, L. H. Wojcik, and J. H. Futrell, *Biochem. Biophys. Res. Commun.* **41**, 1111 (1970).

[32] P. A. Leclercq and D. M. Desiderio, *Anal. Lett.* **4**, 305 (1971).

[33] P. A. Leclercq, L. C. Smith, and D. M. Desiderio, *Biochem. Biophys. Res. Commun.* **45**, 937 (1971).

[34] H. R. Morris, D. H. Williams, and R. P. Ambler, *Biochem. J.* **125**, 189 (1971).

[35] H. U. Winkler and H. D. Beckey, *Biochem. Biophys. Res. Commun.* **46**, 391 (1972), and references cited therein.

[36] M. A. Baldwin and F. W. McLafferty, *Org. Mass Spectrom.* **7**, 1353 (1973).

[37] A. J. Geddes, G. N. Graham, H. R. Morris, F. Lucas, M. Barber, and W. A. Wolstenholme, *Biochem. J.* **114**, 695 (1969).

[38] F. W. McLafferty, R. Venkataraghavan, and P. Irving, *Biochem. Biophys. Res. Commun.* **34**, 774 (1970).

[39] H. K. Wipf, P. Irving, M. McCamish, R. Venkataraghavan, and F. W. McLafferty, *J. Amer. Chem. Soc.* **95**, 3369 (1973).

[40] P. Roepstorff, R. K. Spear, and K. Brunfeldt, *FEBS (Fed. Eur. Biochem. Soc.) Lett.* **15**, 237 (1971).

[41] P. Roepstorff and K. Brunfeldt, *FEBS (Fed. Eur. Biochem. Soc.) Lett.* **21**, 370 (1972).

[42] N. Ling, J. Rivier, R. Burgus, and R. Guillemin, *Biochemistry* **12**, 5305 (1973).

[43] H. J. Forster, J. A. Kelley, H. Nau, and K. Biemann, *in* "Techniques of Combined GC-MS" (W. H. McFadden, ed.), p. 385. Wiley (interscience), New York, 1973; also *in* "Chemistry and Biology of Peptides" (J. Meienhofer, ed.), p. 679. Ann Arbor Sci. Publ., Ann Arbor, Michigan, 1972.

[44] D. H. Calam, *J. Chromatogr.* **70**, 146 (1972).

[45] Yu. A. Ovchinnikov and A. A. Kiryushkin, *FEBS (Fed. Eur. Biochem. Soc.) Lett.* **21**, 300 (1972).

[46] R. M. Caprioli, W. E. Seifert, and D. E. Sutherland, *Biochem. Biophys. Res. Commun.* **55**, 67 (1973).

[47] M. A. Baldwin and F. W. McLafferty, *Org. Mass Spectrom.* **7**, 1111 (1973).

[47a] P. Arpino, M. A. Baldwin, and F. W. McLafferty, *Biomed. Mass Spectrom.* **1**, 80 (1974).

[48] F. F. Richards, W. T. Barnes, R. E. Lovins, R. Salomone, and M. D. Waterfield, *Nature (London)* **221**, 1241 (1969).

[49] T. Fairwell, W. T. Barnes, F. F. Richards, and R. E. Lovins, *Biochemistry* **9**, 2260 (1970).

[50] T. Fairwell, S. Ellis, and R. E. Lovins, *Anal. Biochem.* **53**, 115 (1973).

[51] T. Fairwell and H. B. Brewer, *Fed. Proc., Fed. Amer. Soc. Exp. Biol.* **32**, 648 (1973).

[52] E. L. Cannon and R. E. Lovins, *Anal. Biochem.* **46**, 33 (1972).

[53] S. Ellis, T. Fairwell, and R. E. Lovins, *Biochem. Biophys. Res. Commun.* **49**, 1407 (1972).

[54] T. Sun and R. E. Lovins, *Anal. Biochem.* **45**, 176 (1972).

[55] R. E. Lovins, J. Craig, F. Thomas, and C. McKinney, *Anal. Biochem.* **47**, 539 (1972).

[55a] F. Weygand and R. Obermeier, *Eur. J. Biochem.* **20**, 72 (1971).

[55b] H. B. Brewer, T. Fairwell, R. Ronan, G. W. Sizemore, and C. D. Arnaud, *Proc. Nat. Acad. Sci. U.S.* **69**, 3585 (1972).

[56] H. Tschesche, M. Schneide, and E. Wachter, *FEBS (Fed. Eur. Biochem. Soc.) Lett.* **23**, 367 (1972).

[57] H. Tschesche and E. Wachter, *Eur. J. Biochem.* **16**, 187 (1970).

[58] H. Tschesche and E. Wachter, *Hoppe-Seyler's Z. Physiol. Chem.* **351**, 449 (1970).

[59] H. Hagenmaier, W. Ebbighaussen, G. Nicholson, and W. Votsch, *Z. Naturforsch. B* **25**, 681 (1970).

[60] N. A. Aldanova, E. I. Vinogradova, S. A. Kazaryan, B. V. Rosinov, and M. M. Shemyakin, *Biokhimiya* **35**, 854 (1970).

[61] J. J. Maher, M. E. Furey, and L. J. Greenberg, *Tetrahedron Lett.* 29 (1971).

[62] G. C. Barret and J. R. Chapman, *Chem. Commun.* 335 (1968).

[63] V. Bacon, E. Jellum, W. Patton, W. Pereira, and B. Halpern, *Biochem. Biophys. Res. Commun.* **37**, 878 (1969).

[64] W. R. Gray and V. D. del Valle, *Biochemistry* **9**, 2134 (1970).

[65] G. Prota, F. Chioccara, and A. Previero, *Biochimie* **53**, 51 (1971).

[65a] V. M. Orlov, Ya. M. Varshavsky, and A. A. Kiryushkin, *Org. Mass Spectrom.* **6**, 9 (1972).

[65b] P. Brown and G. R. Pettit, *Org. Mass Spectrom.* **3**, 67 (1970).

[66] F. Weygand, A. Prox, H. H. Fessel, and K. K. Sun, *Z. Naturforsch. B* **20**, 1169 (1965).

[67] A. Day, H. Falter, J. P. Lehman, and R. E. Hamilton, *J. Org. Chem.* **38**, 782 (1973).

[68] I. Lengyel, R. A. Salomone, and K. Biemann, *Org. Mass Spectrom.* **3**, 789 (1970).

[69] D. Stehelin, Ph.D. Thesis, Université Louis Pasteur de Strasbourg, France, 1972.

[69a] K. M. Baker, M. A. Shaw, and D. H. Williams, *Chem. Commun.* 1108 (1969).

[70] F. Franěk, B. Keil, D. W. Thomas, and E. Lederer, *FEBS (Fed. Eur. Biochem. Soc.) Lett.* **2**, 309 (1969).

[71] M. Senn and F. W. McLafferty, *Biochem. Biophys. Res. Commun.* **23**, 381 (1966).

[72] J. Van Heijenoort, E. Bricas, B. C. Das, E. Lederer, and W. A. Wolstenholme, *Tetrahedron* **23**, 3403 (1967).

[73] E. Hunt and H. R. Morris, *Biochem. J.* **135**, 833 (1973).

[74] L. I. Rostovtseva and A. A. Kiryushkin, *Org. Mass Spectrom.* **6**, 1 (1972).

[75] C. N. C. Drey and J. Lowbridge, *Org. Mass Spectrom.* **7**, 779 (1973).

[76] M. M. Shemyakin, Yu. A. Ovchinnikov, A. A. Kiryushkin, E. I. Vinogradova, Yu. B. Alakhov, V. M. Lipkin, and B. V. Rosinov, *J. Gen. Chem. USSR* **38**, 765 (1968).

[77] K. Heyns and H.-F. Grützmacher, *Fortschr. Chem. Forsch.* **6**, 536 (1966).

[78] B. C. Das and R. D. Schmid, *FEBS (Fed. Eur. Biochem. Soc.) Lett.* **25**, 253 (1972).

[78a] G. Brignon, J. C. Mercier, B. Ribadeau-Dumas, and B. C. Das, *FEBS (Fed. Eur. Biochem. Soc.) Lett.* **27**, 301 (1972).

[79] A. Prox and K. K. Sun, *Z. Naturforsch. B* **21**, 1028 (1966).

[80] J. P. Flikweert, W. Heerma, T. J. Penders, G. Dijkstra, and J. F. Arens, *Rec. Trav. Chim. Pays-Bas* **86**, 293 (1967).

[81] E. Stenhagen, *Z. Anal. Chem.* **181**, 462 (1961).

[82] B. A. Andersson, *Acta. Chem. Scand.* **21**, 2906 (1967).

[83] A. Prox and F. Weygand, *in* "Peptides" (H. C. Beyerman, A. Van de Linde, and W. Maassen Van den Brink, eds.), p. 158. North-Holland Publ., Amsterdam, 1967.

[84] K. Levsen, H. K. Wipf, and F. W. McLafferty, *Org. Mass Spectrom*, **8**, 117 (1974).

[85] C. Bogentoft, J.-K. Chang, H. Sievertsson, B. Currie, and K. Folkers, *Org. Mass Spectrom.* **6**, 735 (1972).

[86] R. T. Aplin, I. Eland, and J. H. Jones, *Org. Mass Spectrom.* **2**, 795 (1969).

[87] R. T. Aplin, J. H. Jones, and B. Liberek, *Chem. Commun.* 794 (1965).

[88] K. Biemann, *Chimia* **14**, 393 (1960).

[89] K. Heyns and H.-F. Grützmacher, *Justus Liebig's Ann. Chem.* **669**, 189 (1963).

[90] E. Bricas, J. Van Heijenoort, M. Barber, W. A. Wolstenholme, B. C. Das, and E. Lederer, *Biochemistry* **4**, 2254 (1965).

[91] Yu. A. Ovchinnikov, A. A. Kiryushkin, E. I. Vinogradova, B. V. Rosinov, and M. M. Shemyakin, *Biokhimiya* **32**, 427 (1967); *Biochemistry (USSR)* **32**, 351 (1967).

[92] F. W. McLafferty, "Interpretation of Mass Spectra," 2nd ed. Benjamin, Reading, Mass., 1973.

[93] M. M. Shemyakin, Yu. A. Ovchinnikov, A. A. Kiryushkin, E. I. Vinogradova, A. I. Viroshnikov, Yu. B. Alakhov, V. M. Lipkin, Yu. B. Shvestov, N. S. Wul'fson, B. V. Rosinov, V. N. Bochkarev, and V. M. Burikov, *Nature* (*London*) 211, 361 (1966).

[94] P. Roepstorff, K. Norris, S. Severinsen, and K. Brunfeldt, *FEBS* (*Fed. Eur. Biochem. Soc.*) *Lett.* 9, 235 (1970).

[95] H. R. Morris, R. J. Dickinson, and D. H. Williams, *Biochem. Biophys. Res. Commun.* 51, 247 (1973).

[96] P. A. Leclercq and D. M. Desiderio, *Biochem. Biophys. Res. Commun.* 45, 308 (1971).

[97] M. L. Polan, W. J. McMurray, S. R. Lipsky, and S. Lande, *Biochem. Biophys. Res. Commun.* 38, 1127 (1970).

[98] A. A. Kiryushkin, V. A. Gorlenko, Ts. E. Agadzhanyan, B. V. Rosinov, Yu. A. Ovchinnikov, and M. M. Shemyakin, *Experientia* 24, 883 (1968).

[99] E. Bayer and W. A. König, *J. Chromatogr. Sci.* 7, 95 (1969).

[100] E. Bayer, E. Hagenmayer, W. König, H. Pauschmann, and W. Sautter, *Z. Anal. Chem.* 243, 670 (1968).

[101] N. S. Wul'fson, V. N. Bochkarev, B. V. Rosinov, M. M. Shemyakin, Yu. A. Ovchinnikov, A. A. Kiryushkin, and A. I. Miroshnikov, *Tetrahedron Lett.* 39 (1966).

[102] M. M. Shemyakin, Yu. A. Ovchinnikov, A. A. Kiryushkin, A. I. Miroshnikov, and B. V. Rosinov, *Zh. Obshch. Khim.* 40, 407 and 443 (1970).

[103] A. A. Kiryushkin, A. I. Miroshnikov, Yu. A. Ovchinnikov, B. V. Rosinov, and M. M. Shemyakin, *Biochem. Biophys. Res. Commun.* 24, 943 (1966).

[104] P. B. Lodler and E. P. Abraham, *Biochem, J.* 123, 471 (1971).

[105] G. H. de Hass, B. Franĕk, B. Keil, D. W. Thomas, and E. Lederer, *FEBS* (*Fed. Eur. Biochem. Soc.*) *Lett.* 4, 25 (1969).

[105a] J. Novotny, L. Dolejs, and F. Franĕk, *Eur. J. Biochem.* 31, 277 (1972).

[106] H. C. Beyerman, P. Kranemberg, and J. L. M. Syrier, *Rec. Trav. Chim. Pays-Bas* 90, 791 (1971).

[107] R. Burgus, T. F. Dunn, D. M. Desiderio, and R. Guillemin, *C.R. Acad. Sci., Ser. D* 269, 1870 (1969).

[108] R. Burgus, T. F. Dunn, D. M. Desiderio, D. N. Ward, W. Vale, and R. Guillemin, *Nature* (*London*) 226, 321 (1970).

[109] R. Guillemin, M. Amoss, R. Blackwell, J. Rivier, N. Ling, and W. Vale, *Biochem. Biophys. Res. Commun.* 48, 1093 (1972).

[110] J.-K. Chang, H. Sievertsson, C. Bogentoft, B. Currie, K. Folkers, and G. D. Daves, *J. Med. Chem.* 14, 481 (1971).

[111] M. Matsuo, Y. Baba, R. M. G. Nair, A. Aimura, and A. V. Schally, *Biochem. Biophys. Res. Commun.* 43, 1334 (1971).

[112] B. L. Currie, H. Sievertsson, G. Bogentoft, J.-K. Chang, K. Folkers, C. Y. Bowers, and R. F. Doolittle, *Biochem. Biophys. Res. Commun.* 42, 1180 (1971).

[113] P. Ferlund and L. Josefsson, *Science* 177, 173 (1972).

[114] H. Kato, T. Suzuki, K. Okada, T. Kimura, and S. Sakakibara, *Experientia* 29, 574 (1973).

[115] J. Rivier, W. Vale, R. Burgus, N. Ling, M. Amoss, R. Blackwell, and R. Guillemin, *J. Med. Chem.* 16, 545 (1973).

[116] P. Pfaender, *Justus Liebig's Ann. Chem.* 707, 209 (1967).

[117] H. R. Morris, *FEBS* (*Fed. Eur. Biochem. Soc.*) *Lett.* 22, 257 (1972).

[118] M. L. Polan, W. J. McMurray, S. R. Lipsky, and S. Lande, *J. Amer. Chem. Soc.* **94**, 2847 (1972).

[119] N. Ling, R. Burgus, J. Rivier, W. Vale, and P. Brazeau, *Biochem. Biophys. Res. Commun.* **50**, 127 (1973).

[120] H. R. Morris and D. H. Williams, *Chem. Commun.* **114** (1972).

[121] G. W. A. Milne, A. A. Kiryushkin, Yu. B. Alakhov, V. M. Lipkin, and Yu. A. Ovchinnikov, *Tetrahedron* **26**, 299 (1970).

[122] D. W. Thomas, B. C. Das, S. D. Géro, and E. Lederer, *Biochem. Biophys. Res. Commun.* **32**, 519 (1968).

[123] M. M. Shemyakin, E. I. Vinogradova, Yu. A. Ovchinnikov, A. A. Kiryushkin, M. Yu. Feigina, N. A. Aldanova, Yu. B. Alakhov, V. M. Lipkin, B. V. Rosinov, and L. A. Fonina, *Tetrahedron* **25**, 5785 (1969).

[124] F. H. Field, in "Mass Spectrometry" (A. Maccoll, ed.), p 133, Butterworth, London, 1972.

[124a] G. W. A. Milne and M. J. Lacey, *Crit. Rev. Anal. Chem.* **4**, 45 (1974).

[125] R. J. Beuhler, E. Flanigan, L. J. Greene, and L. Friedman, *Biochem. Biophys. Res. Commun.* **46**, 1082 (1972).

[125a] R. J. Beuhler, E. Flanigan, L. J. Greene, and L. Friedman, *J. Amer. Chem. Soc.* **96**, 3990 (1974).

[126] A A. Kiryushkin, H. M. Fales, T. Axenrod, E. J. Gilbert, and G. W. A. Milne, *Org. Mass Spectrom.* **5**, 19 (1971).

[127] D. V. Bowen and F. H. Field, *Org. Mass Spectrom.* **9**, 195 (1974).

[128] D. V. Bowen and F. H. Field, *Int. J. Peptide Protein Res.* **5**, 435 (1973).

[129] D. M. Desiderio, R. Burgus, T. F. Dunn, W. Vale, R. Guillemin, and D. N. Ward, *Org. Mass Spectrom.* **5**, 221 (1971).

[130] E. Flanigan, R. J. Beuhler, L. Friedman, and L. J. Greene, *Fed. Proc., Fed. Amer. Soc. Exp. Biol.* **31**, A446 (1972).

[131] G. W. A. Milne, private communication (1973).

[131a] H. M. Fales, private communication (1973).

[132] M. Meot-Ner and F. H. Field, *J. Amer. Chem. Soc.* **95**, 7207 (1973).

[133] L. Friedman, private communication (1974).

[134] G. A. Junk and H. J. Svec, *Anal. Biochem.* **6**, 199 (1963).

[134a] H. J. Svec and G. A. Junk, *J. Amer. Chem. Soc.* **86**, 2278 (1964).

[135] C. Pasquale, F. W. McLafferty, and W. Simon, unpublished (1968).

[135a] W. A. Wolstenholme and L. C. Vining, *Tetrahedron Lett.* 2785 (1966).

[136] K. Levsen and H. D. Beckey, *Org. Mass Spectrom.* **9**, 570 (1974).

[137] K. L. Rinehart, Jr., R. T. Hargreaves, J. C. Cook, Jr., R. M. Milberg, and K. L. Olson, *168th Meet., Amer. Chem. Soc.* Paper ANAL-58 (1974).

[137a] C. C. Sweeley, J. N. Gerber, B. Soltman, and J. F. Holland, *168th Meet., Amer. Chem. Soc.* Paper ANAL-57 (1974).

[138] S. Asante-Poku, G. W. Wood, and D. E. Schmidt, Jr., submitted for publication.

[139] G. W. Wood, *Biomed. Mass. Spectrom.* **1**, 206 (1974).

[140] C. O. Andersson, *Acta Chem. Scand.* **12**, 1353 (1958).

[141] K. L. Rinehart and G. E. Van Lear, in "Biomedical Application of Mass Spectrometry" (G. R. Waller, ed.), p. 449. Wiley (Interscience), New York, 1972.

[142] D. W. Bishay, Z. Kowalewski, and J. D. Phillipson, *Phytochemistry*, **12**, 693 (1973).

[143] H-W. Fehlhaber, J. Uhlendorf, S. I. Davis, and R. Tschesch, *Justus Liebig's Ann. Chem.* **759**, 195 (1972).

[144] O. A. Mascaretti, V. M. Merkuza, G. E. Ferraro, E. A. Ruveda, C. Chang, and E. Wenkert, *Phytochemistry* **11**, 1133 (1972).

[145] J. Marchand, X. Monseur, and M. Dias, *Ann. Pharm. Fr.* **26**, 771 (1968).

[146] D. A. Holwerda, *Eur. J. Biochem.* **28**, 340 (1972).

[147] J. F. G. Vliegenthart and L. Dorland, *Biochem. J.* **117**, 31 (1970).

[148] R. Toubiana, J. E. G. Barnett, E. Sach, B. C. Das, and E. Lederer, *FEBS (Fed. Eur. Biochem. Soc.) Lett.* **8**, 207 (1970).

[149] A. A. Kiryushkin, V. A. Gorlenko, B. V. Rosinov, Yu. A. Ovchinnikov, and M. M. Shemyakin, *Experientia* **25**, 913 (1969).

[150] S. Takeuchi, M. Senn, R. W. Curtis, and F. W. McLafferty, *Phytochemistry* **6**, 287 (1967).

[151] M. M. Shemyakin, Yu. A. Ovchinnikov, E. I. Vinogradova, M. Yu. Feigina, A. A. Kiryushkin, N. A. Aldanova, Yu. B. Alakhov, V. M. Lipkin, and B. V. Rosinov, *Experientia* **23**, 428 (1967).

[152] T. P. King, *Biochemistry* **5**, 3454 (1966).

[153] H. Vetter-Dietchtl, W. Vetter, W. Richter, and K. Biemann, *Experientia* **24**, 340 (1968).

[154] K. Toi, E. Bynum, E. Norris, and H. A. Itano, *J. Boil. Chem.* **242**, 1036 (1967).

[155] J. Lenard and P. M. Gallop, *Anal. Biochem.* **29**, 203 (1969).

[156] A. A. Kiryushkin, Yu. A. Ovchinnikov, M. M. Shemyakin, V. N. Bochkarev, B. V. Rosinov, and N. S. Wul'fson, *Tetrahedron Lett.* 33 (1966).

[157] M. Barber, P. Jolles, E. Vilkas, and E. Lederer, *Biochem. Biophys. Res. Commun.* **18**, 469 (1965).

[158] P. A. Cruickshank and J. C. Sheehan, *Anal. Chem.* **36**, 1191 (1964).

[159] G. M. Schier and B. Halpern, *Aust. J. Chem.* **27**, 393 (1974).

[160] G. Ungar, D. M. Desiderio, and W. Parr, *Nature (London)* **238**, 198 (1972).

[161] D. M. Desiderio, G. Ungar, and P. A. White, *Chem. Commun.* 432 (1971).

[162] E. Bayer, H. Eckstein, K. Hägele, W. A. Lönig, W. Brüning, H. Hagenmaeir, and W. Parr, *J. Amer. Chem. Soc.* **92**, 1735 (1970).

[163] A. Prox, J. Schmid, and H. Ottenheym, *Justus Liebig's Ann. Chem.* **722**, 179 (1969).

[164] T. Wieland, G. Lueben, H. Ottenheym, J. Faesel, J. X. de Vries, W. Konz, A. Prox, and J. Schmid, *Angew. Chem.* **80**, 209 (1968); *Angew. Chem., Int. Ed. Engl.* **7**, 204 (1968).

[165] R. B. Fairweather, M. L. Tanzer, and P. M. Gallop, *Biochem. Biophys. Res. Commun.* **48**, 1311 (1972).

[166] W. W. Stewart, *Nature (London)* **238**, 202 (1972).

[167] M. M. Shemyakin, Yu. A. Ovchinnikov, E. I. Vinogradova, A. A. Kiryushkin, M. Yu. Feigina, N. A. Aldanova, Yu. B. Alakhov, V. M. Lipkin, A. I. Miroshnikov, B. V. Rosinov, and S. A. Kazaryan, *FEBS (Fed. Eur. Biochem. Soc.) Lett.* **7**, 8 (1970).

[168] A. Kakinuma, A. Ouchida, T. Shima, H. Sugino, M. Isano, G. Tamura, and K. Arima, *Agr. Biol. Chem.* **33**, 1669 (1969).

[169] M. Hiramoto, K. Okada, and S. Nagai, *Tetrahedron Lett.* 1087 (1970).

[170] M. Barber, W. A. Wolstenholme, M. Guinand, G. Michel, B. C. Das, and E. Lederer, *Tetrahedron Lett.* 1331, (1965).

[171] M. Guinand, M. J. Vacheron, G. Michel, B. C. Das, and E. Lederer, *Tetrahedron, Suppl.* **7**, 271 (1966).

[172] F. Laneelle, J. Asselineau, W. A. Wolstenholme, and E. Lederer, *Bull. Soc. Chim. Fr.* 2133 (1965).

[173] E. Vilkas, A. Rojas, B. C. Das, W. A. Wolstenholme, and E. Lederer, *Tetrahedron* **22**, 2809 (1966).

[174] H. Morishima, T. Takita, T. Aoyagi, T. Takeuchi, and H. Umezawa, *J. Antibiot.* **23**, 263 (1970).

[175] T. Miyanu, M. Tomigasu, H. Iizuka, S. Tomisaka, T. Takita, T. Aoyagi, and H. Umezawa, *J. Antibiot.* **25**, 489 (1972).

[176] D. W. Thomas and T. Ito, *Tetrahedron* **25**, 1985 (1969).

[177] D. W. Thomas, E. Lederer, M. Bodanszky, J. Izdebski, and I. Muramatsu, *Nature (London)* **220**, 582 (1968).

[178] A. Voiland, M. Bruneteau, and G. Michel, *Eur. J. Biochem.* **21**, 285 (1971).

[179] M. Bodanszky, A. A. Bodanszky, C. A. Ralofsky, R. C. Strong, and R. L. Foltz, *J. Antibiot.* **24**, 294 (1971).

[180] J. P. Kamerling, W. Heerma, and J. F. G. Vliegenthart, *Org. Mass Spectrom.* **1**, 351 (1968).

[181] R. T. Aplin, J. H. Jones, and B. Liberek, *J. Chem. Soc.,* C 1011 (1968).

[182] R. T. Aplin, and J. H. Jones, *J. Chem. Soc.,* C 1770 (1968).

[183] J. P. Kamerling, W. Heerma, T. J. Penders, and J. F. G. Vliengenthart, *Org. Mass Spectrom.* **1**, 345 (1968).

[184] T. J. Penders, H. Copier, G. Dijkstra, and J. F. Arens, *Rec. Trav. Chim. Pays-Bas* **85**, 879 (1966).

[185] R. C. Hamilton, G. V. Patil, K. Jayasimhulu, and R. A. Day, *Org. Mass Spectrom.* **9**, 211 (1974).

[186] E. Wünsch and E. Jaeger, *Hoppe Seyler's Z. Physiol. Chem.* **352**, 1584 (1971).

[187] M. R. Altamura, R. E. Andreotti, M. L. Bazinet, and L. Lony, *J. Food Sci.* **35**, 134 (1970).

[188] N. Van Chuyen, T. Kurata, and M. Fujimaki, *Agr. Biol Chem.* **37**, 1613 (1973).

[189] G. V. Patil, R. E. Hamilton, and R. A. Day, *Org. Mass Spectrom.* **7**, 817 (1973).

[190] R. Hodges, S. B. H. Kent, and B. C. Richardson, *Biochim. Biophys. Acta* **257**, 54 (1972).

[191] H. Horman, W. Grassman, E. Wünsch, and H. Preller, *Chem. Ber.* **89**, 993 (1956).

[192] D. W. Thomas, *Biochem. Biophys. Res. Commun.* **33**, 483 (1968).

[193] D. W. Thomas, *FEBS (Fed. Eur. Biochem. Soc.) Lett.* **5**, 53 (1969).

[194] K. L. Agarwal, R. A. W. Johnstone, G. W. Kenner, D. S. Millington, and R. C. Sheppard, *Nature (London)* **219**, 498 (1968).

[195] J. Halstrom, K. Brunfeldt, and K. Kovacs, *Experientia* **27**, 17 (1971).

[196] G. M. Schier, J. Korth, and B. Halpern, *Tetrahedron Lett.* 4621 (1972).

[197] W. R. Gray, *in* "Methods in Enzymology" (C. H. W. Hirs, ed.), Vol. II, p. 417. Academic Press, New York, 1967.

[198] D. G. Hoare and D. E. Koshland, *J. Amer. Chem. Soc.* **88**, 2057 (1966).

[199] B. C. Das, S. D. Géro, and E. Lederer, *Biochem. Biophys Res. Commun.* **29**, 911 (1967).

[199a] B. C. Das, S. D. Géro, and E. Lederer, *Nature (London)* **217**, 547 (1968).

[200] G. Marino, L. Valente, R. A. W. Johnstone, F. Mohammedi-Tabrizi, and G. C. Sodine, *Chem. Commun.* 357 (1972).

[201] S. I. Hakomori, *J. Biochem. (Tokyo)* **55**, 205 (1964).

[202] E. Vilkas and E. Lederer, *Tetrahedron Lett.* **26**, 3089 (1968).

[203] C. Cone and R. G. Oldham, *Abstr. 162nd Meet., Amer. Chem. Soc.* ORGN-94 (1971).

[204] K. Jayasimhulu and R. A. Day, *Abstr. 166th Meet., Amer. Chem. Soc.* Biology, p. 122 (1973).

[205] J. R. Coggins and N. L. Benoiton, *Abstr. 156th Meet., Amer. Chem. Soc.* Biology, BIOL-18 (1968).
[206] P. A. White and D. M. Desiderio, *Anal. Lett.* **4**, 305 (1971).
[207] H. R. Morris, K. E. Batley, N. G. L. Harding, R. A. Bjur, J. Dann, and R. W. King, *Biochem. J.* **137**, 409 (1974).
[208] H. R. Morris, D. H. Williams, G. G. Midwinter, and B. S. Hartley, *Biochem. J.* **141**, 701 (1974).
[209] K. Okada, T. Uyehara, M. Hiramoto, H. Kato, and T. Suzuki, *Chem. Pharm. Bull.* **21**, 2217 (1973).
[210] J. Wietzerbin-Falszpan, B. C. Das, C. Gros, J. F. Petit, and E. Lederer, *Eur. J. Biochem.* **32**, 525 (1973).
[211] W. E. Reynolds, V. A. Bacon, J. C. Bridges, T. C. Coburn, B. Halpern, J. Lederberg, E. C. Levinthal, E. Steed, and R. B. Tucker, *Anal. Chem.* **42**, 1122 (1970).
[212] T. Haylett, L. S. Swart, and D. Parris, *Biochem. J.* **123**, 191 (1971).
[213] J. Wietzerbin-Falszpan, B. C. Das, I. Azuma, A. Adams, J. F. Petit, and E. Lederer, *Biochem. Biophys. Res. Commun.* **40**, 57 (1970).
[214] R. Weyers, P. Hagel, B. C. Das, and C. Van der Meer, *Biochim. Biophys. Acta* **279**, 331 (1972).
[215] F. Grosclaude, M. F. Mahe, J. C. Mercier, and B. Ribadeau-Dumas, *FEBS (Fed. Eur. Biochem. Soc.) Lett.* **11**, 109 (1970).
[216] L. B. Sandberg, N. Weissman, and W. R. Gray, *Biochemistry* **10**, 52 (1971).
[217] W. O. Godtfredsen, S. Vangedal, and D. W. Thomas, *Tetrahedron* **26**, 4931 (1970).
[218] G. Bohman-Lindgren, *Tetrahedron Lett.* **28**, 4625 (1972).
[219] Y. Okumura and A. Sakuri, *Bull. Chem. Soc. Jap.* **46**, 2190 (1973).
[220] C. H. Hassal, R. B. Morton, Y. Ogihara, and D. A. S. Phillips, *J. Chem. Soc.*, C 526 (1971).
[221] M. Koncewicz, P. Mathiaparanam, T. F. Uchytil, L. Sparapano, J. Tam, D. H. Rich, and R. D. Durbin, *Biochem. Biophys. Res. Commun.* **53**, 653 (1973).
[222] A. Suzuki, N. Takahashi, and S. Tamura, *Org. Mass Spectrom.* **4**, 175 (1970).
[223] F. Compernolle, H. Vanderhaeghe, and G. Janssen, *Org. Mass. Spectrom.* **6**, 151 (1972).
[224] V. Ludescher and R. Schwyzer, *Helv. Chim. Acta* **55**, 2052 (1972).
[225] N. S. Wul'fson, V. A. Puchkov, V. N. Bochkarev, B. V. Rosinov, A. M. Zyakoon, M. M. Shemyakin, Yu. A. Ovchinnikov, V. T. Ivanov, A. A. Kiryushkin, E. I. Vinogradova, M. Yu. Feigina, and N. A. Aldanova, *Tetrahedron Lett.* 951 (1964).
[226] R. D. Grigsby, C. O. Hansen, D. G. Mannering, W. G. Fox, and R. H. Cole, *Anal. Chem.* **43**, 1135 (1971).
[227] M. Rangarajan, R. E. Ardey, and A. Darbre, *J. Chromatog.* **87**, 499 (1973).
[228] J. J. Pisano, T. Bronzert, and H. B. Brewer, Jr., *Anal. Biochem.* **45**, 43 (1972).
[229] C. Merritt and D. H. Robertson, *J. Gas Chromatogr.* **5**, 96 (1967).
[230] H. Nau, J. A. Kelley, and K. Biemann, *J. Amer. Chem. Soc.* **95**, 7162 (1973).
[231] H. Nau, *Biochem. Biophys. Res. Commun.* **59**, 1088 (1974).
[232] F. W. McLafferty, J. A. Michnowicz, R. Venkataraghavan, P. Rogerson, and B. G. Giessner, *Anal. Chem.* **44**, 2282 (1972).
[233] A. L. Burlingame, R. E. Cox, B. J. Kimble, R. V. McPherron, R. W. Olsen, and F. C. Walls, *J. Chromatogr. Sci.* **12**, 598 (1974).
[234] H. Nau and K. Biemann, *Anal. Lett.* **6**, 1071 (1973).

235 H. Nau and K. Biemann, *Anal. Chem.* **46**, 426 (1974).
236 H. Nau and K. Biemann, *168th Meet., Amer. Chem. Soc.* Paper ANAL-62 (1974).
237 H. Lindley, *Biochem. J.* **126**, 683 (1972).
238 P. Arpino, B. Dawkins, and F. W. McLafferty, *J. Chromatogr. Sci.* **12**, 574 (1974).
239 F. W. McLafferty, R. Knutti, R. Venkataraghavan, B. G. Dawkins, and P. Arpino, *Anal. Chem.* **47**, 1503 (1975.)
240 M. Barber, P. Powers, M. J. Wallington, and W. A. Wolstenholme, *Nature (London)* **212**, 784 (1966).
241 K. Biemann, C. Cone, B. R. Webster, and G. P. Arsenault, *J. Amer. Chem. Soc.* **88**, 5598 (1966).
242 H. Van't Klooster, Ph.D. Thesis, University of Utrecht, 1974; we are indebted to Professor G. Dijkstra for a copy of this reference.
243 K. Biemann, C. Cone, and B. R. Webster, *J. Amer. Chem. Soc.* **88**, 2592 (1966).
244 K. D. Hägele, G. Holzer, W. Parr, S. H. Nakagawa, and D. M. Desiderio, *Biomed. Mass Spectrom.* **1**, 175 (1974).
245 F. W. McLafferty, *Pure Appl. Chem.* **7**, 61 (1971).
246 M. O. Dayhoff and R. V. Eck, Natl. Biomed. Res. Foundation, Silver Spring, Md., Report No. 08710–681115, Part 2, 1968; *Comput. Biol. Med.* **1**, 5 (1970).
247 A. Kanderd, R. B. Spencer, and W. L. Budde, *Anal. Chem.* **43**, 1086 (1971).
248 V. Yu. Gavrilov, A. D. Frank-Kamenetskii, and M. D. Frank-Kamenetskii, *Biokhimiya* **81**, 799 (1966); *Biochemistry (USSR)* **81**, 689 (1966).
249 H. A. Van Klooster, J. S. Vaarkamp-Lijnse, and G. Dijkstra, *Adv. Mass Spectrom.* **6**, 1027 (1974).
250 C. L. Weise and D. M. Desiderio, *Comput. Biol. Med.* **3**, 437 (1973).
251 D. F. Torgerson and R. D. MacFarlane, *Conf. Amer. Soc. Mass Spectrom.* (1975).

Computerized Structure Retrieval and Interpretation of Mass Spectra

2

GAIL M. PESYNA AND F. W. McLAFFERTY

I. INTRODUCTION

The modern scientific investigator has found that the advent of the digital computer has caused profound changes in his research laboratories. It is of little or no practical value to ask which field of scientific investigation has benefited *most* from "the computer revolution," but those working in areas characterized by an enormous amount of data might readily exercise their prejudice by claiming a major portion of the benefit. Mass spectrometry is one such area, and this review is concerned with one aspect of the information problem in this field: the retrieval and interpretation of mass spectra for chemical structure elucidation.

That a single mass spectrum provides a wealth of information on chemical structure is well established.[1-4] This information, consisting of the relative abundances of scores of peaks of unique mass values, can be obtained in seconds using subnanogram samples; the combined techniques of gas chromatography[5-8] and liquid chromatography–mass spectrometry[9,10] often produce hundreds of mass spectra per hour from only microgram samples of complex mixtures. For these reasons, there has been an explosive growth in the applications of mass spectrometry in the last decade, particularly in the areas of forensic, clinical, and environmental chemistry. The magnitude of the resulting information problem reveals the timeliness of the parallel growth of computer technology.

In order to alleviate the enormous interpretive problem generated by this rapid growth, a correspondingly large number of computer systems have been devised. These can be grouped into a small number of classes of systems, within which each approaches the problem of interpretation in a similar way. Between classes, however, the nomenclature and conceptual frameworks used are often very different, sometimes appearing to create illusory differences between the various approaches and obscuring similarities and subtle relationships.

A number of reviews have been written on the computer systems that have been developed for mass spectral interpretation.[11-15] The approach of this review is somewhat different in its deliberate emphasis on the similarities between various methods rather than on the differences between them. Consequently, there has been an attempt wherever possible to make the terminology uniform and the conceptual schemes consistent or complementary. As one might expect, however, any clear-cut division of the computer systems into some small number of classes will be accompanied by more or less difficulty, depending primarily on the number of systems that are left over after all the *obvious* classifications have been successfully made. Although Occam's Razor may provide the strongest justification for the simple division into two classes that is used in this review, it is also felt that forming a dichotomy between "inductive" and "deductive" approaches offers a unique perspective on the similarities within each group.

In the development of an interpretive computer system, the mass spectrometrist sets up what is inevitably a human model of reasoning from a mass spectrum to molecular structure. If strongly influenced by the computer's ability to process large quantities of data in short periods of time, he may set up a model that is primarily "inductive" in its approach. In such a system the computer is given little or no specialized knowledge about mass spectral fragmentation; instead it is given a large number of spectra from which the computer makes empirical correlations, generalizations or statistical inferences *on its own* about mass spectral fragmentation patterns. By applying this newly acquired "knowledge" to an unknown spectrum, the computer predicts a complete or partial molecular structure for the unknown compound.

On the other hand the mass spectrometrist may decide not to deprive the computer of all the accumulated knowledge of mass spectral behavior and may choose instead to set up a "deductive" model. The computer is given an extensive set of mass spectral rules and correlations, which are assumed to be true as well as useful, and by the logical application of these rules the computer is able to deduce the molecular structure from the mass spectrum.

Making such distinctions between inductive and deductive methods of reasoning is difficult primarily because "both in our daily life and in the sciences the two methods are complementary, and it is often difficult to know where one leaves off and the other begins."[16] Likewise, many of these computer systems combine elements of both approaches, but for most of these cases either one of the approaches predominates, or the two are easily separable and therefore can be discussed individually with no loss of the essential nature of the system. This is not true, however, for the two systems discussed in the final section, STIRS and DENDRAL. These systems resist all efforts to separate clearly those elements that are inductive from those that are deductive while still maintaining the integrity of the systems, a condition necessary for a complete understanding of how the computer imitates the mass spectrometrist.

As previously emphasized, the ultimate goal of this review is to point out the underlying similarities among the various computer systems—not to argue the epistemology of human modes of reasoning. The classification of "deductive approaches" should generate little controversy; the name "Deduction Programming Approach" was in fact given to this group of systems by Delfino and Buchs in an earlier review.[15] The use of the classification of "inductive approaches," however, may appear to many to be imprecise; but if this is so, it is due at least in part to the fact no one seems to be in agreement concerning the true nature of scientific induction and, in fact, as to whether induction really *is* a mode of human reasoning at all.* Nevertheless, the organizational framework provided by this classification scheme and the insight into the similarities among the various approaches obtained through its use appear to outweigh the benefits gained by remaining a philosophical purist.

II. INDUCTIVE APPROACHES

The rather liberal definition of induction used here is "that process of reasoning from some observed cases to a universal conclusion regarding all similar cases, some of which are unobserved."[19] Central to this notion of induction is that "inductive generalizations . . . are only probable, ranging from low probability to high."[20] A further qualification on the use of the word "induction" is imposed by the topic of this discussion: the computer,

---

* As an example, we are trained as scientists to believe that induction "is the basis of all experimental science."[17] Yet many philosophers of science believe, as Popper, that "induction, i.e., inference based on many observations, is a myth. It is neither a psychological fact nor a fact of ordinary life nor of scientific procedure."[18]

not the mass spectrometrist, is forming the "conclusions" from the data. Three types of systems are included in this classification: retrieval systems, clustering methods, and nonparametric "learning machines."

There is an undeniable emphasis on retrieval systems in this section for several reasons. The first is simply the tremendous quantity of research that has been done in this area. Retrieval systems are also important in this study both because they are the most commonly used method of computerized structure determination, and because an understanding of the retrieval process provides a solid foundation for all the other techniques that utilize libraries of mass spectral data.

It may seem at first rather pretentious to classify retrieval systems as "inductive approaches." To understand the reasons for this, it is necessary to realize that "perfect matches," in which every peak and every intensity value are identical in both the unknown and the reference spectra, are very seldom found in real situations. Differences among spectra due to instrumental variations, low or changing concentrations of sample while recording, or contamination of the sample, imply that in general the computer will be searching for "best matches." The computer essentially does this by "counting" the number of matching peaks in the unknown spectrum and each reference spectrum, and "in its simplest form, induction may be illustrated by the simple process of counting."[21] Furthermore, there is associated with each match some probability that it is indeed the spectrum of the same compound, and whereas a probability of 1.0 is hoped for, it is frequently not obtained.

A. Retrieval Methods

Retrieval methods (also referred to as "library-" or "file-searching" or "matching" methods) are by far the easiest methods to conceptualize and to program. If a spectrum of an "unknown" compound is contained in the library and the two spectra are identical, it is guaranteed that the correct structure identification will be made within the limitations of mass spectrometry to distinguish molecular structures. Thus the library is the most important element of any retrieval system and, in general, the larger the library, the better the system. Several sources of data for mass spectral files can be found in an extensive survey of analytical data sources by Gevantman,[22] and since this publication the *Registry of Mass Spectral Data* has also been made available.[23]

The methods used by various researchers to correlate the unknown mass spectrum with those in the library vary, but the theoretical framework for all systems is the same. The retrieval of information from a numeric data base such as a file of tabulated mass spectra is a part of the field of information science—a field that is concerned with the structure, analysis, organization, storage, searching and retrieval of information.[24] The bulk of research in this field deals with natural language documents in an attempt to find a solution to "the information problem."* As opposed to processing documents, however, the handling of numeric, scientific data is perhaps one of the simpler problems in the field for several reasons. First of all, the data exist in a form readily compatible with digital computers. Second, the data are highly specific and the range of the data is limited. (This is obviously not the case when the data base is the English language.) Third, simple rules for choosing "important" information from a mass spectrum can be easily defined using the intensities of peaks or easily calculated statistical information.

In the semiautomatic document retrieval systems currently in existence, content identifiers, or "terms," are generally assigned to each document; user requests, or "queries," are subsequently drawn up from a list of allowable terms. If the total number of allowable terms is said to be N and some number m of these terms are used to specify a document (or a query), then one may consider a "term vector" containing N elements, m of which are nonzero, that identifies or represents that document.† One may further conceptualize the document, represented by its term vector, as a point in N-dimensional space. Finding the document or documents whose content most closely matches that of the posed query can then be thought of as finding those points nearest the query in N-dimensional space. There are many distance measures employed to judge "nearness"; two of these that are easy to visualize are the Euclidean distance in N-space and the cosine of the angle between the document and query. In addition many names are given to these similarity measures, some of which are more common than others; "similarity index," "correlation coefficient," "criterion of agreement," and "match index" are several of these.‡

* "The information problem" is a catchphrase that has been assimilated into our language and has come to mean that situation in which "more and more information is generated and put into circulation; the existing tools, classification schedules, and storage arrangements are often inadequate, particularly in the newer fields; and it generally becomes more difficult and more expensive to get to know what one needs to know."[25]

† As an illustration, consider the following "document," which is a sentence from an abstract:

"The problem of the identification of low resolution mass spectra has been approached from the viewpoint of statistics and information theory."[26]

If the major concepts are chosen to be "identification," "low resolution," "mass spectra," "statistics," and "information theory," one could construct an m-tuple ($m = 5$) representing this "document." However, if our list of allowable terms is controlled and N arbitrarily equals 100, and if the above terms are assigned the numbers, 1, 2, 5, 98, and 99 within this set of N terms, then the binary term vector, (1, 1, 0, 0, 1, 0, 0, ..., 1, 1, 0), containing 100 elements, 5 of which are nonzero, can be constructed.

‡ In many mass spectral search systems, "dissimilarity measures" are calculated. These are identical in purpose, if not in orientation, with "similarity measures"; one can be easily converted to the other by appropriate renormalization, inversion, or subtraction from a constant.

Within the same conceptual framework, each mass spectrum can be represented by a vector or a point in N-dimensional space where N equals the m/e value of the highest peak recorded for any spectrum. Each of the m peaks in an individual mass spectrum is a nonzero element of the vector and determines the locus of the point. When presented with an unknown mass spectrum, the computer retrieves as "relevant" to the mass spectroscopist's request those spectra in the collection that are closest to the unknown in N-space.

So far this discussion has been concerned only with binary term vectors, e.g., presence or absence of a peak at a particular m/e value. Most document and mass spectral retrieval systems, however, make use of "weighted" term vectors in which each term is assigned a positive weight that reflects the relative importance of the term in the vector. As mentioned previously, the most straightforward means of assigning weights for mass spectra is to use the peak intensity data directly; other weighting systems are also in use and will be discussed further. In general, it has been shown[27] that weighted term vectors produce better retrieval results than nonweighted ones. This result has been corroborated for the special case of mass spectral retrieval.[26, 28]

The most accurate method of "determining the locus of the point (i.e., spectrum) in N-dimensional mass spectral space" and retrieving unambiguously the spectrum (or spectra) identical to that of the unknown (i.e., an "exact" match) is by using all of the peak and intensity data in each spectrum. There are several disadvantages to this approach, however. One important consideration is the difficulty of reproducing mass spectra due to instrument variations. Any retrieval system should be designed to minimize these effects on the search, and a peak-by-peak comparison for all peaks in a spectrum may give misleading results or none at all. Small impurity peaks in a spectrum may also unnecessarily worsen the search results in such a system. As it is only necessary to store the nonzero elements of the mass spectral "vectors," the more peaks used per spectrum, the greater the amount of storage space required. Also, if the computer is not bound by the speed at which it can read in the data, the search time will depend on the number of peaks that must be compared.* Furthermore, the large volume of information contained in a mass spectrum is often much more than is needed to obtain a reliable structure identification.[26, 28] Consequently, most of the mass spectral search systems that have been developed and are discussed below use "abbreviated" spectra. Research in the field of mass spectral retrieval is generally directed toward the development of systems that will identify structures unambiguously in the shortest possible search times within the limitations of the available computer hardware. The search algorithms developed differ from one another in any of four ways: the number and method of choosing the peaks

---

* And, if the computer *is* input/output bound, reducing the number of peaks used for matching may decrease search times by reducing the number of disk accesses required to read the entire file, for example.

used for matching (i.e., the number of nonzero terms in the term vector and the criteria used to determine which of the "concepts" are most important), the weighting system (if any) that is used, and the form of the similarity measure. Optimal file organization and techniques of prefiltering may be incorporated into the system to decrease search times, but these do not substantially change the nature of the search itself.

Another desirable characteristic of these search systems is that if the spectrum of the unknown compound is not contained in the library, the computer should retrieve spectra of compounds whose structures are *similar* to that of the unknown compound (i.e., items relevant to the query). In addition, there should be some way for the chemist to know how close the matches are to the spectrum of his unknown, and whether the best match is in fact the correct compound or one of similar structure.

1. Predecessors of the Computer System

An early approach to mass spectral retrieval systems formulated by Zemany[29] is in fact a manual method, but this work is of historical interest in that it is the forerunner of all the later machine-assisted and computer-based techniques. The method utilizes 5 × 8 inch McBee Keysort cards onto which the complete mass spectrum is pasted or transcribed along with any pertinent experimental data. The cards are punched under two headings for retrieval purposes—the molecular weight of the compound and the five or six most prominent peaks. No intensity information is included, and a simple sorting procedure retrieves the spectra of those compounds containing the combination of peaks of interest. Perhaps one of the most significant statements in this paper is that "analysis with the mass spectrometer depends ultimately on an adequate library of reference spectra, for there exists at present [1950] no adequate theory for predicting these spectra." Not only was this statement certainly true for mass spectrometry in general, but it was also prophetic with respect to the development of computer interpretation systems.

The next system to appear that is of importance in tracing the development of mass spectral retrieval systems is that of McLafferty and Gohlke.[30] This system utilizes the machine filing and manual sorting of 80-column IBM cards. On each card (one card per spectrum for a library of 4000 spectra) are recorded the m/e values of the ten most intense peaks in the spectrum and five "peculiar" peaks that may stand out in the spectrum of a mixture. A second part of the system includes information on masses at which there are no peaks in the spectrum. By assigning a particular column on each IBM card to an individual m/e value, each spectrum is completely specified with respect to mass, using as many cards as necessary. An IBM collator is then used to search the card collection.

TABLE I

A Comparison of the Search Strategies Employed in Various "Traditional" Retrieval Systems

Reference	Prefilter	Search	Similarity measure
Abrahamsson et al.[31-33]	Molecular weight	1. 5 most intense peaks 2. 5 peaks of greatest weighted intensity 3. 5 most intense peaks with $m/e > 50$ amu 4. 5 peaks of greatest weighted intensities with $m/e > 50$ amu 5. n most intense peaks in particular mass ranges	$\text{INDEX} = \sum_{m/e=1}^{M} \lvert I_1 - I_2 \rvert_{m/e}$ where M = all peaks in spectrum I_1, I_2 = intensities of matching peaks in unknown and reference
Pettersson and Ryhage[38]	Molecular weight	1. 6 most intense peaks 2. 6 most intense peaks if $MW \leq 200$ amu; 10 most intense peaks if $MW > 200$ amu 3. 6 most intense peaks if $MW \leq 200$ amu; 2 most intense peaks every 25 mass units if $MW > 200$ amu	1. None 2. Abundances must agree to within $\pm 15\%$ relative intensity 3. Abundances must agree to within $\pm 5\%$ relative intensity for 60% of *all* peaks with abundances $> 3\%$

Ridley et al.[39,40]	Either or both of the following:
	
	a. Molecular weight
	
	b. Base peak in unknown must be one of 6 most intense in reference
		1. n most intense peaks
		2. n most intense peaks every m mass units
			1. a. $P = \dfrac{A}{n}$
			   b. $P = \dfrac{1}{n^2} \sum_{k=1}^{A} (n - \lvert i - j \rvert_k)$
			2. a. $P = \dfrac{1}{R} \sum_{r=1}^{R} \dfrac{A_r}{n}$
			   b. $P = \dfrac{1}{R} \sum_{r=1}^{R} \dfrac{1}{n^2} \sum_{k=1}^{A_r} (n - \lvert i - j \rvert_k)$
			  A = number of agreements
			  n = number of peaks used in matching

i, j = relative "positions" of matching peaks in unknown and reference
R = number of mass ranges

Crawford and Morrison[41]

Any of the following:

a. The 2, 3 or 4 most intense unknown peaks must be the same as the 2, 3 or 4 most intense reference peaks

b. Molecular weight and 2 most intense peaks the same

c. Either of the 2 most intense unknown peaks must be the reference base peak

d. Either of the 2 or 3 most intense unknown peaks must be one of the 2 or 3 most intense reference peaks

e. Molecular weight

1. Entire mass spectrum

2. 6 most intense peaks

1. a. $D = \sum_{n=1}^{k} |Pn_{ref} - Pn_{unk}|$

where $\sum_{n=1}^{k} Pn = 1$

b. $D = \sum_{n=1}^{k} |\sqrt{Pn_{ref}} - \sqrt{Pn_{unk}}|$

where $\sum_{n=1}^{k} \sqrt{Pn} = 1$

c. $D = \sum_{n=1}^{k} (Pn_{ref} - Pn_{unk})^2$

where $\sum_{n=1}^{k} Pn^2 = 1$

2. $D = \sum_{n=1}^{k} |Pn_{ref} - Pn_{unk}|$

where strongest peak has intensity = 100, and Pr = intensity of peaks in unknown or reference

TABLE I (*Continued*)

Reference	Prefilter	Search	Similarity measure
Biemann *et al.*[50]	All of the following: a. Base peak in reference spectrum must be $\geq 25\%$ relative intensity in unknown and vice versa b. The mass ranges covered by the known and unknown spectra must differ by $\leq$ a factor of 3 c. Total abundance of homologous series of ions must be similar	2 most intense peaks in every 14 mass unit interval	$$\text{INDEX} = \frac{1/n \sum_{i=1}^{n} [(R_n/\bar{R}_{>10\%})W]}{F + 1}$$ n = (total number of peaks which are contained in both the unknown and the reference spectra) + (total number of peaks either in the unknown *or* in the reference, but *not* in both) R_n = the ratio of the intensities of unknown and reference peaks $\bar{R}_{>10\%}$ = the average intensity ratio for those peaks $>10\%$ abundance W = weighting factor = 12, 4, or 1 if intensity is $>10\%$, $>1\%$, or $<1\%$ F = fraction of unmatched intensities
Grotch[65-69]	All or none of the following: a. The base peak in the unknown must be one of the 5 most intense reference peaks and vice versa b. Molecular weight	1. a. Entire mass spectrum, no intensity data b. 20 most intense peaks, no intensity data	1. $C = \text{XOR}(A, B)$, or $C = \text{AND}(A, B)$, or $$C = \mu N \sum_{i=1}^{M} [(\text{XOR})_i - (\text{AND})_i]$$ A, B = binary representations of unknown and reference spectra μ = weighting factor N = number of peaks in the unknown M = highest m/e value in either the unknown or reference

2. Entire mass spectrum, 8 or 10^4 intensity levels

2. $C = \sum_{i=1}^{M} |x_i - l_i|$

   $x, l =$ intensities of peaks in unknown and reference

3. Most intense peak in every 14 mass unit interval, no intensity data

3. $C = \mu N \sum_{i=1}^{M} [(XOR)_i - (AND)_i]$

4. Most intense peak in every 14 mass unit interval, 4 intensity levels

4. a. $C = \dfrac{\sum (x_i - l_i)^2}{\sum (x_i^2 + l_i^2)}$

   b. $C = \dfrac{\sum x_i - l_i}{\sum (x_i + l_i)}$

One of the significant aspects of this system is the inclusion of five "peculiar" peaks (i.e., peaks that are found to occur infrequently in mass spectra). This system is consequently the first to incorporate both m/e "uniqueness" and peak intensity as criteria for choosing the data used in spectral identification. The second part is the direct predecessor of the 1-bit, peak–no peak coding techniques (*vide infra*). Both the system described by McLafferty and Gohlke and the Keysort system of Zemany were used successfully for a number of years in the identification of unknown spectra of pure compounds and mixtures.

2. The Digital Computer Systems

The first published information concerned with a mass spectral retrieval system operating on a digital computer appeared as a brief note by Abrahamsson *et al.* in the 1964 *Proceedings of the Biochemical Society*,[31] with details of this work appearing in the later literature.[32, 33] These researchers focus their attention on different types of spectral abbreviation. Five different libraries are created and are available for searching. Method 1 employs only an abundance criterion for choosing peaks by using the five peaks of greatest intensities. The second method uses the five peaks of greatest weighted intensities, where the weighted intensity is equal to the original intensity multiplied by the m/e value of the peak. Since in general the frequency of occurrence of peaks at a particular m/e decreases with increasing mass,[34–36] this second method is seen to employ both abundance and uniqueness criteria. These two methods are also modified by restricting the mass range to only those peaks with masses greater than 50 amu. A final method uses the most intense peak in different m/e intervals of the spectrum. These methods are summarized in Table I.

The sets of peaks generated by each of these five methods are referred to as "keys," and the procedure may be terminated after a "key-search" (Table II). Alternatively, a comparison program may be executed after a key-search to calculate for each retrieved spectrum a "disagreement index" (see Table I). The list of retrieved spectra is then ordered by increasing disagreement indices—a perfect match has an index of zero (Table III). The calculation of this disagreement index on the basis of intensity differences establishes Abrahamsson's system as the first to employ weighted term vectors (*vide supra*). One of the major disadvantages of this unnormalized disagreement index, however, is that the measure of disagreement is unbounded, i.e., the largest value that the index may attain will differ from one spectrum to another.*

The details of the search systems investigated by Pettersson and Ryhage[38] are very similar to those of Abrahamsson *et al.*, but the later research includes

---

* As will be discussed further, recent work[37] has suggested that such unnormalized similarity measures may be less tolerant of spectral errors than normalized measures.

TABLE II

Programs (in Algol) for Mass Spectrometry Information Retrieval System Implemented by Abrahamsson et al.[32]

Program	Function
1. READ	Reads and checks mass spectrum from punched cards or paper tape, normalizes the intensities, and converts the spectrum into the correct magnetic tape format.
2. ADD	Adds spectrum to master file.
3. PRINT	Prints all information stored for any desired spectrum.
4. PLOT	Plots bar diagram of spectrum.
5. KEYSEARCH (K, D, N, M)	Searches through master file using key number K. For parameters D, N, and M, see references.
6. COMPARE	Compares in detail unknown spectrum with spectra extracted from master file by KEYSEARCH. To each extracted spectrum is attached a "disagreement index" (INDEX in Table III) which is proportional to the sum of the absolute values of the differences in intensities between the unknown and extracted spectrum: $$\sum_{m/e=1}^{M} |I_1 - I_2|_{m/e}$$
7. LINESEARCH (P_i, R_i, P_{i+1}, etc.)	Extracts from master file spectra with lines of $m/e = P_i, P_{i+1}$, etc., of intensity greater than R_i, R_{i+1}, etc., respectively.
8. SORT (S)	Sorts spectra in order after item S (S can specify name chemical type, molecular weight, disagreement index, etc.).
9. LIST	Lists the result of programs 5–8.

a qualitative evaluation of the effectiveness of the various methods of spectral abbreviation. The first method searches the file using the six most intense peaks for compounds with molecular weights up to 200 amu and the ten most intense peaks for compounds with molecular weights above 200 amu (cf. Abrahamsson's methods 1 and 5 mentioned above). A criterion of similarity is placed on those spectra retrieved using method 2: The intensities of the peaks in the reference compound matching those in the unknown must agree within $\pm 15\%$ relative abundance. Pettersson reports[38] that this second method gives more satisfactory retrieval results than the first, especially for compounds with molecular weights over 200. Since in general the more intense peaks in a mass spectrum tend to occur at lower m/e values, while the more characteristic peaks tend to be found at higher m/e values,[1] it is not surprising that the use of more peaks from the spectra of the higher molecular weight compounds produces better results. Pettersson and Ryhage's third approach (similar to the combination of KEYSEARCH and COMPARE in Abra-

TABLE III

Result from Automatic Identification (Abrahamsson) Using the Following Directives: READ, KEYSEARCH (1, 0, 4, 300), COMPARE, SORT (5), LIST[a]

Molecular weight	Compound name	Series	Number	NOL	INDEX
106.160	1,2-Dimethylbenzene (o-xylene) (gas)	API	420	71	0
106.160	1,2-Dimethylbenzene (o-xylene) (gas)	API	178	76	10
106.160	1,2-Dimethylbenzene (o-xylene) (gas)	API	253	84	10
106.160	1,3-Dimethylbenzene (m-xylene) (gas)	API	254	85	16
106.160	1,4-Dimethylbenzene (p-xylene) (gas)	API	180	77	16
106.160	1,3-Dimethylbenzene (m-xylene) (gas)	API	179	75	18
106.160	1,4-Dimethylbenzene (p-xylene) (gas)	API	255	73	23
106.160	m-Xylene	DOW	313	70	33
106.160	p-Xylene	DOW	314	75	34
106.160	Ethylbenzene (gas)	API	252	80	47
106.160	Ethylbenzene (gas)	API	177	73	48
106.160	1-4,Dimethylbenzene (p-xylene) (gas)	API	422	67	49
106.160	1,3-Dimethylbenzene (m-xylene) (gas)	API	421	67	50
106.160	Ethylbenzene (gas)	API	419	68	73
106.160	o-Xylene	DOW	312	76	80
106.160	1,3-Dimethylbenzene (m-xylene) (gas)	API	308	49	125
106.160	1,4-Dimethylbenzene (p-xylene) (gas)	API	309	52	127
106.160	1,2-Dimethylbenzene (o-xylene) (gas)	API	307	49	151

[a] In this case, there is in practice no molecular weight discrimination ($M = 300$). The extracted spectra are sorted in order of increasing "disagreement index." NOL gives the number of peaks in each compound.[33]

hamsson's work) was found to give the best results overall, however. The six most intense peaks (or the two most intense peaks in every 25 mass unit interval) are essentially used as a presearch. All peaks with intensities greater than 3% in the unknown spectrum are then compared with the reference spectrum, and if 60% of these peaks agree within ± 5% relative intensity, the reference compound is retrieved. Thus this peak intensity comparison is a slightly less formal calculation of similarity than the "disagreement index" proposed by Abrahamsson; but both are measures based upon intensity differences between the unknown and the reference spectra.

Ridley and co-workers reported the results of a more systematic and complex retrieval evaluation.[28, 39, 40] In addition to varying the number and method of choosing the peaks to be used in matching, which were the primary areas of investigation of both Abrahamsson and Pettersson, Ridley also studied the retrieval characteristics of several new similarity measures.

Ridley's methods of choosing peaks are identical to two discussed previously: the n most intense peaks in each spectrum and the n most intense peaks in every m mass unit interval. Both an unweighted system (i.e., a binary system as discussed above) and a weighted system are examined. Unlike Abrahamsson and Pettersson, however, Ridley et al. do not use peak intensity data directly as weights: they use instead the relative position of the n peaks when ordered by decreasing intensity. The forms of the similarity measures employed are listed in Table I.

The conclusions of these researchers are extremely important for the study and development of mass spectral retrieval systems. For the tests run on the methods of choosing peaks, variations in n and m were not found to produce significant differences in retrieval results. In the method of choosing the n most intense peaks, n ranged from 5 to 20; in the study of n peaks every m mass units, combinations such as $n = 2$, $m = 14$; $n = 3$, $m = 20$; and $n = 3$, $m = 45$ were investigated. These workers also showed that, for the specific case of the retrieval of mass spectra, weighted systems are generally superior to unweighted systems (i.e., similarity measures 2 and 4 were more successful than measures 1 and 3; see Table I). Finally, it was discovered that the use of n peaks every m mass units generally gives better retrieval results than the use of approximately the same number of "most intense peaks" in the spectrum.

Ridley et al. demonstrate that using peak intensity data directly in the calculation of similarity is not necessary for satisfactory mass spectral retrieval. Crawford and Morrison argue that it may not even be desirable if the common normalization procedure of setting the base peak equal to 100% is employed.[41] They use as the major arguments against this normalization procedure the facts that any error in the recording of the base peak height will affect the entire normalized mass spectrum, and that the base peak is often not the most significant piece of spectral information for identification. As a consequence, a primary direction of their research is toward an examination of the effects of random errors in peak heights on the quality of retrieval when various normalization procedures are employed.

The methods of normalization and the corresponding forms of the similarity measures are given in Table I. The second method in which the sum of the square roots of all the peak heights is set equal to unity is found to give the most satisfactory results when random errors of $\pm 20\%$ on all peaks are introduced. This procedure is also found to be nearly as satisfactory as the more complicated and time-consuming method of normalizing groups of peaks within specific mass ranges—an approach that might be taken, for example, to minimize the effects of quadrupole "roll-off."[42]

Crawford and Morrison make the important point that the identification of a mass spectrum that is contained in a file is an almost trivial problem if the entire mass spectrum is used. As the time requirements for such a system are prohibitive for real-time gas chromatography–mass spectrometry systems, however, they undertook extensive testing of a 6-peak search preceded by various prefiltering techniques. Depending on the ordering of spectra in the reference file, increases in speed of up to a factor of 7 were obtained. An example of the search results obtained for two unknown spectra previously published by Pettersson and Ryhage[38] demonstrates that even this simple system can be useful.

Biemann and co-workers are primarily concerned with the development of a retrieval system to be used in conjunction with gas chromatograph–mass spectrometer combinations.[43-53] The system designed at the Massachusetts Institute of Technology has apparently progressed through three major stages of development. In the first stage,[43-49] three independent measures of similarity were computed; these were later modified and combined (stage 2)[50] into a single measure, which is discussed below.

As can be seen in Table I, this system also uses the n most intense peaks in every m mass unit interval, but in this case n and m are fixed values of 2 and 14, respectively. The choice of these values is defended on the basis of a general knowledge of mass spectral fragmentation patterns; as discussed above, the findings of Ridley et al. support this choice.[28]

The similarity measure is considerably more complex than the others examined up to this point. The major difference that should be noted is the use of intensity ratios rather than intensity differences in the computation. Intensity values are thus used directly as one set of weights, but a second set of weights based on the intensities is also established (W in Table I). For a detailed explanation of the terms in this similarity measure, direct reference should be made to the paper of Hertz et al.[50]

The most recent program development[51-53] takes into consideration the gas chromatographic retention index of the unknown compound. One suggestion made by the authors is that, for a file containing the retention indices of each reference compound, the retention index be used as a prefilter. They believe it preferable, however, to use the retention index as an independent parameter reflecting the degree of similarity between the unknown compound and those whose spectra are retrieved from the reference file.[51] The difference between the retention indices of the unknown and each reference compound is computed and printed at the time of output next to the similarity measures, calculated as in Table I. Use of retention indices in the manner suggested by these authors is particularly valuable for determining whether the best match is indeed the same compound or merely one of similar structure. The main disadvantage to the technique is its lack of generality, but it appears to hold great promise for laboratories routinely investigating limited sample or compound types by gas chromatography–mass spectrometry (GC/MS).

Entire systems fashioned after that of Biemann *et al.* have appeared in the literature in reports of environmental[54] and biological[55] research; an interesting modification of the algorithm is reported by Bell.[56] This modification uses the three most intense peaks in every 14 amu interval to calculate two ratios, the abundance of the most intense peak to both the abundance of the second and third most intense peaks, for both the unknown and reference spectra. These 2 *ratios* in every 14 amu interval are substituted for the 2 most intense peaks used in the normal Biemann search method. As a separate consideration, however, it must be noted that the selection of the 2 most intense peaks in every 14 mass unit interval has perhaps become the most widely used method of abbreviation in computerized retrieval systems.* The most highly publicized of these is the National Institutes of Health mass spectral search system,[57-63] which is available internationally over the General Electric computer network.[62, 63] As seen in Table IV, a variety of different search systems are available to the chemist using this system; all but one of the peak search systems use the Biemann method of spectral abbreviation. This one, the "dissimilarity comparison," uses all the peaks of the reference spectra and searches the entire file of 8782 mass spectra. This method is also the only one that involves the calculation of a similarity measure.[63] The authors do not give an estimate of the time required for such a search; the other search routines use only on the average of 2–6 seconds of computer central processing unit (CPU) time per unknown. The entire process of requesting searches, specifying search parameters, typing in data and any other information, receiving output, etc., takes the average user 5–15 minutes at his remote terminal. These times should be compared with those of the original Abra-

---

* This method of spectral abbreviation is often referred to as "the Biemann system." One should keep in mind that this phrase, when commonly used in such a context, does not generally refer to *all* the aspects of the entire Biemann retrieval system discussed above.

TABLE IV

Options of the National Institutes of Health
(Division of Computer Research and Technology) Mass Spectral Search System[63]

---

1. Peak and intensity search
2. Molecular weight search
3. Molecular formula search
   a. Complete
   b. Imbedded
4. Molecular weight and peak search
5. Molecular formula and peak search
6. Molecular weight and molecular formula search
7. Dissimilarity comparison
8. Spectrum printout
9. Microfiche display of spectrum
10. Display of spectrum
11. Plotting of spectrum
12. CRAB—Comments and complaints
13. HARVEST—Entering new data
14. NEWS—News of the system
15. MSDC classification code list

---

hamsson system (*vide supra*), which required 5 minutes of computer search time *alone* for a file of comparable size.

Crawford and Morrison pointed out that the identification of an unknown that is present in the reference file is an almost trivial problem if the masses and intensities of all peaks in the spectra are used. Grotch shows in his studies on binary retrieval systems (systems in which no weights, such as peak intensities, are used) that reasonably good retrieval is obtained when only the presence or absence of a peak at each m/e value is considered.[64-68] Inclusion of 2, 3, or 32 bits* of intensity information does improve the retrieval results; but it does so at the expense of storage requirements and search times (time increases of approximately a factor of 3 for the 3-bit case, and of an order of magnitude for the 32-bit case). Since Grotch, in his work in space applications at the Jet Propulsion Laboratory, is primarily concerned with limitations of time and storage on searching, a sacrifice of high retrieval precision in exchange for minimum storage and time requirements may well be justified. Grotch's results should not be interpreted as a legitimatization of the total abandonment of intensity information in the design of retrieval systems, however. He clearly states that when only the 20 most intense peaks are used and no intensity information is included, "the performance here is significantly

---

* Note that although Grotch refers to "32 intensity bits," he uses only 14 bits to code the entire range of abundances from 0.01–100% (or 1–10,000).

poorer than any of the other techniques [matching complete spectra using only presence or absence of each peak, and matching complete spectra with 3 or 32 bits of intensity data included]"; and furthermore, that while the "results for $[N] = 8$ were comparable to that of $N = 20, \ldots$ those for $N = 4$ were decidedly poorer."[68] Table V presents some typical results from an "unrestricted" search of the entire file and a "restricted" search which employs prefiltering techniques.

TABLE V

Comparison of Grotch's Results[67, 68] Using 1, 3, and 32 Bits of Intensity Information

Unrestricted search: all members of the library compared with unknown				
	% of unknowns identified with confusion $\leq C_0$			
	1 Bit		3 Bits	32 Bits
Confusion[a] (C_0)	20 Most intense peaks	All peaks in spectrum	All peaks in spectrum	All peaks in spectrum
0	48.0	64.8	73.6	80.8
1	54.2	76.0	84.0	88.0
5	71.2	82.4	91.2	96.0
10	77.6	86.4	93.6	97.6

Restricted search: prefiltering employed (see Table I)				
	% of unknowns identified with confusion $\leq C_0$			
	1 Bit		3 Bits	32 Bits
Confusion (C_0)	20 Most intense peaks	All peaks in spectrum	All peaks in spectrum	All peaks in spectrum
0	60.0	70.4	80.8	84.0
1	71.2	79.2	88.0	94.4
5	85.6	95.2	97.6	98.4
10	92.0	99.2	100.0	99.2

[a] "Confusion" is defined as the number of library compounds which match the unknown as well as or better than the correct answer. Thus using only the 20 most intense peaks in the unknown and reference spectra and 1 bit of intensity information (i.e., "peak–no peak" system) 48% of the unknowns are identified unambiguously ($C_0 = 0$), 54.2% appear as one of the top 2 compounds retrieved, etc. Using 32 bits of intensity information and all the peaks in the spectra, 80.8% of the unknowns are identified unambiguously.

In Grotch's "1-bit encoding" of mass spectra, 10 computer words of 32 bits each are all that are needed to code all peak data over the mass range 12–331 amu. Each of the 320 computer bits represents a discrete mass value;

if a peak is present at a particular mass, the corresponding bit $= 1$, otherwise it is zero. (Efforts to combine several peak positions into a single bit with a minimal loss of important information have been reported by Wangen *et al.*[34]) To compare an unknown spectrum compressed in this manner with that of a similarly compressed reference spectrum, 10 computer operations can be used to replace the 320 (maximum) mass-by-mass comparisons if the similarity measure is composed of logical operators such as XOR ("exclusive or") and AND ("logical and"). Performing an XOR function on two words of compressed mass spectral data and counting the resulting number of "ones" in the result is equivalent to determining the number of peak disagreements between the two spectra over a range of 32 amu; likewise, the AND function will determine the number of co-occurring peaks. In an investigation of three similarity measures thus constructed, Grotch determined that counting only the number of agreements (AND) between two spectra produces retrieval results (cf. the work of Ridley *et al.*) worse than counting only the number of disagreements (XOR) between the two spectra. Best results are obtained when the measure takes into account both the number of disagreements *and* the number of agreements, with the agreements weighted approximately twice as heavily in the calculation. Normalization of spectra on the basis of total ionization, Grotch reports, produces slightly better retrieval than normalization on the basis of the most intense peak—a finding consistent with the arguments put forward by Crawford and Morrison.[41]

In an effort to reduce storage and search time for those systems using n peaks in 14 mass unit intervals, Grotch devised a compressed spectral code using only 4 bits per mass value.[26, 69] A similar approach, which uses 3 bits* to represent the most intense peak in every 7 amu interval, has been examined by Robertson *et al.*[70–72]

The aspect of mass spectral retrieval systems most recently investigated by Grotch is that concerned with the distribution of mismatches obtained for an unknown matched against a particular reference file.[37, 73, 74] After demonstrating that this distribution can be predicted theoretically, Grotch shows that it determines the maximum number of spectral errors (e.g., in recording, measuring, counting, inputting, etc., the spectrum) that can be tolerated in the unknown before misidentification occurs. For the general case in mass spectra, where the probability of finding a peak at any one particular mass is almost always less than 0.50,[34–36] it is shown that spectral errors caused by the addition of incorrect data (e.g., impurities) present a far less serious problem than errors resulting from the *omission* of important information. Grotch therefore suggests a simple coding maxim: "When in doubt, code a peak." Preliminary studies[37] seem to indicate that the use of normalized, rather than

---

* This method is given the name "octal coding", since 3 binary digits can be combined to form one octal digit. Analogously, Grotch's 4-bit representation might be christened "hexadecimal coding."

unnormalized, measures reduces the severity of this "problem of omission." If this is the case, it is therefore more likely that a normalized similarity measure will produce a correct identification for a poorly recorded (e.g., a very weak GC peak) or badly coded mass spectrum.

The mass spectral retrieval systems discussed so far have been the major efforts in the development of traditional systems designed to retrieve from a reference file the exact or best match to an unknown spectrum. A criterion essential to the reasonable performance of any of these systems is that the unknown spectrum must be of a single, *relatively pure* compound. These features and requirements of the traditional retrieval systems have prompted several researchers to ask such questions as "Can a faster system be devised if one is interested only in *exact* matches and not in *best* matches?" and "How might a system be designed to insure the identification of components in highly contaminated spectra or in the spectra of mixtures?"

In answer to the first of these questions, Robertson et al. calculate a single number, the Khinchine entropy function, which represents an entire spectrum.[75, 76] The function η is defined as

$$-\eta = \sum_{i=1}^{n} p_i \log p_i$$

where p is the fraction of the total ion current for each ion, i, summed over the entire mass range of the spectrum. This approach is *not* recommended in general since this function is extremely sensitive to peak intensity differences, and "the variability in the values of relative ion abundances with variations in mass spectrometer design and operation produces considerable uncertainty in the reliability of diagnostics based on spectral intensity factors."[76] (The use of the "divergence function" to calculate dissimilarity between two spectra is also proposed by these researchers[75-77] and appears to be a more successful, albeit traditional, method of retrieval.)

Jurs suggests[78] that a very rapid search system which is totally independent of the size of the reference file can be developed using hash coding techniques.[79] This researcher chooses specifically to examine infrared data in this paper, but it is implied that the technique is generally applicable to other types of spectrometric data. In a communication issued in response to Jurs' publication,[80] Lytle points out that the "exact matching" obtained in a hash coding system may prove useful in mass spectrometry for binary systems such as those investigated by Grotch (*vide supra*). It is emphasized that a retrieval system based on hash coding will *only* be successful for perfect matches— every peak that is present in the unknown spectrum *must* occur in the reference spectrum and vice versa. Lytle also shows that for a particular library size, hash coding may or may not possess memory and time advantages over other methods of exact matching.

The problem of identification of compounds in the spectra of mixtures is one that has plagued mass spectrometrists since the earliest analytical applications of the technique. The problem originally received great attention from researchers in the petroleum industry[81] who were using mass spectrometry to provide "hydrocarbon type" analyses of petroleum fractions. Despite rapid advances in gas chromatography–[5–8] and liquid chromatography–mass spectrometry[9, 10] systems in recent years, the "mixture problem" is still significant, if for no other reason than the exponential growth in the application of mass spectrometry to analytical problems over the past decade.

One of the earliest computer techniques for determining the qualitative (as well as the quantitative) composition of a mixture from a single mass spectrum is that of Tunnicliff and Wadsworth.[82] A stepwise regression procedure chooses from the library of reference spectra that combination of spectra that, when each is multiplied by an appropriate concentration factor, will give the best least-squares fit to the sample spectrum. One of the major limitations of this method is the fact that the computer time is significant and is dependent upon the number of reference spectra; the file used here contains only 129 spectra and the computer time required is 2.5 minutes. For reliable results, the spectra of all components of the mixture must be contained in the reference file.

A considerably less complicated method that utilizes small numbers of characteristic peaks instead of all peaks in the spectra has been developed and is currently under investigation by several researchers.[35, 36, 83–86] This procedure has been given the name "reverse search," since, in contrast to the traditional "forward" searches that attempt to determine whether the spectrum of a particular unknown is contained in the reference file, the "reverse search" attempts to determine which (if any) of the spectra in the library are contained in the spectrum of the unknown. In other words, "rather than selecting a characteristic set of masses and intensities from an unknown spectrum and searching an equivalently reduced library, [it] selects a reduced library spectrum, extracts from the unknown spectrum just those intensities corresponding to the library spectrum masses and then compares the intensities."[84] A list of reference compounds, all of which are possibly contained in the mixture, is produced as output; if the unknown spectrum is that of a pure compound (whose spectrum is contained in the library, of course), this list should contain only one member—the compound that provides the correct identification of the unknown.

The greatest advantage of the reverse search over the forward searches discussed previously is that most interferences in the unknown spectrum are ignored; this often allows clear identification of contaminated spectra. For the reverse search to be totally effective it is necessary that there be minimal interference between the peaks of various components. Peaks that are more abundant than would be expected, however, are recognized as possibly contaminated and will not "count against" the match if one has sufficient confidence in an adequate number of the other peaks used for identification.

The work of Grotch, Ridley, and others has repeatedly shown that the method of choosing the peaks used in the matching process is an important factor in the performance of mass spectral retrieval systems. In order to minimize the handicaps associated with mixture interpretation, this aspect is extremely important to reverse searches. In these systems it is essential that the peaks used to characterize each reference spectrum actually be most characteristic of the reference compound. In order to reduce the ambiguities introduced by contamination, it is also desirable that the peaks which are used be the least likely to be contaminated.

Abramson's system, which normally uses the ten most intense peaks in each reference spectrum for matching, attempts to circumvent the second of these two problems by severely restricting both the size and the composition of the reference file.[86] He suggests creating small libraries of no more than approximately 100 spectra, all easily distinguishable from one another and specifically chosen in view of the problem at hand (pesticides, illicit drugs, etc.).

The "probability-based matching" system (PBM) of McLafferty, Hertel, and Villwock[35, 36] provides a systematic, automated method of selecting the peaks in each reference spectrum that are both most characteristic and least subject to contamination. At the same time, it introduces a radical departure from other search systems by attempting to provide a statistically meaningful way of expressing the similarity between the abbreviated spectrum of the reference compound and the matching peaks extracted from the unknown spectrum.

Mass spectrometrists have long been aware of the fact that peaks at certain m/e values occur far less frequently than peaks at other m/e values, and that abundant peaks are less common than peaks of low intensities; for example, a peak of 1% abundance at m/e 343 is less common than a 1% peak at m/e 43, and a 100% peak at m/e 343 is less common still. [34, 64, 87] McLafferty and co-workers used a compilation of the eight largest peaks in 17,124 spectra[88] to estimate mass value probabilities for the three most abundant peaks in each spectrum. The peaks then chosen as the most characteristic and least subject to interferences by contaminants are those with the *smallest probability* of occurrence. In other words, a peak is determined to be important in the identification of a compound either because it occurs at an unique m/e value, or because it is a very intense peak; or, as is most likely, because it has some desirable combination of both mass and abundance characteristics.

The retrieval system itself is based upon the "General Rule of Multiplication" of probability theory,[89] which states that if n independent events occur with probabilities $p_1, p_2, \ldots, p_n$, then the probability of all n of these events occurring is

$$\prod_{i=1}^{n} p_i$$

These researchers apply this to the identification of mass spectra in the following way: Assume that a peak at some m/e value m_1 with intensity i_1 has an independent probability of occurrence in a mass spectrum of p_1; and likewise that another peak (m_2, i_2) has a probability of occurrence of p_2. Both of these peaks are contained in the spectrum of a particular reference compound Z. The occurrence of this combination of peaks m_1, m_2 with intensities i_1 and i_2 in an unknown spectrum can be interpreted as due either to the presence of reference compound Z or to some random combination of spectra (one or more), the probability of which is $p_1 \times p_2$. For an unknown spectrum that one knows is of a pure compound, this latter possibility can be taken to mean that $(p_1 \times p_2)^{-1}$ spectra would have to be examined before a spectrum would be found that matches the unknown as well as (or better than) that of compound Z.

   Although this retrieval approach surpasses all others in its attempt to provide a statistically meaningful measure of the similarity between two mass spectra, these researchers do not claim that the imposed meaning is absolute, nor that the attempt is theoretically perfect. The assumption that mass spectral peaks are totally independent events is central to this approach, and the validity of this assumption is essential if the General Rule of Multiplication is formulated as above. It is a well-known fact of mass spectrometry, however, that this is *not* the case—that mass spectral peaks *are* highly correlated.[1-4, 35, 66, 87] Strong correlations between peaks 14 mass units apart provide one example; those between certain neutral losses and/or ions characteristic of particular compound classes (where a class may be very large, such as the class of all hydrocarbons; or very small, such as the class of all xylene isomers) provide many more examples. A rigorous approach to probability-based matching would have to consider the probability of occurrence of any peak given that any other combination of (one or more) peaks is known to occur. The evaluation of all these dependent probabilities can hardly be considered a reasonable task, but future refinements of the system using some of the most common correlations should be considered as a possibility. However, as long as the *degree* of dependence is approximately the same for the data of different spectra, the resulting measure of similarity of spectra will be valid on a relative scale; in support of this the results obtained so far do not appear misleading when a sufficient number of peaks are used to characterize each reference spectrum. Distinction among the spectra of closely related isomers, several of which may be retrieved by the system, can often be made within the normal limitations of distinguishability among the spectra.[36]

McLafferty *et al.* express the similarity measure ("confidence value" or "K value"), as well as all the individual probabilities, as the corresponding base 2 logarithms for convenience of calculation; inverse probabilities are also used to simplify the calculations and to produce a final result which is a direct measure of "confidence," In this reverse search, there is computed for each abbreviated reference spectrum matched against the unknown a confidence value K,

$$K = \sum_{j=1}^{n} U_j - A_j + W_j - D$$

where n is the number of peaks in the reference spectrum, U the contribution to the probability of the "uniqueness" of the m/e value of the peak, and A the contribution to the probability of the abundance value of the peak as it appears in the reference spectrum; W, the "window factor," is a measure of the agreement between the abundance of the peak in the reference and in the unknown, and D, the "dilution factor" for mixture spectra, is a measure of the overall reduction of peak intensities in the unknown due to the presence of other components—If the unknown spectrum is of a pure compound, $D = 0$. As in any reverse search, the n peaks of a particular reference compound are extracted from the unknown spectrum (which may be either the spectrum of a mixture *or* of a pure compound) at the time of matching. In general, $n = 10$–15; these peaks are chosen in order from the peak with the highest value of $U - A$ to that with the next highest $U - A$ value, etc. Those peaks in the unknown whose intensities agree within a predetermined range each contribute increments of U, A, and W to the confidence value; those peaks which do not agree because they are *more* intense than would be expected are termed "contaminated" and are ignored in the cumulative calculation of confidence. Since all the reference spectra were recorded on the same instrument and the background level is known for each unknown sample, peaks which do not agree because they are *less* intense than would be expected rule out the possibility of a match corresponding to this particular reference compound. Either all compounds whose K values exceed some preset threshold can be displayed, or particular identifications can be sought and the K values listed for the comparisons.

This probability-based matching system is able to give quantitative as well as qualitative indications of the components in the mass spectrum. Reduction in the quantity of a sample component is found to produce a regular decrease in the K value both because of the dilution factor D and because of the "disappearance" of normally low-intensity peaks into the background level of the spectrum. Figure 1 clearly illustrates this effect in addition to the power of PBM to distinguish between the spectra of structurally similar compounds unambiguously.

The concept of probability-based matching has been generalized by researchers at Cornell to a retrieval system that uses large libraries of mass

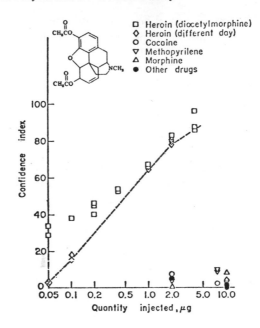

FIG. 1. Heroin identification algorithm for the probability-based matching system (PBM)[36]: Effect on confidence index of the amount of heroin and of other drugs. Other drugs tested include quinine sulfate, procaine hydrochloride, ephedrine, strychnine sulfate, amphetamine sulfate, metamphetamine hydrochloride, secobarbital, phenobarbital, amobarbital, and pentobarbital. The 2 in the lower left corner indicates the number of determinations yielding the same value. Structure given is that for heroin.

spectra originating from diverse sources. For a library of nearly 24,000 mass spectra, the base peak, the molecular ion peak (if one is present), one (or two, if there is no molecular ion) neutral molecular loss peak with the highest $U + A^*$ of all such losses, and the 12 other peaks of highest $U + A$ in each reference spectrum are used for matching in the reverse sense.

Tests on approximately 900 spectra of pure compounds and 200 synthetic mixture spectra have shown that this system is superior to commonly used forward search techniques. Probability-based matching, which uses both abundance and uniqueness criteria in a statistically meaningful relation to each other in order to determine the importance of each peak in the spectrum, is a refinement on all previous methods of weighting. Thus, as dependent probabilities are incorporated into the system, not only will the computation

---

* As opposed to the specialized PBM system discussed above, A values in this generalized system are determined at the 1% abundance level, thus assuming positive values at greater abundances; the corresponding decrease in occurrence probability is therefore expressed by $\frac{1}{2}^{U+A}$.

of K become more valid statistically, but it is also expected that the overall system performance will improve as the weighting system is further refined.

B. Clustering Methods: Empirical Identification of Compound Class

When mass spectra, represented by the appropriate term vectors containing mass and abundance information, are conceptualized as points in N-dimensional mass spectral space, the retrieval process discussed in the previous section reduces to that of finding the point representing the reference spectrum which is identical to or nearest that point representing the unknown spectrum in N-space. Identification of an unknown mass spectrum using a pure retrieval system can be thought of, then, as a special and limiting case of "pattern recognition," a set of methods which comprise "the detection, perception, and recognition of invariant properties among sets of measurements of objects or events."[90] When the distribution of all these points in N-dimensional space is examined, however, some more general propositions about pattern similarities can be made. It will be found, for example, that these points are not distributed evenly throughout the space but instead tend to cluster in certain regions due to the presence of similar terms in the vectors or, in other words, due to the existence of pattern similarities among the spectra of different compounds. Mass spectrometrists have been aware of this phenomenon since the first serious application of the technique to molecular structure elucidation.[1-4] The determination of which of the terms are common to a cluster of points is referred to as the process of finding mass spectral correlations.[87]

The presence of certain chemical functional groups in a molecule affects the mass spectral fragmentation pattern to some greater or lesser extent, producing ions that are characteristic of the mass spectra of members of that compound class. Thus these pattern similarities that cause clustering in the universe of points are the result of structural similarities among the compounds whose spectra are represented by these points. It is, of course, this feature that makes it possible to retrieve compounds with structures similar to that of the unknown when the spectrum of the unknown is not contained in the file.

In document retrieval systems, the phenomenon of clustering is generally exploited to speed up the search; by matching the query first with the centroids of all the clusters, the search space may be reduced dramatically. This technique could be applied to mass spectral retrieval systems as well if one so desired—if the unknown is first matched with the cluster centroids (each of which can be thought of as a sort of "average spectrum" for a particular compound class), the only spectra in the file that would be individually searched would then be the members of the (one or more) compound classes whose centroids best match the unknown spectrum.

The research done in mass spectrometry so far in the area of cluster analysis has instead taken advantage of the predictive ability obtained by the clustering process. Provided that there is *minimal overlap* between the clusters in *N*-dimensional space, the chemical class of a monofunctional compound is relatively easy to determine (within appropriate confidence limits) by determining in which cluster its spectrum is found. If considerable overlap occurs, however, the resulting classification will be extremely unreliable. Multifunctional compounds *may* be found to be "closest to" the centroids corresponding to each of the functional groups present; if the fragmentation pattern is dominated by one of the groups, however, the effects of other groups on the spectrum may go unnoticed, as might be expected in manual interpretation as well. A third possibility for the case of multifunctional compounds is, of course, that the interaction between the groups is such that the spectrum does not strongly resemble the spectra of any of the individual classes. The resulting classification would be ambiguous at best and totally incorrect in the worst case.

Depending on the type of classification sought, it may be advantageous to construct the term vectors by criteria other than the direct use of all *m/e* and intensity values in the spectrum. For example, neutral loss data may prove useful for the classification of compounds with small electronegative functionalities. Reduction of the term vector to 14 dimensions where each term is weighted in correspondence with the relative importance of each of the 14 homologous ion series may not only increase the speed of the classification, but also provide equally reliable classification in many cases because of the retention of structurally significant information.[1, 91]

Although the same information concerning compound class might be obtained using a pure retrieval system, the cluster approach has two advantages over retrieval systems for this type of application. First, the method is guaranteed to be faster than using a pure retrieval method since the library is reduced to a much smaller collection of "average spectra" (i.e., the centroids). Further reductions in time can be obtained by the construction of a hierarchical system of clusters. The first level of such a system, for example, might be designed to establish the presence or absence of various heteroatoms in the molecule by determining the centroids of such large classes as "oxygen-containing" and "non-oxygen-containing" compounds. The computer might thus classify the compound as "oxygen containing" and then proceed to compare the unknown to the centroids for "alcohols," "ketones," and so on; perhaps having classified the compound as "alcohol," the third hierarchical level would determine "primary alcohols," "secondary alcohols," etc.

A second advantage of the clustering approach is that the construction and use of the cluster centroid "smoothes out" individual differences among the spectra of members of the same class; and, since a centroid is determined specifically for each class, there is no confusion in interpreting the output from the matching process. The results obtained from a retrieval system when it is used for compound classification may often be difficult to interpet if no single functional group positively dominates the list of retrieved compounds. Also, the appearance of anomalous compounds in the list may be due to poor recording or coding of the reference spectra or to individual peculiarities of the mass spectral behavior of a particular compound and *not* to true structural similarities among the compounds.

Empirical compound classification is an interpretive technique used to obtain structural information for a compound provided that the reference file contains at least one member of the same compound class. Its accuracy is

TABLE VI

Distances between Centroids of Each Compound Class and Surface of Hypersphere as Determined by Crawford and Morrison[91][a, b]

Aromatics	14.6	Esters	26.8
Dienes	18.5	Alkanes	29.5
Alkenes	20.9	Alcohols	30.4
Ketones	21.4	Amines	30.6
Cycloalkanes	24.6	Ethers	33.0
Aldehydes	25.8	Acids	37.5

[a] The tighter the class cluster, the closer will the centroid (i.e., center of gravity) lie to the surface of the hypersphere.

[b] Radius of hypersphere = 100.

primarily dependent upon the degree of overlap between the clusters used for classification. Unless the classification process is used as a prefilter for a retrieval system or is combined with the approaches discussed in Section III (*vide infra*), a positive identification of the complete structure of the compound will *not* be obtained even if the actual spectrum of the compound is contained in the original library.

Crawford and Morrison[91] first reported their investigations on the properties of clusters in 1968.* Three separate vector spaces are considered. The first of these uses all peaks in the spectrum from m/e 1 to 500; the intensities of the peaks, normalized such that

$$\sum_{n=1}^{500} P_n^2 = 1$$

are used as weights in the term vector. For each of twelve compound classes the "tightness" (or "spread") of the clusters is determined by computing the

* Raznikov and Tal'roze[92] proposed a similar approach two years earlier; however, this work will be considered in the next section on "learning machines," to which it is more closely related.

distances of the respective centroids from the surface of an N-dimensional hypersphere with radius 100. Table VI lists the classes in order of decreasing tightness of clusters; as can be seen, the "aromatic" cluster is the tightest, indicating very high similarity and strong correlations among the spectra of all aromatic compounds. At the other extreme, the "acid" cluster is the most diffuse of the twelve, indicating that the spectra of acids are less similar from one to another in the group and less highly correlated than the other compound types.

The overlap among clusters is then approximated from the distances between the centroids projected onto the hypersphere. These distances are presented in Table VII. The clusters that are closest (35 units) to each other in the vector space are those of ketones and esters; an unknown spectrum of a compound of either of these two classes might therefore prove difficult to classify unambiguously unless the clusters are each sufficiently tight (Table VI) to reduce the amount of overlap. The distance between the ether and alcohol centroids (50 units), although slightly larger than that between ketones and esters, may in fact be more open to ambiguous results since the ether and alcohol clusters are much looser. It is interesting to note in Table VII that the centroids of the "aromatic" cluster and the "amine" cluster are quite distant from all other centroids; this is generally true of the "diene" cluster as well.

In addition to forming class clusters, Crawford and Morrison constructed and examined the properties of clusters formed by compounds with equivalent numbers of rings, double bonds, oxygen atoms, and nitrogen atoms. One compound from each of the twelve compound classes was used as an unknown and the distances between these points and each of the centroids were computed. The identification of compound class using the actual class clusters is found to be more accurate than that information obtained by using the clusters based on rings, double bonds, etc. (two misclassifications and two ambiguous classifications as opposed to five misclassifications), although the information obtained using *both* methods is "frequently complementary."[91]

The "mass periodicity spectrum," calculated by using the formula

$$Q(m) = \sum_{n=1}^{500} P(n) \times P(n + m)$$

where $Q(m)$ equals the relative probability of periodicity m ($m = 1, 2, \ldots, 50$), n the m/e values of all peaks in the spectrum, and P the intensity of an ion (at mass n or mass $n + m$), can be thought of as a point in 50-dimensional space and the *average* mass periodicity spectrum of a particular compound class as the centroid of that class cluster. The dimensionality of the vector space (and thus the number of calculations that must be performed for each unknown) is reduced by an order of magnitude from the 500-dimensional space considered above. This method of identification produced only one mis-

TABLE VII

Distances between Projections on the Hypersphere of the Centroids of Each Class Cluster[a,b]

	Aromatic	Ether	Acid	Ketone	Aldehyde	Ester	Alkene	Alkane	Alcohol	Cyclo-alkane	Diene	Amine
Aromatic	0	129	134	125	128	130	120	128	129	121	115	131
Ether	129	0	84	91	85	77	105	88	50	117	129	115
Acid	134	84	0	108	96	97	114	104	96	120	132	110
Ketone	125	91	108	0	56	35	85	42	81	106	124	118
Aldehyde	128	85	96	56	0	59	79	49	74	96	124	103
Ester	130	77	97	35	59	0	91	46	74	110	127	120
Alkene	120	105	114	85	79	91	0	75	86	38	102	116
Alkane	128	88	104	42	49	46	75	0	78	94	124	114
Alcohol	129	50	96	81	74	74	86	78	0	100	129	115
Cycloalkane	121	117	120	106	96	110	38	94	100	0	93	118
Diene	115	129	132	124	124	127	102	124	129	93	0	129
Amine	131	115	110	118	103	120	116	114	115	118	129	0

[a] See Table VI and Crawford and Morrison.[91]
[b] Radius of hypersphere = 100.

classification and one ambiguous classification, although the unknown compounds used were different from those used in the previous classification schemes, thus making an absolute comparison of the methods impossible.

Using the simple intensity normalization

$$\sum_{n=1}^{500} P(n) = 1$$

and the reduction formula

$$T(m) = \sum_{n=0}^{35} P(m + 14n) \quad m = 1, 2, \ldots, 14$$

where $T(m)$ is the probability of the "condensed mass" m (i.e., one of the 14 homologous ion series), and n is a sequential number indicating a specific 14 mass unit interval, Crawford and Morrison create a "reduced mass spectrum" which is an ion series summation over the mass range 1–504 amu. The dimensionality of the vector space is reduced to 14, and the centroid of each of the clusters is the average ion series spectrum for all compounds of that class. As mentioned in the discussion above, this technique is found to produce fairly reliable identifications (3 misclassifications and 3 ambiguous classifications out of 18 unknowns).

Smith and Eglinton[93, 94] also developed a classification system based on ion series spectra and reported excellent results. Compound classes are more rigorously defined in this work and are considerably smaller than the broad classes examined by Crawford and Morrison: for example, while Crawford and Morrison consider one cluster of esters, Smith creates ten very specific ester clusters. The centroids of sixty-five clusters of compounds of geochemical and environmental interest are calculated using all masses greater than 35 amu.

A somewhat different approach to cluster generation based on neutral loss data has been considered by Ishida et al.[95] Each vector is composed of 50 neutral loss terms, and the clusters formed by the points representing compounds that contain a particular functional group are examined. Rather than determining the centroid of each cluster, Ishida et al. use factor analysis[96] to extract the terms that are correlative in the "neutral loss spectra" of the cluster. A vector composed only of these terms (i.e., neutral losses) is thus constructed for each class and used instead of centroid matching to determine the "plausibility" of a particular group's presence in the unknown compound.

Both neutral loss data *and* characteristic ion data were used by Heller et al. in their approach to cluster analysis.[62, 97] A vector in 227-dimensional space whose terms include all peaks from m/e 14 to 140 and all losses from the molecular ion to $M - 99$ is constructed for each spectrum in a reference file which is exclusively composed of the spectra of compounds containing exactly one sulfur atom. The major distinction of this work is that the computer is not told which compounds to form into clusters; rather the computer is free to form the tightest clusters that it can using a graph-theoretical method called the "Shortest Spanning Path" (SSP). To do this, the computer is first given a list of the points as represented by their term vectors; then, using an algorithm which iteratively reorders the list, the computer constructs an ordered list such that there is a minimum sum of distances between all points adjacent in the list. This final list can be segmented, and the compounds in any segment are thus members of the same empirical class cluster. The segmentation can be done either manually or by machine; these researchers performed the segmentation manually, using as criteria the chemists' "intuitive judgment."[07] Heller et al. anticipate that future work will include automated segmentation procedures; this addition, plus an automated method of factor analysis and subsequent correlation of these terms with the structural features common to members of the clusters should be an interesting study.

Chu uses four additional types of cluster analysis techniques in an investigation of the pharmacological activity of certain drugs, based on their structure-activity relationships.[98] Since the mass spectrum of a drug molecule is directly related to its structure and its structure to its pharmacological activity, the clustering properties of the mass spectra of sedatives and tranquilizers are examined in hopes of predicting the biological properties of other as yet unclassified organic compounds.

One of these methods, the "K-Nearest-Neighbor" (KNN) approach, is the offspring of simple retrieval systems. The classification of the unknown mass spectrum, represented by a point in N-dimensional mass spectral space, is made on the basis of a "majority vote" of its K nearest neighbors. In other words, the distance between the unknown mass spectrum and other members of the library is computed, often using the Euclidean distance measure, and, if K equals one, for example, the unknown is put in the same class as the reference spectrum that lies closest to it in N-space. The reference "library" or "training set" is limited to only the spectra of compounds that are known to belong to the classes of interest.

The "Nonlinear Map" (NLM) technique maps the spectral points in N-dimensional space onto 2-dimensional space so that visual determination of clusters is possible, and an unknown may be identified as belonging to one class or another by human inspection. The "Minimal Spanning Tree" (MST) and "Hierarchical Clustering Algorithm" (HAR), like the SSP technique employed by Heller et al., are graph-theoretical methods in which the computer forms the best clusters it can. Examination of these clusters then produces the classification of the spectra.

Since over half the spectra of sedatives and tranquilizers used in this study are barbituric acid and phenothiazine derivatives, respectively, it is not surprising that reasonably tight clusters and consequently correct pharmacological classifications were obtained by these researchers. A learning machine approach to this problem is described in the following section, and some of the criticisms and controversies surrounding this work will be discussed concurrently (*vide infra*).

C. The Learning Machine Approach to Pattern Recognition

Once it is recognized that mass spectral patterns, represented by points in N-dimensional mass spectral space, tend to form clusters based on structural similarities among the compounds, it becomes possible both to visualize and to construct surfaces that separate the clusters from one another. The "learning machine"[99] approach to mass spectral classification and interpretation involves the attempt to construct such surfaces. It is very closely related to cluster analysis, discussed in the previous section; indeed, cluster analysis has occasionally been referred to as "learning without a teacher" or "unsupervised learning."[97] The separation of the two techniques in this discussion is in fact rather arbitrary and is motivated in part by the indebtedness of the more recent methods discussed in this section to the influential monograph of Nils Nilsson,[99] both with regard to vocabulary and conceptual framework. A learning machine is defined by Nilsson to be "any device whose actions are influenced by past experiences,"[100] although he and most of the mass spectrometrists working with learning machines are specifically concerned with "a subclass of learning machines, those which can be *trained* to recognize patterns."[100]

The construction of the surface separating classes of data is accomplished through the use of a "discriminant function" which defines that surface in N-dimensional space. Specifically, for the mass spectrum X consisting of the mass positions ("terms") $x_1, x_2, \ldots, x_d$, R functions are sought for which $g_i(X) > g_j(X)$, where $i, j = 1, \ldots, R, j \neq i$, and R is the number of distinct classes. In other words, if X is in class i, the value of $g_i(X)$ will be the largest of the R values. The number of discriminant functions can be reduced from R to $R - 1$ by selecting one of them and subtracting it from all the others; thus, for the special case of $R = 2$ [a "binary pattern classifier" (BPC)], only one discriminant function is necessary. It is especially important to note this characteristic in a discussion of the learning machines currently being applied to chemical problems, since much of this research has dealt with the special case of binary pattern classifiers. In general, then, for a BPC X is placed in class 1 if $g(X)$ is positive (and consequently X lies on one side of the surface in N-space) and in class 2 if $g(X)$ is negative. Thus $g(X) = 0$ defines the actual decision surface.

The publication of Raznikov and Tal'roze in 1966[92] is believed to be the earliest application of trainable pattern classifiers to any chemical problem.

These researchers began with the attempt to classify saturated hydrocarbons and monoolefins using only the peaks at m/e 41, 43, 55, and 57. Since it is known that the intensities of the $(C_nH_{2n+1})^+$ ions in the spectra of saturated hydrocarbons predominate over the intensities of the $(C_nH_{2n-1})^+$ ions,[1] and that for olefins the opposite tends to be true, the declared interest of these researchers was to determine whether the computer could discover a better or more precise criterion for differentiating these two classes.

The 4-dimensional vectors generated for the members of the training set were normalized to unit length. Then, using several different algorithms, "training" of the system was accomplished by the construction of a system of cones with their apexes at the origin of the hypersphere such that (1) for one of the classes of compounds, each vector representing the mass spectrum of such a compound falls within one or more of the cones, and (2) none of the vectors associated with the other class of compounds falls within any of the cones. The computer was indeed able to discover a more quantitative relationship among the intensities of these four peaks to use in distinguishing saturated hydrocarbons and monoolefins, and correct recognition of training set members and prediction of unknown spectra was obtained 94% of the time.

With such a successful first experiment, Raznikov and Tal'roze attempted to use the peaks at m/e 41, 42, and 43 to make the same classification; no empirical rules were known that would successfully distinguish these two classes using these peaks, however. Using several different algorithms, recognition and prediction rates of 84% to 91% were obtained.

Finally, an attempt was made to distinguish between monoolefins and cycloparaffins, a classification problem for which no empirical rules at all were known. Peaks at m/e, 39, 41, 42, 43, 55, and 57 were chosen, and, depending upon the algorithm used, a recognition-prediction rate of about 78% could be achieved.

Despite the limitations of training set size on this work and the state of computer technology in 1966, and regardless of the simplicity of the three problems investigated, the research of Raznikov and Tal'roze is undeniably of great historical significance in the field of pattern recognition as applied to chemical problems. Their work was the precursor of other extremely elegant and exciting techniques for the interpretation of mass spectra.

The early work of Jurs, Kowalski, and Isenhour, however, is the true fountainhead of the intense interest in the application of learning machine techniques to chemical problems.[101] This original paper applied learning machines to the determination of molecular formulae from low resolution mass spectra. Since then, there has been a profusion of work in many areas of chemistry, notably in other areas of spectroscopy; the ensuing discussion will be concerned only with the work that has appeared in the area of mass spectrometry.

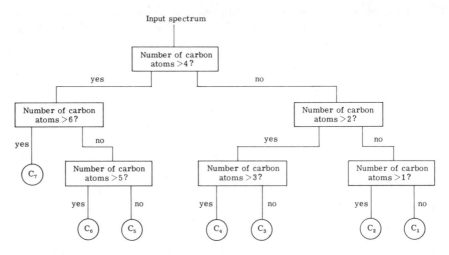

FIG. 2. Branched tree structure of binary pattern classifiers used to determine carbon number (C_1-C_7).[101] A positive value of $g(X)$ corresponds to a left branch.

The research on molecular formula determination used a number of binary pattern classifiers arranged in a "binary tree"[102] structure. The part of this tree used in the determination of carbon number is presented in Fig. 2. It is seen that 6 binary pattern classifiers are needed to determine the number of carbons (from 1 to 7) that the compounds contain; 15 are used to determine the number of hydrogens (H_1-H_{16}), 3 to determine the oxygen content (O_0-O_3), and 2 to determine nitrogen, N_0-N_2.

This learning machine uses a linear discriminant function of the form:

$$g(X) = w_1x_1 + w_2x_2 + \cdots + w_dx_d + w_{d+1}$$

where X is the mass spectral vector defined by $x_1, x_2, \ldots, x_d$, and $w_1, w_2, \ldots, w_d, w_{d+1}$ are the parameters on which the function depends. If these parameters, or "weights," were known *a priori*, it would be a trivial task to construct the decision surface. In other words, if a mass spectrometrist knew precisely the usefulness and the importance of finding a peak at any particular m/e value for distinguishing between two classes of compounds, there would be no need to use a computer to assist him in this step of the interpretation—except, perhaps, as a high-speed calculator. Not having this *a priori* knowledge, however, mass spectrometrists may design learning machines to uncover these relationships and to predict from these relationships the class of an unknown compound.

The "Threshold Logic Unit" (TLU) describes in physical terms this process of pattern classification; a TLU is presented in Fig. 3. Although pictured as a machine, the TLU is implemented using computer software.

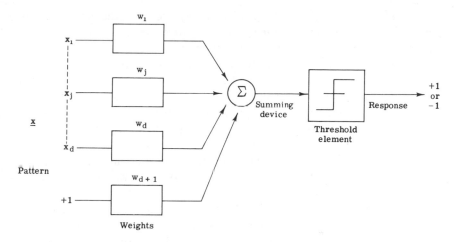

FIG. 3. A threshold logic unit (TLU)[99, 101] for the binary pattern classifier used for determining the class of mass spectrum X. The discriminant function is

$$g(X) = \left[\sum_{j=1}^{d} w_j x_j \right] + w_{d+1}$$

The parameters w_j are set to any arbitrary value (usually $+1$ or -1) to start. The first mass spectrum in the training set, whose classification is thereby known, is presented and the response of the TLU is examined. If the response is the desired one, with a $+1$ indicating membership in one of the two classes under investigation and a -1 indicating membership in the other, the machine proceeds to the next mass spectrum. If the desired response is *not* obtained, however, the parameters w_j are adjusted so that the spectrum falls on the correct side of the decision surface. The next mass spectral pattern is presented, and this "training" continues either until all training set spectra are correctly classified, in which case the clusters are said to be linearly separable, or until some maximum number of iterations has been performed.* Now the machine has learned, through this error correction process, something about the relationship of the peaks in a mass spectrum to the structure of the compound. Unknown spectra may then be presented and, as the weights w_j are determined, prediction of the class of the compound is extremely rapid.

Jurs *et al.* used the m/e values of all peaks in the mass spectra having abundances greater than or equal to 1%. In another test, the square roots of

---

* This latter case is not proof of cluster overlap or linear inseparability but strongly indicates that this may be the case.

the abundances were used, which substantially decreases the amount of time required for training by decreasing the "dynamic range" of each element x_j in the pattern vector by an order of magnitude.

The same basic method was used again in another study by Jurs and co-workers on the convergence rate and predictive ability of BPC's.[103] The discovery of the previous paper that the convergence rate increases as the dynamic range of the terms decreases was found to be generally true. For the special case in which the mass spectra are reduced to binary vectors in which all peaks of 1% abundance or greater are given the value 1, a prediction rate of 91% is obtained for the determination of oxygen presence. Once again the overdetermined nature of low-resolution mass spectra for particular problems is demonstrated (cf. Grotch).

Various error correction or "feedback" techniques are compared, as are the convergence rates and predictive abilities obtained by a reduction in dimensionality of the vectors. This latter aspect of the work is significant since the number of calculations that must be made in training is linearly related to the number of terms in the vectors, and training is the time-consuming part of the learning machine approach. Table VIII shows that elimination of the

TABLE VIII

Convergence Rate and Predictive Ability of Learning Machine to Determine Oxygen Presence as a Function of Number of Terms[103, 105]

Number of terms	Average recognition	Average % predicted
155	300[a]	90.6
95	300	91.0
65	300	89.9
50	300	90.2
35	300	90.2
20	284	86.2
10	245	72.2

[a] Out of a training set of 300 spectra.

appropriate terms from the pattern vector is able to reduce the dimensionality of the "oxygen presence problem" from 155 terms to 35 terms without significantly altering the predictive ability of the system, providing further proof of the overdeterminedness of mass spectra for solving certain problems. More detailed studies of "feature extraction," or this reduction in dimensionality that preserves the features of the mass spectra most significant for classification, have also been reported.[104, 105]

The first work on multicategory pattern classifiers was undertaken by these same researchers using a least-squares method.[106] This technique avoids having to calculate R individual discriminant functions (*vide supra*) by developing a single weight vector that produces a value having both a sign *and a magnitude* indicative of the class to which the pattern belongs. For a small number of categories, this method is found to be usually much slower than a series of binary pattern classifiers, but it becomes much more practical as the number of categories increases. In general, the predictive ability of the least-squares multicategory classifier is less than for the linear learning machine although it is often able to produce classifications when the clusters are not linearly separable. In addition, the reduction of the dimensionality of the vectors becomes more pressing for the least squares procedure since calculation time increases as the *square* of the number of dimensions.

One of the most interesting aspects of pattern recognition techniques is that the data used in the construction of the pattern vectors need not come from a single source. Thus Jurs *et al.* combined mass spectral data with infrared spectral data and melting and boiling points in a later study.[107] The learning machine approach is thus able not only to evaluate which analytical method may be most useful in a particular situation, but also the optimum combination of techniques in the solution of certain problems.

A parallel arrangement of BPC's, rather than the branched binary tree discussed above, has been used by these authors in the determination of structural features of a molecule from its low-resolution mass spectrum.[108] The predictive ability of this system is quite high for individual structural features, averaging 90% for hydrocarbons and 88% for compounds containing either oxygen or nitrogen or both; however, the ability of the system to correctly predict *all* aspects of the molecular structure of a compound is quite limited, making this approach unreasonable for the positive identification of an unknown mass spectrum. Additional work on the determination of partial molecular structures using both the ten most intense ions in the spectra and neutral loss data has been reported by Sasaki and Abe.[109]

Learning machines have also been used in the reverse sense of that described above: reasoning from molecular structure to mass spectra.[90, 105, 110] Structural features are represented through use of a computer-compatible fragmentation code; training sets can then be formed by collecting the spectra containing the same structural feature. Discriminant functions are calculated for 60 m/e values such that each weight vector has the ability to classify the compound into category 1 if its spectrum should contain a peak of at least 0.5% of the total ion current at that m/e value or into category 2 if it should not. For 11 of these m/e values, additional weight vectors were determined for the 0.1% and 1% threshold values.

Lytle[111] has suggested using error-correcting Hamming codes[112] instead of branched tree or parallel computation in order to minimize the number of

TLU's that are needed to make multicategory classifications. Felty and Jurs[113] then compare the three methods for the multicategory problem of carbon-number determination. Two significant findings are reported: first, that the individual clusters formed by the C_4–C_{10} data sets are linearly separable since all methods, including the binary code schemes, are able to be completely trained within a reasonable number of feedbacks; and second, that the branching tree arrangement of binary pattern classifiers (Fig. 2) resulted in the highest prediction rates.

The representation of the data used in training a learning machine is extremely important. Just as it was shown that particular normalization procedures can have a profound effect upon the rate of convergence of the machine,[103] it has also been demonstrated that clusters that appear to be linearly inseparable using previously discussed forms of data representation (e.g., spectral intensities normalized to 100% base peak, square root intensities, etc.) may become separable if particular transformations are applied to the data.[114-118] The use of Fourier[114, 115] and Hadamard[116] transforms, as well as new feature generation from trained weight vectors,[117] spectral moments, and intensity histograms,[118] has been found to give improved results for particular classifications. These methods also have the added benefit of possibly reducing the dimensionality of many of the problems without significant losses of predictive ability.

In all the above examples of binary pattern classifiers, the discriminant function $g(X) = 0$ defines a surface on one side of which ($g(X) > 0$) spectra are classified as belonging to category 1 and on the other side ($g(X) < 0$) to category 2. The addition of a "width parameter" or "dead zone," $\pm \Delta$, such that if $g(X) > +\Delta$ the spectrum is classified in category 1 and in category 2 if $g(X) < -\Delta$ [if $-\Delta < g(X) < +\Delta$ no classification is made] is also found to give improved results for linearly inseparable clusters.[119, 120] Jurs examines the effect of using such a dead zone in a general and empirical study of prediction and separability.[120]

A later study on prediction and separability using TLU's has been made by Anderson and Isenhour.[121] It has been noted[122, 123] that essentially meaningless results may be obtained by a pattern classifier if the number of objects to be classified (N) does not exceed the number of terms used (D, the dimensionality of the vector) by a considerable margin, e.g., $N/D > 2$. By creating vectors filled with random numbers of Gaussian distribution, Anderson et al. could train a BPC and collect meaningful statistics on the training. Not only is the expected result, that predictive ability increases as the ratio N/D increases (for constant D), found to be true, but it is also found that early indications of the separability of two sets of data may be obtained if N/D is greater than 2.5.

Another method of improving the results obtained using a linear learning machine is to improve the discriminant function. Attempts to do this using an iterative least-squares procedure have been made.[105, 124] As opposed to error-correcting feedback methods, this approach searches for the best fit of patterns to a nonsingular function with linear parameters. Justice *et al.* demonstrate the feasibility of using a complex-valued nonlinear discriminant function for predicting molecular structural features from mass spectra.[125]

Another "nonlinear approach" to learning machine classification has been investigated by Jurs in a study concerned with the development of cross terms to improve the system.[126] Graph-theoretical procedures are used to determine both intraset and interset cross terms, i.e., those pairs of terms that occur together sufficiently often in the spectra of compounds of the same class, or sufficiently seldom in spectra from the two different classes. It is found that the intraset cross terms are not particularly helpful for pattern classification by the learning machine but that interset cross terms can be useful both in decreasing the convergence rate of the machine and reducing the dimensionality of the vectors.

In addition to being theoretically interesting, learning machines can be used in the solution of practical problems. Tunnicliff and Wadsworth[127] used the ability of BPC's to categorize the structural features of hydrocarbons and hydrocarbon types in general in the analysis of gasoline samples. Using carefully selected training sets composed of the spectra of compounds likely to be contained in gasoline, excellent quantitative results are obtained for these extremely complex mixtures.

The application of cluster analysis to the determination of structure–activity relationships of drugs using mass spectral data was discussed previously; the same authors have also applied learning machine techniques[98, 128, 129] to the data sets of sedatives and tranquilizers and obtain satisfactory results in terms of linear separability and predictive ability. The bias of the data sets towards the spectra of barbituric acid and phenothiazine derivatives has already been mentioned (*vide supra*). This lack of independence is noted in a response to the work of Ting *et al.* by Perrin.[130] Clerc *et al.* also raise a criticism to the work on the basis of the low N/D ratio.[123] Both sets of critics demonstrate that these shortcomings may result in verification of absurd hypotheses, and both issue stern warnings to all who would overlook important criteria in the construction of learning machines.

An interesting comparison of cluster analysis methods and learning machine techniques is made by Justice and Isenhour.[131] In a study of the determination of structural features from low-resolution mass spectra, the following relationships between the increase in predictive abilities of the various methods were observed:

Sum spectra < binary spectra < normalized sum spectra < nonlinear transform = learning machine < nearest neighbor

The first four methods involve computing a centroid for the class clusters. The "sum spectra" are the average spectra of the classes where each spectrum is normalized to the base peak equal to 100% prior to calculation. *This gave the poorest results of all methods*—a caveat for those who are tempted carelessly to determine "average" spectra for a given compound class. The averaged binary spectra ("peak–no peak") produced better results, presumably by lessening the effects of peaks irrelevant to the classification of interest. Normalization based on percent total ionization produces slightly better results, and the application of a nonlinear transformation[125] before averaging produces the best results for the four centroid type determinations. This latter method gives approximately equivalent results when compared with the linear nonparametric learning machines of the type discussed in this section. The most interesting result is that the nearest-neighbor method results in *the highest predictive ability of all the methods*. As mentioned previously, the nearest-neighbor approach reduces to a normal library search technique if both the similarity measure and the library of spectra (e.g., only two classes of compounds) are carefully selected. Other explicit relationships between learning machines and information retrieval systems can be found in a study by Negoita.[132]

Several general articles on the application of learning machines to chemical problems are available[133-135] as is a demonstration of the utility of various display methods to the field of pattern recognition in general.[136] For an excellent comprehensive review of all the work mentioned in this section, that of Isenhour *et al.*[135] is highly recommended. Nilsson,[99] Jurs *et al.*,[101] and others cited therein should be consulted for a much more detailed treatment of learning machine theory.

III. DEDUCTIVE APPROACHES

Having first considered the inductive approaches, it has been shown how empirical correlations of mass spectra can be made by the computer and used to predict the structure of an unknown compound. This prediction is associated with a probability of its being correct, regardless of whether the probability can be accurately determined and confidence limits known precisely. If, however, these correlations are assumed to be *true*, they can also be used in combination with the knowledge accumulated over the years by mass spectrometrists to *deduce* molecular structure.

In deductive reasoning "one point that is interesting to note, especially in connection with scientific work, is that a false conclusion cannot follow logically from true premises."[137] Likewise in these deductive interpretation systems, an assigned structure is guaranteed to be as reliable as that obtained by a skilled interpreter provided that all the rules and correlations given to the computer are true for every case.

Unfortunately there are numerous disadvantages to these systems.

Although much of what is known about mass spectra is known to be "true" (e.g., the constancy of the ratios of naturally occurring isotopes), our knowledge of mass spectral behavior is far from complete. Therefore, the choice must often be made between a partial structure in which one has 100% confidence and a more "complete" structure, accompanied by an associated risk, obtained by the "computer induction" of correlations which have not been verified or confirmed. These methods are also, in general, among the most difficult to program, especially if any degree of sophistication or flexibility is desired in the types of problems which the system will solve.

A number of manual and computer techniques originally designed as aids for the human interpretation of mass spectra have strongly influenced the direction taken in this area of computer interpretation. The "rectangular array coupled with identifying charts,"[138] a direct descendant of the "matrix form"[139] of representing mass spectra, was proposed by Hamming and Grigsby and demonstrated the utility of ion series summations in the identification of compound class. In order to simplify structure determination, various means of displaying elemental compositions[140-142] obtained by high-resolution mass spectrometry were devised; these later provided models for the computer examination of these data. Methods for determining the molecular ion in a mass spectrum[143,144] were automated, and several methods of computer calculation of monoisotopic spectra[145,146] from known elemental compositions were developed.

The inductive systems discussed in Section II do not necessarily require all the information contained in a mass spectrum. The mass spectrometrist has a good deal of freedom in his choice of which data from the mass spectrum to use in retrieval systems since identification is a relatively easy task; clustering methods and learning machines are not designed to determine complete structures by themselves and as a consequence partial information may be all that is needed. The deductive approaches, however, require the use of *all* the data in the mass spectrum since they do attempt to determine the structure of the unknown as completely as possible. The more information that one has about the unknown, the greater the possibility of complete structure determination. Since high-resolution mass spectra contain far more information about elemental composition than low-resolution mass spectra, deductive approaches have developed in both these directions.

A. Low-Resolution Spectra

The spectra of straight-chain and monomethyl-substituted aliphatic hydrocarbons and methyl esters of fatty acids are the interpretive problems explored by Pettersson and Ryhage.[39,147] Using ion series summations and with the *a priori* knowledge of which series predominate in the spectra of each class of compounds, the spectra of saturated and unsaturated hydrocarbons and fatty acids are distinguished from other compound types and from each other. The presence of some small number of peaks characteristic of the particular class must then be verified in order to confirm the assignment. By a detailed comparison of the intensities of structurally significant homologous ions, the molecular weight and the degree and site (if monomethyl) of branching is determined. The rules established for these limited compound types produce very reliable results; correct identifications were made for all the fatty acid methyl ester spectra tested, and only one out of the fifty aliphatic hydrocarbon spectra tested was identified incorrectly.

Crawford and Morrison[41] used the correlations of characteristic ions with compound class tabulated by McLafferty[87] to determine the probability that an unknown compound contains a particular structural feature. The four most intense peaks in the unknown spectrum are used for this classification, and the results appear to be comparable to those obtained by these researchers using the approaches discussed in Section II,B. All of these classification methods were further employed by Crawford and Morrison in the design of a system for complete spectral interpretation.[148, 149] After the compound class has been determined, a molecular weight and formula consistent with this classification is sought. Special chemical class routines for the actual structure determination are available for alkyl benzenes, alkanes, esters, amines, alcohols, ethers, aldehydes, ketones, and acids. If these are not adequate for a complete determination of structure, a more general routine that resembles, according to the authors, the "inspired guess-work" of the mass spectrometrist in such a situation, is employed. By the examination of characteristic ions and neutral losses, a list of most probable fragments is created; the combination of fragments that satisfactorily accounts for the molecular formula is proposed as the structure of the molecule. The computer's performance in the interpretation of the mass spectra of primarily monofunctional compounds is approximately equal to that of a group of undergraduate chemistry majors.

In a similar manner, Smith combined the classification approach discussed in Section II,B with the mass spectrometrists' knowledge of mass spectral fragmentation patterns[93, 94] in the design of a system for complete spectral interpretation. The "correlation set" of ion series mass spectra are used to obtain a preliminary identification of compound class. Assuming the correctness of this classification, molecular weight and formula are determined; specialized subroutines for a number of the compound classes are available to complete the details of structure.

B. High-Resolution Spectra

As seen in the study of low-resolution systems, *a priori* knowledge of compound type greatly simplifies the interpretive problem. If this knowledge is combined with the wealth of information contained in a high-resolution mass spectrum, a complete structure determination becomes even more likely. Thus it is not surprising that such specialized systems for classes of "important" compound types have been developed and extensively tested. The most successful of these are the "peptide sequencing" programs for the interpretation of the spectra of both pure oligopeptides and mixtures of oligopeptides.[150-155] These programs are discussed in a recent publication by Arpino and McLafferty.[156]

Of more general use is a system developed by Venkataraghavan et al.[157, 158] which performs a computer interpretation of a mass spectrum following a generally accepted approach.[1] This system is applicable to ketones, esters, alcohols, amines, and keto alcohols containing saturated acyclic hydrocarbon moieties. The elemental composition obtained from the high-resolution data for each of the ions is assumed to be correct—even if alternative compositions exist, none are considered as possibilities in the interpretation procedure.

Having at hand, then, the elemental compositions of all ions in the spectrum, the program proceeds to postulate molecular ions for the unknown spectrum and to test all hypotheses by applying a number of common criteria for the molecular ion.[1] Using the molecular ion that has been determined to be the most likely candidate, all primary neutral fragments lost are determined and possible functional groups are suggested. Futher indication of compound class is obtained by an examination of the 14 homologous ion series and important nonhomologous ion series such as the aromatic and perfluoro series. Subroutines, which are available for the various compound classes, may give additional indication of a particular compound type.

The information obtained up to this point is examined for consistency and, depending on the predicted compound class, a final computer subroutine is called that contains the characteristic fragmentation patterns of that class. These constitute the "rules" of mass spectral behavior that are then used to postulate the most likely structure of the unknown compound.

A similar system, applicable to ketones, amides, and amines, has been developed by Biemann et al.[159]

IV. THE COMBINATION APPROACHES

A. STIRS

The Self-Training Interpretive and Retrieval System, or STIRS,[160-167] resembles to some extent all of the methods discussed so far; in fact, STIRS seeks to combine the advantageous features of all these approaches. The direct predecessor of STIRS is the interpretive system for high-resolution mass spectra also developed by McLafferty et al. and described in Section III,B. The STIRS routines for identification of the molecular ion have their origin in this earlier work, and the choice and definition of the STIRS data classes are the result of the testing and application of this earlier system. Unlike its progenitor, however, STIRS utilizes a large library of reference spectra (at present nearly 24,000); thus STIRS resembles those systems discussed in Section II, and likewise its performance depends to a great extent upon the size and the quality of the reference file.[165] The reference and the unknown spectra are abbreviated or "condensed" by the extraction of those data belonging to the data classes listed in Table IX.

TABLE IX

Spectral Data Classes and Match Factors Utilized by STIRS[162]

Class of spectral data	Match factor
1. Ion series	MF1
2. Low mass characteristic ions	MF2
3. Medium mass characteristic ions	MF3
4. High mass characteristic ions	MF4
5. Small primary neutral losses	MF5
6. Large primary neutral losses	MF6
7. Secondary neutral losses from the most abundant odd mass loss	MF7
8. Secondary neutral losses from the most abundant even mass loss	MF8
Class 8 data of the unknown spectrum matched against Class 5 data of the reference spectrum	MF9
9. Fingerprint ions	MF10
Overall match factor (MF1 + MF2 + 2MF3 + 2MF4 + 4MF5 + 2MF6)/12	MF11

Data class 9, composed of the fingerprint ions, was designed to perform the function of a retrieval system; indeed the selection of data strongly resembles that of other retrieval systems discussed previously.[43-73] The similarity measure, or "match factor," in this case utilizes both the ratios and sums of the intensities of the matching ions which have been normalized by setting the sums of the square roots of the intensities equal to 1—a method proposed by Crawford and Morrison.[41] If a match factor of 900 (out of a possible 1000) or greater is obtained using the fingerprint ions, the unknown spectrum is considered to be identified by simple retrieval, and no further interpretation by STIRS is considered necessary.

If the unknown spectrum is that of a truly "unknown" compound, i.e., one whose spectrum has not been previously recorded and stored in the library, or if the spectrum is significantly different from that of the correct compound contained in the library and, as a consequence, the fingerprint ions cannot unambiguously identify the spectrum, then the full power of the Self-

Training Interpretive and Retrieval System is necessary. The interpretation begins in the same way a mass spectrometrist would begin—with the identification of the molecular ion using the reasoning procedure and associated tests normally employed by the human interpreter.[1]

The peaks in each of the nine individual classes of data specify the locus of points in N-dimensional "ion series" space, "characteristic ion" space, "neutral loss" space, etc. Since these classes of data were chosen because of their known significance in the assignment of molecular structure, it is anticipated that the clusters formed by compounds of similar structure should be relatively tight, to the degree that a particular data class is sensitive to that structural feature. Thus STIRS makes use of the clustering phenomenon to determine the compound class or classes to which a particular unknown belongs, but with one important difference with regard to some other clustering methods: no initial delimiting of classes (generally monofunctional) is required. STIRS is free to discover its own correlations between molecular structure and the mass spectral data; in effect, then, STIRS forms its own clusters, and those 15 reference compounds which are the "nearest neighbors" of the unknown in each N-dimensional space, hopefully composed of tight clusters, are presented to the chemist. Significant overlap among clusters or the lack of sensitivity of a data class to a particular structural feature generally results in the retrieval of compounds with structures seemingly unrelated to each other. Consequently, if tight clusters are not formed, STIRS provides no structural information. Only the *presence* of structural features can be determined—if a feature is *not* found, this does not imply that it is not contained in the molecule.

In order to determine the significance of the appearance of a particular structural feature several times in the output list, criteria for both manual and automated analyses have been devised. Originally the STIRS results of 110 unknown compounds were examined, and criteria for manual analysis of the results were empirically determined (Table X). It is required that, for each class of data in the unknown, some minimum number of entries ("peaks") be used for matching. Thus, for example, if there are at least three ion series present in an unknown spectrum, and a particular structural feature is contained in at least seven of the top ten different reference compounds retrieved with match factor $1 \geq 800$, then that feature is indicated by STIRS to be present in the unknown molecule. For several of the match factors, the criteria are relaxed with an increase in the relative size of the common feature.

The set of criteria for manual analysis does not take into consideration the fact that certain classes of data perform better than others in distinguishing particular structural features—for example, the neutral-loss data may give a much stronger indication of chlorine than the low mass ion series. The criteria for the automated analysis of STIRS results does take this into account for 27 commonly occurring functionalities. Based on the distributions of match

TABLE X

Criteria for Structural Feature Identification[162]

	Minimum number	Minimum MF value	Number of WLN symbols	Minimum number of entries
MF1	7/10	800		3 ion series
MF2	5/10	500		1 odd-mass and 1 even-mass ion
MF3, MF4	2/10	700	≥ 5	3 ions
	3/10	700	3, 4	
MF5	5/10	500	≥ 3	3 neutral losses
	6/10	500	2	
	7/10	500	1	
MF6	4/10	600		3 neutral losses
MF10	2/10	600	≥ 4	6 ions
	3/10	600	3	
	4/10	600	2	
MF11[a]	2/5	350	≥ 4	
	3/5	350	3	
	4/5	350	2	

[a] For MF11 the identified structural feature may be an isomer or homolog of the true feature.

factor values for the "identification" of each structural feature both when it is and when it is not contained in an unknown molecule, a meaningful percent confidence that any of the 27 structural features are contained in a particular unknown compound can be calculated by the computer. These two distributions for phenyl using match factor 11 are presented in Fig. 4. The computer also determines whether the repeated occurrence of any of 179 additional structural features in the output list is statistically significant based on the percent of the reference file containing each feature and on a random drawing model. Examples of this automated analysis are given in Tables XI and XII.

Of all the match factors, the overall match factor (MF11), a weighted combination of match factors 1 through 6, generally gives the most complete indication of the structure of the unknown. When 3-*p*-nitrophenylsydnone is run as an unknown, the low mass characteristic ions indicate the presence of phenyl, carbonyl, and nitrogen; Table XIII shows that the combination of match factors clearly uncovers the gross structure of the molecule, which is not revealed by any of the individual match factors.

As a retrieval system STIRS seems to perform as well as or better than the systems discussed in Section II,A; but, more importantly, as an *interpretive* system STIRS is able to provide the chemist with complete or partial structural information with approximately 95% reliability. A STIRS run, including a complete statistical analysis of the data, requires 60 seconds. Because of the

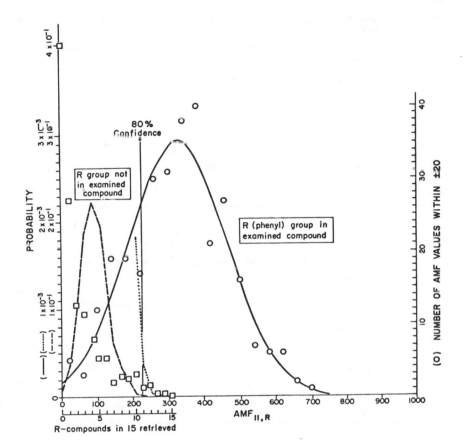

FIG. 4. Distribution of average match factor values $AMF_{11,R}$ for data class MF11 (overall match factor) and functional group R (phenyl), assuming a Gaussian distribution (solid line), based on 300 examined spectra of compounds selected at random whose WLN structure codes contain the symbol R.[167] The experimental values (circles, referred to right-hand ordinates) represent the number of spectra found with $AMF_{11,R}$ values within ± 20 units of the abscissa value shown. The dashed line (abscissa "R-compounds in 15 retrieved") represents the probability of drawing a particular number of R-containing compounds at random from a file with 28.6% such compounds out of 15 compounds drawn. For the dotted line these probability values are shown at $100 \times$ sensitivity, the same scale as used for the $AMF_{11,R}$ values above. This abscissa scale is converted to $AMF_{11,R}$ values by multiplying the number of retrieved compounds containing R by $\overline{MF}_{k,-1}/15$ (see text). The squares represent the proportion of examined compounds not containing phenyl out of 300 total (ordinate) for which a particular number of the 15 selected compounds (abscissa) do contain phenyl.

TABLE XI

Structural Group Probabilities from STIRS for the Mass Spectrum of Tryptophan Ethyl Ester, T56 BMJ D1YZVO2

WLN	Functional group	Match factor, % confidence
1	—CH$_2$—, —CH$_3$	MF5, >99
V	Carbonyl	MF3, 98
VO	Ester, anhydride	MF6, >99
Z	Primary amine	MF11, >99
M	Secondary amine	MF3, MF10, >99; MF11, 98
VO2	Ethyl ester	MF5, MF11, 99
T.J	Heterocycle	MF3, 97; MF10, >99
Y	Single branch	MF6, 98
T56 BMJ	Indole	MF3, MF10, MF11, >99

TABLE XII

Structural Group Probabilities from STIRS for the Mass Spectrum of (CH$_3$)$_2$CHCH(COCH$_3$)COOC$_2$H$_5$, 2OVYV1&Y[167]

WLN	Functional group	Match factor, % confidence
1	—CH$_2$—, —CH$_3$	MF2, MF11, 99
2	—CH$_2$CH$_2$—, —C$_2$H$_5$	MF5, 99; MF11, >99
≥ 3	Alkyl ≥ C$_3$	MF1, MF4, >99
V	Carbonyl	MF4, >99; MF5, 81; MF6, 99; MF11, >99
VO	Ester anhydride	MF11, >99
O2	Ethoxy	MF5, MF11, >99
Y	Single branch	MF1, 98; (MF11, 69)

TABLE XIII

STIRS MF11 (Overall Match Factor) Results for 3-p-Nitrophenylsydnone (T5NNOVJ AR DNW)

Compound name	Wiswesser line notation	MF11
3-p-Bromophenylsydnone	T5NNOVJ AR DE	559
3-p-Fluorophenylsydnone	T5NNOVJ AR DF	516
3-(2-Naphthyl)sydnone	L66J C- AT5NNOVJ	511
3-p-Carboxyphenylsydnone	T5NNOVJ AR DVQ	504
3-m-Methoxyphenylsydnone	T5NNOVJ AR CO1	494
3-p-Methoxyphenylsydnone	T5NNOVJ AR DO1	494
3-p-(Carboxymethylphenyl)sydnone	T5NNOVJ AR D1VQ	481
3-o-Methoxyphenylsydnone	T5NNOVJ AR BO1	474
3-(4-Nitrophenyl)-5-oxozolidine	T5NMOV EHJ AR DNW	468
3-Phenyl-4-acetylsydnone	T5NNOVJ AR & EV1	467

speed of the system, it is feasible to use STIRS routinely in the mass spectrometry laboratory and from remote terminals, as STIRS is also available over telephone lines.

B. Heuristic DENDRAL

The most elegant and sophisticated interpretive system for mass spectral data is Heuristic DENDRAL, the creation of Joshua Lederberg, Edward Feigenbaum, Carl Djerassi et al. of the Departments of Genetics, Computer Science, and Chemistry at Stanford University. DENDRAL defies the elementary categorization used in this review primarily because it was originally designed to model the complex processes of scientific reasoning in toto. In an illustration of the complementarity of inductive and deductive processes, Herbert Searles has remarked:

> In science, the inductive and observational approach to the materials provides the starting point and ground work for the making of hypotheses, and deductions draw out and explore the logical consequences or implications of these hypotheses in order to eliminate those that are inconsistent with the observed facts, while induction again contributes to the verification of the remaining hypothesis.[168]

As if to corroborate this assertion with a specific example, Lederberg states that

> the primary motivation of the Heuristic DENDRAL project is to study and model processes of inductive inference in science, in particular, the formation of hypotheses that best explain given sets of empirical data ... Most of our work is addressed to the following problem:

> Given the data of the mass spectrum of an unknown compound, induce a workable number of plausible solutions, i.e., a small list of candidate molecular structures. In order to complete the task, the DENDRAL program then deduces the mass spectrum predicted by the theory of mass spectrometry for each of the candidates and selects the most productive hypothesis, i.e., the structure whose predicted spectrum most closely matches the data.[169]*

Interpretive systems that intend to deal seriously and flexibly with the analysis of mass spectral data for chemical structure determination must furnish the computer with the means of recognizing and manipulating structural information. For example, those conducting research in the area of learning machines have chosen to use fragmentation codes,[110] while the

---

* Feynman would also concur. He writes, "In general we look for a new law by the following process. First we guess it. Then we compute the consequences of the guess to see what would be implied if this law that we guessed is right. Then we compare the result of the computation to nature, with experiment or experience, compare it directly with observation, to see if it works. If it disagrees with experiment it is wrong. In that simple statement is the key to science."[170]

group at Cornell who developed STIRS has opted for Wiswesser Line Notation.[165] The method of structure representation used by the DENDRAL researchers is based upon topological graph theory and was developed by Lederberg.[171] The DENDRAL algorithm is inseparable from this notation. Using this notation the DENDRAL algorithm is capable of generating exhaustive, nonredundant, and ordered lists of all structures having the same empirical formula, i.e., the set of all isomers having a particular elemental composition. The general procedure involves considering each chemical structure as a tree-graph which has been shown to contain a unique centroid.[172] For an acyclic structure with an odd number of atoms, excluding hydrogen, this centroid is the atom from which each branch carries less than half of the remaining skeletal atoms; for a structure with an even number of skeletal atoms this centroid is the bond that evenly divides the number of atoms, one-half on either side. (Ring structures, which pose ambiguities due to symmetries within the molecules, are much more difficult to represent uniquely. For a detailed discussion of the classification of cyclic structures, the reader should refer to Lederberg.[169,171]) This centroid provides the starting point for a canonical mapping of the tree, which can be represented in a linear, and thus computer-compatible, notation. The rules of precedence that specify a unique representation of a single isomer also provide the means for ordering all possible isomers of a given elemental composition and thereby permit this generation of an exhaustive, nonredundant list of isomers. This *dendr*itic *al*gorithm is the heart of the mass spectral interpretation system as originally conceptualized and is independently known as DENDRAL. An example of this notation is given in Table XIV. A detailed explanation of the notation

TABLE XIV

Examples of DENDRAL Notations[173]

	Linear DENDRAL notations	
Molecule	Extended	Compressed[a]
Ethane	$.CH_3\,CH_3$	$.CC$
Propane	$CH_2..CH_3\,CH_3$	$C..CC$
n-Butane	$.CH_2.CH_3CH_2.CH_3$	$.C.CC.C$
Isobutane	$CH...CH_3CH_3CH_3$	$C...CCC$
n-Pentane	$CH_2..CH_2.CH_3CH_2.CH_3$	$C..C.CC.C$
Isopentane	$CH...CH_3CH_3CH_2.CH_3$	$C...CCC.C$
Neopentane	$C....CH_3CH_3CH_3CH_3$	$C....CCCC$

[a] In the compressed formulas it is useful to remember that adjacent letters (i.e., without dots in between) represent atoms not connected to each other (except for ethane). Notations beginning with a letter are atom centered; those with a dot are bond centered. The single dot represents a single bond.

has been presented in an excellent review by Buchanan *et al.*,[173] and the DEN-DRAL algorithm is the subject of the first paper in the series.[174]

The basic approach to the interpretation of mass spectra is shown in Fig. 5. This scheme is known as "Heuristic DENDRAL" since it utilizes the

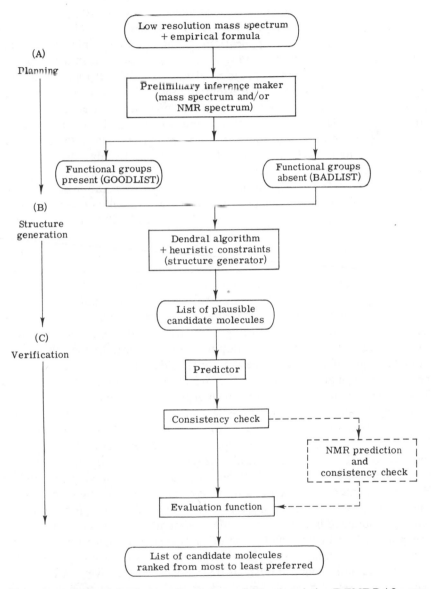

FIG. 5. The original overall design of the heuristic DENDRAL program.[173, 175-177]

DENDRAL algorithm supplemented by "heuristics" or rules that assist the computer in making decisions which will enable it to deduce the molecular structure. As seen in Fig. 5, there are three main parts to the program: Planning, Structure Generation, and Verification.

The Preliminary Inference Maker (PIM) is the core of the Planning subprogram. The first compounds for mass spectral interpretation investigated by these researchers were acyclic ketones[175] and ethers,[176] and in these relatively simple, well-defined applications the Preliminary Inference Maker determined only the nature of the functional group present. In later applications, PIM was expanded to contain all theoretical input (*vide infra*). The rules contained in this section are essentially those that a mass spectrometrist would use to determine functional group presence—since originally the empirical formula of the parent compound was supplied, this presented no obstacle. Functional groups thought to be present are put on GOODLIST; those absent or unlikely for chemical reasons (e.g., unstable structures) are put on BADLIST. GOODLIST and BADLIST entries are then used to restrict the DENDRAL algorithm in the Structure Generation subprogram. Any isomers containing a functionality on BADLIST will not be produced—only those containing functionalities on GOODLIST will be generated. This limited list of plausible candidate molecules is input to the Verification subprogram in which the Predictor, using the rules of organic mass spectrometry, predicts the significant spectral features of each candidate molecule. Each predicted spectrum is compared with the actual spectrum of the unknown, and any inconsistencies between the two result in the elimination of that possible molecular structure. All structures that are retained in this consistency check are ranked by a scoring or evaluation function, and the ranked list of more likely structures is output to the chemist. In the ether study, optional input of nuclear magnetic resonance data in the verification phase permits additional truncation of the list of possible molecular structures. Some typical results of the computer interpretation of ketones are presented in Table XV, and the heuristics for ether interpretation are shown in Fig. 6.

The number of saturated amine isomers that can be generated is much larger than the numbers of either aliphatic ketones or ethers having the same number of carbon atoms.[174] Thus in the study of the interpretation of saturated amines,[177, 178] optional NMR data is used in the Planning phase, and the Preliminary Inference Maker as a whole is expanded to contain much more of the mass spectrometric theory formerly contained in the Predictor. The efficiency obtained in terms of restricting the number of possible structures by this approach frequently makes the other two stages, Structure Generation and Verification, unnecessary. The concept of the "superatom," defined as a structural subunit having at least one free valence, is extensively used to define the complete set of possible amine subgraphs. Thirty-one such superatoms completely specify this set; several examples of these are presented in Table

Results of the Computer Interpretation of Some Ketone Mass Spectra[173]

Ketone spectrum	Number of acyclic structures				Ranking of candidates
	DENDRAL		Heuristic DENDRAL		
	Total isomers[a]	Total ketones[a]	After structure generation	After consistency check	
2-Butanone[b]	11	1	1	1	1st, 2-butanone
3-Pentanone[b]	33	3	1	1	1st, 3-pentanone
3-Hexanone[c]	91	6	1	1	1st, 3-hexanone
2-Methylhexan-3-one[c]	254	15	1	1	1st, 2-methylhexan-3-one
3-Heptanone[b]	254	15	2	2	Tie for 1st; 3-heptanone and 5-methylhexan-3-one
3-Octanone[b,c]	698	33	4	4	1st, 3-octanone
4-Octanone[c]	698	33	2	1	1st, 4-octanone
2,4-Dimethylhexan-3-one[c]	698	33	4	3	Tie for 1st; 2,4-dimethylhexan-3-one and 2,2-dimethylhexan-3-one
6-Methylheptan-3-one[b]	698	33	4	4	1st, 3-octanone; tied for 2nd, 6-methylheptan-3-one, 5-methylheptan-3-one, and 5,5-dimethylhexan-3-one
2-Nonanone[c]	1936	82	7	7	1st, 3-nonanone
3-Methyloctan-3-one[c]	1936	82	4	3	Consistency check eliminated correct structure because no McLafferty +1 peak was present in original mass spectrum.
4-Nonanone[c]	1936	82	4	1	1st, 4-nonanone. The program was revised after results were first published[174] showing the correct structure eliminated.

[a] Total isomers were computed with enol on BADLIST; the numbers therefore include ketones, aldehydes, unsaturated ethers (including enol ethers), and all unsaturated alcohols except enols. Total ketones were computed by putting ketone (only) on GOODLIST.

[b] Literature mass spectrum: A. G. Sharkey, J. L. Shultz, and R. A. Friedel, Anal. Chem. 28, 934 (1956).

[c] Mass spectrum from Stanford University Chemistry Department.

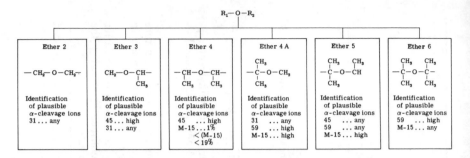

FIG. 6. Heuristics used by heuristic DENDRAL in ether identification. The numbers in the names (e.g., Ether 2) serve as reminders of the number of carbon atoms in the ether subgraphs.

TABLE XVI

Several Amine "Superatoms"^{a} Used in Heuristic DENDRAL[173, 177, 179]

$H_2N{-}CH_2{-}$	$H_2N{-}CH\overset{/}{\underset{\backslash}{}}$	$H_2N{-}\overset{\vert}{\underset{\vert}{C}}{-}$
${-}CH_2{-}NH{-}CH_2$	${-}CH_2{-}NH{-}CH\overset{/}{\underset{\backslash}{}}$	${-}CH_2{-}NH{-}\overset{\vert}{\underset{\vert}{C}}{-}$
${-}CH_2{-}N\overset{/CH_3}{\underset{\backslash CH_3}{}}$	${-}CH_2{-}\overset{\vert}{\underset{\vert}{N}}{-}CH_2{-}\atop CH_2$	$\overset{}{\underset{/}{}}CH{-}N{-}CH\overset{}{\underset{\backslash}{}}\atop \underset{/\backslash}{CH}$

^{a} A "superatom" for an amine is defined as the structural unit containing the nitrogen atom and having at least one free valence. Thirty-one amine superatoms completely specify the set of all possible subgraphs.

XVI. For a thorough discussion of this concept, the reader should refer to papers by Delfino *et al.*[15, 177, 179]

This approach was generalized to interpretation of all "saturated acyclic monofunctional" compounds[179] of the formula $C_nH_{2n+v}X$, where X = O, S, or N and v is the valence of X, after a brief excursion into the investigation of a group of cyclic structures—the cyclic ketones.[180] This generalized approach need not be provided with the empirical formula of the molecular ion; in fact, a molecular ion need not even be present in the spectrum. The program determines which of the heteroatoms could be contained in the molecule and by closely examining the peaks at the high mass end of the spectrum calculates the lowest possible molecular weight. If the interpretation goes poorly, the predicted molecular weight is increased in successive steps of 14 mass units, and if this still fails to account for the origin of the spectrum, DENDRAL will try the interpretation again using the next most plausible heteroatom.

After these initial studies, no further investigations using the DENDRAL system appeared for nearly two years. In that period of time, DENDRAL matured to the interpretation of the high-resolution mass spectra of estrogenic steroids[181] and within another year was solving problems in the analysis of mixtures of estrogenic steroids.[182] The use of metastable ion and low ionizing voltage data facilitates the determination of candidate molecular ions. Heuristic DENDRAL has been provided with the estrogen "superatom" and the fragmentation mechanisms common to estrogens. The approach may easily be generalized to other complex molecules, however.

The difference between the empirical formula(s) of the molecular ion(s) and the estrogen skeleton determines the number and composition of substituents. (Unsaturation is also considered as a substituent.) Evidence is sought for the various characteristic fragmentations of the molecule, and the possible substituents and their placement on the skeleton are determined. These pieces of evidence are combined, candidate molecular structures determined and consistency checks made.

DENDRAL has branched out in four other extensions and modifications of note. The first of these developed from another study of estrogenic steroids.[183] INTSUM, a program for data interpretation and study, explored the fragmentation processes for classes of compounds whose mass spectra have not been extensively correlated manually. Given the "superatom" of the class of compounds, a nonredundant list of all possible fragmentations of this skeleton is generated. A set of structure–spectrum pairs of these compounds is examined for evidence of these fragmentations. This evidence is collected, correlated, and finally summarized for the class. The extension of this work is Meta-DENDRAL,[15, 184, 185] which, given such sets of structure–spectrum pairs, will be able to discover the parts of a unifying theory of mass spectrometry.

The amine problem has been extended using Heuristic DENDRAL to interpretation using carbon-13 NMR data.[186] The DENDRAL structure generation is now complete for the generation of cyclic structures.[184]

Finally, just as in the learning machine approach to artificial intelligence,[90, 105, 110] the heuristic programming approach can be turned around so that reasoning from chemical structure to mass spectrum is achieved.[15, 187] The "Ion Generator" program of Delfino and Buchs at the University of Geneva is an attempt to simulate the formation of ions in the ion source of the mass spectrometer. Unlike INTSUM, which examines a number of structure–spectrum pairs for a particular class of compounds in order to determine which fragmentation mechanisms are most common to that class, the "Ion Generator" accepts as input a single structure for which it predicts a mass spectrum. Its prediction is based on the method of "electron book-keeping," of moving "positive charges" and "radical sites" around within ions to effect fragmentations and rearrangements. As this is one of the first methods used by mass spectrometrists to gain insight into mass spectral fragmentation processes,[1-4] it appears that we have come full circle. Meta-DENDRAL and the "Ion Generator" may begin a new era in the computer interpretation of mass spectra.

Until these products of man's intelligence, these computer systems for the retrieval and interpretation of mass spectra, are sufficiently capable of imitating or going beyond the traditional modes of human reasoning, we must recognize that it will be necessary for a time to let them and the mass spectrometrist interact in order to obtain the best solutions to our temporal problems. The combination of two or more of the techniques which are currently available could be of tremendous utility. Using a retrieval system as a "front-end" for an interpretive program is a most obvious combination of techniques. One can also imagine using a learning machine or cluster analysis to determine which features in the spectra of a class of compounds are most useful in distinguishing this class from others, then using this information to develop new match factors for STIRS that would be sensitive to these functions. Another powerful combination would be the use of STIRS for uncovering the structural features of the molecule, followed by DENDRAL for "putting the pieces back together" in the order that is the most logical and consistent with the entire mass spectrum.

But as Kowalski and Bender have stated with regard to their field of interest and expertise,

> It would be erroneous to infer that pattern recognition removes the scientist from data analysis. Man–machine interaction is currently in vogue in chemistry and other fields for a good reason: man is the best pattern recognizer known today. . . . At this point computer techniques should be used but carefully supervised by the scientist.[134]

As long as scientists maintain this realistic attitude, the entry of the digital computer into the areas of mass spectral interpretation and other scientific investigations will produce numerous future benefits, not the least of which may be a deeper understanding of scientific methods.

References

[1] F. W. McLafferty, "Interpretation of Mass Spectra," 2nd ed. Benjamin, New York, 1973.
[2] K. Biemann, "Mass Spectrometry: Organic Chemical Applications." Mc-Graw-Hill, New York, 1962.
[3] H. Budzikiewicz, C. Djerassi, and D. H. Williams, "Mass Spectrometry of Organic Compounds." Holden-Day, San Francisco, California, 1967.

[4] J. H. Beynon, R. A. Sanders, and A. E. Williams, "The Mass Spectra of Organic Molecules." Elsevier, Amsterdam, 1968.

[5] G. R. Waller, ed., "Biochemical Applications of Mass Spectrometry." Wiley, New York, 1972.

[6] G. W. A. Milne, ed., "Mass Spectrometry: Techniques and Applications." Wiley (Interscience), New York, 1971.

[7] W. H. McFadden, "Techniques of Combined Gas-Chromatography/Mass Spectrometry: Applications in Organic Analysis." Wiley (Interscience), New York, 1973.

[8] R. Venkataraghavan and F. W. McLafferty, Chem. Tech. p. 364 (1972).

[9] M. A. Baldwin and F. W. McLafferty, Org. Mass. Spectrom. 7, 1111 (1973).

[10] P. Arpino, M. A. Baldwin, and F. W. McLafferty, Biomed. Mass Spectrom. 1, 80 (1974).

[11] P. V. Fennessey, in "Mass Spectrometry: Techniques and Applications" (G. W. A. Milne, ed.), p. 77. Wiley (Interscience), New York, 1971.

[12] R. G. Ridley, in "Biochemical Applications of Mass Spectrometry" (G. R. Waller, ed.), p. 177. Wiley, New York, 1972.

[13] S. L. Sasaki, in "Determination of Organic Structures by Physical Methods" (F. C. Nachod and J. J. Zuckerman, eds.), Vol. 5, p. 285. Academic Press, New York, 1973.

[14] T. Clerc and F. Erni, Fortschr. Chem. Forsch. 39, 91 (1973).

[15] A. B. Delfino and A. Buchs, Fortschr. Chem. Forsch. 39, 109 (1973).

[16] H. L. Searles, "Logic and Scientific Methods," p. 5. Ronald Press, New York, 1968.

[17] L. B. Young, "Exploring the Universe," p. 16. McGraw-Hill, New York, 1963.

[18] K. R. Popper, "Conjectures and Refutations: The Growth of Scientific Knowledge," p. 53. Routledge & Kegan Paul, London, 1963.

[19] H. L. Searles, "Logic and Scientific Methods," p. 211. Ronald Press, New York, 1968.

[20] H. L. Searles, "Logic and Scientific Methods," p. 250. Ronald Press, New York, 1968.

[21] H. L. Searles, "Logic and Scientific Methods," p. 207. Ronald Press, New York, 1968.

[22] L. H. Gevantman, Anal. Chem. 44, 30A (1972).

[23] S. Abrahamsson, E. Stenhagen, and F. W. McLafferty, "Registry of Mass Spectral Data." Wiley, New York, 1974.

[24] G. Salton, "Automatic Information Organization and Retrieval," p. v. McGraw-Hill, New York, 1968.

[25] G. Salton, "Automatic Information Organization and Retrieval," p. 1. McGraw-Hill, New York, 1968.

[26] S. L. Grotch, Anal. Chem. 45, 2 (1973).

[27] G. Salton, "Automatic Information Organization and Retrieval," p. 324. McGraw-Hill, New York, 1968.

[28] B. A. Knock, I. C. Smith, D. E. Wright, R. G. Ridley, and W. Kelly, Anal. Chem. 42, 1516 (1970).

[29] P. D. Zemany, Anal. Chem. 22, 920 (1950).

[30] F. W. McLafferty and R. S. Gohlke, Anal. Chem. 31, 1160 (1959).

[31] S. Abrahamsson, S. Stallberg-Stenhagen, and E. Stenhagen, Biochem. J. 92, 2p (1964).

[32] S. Abrahamsson, G. Haggstrom, and E. Stenhagen, *Proc. 14th Annu. Conf. Mass Spectrom. Allied Top.*, *1966* p. 522 (1966).

[33] S. Abrahamsson, *Sci. Tools* 14, 29 (1967).

[34] L. E. Wangen, W. S. Woodward, and T. L. Isenhour, *Anal. Chem.* 43, 1605 (1971).

[35] F. W. McLafferty, R. H. Hertel, and R. D. Villwock, *Proc. 22nd Annu. Conf. Mass Spectrom. Allied Top.*, *1974* p. 452 (1974).

[36] F. W. McLafferty, R. H. Hertel, and R. D. Villwock, *Org. Mass Spectrom.* 9 690 (1974).

[37] S. L. Grotch, *Proc. 22nd Annu. Conf. Mass Spectrom. Allied Top.*, *1974* p. 456 (1974).

[38] B. Pettersson and R. Ryhage, *Ark. Kemi* 26, 293 (1966).

[39] I. C. Smith, W. Kelly, A. Brickstock, and R. G. Ridley, *Proc. 15th Annu. Conf. Mass Spectrom. Allied Top.*, *1967* p. 102 (1967).

[40] B. Knock, D. Wright, W. Kelley, and R. G. Ridley, *Proc. 17th Annu. Conf. Mass Spectrom. Allied Top.*, *1969* p. 398 (1969).

[41] L. R. Crawford and J. D. Morrison, *Anal. Chem.* 40, 1464 (1968).

[42] *Finnigan Spectra* 3, 3 (1973).

[43] C. Cone, P. Fennessey, R. Hites, N. Marcuso, and K. Biemann, *Proc. 15th Annu. Conf. Mass Spectrom. Allied Top.*, *1967* p. 114 (1967).

[44] R. A. Hites and K. Biemann, *Anal. Chem.* 39, 965 (1967).

[45] R. A. Hites, Ph.D. Thesis, Massachusetts Institute of Technology, Cambridge, 1968.

[46] R. A. Hites and K. Biemann, *Anal. Chem.* 40, 1217 (1968).

[47] R. A. Hites and K. Biemann, *Advan. Mass Spectrom.* 4, 37 (1968).

[48] R. C. Murphy, M. V. Djuricic, S. P. Markey, and K. Biemann, *Science* 165, 695 (1969).

[49] R. A. Hites and K. Biemann, *Anal. Chem.* 42, 855 (1970).

[50] H. S. Hertz, R. A. Hites, and K. Biemann, *Anal. Chem.* 43, 681 (1971).

[51] H. Nau and K. Biemann, *Anal. Lett.* 6, 1071 (1973).

[52] H. Nau and K. Biemann, *Anal. Chem.* 46, 426 (1974).

[53] C. E. Costello, H. S. Hertz, T. Sakai, and K. Biemann, *Clin. Chem.* 20, 255 (1974).

[54] J. M. McGuire, A. L. Alford, and M. H. Carter, *Proc. 20th Annu. Conf. Mass Spectrom. Allied Top.*, *1972* p. 366 (1972).

[55] R. Reimendal and J. B. Sjövall, *Anal. Chem.* 45, 1083 (1973).

[56] N. W. Bell, *Proc. 20th Annu. Conf. Mass Spectrom. Allied Top*, *1972* p. 364 (1972).

[57] S. R. Heller, H. M. Fales, and G. W. A. Milne, *Proc. 20th Annu. Conf. Mass Spectrom. Allied Top.*, *1972* p. 353 (1972).

[58] S. R. Heller, H. M. Fales, and G. W. A. Milne, *J. Chem. Educ.* 49, 725 (1972).

[59] S. R. Heller, *Anal. Chem.* 44, 1951 (1972).

[60] S. R. Heller, H. M. Fales, G. W. A. Milne, and R. J. Feldmann, *Proc. 21st Annu. Conf. Mass Spectrom. Allied Top.*, *1973* p. 192 (1973).

[61] S. R. Heller, H. M. Fales, and G. W. A. Milne, *Org. Mass Spectrom.* 7, 107 (1973).

[62] S. R. Heller, *in* "Computer Representation and Manipulation of Chemical Information" (W. T. Wipke *et al.*, eds.), p. 175. Wiley (Interscience), New York, 1974.

[63] S. R. Heller, D. A. Koniver, H. M. Fales, and G. W. A. Milne, *Anal. Chem.* 46, 947 (1974).

64 S. L. Grotch, *Proc. 17th Annu. Conf. Mass Spectrom. Allied Top.*, *1969* p. 459 (1969).

65 S. L. Grotch, *Proc. 18th Annu. Conf. Mass Spectrom. Allied Top.*, *1970* p. B453 (1970).

66 S. L. Grotch, *Anal. Chem.* **42**, 1214 (1970).

67 S. L. Grotch, *Proc. 19th Annu. Conf. Mass Spectrom. Allied Top.*, *1971* p. 72 (1971).

68 S. L. Grotch, *Anal. Chem.* **43**, 1362 (1971).

69 S. L. Grotch, *Proc. 20th Annu. Conf. Mass Spectrom. Allied Top.*, *1972* p. 362 (1972).

70 D. H. Robertson, J. Cavagnaro, J. B. Holz, and C. Merritt, Jr., *Proc. 20th Annu. Conf. Mass Spectrom. Allied Top.*, *1972* p. 359 (1972).

71 D. H. Robertson and C. Merritt, *Proc. 21st Annu. Conf. Mass Spectrom. Allied Top.*, *1973* p. 65 (1973).

72 D. H. Robertson and C. Merritt, *Proc. 22nd Annu. Conf. Mass Spectrom. Allied Top.*, *1974* p. 447 (1974).

73 S. L. Grotch, *Proc. 21st Annu. Conf. Mass Spectrom. Allied Top.*, *1973* p. 69 (1973).

74 S. L. Grotch, *Anal. Chem.* **46**, 526 (1974).

75 D. H. Robertson and R. I. Reed, *Proc. 19th Annu. Conf. Mass Spectrom. Allied Top.*, *1971* p. 68 (1971).

76 C. Merritt, Jr., D. H. Robertson, R. A. Graham, and T. L. Nichols, *Proc. 20th Annu. Conf. Mass Spectrom. Allied Top.*, *1972* p. 355 (1972).

77 S. Farbman, R. I. Reed, D. H. Robertson, and M. E. F. Silva, *Int. J. Mass Spectrom. Ion Phys.* **12**, 123 (1973).

78 P. C. Jurs, *Anal. Chem.* **43**, 364 (1971).

79 C. W. Gear, "Computer Organization and Programming," pp. 333–338. McGraw-Hill, New York, 1969.

80 F. E. Lytle, *Anal. Chem.* **43**, 1334 (1971).

81 S. Meyerson, *Anal. Chem.* **31**, 174 (1959).

82 D. D. Tunnicliff and P. A. Wadsworth, *Anal. Chem.* **37**, 1082 (1965).

83 T. L. Isenhour, *Anal. Chem.* **45**, 2153 (1973).

84 F. P. Abramson, *Proc. 21st Annu. Conf. Mass Spectrom. Allied Top.*, *1973* p. 76 (1973).

85 F. P. Abramson and M. F. Schulman, *Proc. 22nd Annu. Conf. Mass Spectrom. Allied Top.*, *1974* p. 453 (1974).

86 F. P. Abramson, *Anal. Chem.* **47**, 45 (1975).

87 F. W. McLafferty, "Mass Spectral Correlations," Advan. Chem. Ser. No. 40. Amer. Chem. Soc., Washington, D.C., 1963.

88 "Eight Peak Index of Mass Spectra." Mass Spectrometry Data Centre, Aldermaston, 1970.

89 J. E. Freund, "Mathematical Statistics," p. 46. Prentice-Hall, Englewood Cliffs, New Jersey, 1962.

90 P. C. Jurs, *in* "Computer Representation and Manipulation of Chemical Information" (W. T. Wipke *et al.*, eds.), p. 265. Wiley (Interscience), New York, 1974.

91 L. R. Crawford and J. D. Morrison, *Anal. Chem.* **40**, 1469 (1968).

92 V. V. Raznikov and V. L. Tal'roze, *Dokl. Akad. Nauk SSSR* **170**, 379 (1966).

93 D. H. Smith and G. Eglinton, *Nature (London)* **235**, 325 (1972).

94 D. H. Smith, *Anal. Chem.* **44**, 536 (1972).

[95] Y. Ishida, Y. Kudo, M. Sakamoto, H. Abe, and S. Sasaki, *in* "Recent Developments in Mass Spectroscopy" (K. Ogata and T. Hayakawa, eds.), p. 1286. Univ. Park Press, Baltimore, Maryland, 1970.

[96] G. Salton, "Automatic Information Organization and Retrieval," p. 138. McGraw-Hill, New York, 1968.

[97] S. R. Heller, C. L. Chang, and K. C. Chu, *Anal. Chem.* **46**, 951 (1974).

[98] K. C. Chu, *Anal. Chem.* **46**, 1181 (1974).

[99] N. J. Nilsson, "Learning Machines—Foundations of Binary Pattern Classifying Systems." McGraw-Hill, New York, 1965.

[100] N. J. Nilsson, "Learning Machines—Foundations of Binary Pattern Classifying Systems," p. vii. McGraw-Hill, New York, 1965.

[101] P. C. Jurs, B. R. Kowalski, and T. L. Isenhour, *Anal. Chem.* **41**, 21 (1969).

[102] D. E. Knuth, "Fundamental Algorithms," Vol. 1, Chapter 2. Addison-Wesley, Reading, Massachusetts, 1969.

[103] P. C. Jurs, B. R. Kowalski, T. L. Isenhour, and C. N. Reilley, *Anal. Chem.* **41**, 690 (1969).

[104] P. C. Jurs, *Anal. Chem.* **42**, 1633 (1970).

[105] J. Schechter and P. C. Jurs, *Appl. Spectrosc.* **27**, 225 (1973).

[106] B. R. Kowalski, P. C. Jurs, T. L. Isenhour, and C. N. Reilley, *Anal. Chem.* **41**, 695 (1969).

[107] B. R. Kowalski, P. C. Jurs, T. L. Isenhour, and C. N. Reilley, *Anal. Chem.* **41**, 1949 (1969).

[108] B. R. Kowalski, P. C. Jurs, T. L. Isenhour, and C. N. Reilley, *Anal. Chem.* **42**, 1387 (1970).

[109] S. Sasaki and H. Abe, *Shitsuryo Bunseki* **20**, 131 (1972).

[110] J. Shechter and P. C. Jurs, *Appl. Spectrosc.* **27**, 30 (1973).

[111] F. E. Lytle, *Anal. Chem.* **44**, 1867 (1972).

[112] H. Hellermore, "Digital Computer System Principles." McGraw-Hill, New York, 1969.

[113] W. L. Felty and P. C. Jurs, *Anal. Chem.* **45**, 885 (1973).

[114] L. E. Wangen, N. M. Frew, T. L. Isenhour, and P. C. Jurs, *Appl. Spectrosc.* **25**, 203 (1971).

[115] P. C. Jurs, *Anal. Chem.* **43**, 1812 (1971).

[116] B. R. Kowalski and C. F. Bender, *Anal. Chem.* **45**, 2234 (1973).

[117] C. F. Bender and B. R. Kowalski, *Anal. Chem.* **45**, 590 (1973).

[118] C. F. Bender, H. D. Shepherd, and B. R. Kowalski, *Anal. Chem.* **45**, 617 (1973).

[119] L. E. Wangen, N. M. Frew, and T. L. Isenhour, *Anal. Chem.* **43**, 845 (1971).

[120] P. C. Jurs, *Anal. Chem.* **43**, 22 (1971).

[121] D. N. Anderson and T. L. Isenhour, *Pattern Recognition*, **5**, 249 (1973).

[122] J. W. Sammon, Jr. and D. Foley, *Proc. 1970 IEEE Symp. Adapt. Processes, 1970* p. IX. 2.1 (1971).

[123] J. T. Clerc, P. Naegeli, and J. Seibl, *Chimia* **27**, 639 (1973).

[124] L. Pietrantonio and P. C. Jurs, *Pattern Recognition* **4**, 391 (1972).

[125] J. B. Justice, Jr., D. N. Anderson, T. L. Isenhour, and J. C. Marshall, *Anal. Chem.* **44**, 2087 (1972).

[126] P. C. Jurs, *Appl. Spectrosc.* **25**, 483 (1971).

[137] D. D. Tunnicliff and P. A. Wadsworth, *Anal. Chem.* **45**, 12 (1973).

[128] K.-L. H. Ting, R. C. T. Lee, G. W. A. Milne, M. Shapiro, and A. M. Guarino, *Science* **180**, 417 (1973).

[129] K.-L. H. Ting and V. A. Vinton, *Proc. 22nd Annu. Conf. Mass Spectrom. Allied Top., 1974* p. 445 (1974).

[130] C. L. Perrin, *Science* **183**, 551 (1974).
[131] J. B. Justice and T. L. Isenhour, *Anal. Chem.* **46**, 223 (1974).
[132] C. V. Negoita, "Linear and Nonlinear Information Systems." Science Research Council, Atlas Computer Laboratory, Chilton Didcot, Berkshire, England, 1972.
[133] T. L. Isenhour and P. C. Jurs, *Anal. Chem.* **43**, 20A (1971).
[134] B. R. Kowalski and C. F. Bender, *J. Amer. Chem. Soc.* **94**, 5632 (1972).
[135] T. L. Isenhour, B. R. Kowalski, and P. C. Jurs, *CRC Crit. Rev. Anal. Chem.* **4**, 1 (1974).
[136] B. R. Kowalski and C. F. Bender, *J. Amer. Chem. Soc.* **95**, 686 (1973).
[137] L. B. Young, "Exploring the Universe," p. 17. McGraw-Hill, New York, 1963.
[138] M. C. Hamming and R. D. Grigsby, *Proc. 15th Annu. Conf. Mass Spectrom. Allied Top.*, *1967* p. 107 (1967).
[139] *Amer. Petrol. Inst.*, *Res. Proj. 44*.
[140] K. Biemann, P. Bommer, and D. M. Desiderio, *Tetrahedron Lett.* **26**, 1725 (1964).
[141] R. Venkataraghavan and F. W. McLafferty, *Anal. Chem.* **39**, 278 (1967).
[142] A. L. Burlingame and D. H. Smith, *Tetrahedron* **24**, 5749 (1968).
[143] K. Biemann and W. McMurray, *Tetrahedron Lett.* **11**, 647 (1965).
[144] A. Jardine, R. I. Reed, and M. E. S. F. Silva, *Org. Mass Spectrom.* **7**, 601 (1973).
[145] B. Boone, R. K. Mitchum, and S. E. Scheppele, *Int. J. Mass Spectrom. Ion Phys.* **5**, 21 (1970).
[146] L. R. Crawford, *Int. J. Mass Spectrom. Ion Phys.* **10**, 279 (1972/1973).
[147] B. Pettersson and R. Ryhage, *Anal. Chem.* **39**, 790 (1967).
[148] J. D. Morrison, *Proc. 17th Annu. Conf. Mass Spectrom. Allied Top.*, *1969* p. 38 (1969).
[149] L. R. Crawford and J. D. Morrison, *Anal. Chem.* **43**, 1790 (1971).
[150] M. Barber, P. Powers, M. J. Wallington, and W. A. Wolstenholme, *Nature (London)* **212**, 784 (1966).
[151] M. Senn, R. Venkataraghavan, and F. W. McLafferty, *J. Amer. Chem. Soc.* **88**, 5593 (1966).
[152] K. Biemann, C. Cone, B. R. Webster, and G. P. Arsenault, *J. Amer. Chem. Soc.* **88**, 5598 (1966).
[153] P. Irving, Ph.D. Thesis, Cornell University, Ithaca, New York, 1971.
[154] F. W. McLafferty, R. Venkataraghavan, and P. Irving, *Biochem. Biophys. Res. Commun.* **39**, 274 (1970).
[155] H. K. Wipf, P. Irving, M. McCamish, R. Venkataraghavan, and F. W. McLafferty, *J. Amer. Chem. Soc.* **95**, 3369 (1973).
[156] P. Arpino and F. W. McLafferty, *in* "Determination of Organic Structures by Physical Methods" (F. C. Nachod, J. J. Zuckerman, and E. W. Randall, eds.), Vol. 6, p. 1. Academic Press, New York, 1975.
[157] R. Venkataraghavan and F. W. McLafferty, *Proc. 15th Annu. Conf. Mass. Spectrom. Allied Top.*, *1967* p. 102 (1967).
[158] R. Venkataraghavan, F. W. McLafferty, and G. E. VanLear, *Org. Mass Spectrom.* **2**, 1 (1969).
[159] A. Mandelbaum, P. Fennessey, and K. Biemann, *Proc. 15th Annu. Conf. Mass Spectrom. Allied Top.*, *1967* p. 111 (1967).
[160] K.-S. Kwok, R. Venkataraghavan, and F. W. McLafferty, *Proc. 19th Annu. Conf. Mass Spectrom. Allied Top.*, *1971* p. 70 (1971).

[161] R. Venkataraghavan, K.-S. Kwok, G. Pesyna, and F. W. McLafferty, *Proc. 21st Annu. Conf. Mass Spectrom. Allied Top.*, *1973* p. 197 (1973).

[162] K.-S. Kwok, R. Venkataraghavan, and F. W. McLafferty, *J. Amer. Chem. Soc.* **95**, 4185 (1973).

[163] K.-S. Kwok, Ph.D. Thesis, Cornell University, Ithaca, New York, 1973.

[164] F. W. McLafferty, R. Venkataraghavan, K.-S. Kwok, and G. Pesyna, *Advan. Mass Spectrom.* **5**, 999 (1974).

[165] F. W. McLafferty, M. A. Busch, K.-S. Kwok, B. A. Meyer, G. Pesyna, R. C. Platt, I. Sakai, J. W. Serum, A. Tatematsu, R. Venkataraghavan, and R. G. Werth, *in* "Mass Spectrometry and NMR Spectroscopy in Pesticide Chemistry" (F. J. Biros and R. Haque, eds.), p. 49. Plenum, New York, 1974.

[166] G. M. Pesyna, H. Dayringer, R. Venkataraghavan, and F. W. McLafferty, *Proc. 22nd Annu. Conf. Mass Spectrom. Allied Top.*, *1974* p. 451 (1974).

[167] H. E. Dayringer, G. M. Pesyna, R. Vankataraghavan, and F. W. McLafferty, *Org. Mass. Spectrom.* (submitted for publication).

[168] H. L. Searles, "Logic and Scientific Methods," p. 6. Ronald Press, New York, 1968.

[169] J. Lederberg, *in* "Biochemical Applications of Mass Spectrometry" (G. R. Waller, ed.), p. 194. Wiley, New York, 1972.

[170] R. P. Feynman, "The Character of Physical Law," p. 156. MIT Press, Cambridge, Massachusetts, 1965.

[171] J. Lederberg, *Proc. Nat. Acad. Sci. U.S.* **53**, 134 (1965).

[172] H. R. Henze and C. M. Blair, *J. Amer. Chem. Soc.* **53**, 3077 (1931).

[173] B. G. Buchanan, A. M. Duffield, and A. V. Robertson, *in* "Mass Spectrometry: Techniques and Applications" (G. W. A. Milne, ed.), p. 121. Wiley (Interscience), New York, 1971.

[174] J. Lederberg, G. L. Sutherland, B. G. Buchanan, E. A. Feigenbaum, A. V. Robertson, A. M. Duffield, and C. Djerassi, *J. Amer. Chem. Soc.* **91**, 2973 (1969).

[175] A. M. Duffield, A. V. Robertson, C. Djerassi, B. G. Buchanan, G. L. Sutherland, E. A. Feigenbaum, and J. Lederberg, *J. Amer. Chem. Soc.* **91**, 2977 (1969).

[176] G. Schroll, A. M. Duffield, C. Djerassi, B. G. Buchanan, G. L. Sutherland, E. A. Feigenbaum, and J. Lederberg, *J. Amer. Chem. Soc.* **91**, 7440 (1969).

[177] A. Buchs, A. M. Duffield, G. Schroll, C. Djerassi, A. B. Delfino, B. G. Buchanan, G. L. Sutherland, E. A. Feigenbaum, and J. Lederberg, *J. Amer. Chem. Soc.* **92**, 6831 (1970).

[178] A. M. Buchs, A. B. Delfino, C. Djerassi, A. M. Duffield, B. G. Buchanan, E. A. Feigenbaum, J. Lederberg, G. Schroll, and G. L. Sutherland, *Advan. Mass Spectrom.* **5**, 314 (1971).

[179] A. Buchs, A. B. Delfino, A. M. Duffield, C. Djerassi, B. G. Buchanan, E. A. Feigenbaum, and J. Lederberg, *Helv. Chim. Acta* **53**, 1394 (1970).

[190] Y. M. Sheikh, A. Buchs, A. B. Delfino, G. Schroll, A. M. Duffield, C. Djerassi, B. G. Buchanan, G. L. Sutherland, E. A. Feigenbaum, and J. Lederberg, *Org. Mass Spectrom.* **4**, 493 (1970).

[181] D. H. Smith, B. G. Buchanan, R. S. Englemore, A. M. Duffield, A. Yeo, E. A. Feigenbaum, J. Lederberg, and C. Djerassi, *J. Amer. Chem. Soc.* **94**, 5962 (1972).

[182] D. H. Smith, B. G. Buchanan, R. S. Englemore, H. Adlercreutz, and C. Djerassi, *J. Amer. Chem. Soc.* **95**, 6078 (1973).

[183] D. H. Smith, B. G. Buchanan, W. C. White, E. A. Feigenbaum, J. Lederberg, and C. Djerassi, *Tetrahedron* **29**, 3117 (1973).
[184] D. H. Smith, L. M. Masinter, and N. S. Sridharan, *in* "Computer Representation and Manipulation of Chemical Information" (W. T. Wipke *et al.*, eds.), p. 287. Wiley (Interscience), New York, 1974.
[185] D. H. Smith and B. G. Buchanan, *Proc. 22nd Annu. Conf. Mass Spectrom. Allied Top., 1974* p. 440 (1974).
[186] R. E. Carhart and C. Djerassi, *J. Chem. Soc., Perkin Trans.* 2 1753 (1973).
[187] A. B. Delfino and A. Buchs, *Helv. Chim. Acta* **55**, 2017 (1972).

Flash Photolysis and Structure

3

ROBERT L. STRONG

I. INTRODUCTION

Photochemistry is the study of reactions resulting from the absorption of nonionizing radiation. With but few exceptions, the effective range of electromagnetic radiation for photochemistry is from the visible or near-infrared region (1000 nm, 120 kJ) into the vacuum ultraviolet (100 nm, 1200 kJ), and therefore the general effect of the initial act of light absorption is the electronic excitation of the absorbing particle (with, of course, some accompanying rotational and vibrational excitation in most molecules where such motions are possible).

Until 25 years ago, most photochemical studies were limited to times greater than 1 second with low-intensity photolyzing light sources. Conventionally, systems were illuminated (hopefully with a steady monochromatic light beam), concentrations of reactants and/or products were measured as functions of time, and the detailed mechanism was inferred from a kinetic analysis of these data. There were exceptions, of course—rotating sector experiments and fluorescence and phosphorescence measurements to single out only two—but in general the determination of the reaction path was based on the usual steady-state approximation for the undetected but very reactive reaction intermediates, such as higher electronic states or radicals, assumed to be present in very low concentrations. The situation was drastically altered in 1949 with the introduction of the flash photolysis technique to the study of photochemical processes by Norrish and Porter,[1,2] and independently by Herzberg and Ramsay[3] and by Davidson et al.[4] In this technique a short pulse of light of very high intensity, produced by the discharge of a high-voltage capacitor through a gas-discharge tube, is used as the photolyzing source. Intermediates are thus produced in sufficiently high concentrations for direct observation by spectroscopic and other analytical means, and this unique character of the flash photolysis technique has resulted in its extensive use in chemical, physical, and biological research. The development and massive use of this technique by many research groups in a variety of chemical, spectroscopic, and biochemical laboratories have contributed greatly to our understanding of the kinetic and structural behavior particularly of excited-state chemical systems, and led to the awarding of the 1967 Nobel Prize to Norrish and Porter (shared with M. Eigen for his relaxation studies of fast reactions).

Most flash systems using this conventional gas discharge technique cannot get down to times less than approximately 1 μsecond, which however is still sufficient for a variety of different types of physical and chemical processes occurring in a great many organic systems. The introduction of the giant Q-switched pulsed laser independently in several groups[5-10] has extended the study of photochemical intermediates to the nanosecond region, and direct measurements in the picosecond region are now possible using the mode-locked laser.[11]

An extensive chapter on the experimental techniques of the more conventional flash photolysis and these newer applications of pulsed lasers has recently been published,[12] and new developments are updated in annual reviews.[13] Although complete flash apparatus "packages" are available from at least two companies in the United States and England, most flash work reported so far has been done on units constructed from component parts or from scratch to meet specific design requirements for a particular application. A wide variety of equipment designs have therefore emerged, all utilizing however the basic techniques. It is the purpose of this presentation to review briefly these techniques, particularly the experimental aspects that are pertinent to an understanding of transient systems, and then to summarize the applications to structural studies of selected organic systems.

Much of the work involving flash photolysis has been concerned quite naturally with the kinetic behavior of transitory intermediates and has led to better interpretations of reaction mechanisms—i.e., the time structure of reacting systems. In fact, the flash photolysis technique was independently developed and applied by Davidson et al.[4] to the study of one of the conceptually most simple reactions—the recombination of halogen atoms in the presence of gas-phase third bodies and in the liquid phase—and this type of study over the past two decades in several laboratories has been one of the most fruitful applications of the flash technique in determining intermolecular and three-body interactions and solvent cage effects. (In this case, however, the course of the reaction is followed by measurement of the changes in parent molecule concentration rather than by the direct detection of the transient intermediates as is done in most flash experiments.) This interpretation or aspect of structure will not be covered here except as it may apply to conformational, configurational, and electronic structure.

II. INSTRUMENTATION

As mentioned previously, an excellent publication has recently appeared covering in depth the experimental methods and details of conventional and laser flash methods,[12] and the evolution of flash and laser techniques has been developed with a rather complete bibliography (to 1971) in another review chapter.[14] It is therefore the purpose of this section to give a brief picture of the elements common to flash apparatus and the design aspects that need to be considered in developing a complete flash photolysis system, rather than to give all the background needed for actual construction. Many features are common to all flash systems, and this aspect leads to the general applicability of the technique to a wide variety of studies of transient intermediates. However, specific needs for different photochemical studies lead naturally to a variety of custom-designed systems.

In constructing any apparatus of the type described here, it is important to recognize that the combination of high capacitance and high voltage leads to the potential to deliver a very large (and almost certainly fatal) current. It should also be recognized that optical reciprocity effects probably do not hold over a wide time range, and very short pulses may be quite damaging even though the total energy dissipated may be small; in this respect a recent study has shown that ocular damage can result from picosecond pulses at a threshold energy one-fifth that of nanosecond pulses.[15]

A. Activating Light

1. Conventional Flash

The flash photolysis technique is based on the two conditions that (1) a short-lived species is produced by a pulse of light in sufficiently high concentration to be directly observed, and (2) some physical property of the species can be followed during its lifetime. [Alternately, disappearance or decrease of a parent molecular species in its ground state may be followed under the conditions (1) and (2).]

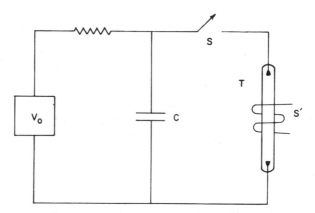

FIG. 1. Simplified schematic flash circuit.

Figure 1 is a very simplified circuit diagram used in virtually all conventional flash apparatuses for producing a single light pulse from gas discharge tubes. The bank of capacitors of capacitance C is charged to a high voltage V_0 by the dc source and discharged through the gas-filled lamp T either by closing a switch S (such as a thyratron or ignitron) or by triggering with a high-voltage pulse from a third electrode S', depending on the breakdown voltage of the discharge tube.

If the impedance of the lamp during discharge is solely resistive (R_T), the voltage across the lamp will decay exponentially:

$$V = V_0 e^{-t/R_T C} \tag{1}$$

and the total energy dissipated by the lamp is

$$E = CV_0^2/2 \tag{2}$$

In this exponential case the discharge time of the lamp may be characterized by the mean or average lifetime τ, which is the time for the discharge to decay to $1/e$ of its initial value, and is equal to $R_T C$. It is apparent, then, that a short discharge time τ for a given discharge energy E results from high-flash voltage (and hence low C) and low R_T (large lamp cross section, short length), and that greater light output is better accomplished by increasing the charging voltage V_0 rather than by increasing capacitance so that τ remains constant and peak light intensity is increased. In addition to problems of insulation, corona discharge, etc., which accompany high-voltage circuitry, it has also been shown[16] that a saturation point is apparently reached where the peak intensity remains roughly constant and τ increases with increasing V_0. Most flash apparatus are designed to operate below 20 kV.

The impedance during discharge is not solely resistive, however, but also includes effects of the associated circuit inductance L (including inductance of the lamp itself) and capacitance C. If $R_T^2 \ll L/C$, which is a condition frequently encountered, then the circuit is underdamped and the voltage across the lamp at time t is

$$V = A\, e^{-R_T t/2L} \sin (t/\sqrt{LC}) \tag{3}$$

This is the equation for a damped oscillatory decay and leads to a flash discharge that has several peaks of diminishing intensities.

The most desirable properties depend on the particular problems to be studied with the flash apparatus. For long-lived intermediate species the main concern may be maximum energy dissipation, and the problems associated with an oscillatory light output consisting of several peaks are only incidental. One of the great advantages of the flash technique, however, is the direct detection of very short-lived species, and in such cases energy dissipated in peaks after the first major peak is wasted (and in fact these secondary peaks usually interfere with the measurement of the transient species). Optimum conditions will occur when the discharge circuit is approximately critically damped:

$$R_T^2 = 4L/C \tag{4}$$

It follows then, that the greater the inductance in the circuit, the greater must be the resistance (and hence the time constant τ) to prevent discharge oscillations. Flash lamp circuitry design therefore in general includes low-inductance (high ringing frequency) capacitors, coaxial leads, and lamps designed to minimize inductance (if straight tubes are used, the inductance increases with increasing length), consistent with the ability to discharge the stored energy, which also increases with length. (Lindqvist[17] has shown that, for an underdamped circuit, the flash duration time is independent of C, being proportional to L/R_T.) In some cases additional resistance has been added to the discharge circuit to approach critical damping, which has only a small effect on the peak light intensity but does greatly reduce the energy dissipated in the secondary peaks.[16] Multiple lamps have also been used to increase the cross sectional discharge area and thereby to decrease lamp resistance,[17] and a currently popular design to minimize lamp inductance is to construct the lamp of concentric tubes with the electrodes at one end, so that the whole assembly is coaxial. Optimum conditions of flash lamp length, discharge, voltage, and capacitance have been evaluated by Vallotton and Wild.[18] More detailed considerations of lamp and flash circuitry design are given by Porter.[12]

Discharge time also depends very significantly on the gas used in the lamp. Probably the most common filling gas is xenon or krypton (xenon is the most efficient gas for converting electrical into visible and ultraviolet light), but in many cases an afterglow results with these rare gases from the light-emitting ions produced by the electrical discharge; this afterglow increases the discharge time beyond the actual power dissipation profile.[19] The afterglow can be reduced somewhat by adding quartz chips to increase the surface area or by adding a quenching gas such as nitrogen (although tailing does exist in pure oxygen- or air-filled lamps[16]). Argon or nitrogen each has appreciable afterglow tailing, but mixtures of the two (for example, 9 nitrogen:1 argon, 2 kJ discharge energy) effectively eliminates tailing and the light output follows the current profile.[17]

The total amount of light emitted and spectral distribution also depend on the nature and pressure of the filling gas. Pure xenon in a quartz lamp at medium pressures (100 Torr) gives a continuum of roughly constant intensity from 200 nm to 700 nm (with superimposed lines primarily in the 400–500 nm region).[20] Pure oxygen has a maximum at ca. 300 nm[21] and at the longer visible-region wavelengths, which gives it a predominantly red discharge, particularly at the lower discharge current densities; the efficiency of energy conversion is also lower for oxygen-filled lamps than with lamps using the rare gases. The light output from a xenon flash is approximately twice that of a discharge in air.[22]

A quite different type of discharge is the theta-pinch flash lamp,[23] which results from the fast magnetic compression of plasma, produced by a high current (of the order of megaamperes) through a one or more turn coil around the electrodeless lamp. Discharge decay lifetimes of the order of 6 μseconds are obtained.[23]

2. Laser Flash

Further refinements of the conventional flash apparatus undoubtedly can lead to somewhat shorter discharge times beyond the microsecond time range, and probably much more time and effort would have been devoted to this aspect of flash equipment development in the past few years were it not for the quantum jump to much shorter pulse times with lasers.[5] In addition to the much shorter flash times possible, the laser also is capable of providing extremely monochromatic flash excitation radiation in contrast to the conventional flash discharge, which is primarily a continuum throughout the visible and ultraviolet regions.

In general the stimulated emission from a pulsed solid-state laser (such as the ruby laser) consists of a series of irregular spikes over a microsecond range, and this irregularity and time length would greatly limit the usefulness of a laser in flash photochemical studies. Q-switching effectively eliminates this, producing single nanosecond pulses of higher peak powers. This technique consists of placing a very fast switching device or "shutter" in front of the laser reflector so that laser action does not occur, thereby developing very large population inversion—much higher than that developed under normal lasing conditions. The shutter is then quickly opened allowing lasing to occur [technically this means that the regenerative power (Q) of the laser cavity is suddenly changed], the result being a single high-intensity pulse of approximately 20 nseconds length[24] that is approximately symmetrical with respect to time, rather than the "tailing off" that is characteristic of many conventional microsecond flash discharges. Some useful switching devices are rotating prisms or mirrors, bleachable dyes, and electro-optical shutters such as Kerr or Pockel cells.

A further time reduction of the order of 10^4 is possible with the use of the output train of a "mode-locked" or "phase-locked" laser.[25] The emission from a Q-switched laser consists of a series of discrete oscillating frequencies, with the separation between modes equal to the reciprocal of the laser cavity round-trip time. In general there is no fixed phase relationship between the discrete frequencies, and the pulse width is limited by the kinetics of the laser to the nanosecond times. However, if these frequencies or modes are "locked" in phase (such as by inserting in the laser cavity a saturable dye, which bleaches in proportion to the laser intensity thereby preferentially transmitting the higher intensities), then there results a series of very narrow (few picoseconds) pulses separated by the cavity round-trip time. Further refinements have permitted selection of a single pulse from the train of mode-locked pulses, thereby now permitting single-flash photolysis experiments in the picosecond region.[26] The experimental techniques and operating procedures have been reviewed by Malley.[27]

The high monochromaticity of lasers in general is an obvious advantage for a photochemical light source. Early high-power pulsed solid-state lasers were limited to the infrared or longer-wavelength (red) visible spectral regions; for example, the two most common solid-state lasers, the ruby and the neodymium lasers, have fundamental harmonics at 694.3 nm and 1060 nm, respectively. However, very efficient (5–25%) frequency doubling and quadrupling with a variety of nonlinear optic crystals (such as potassium or ammonium dihydrogen phosphate tetragonal crystals) is now possible at the peak powers generated by pulsed lasers. The result in the case of the ruby laser is a second harmonic at 347.1 nm, and in the case of the neodymium laser there are significant outputs at 530 and 265 nm. A typical high-powered UV gas laser is the superradiant nitrogen laser involving the second positive band (337.1 nm), and nanosecond flash, even in the vacuum ultraviolet region, is now possible with the molecular hydrogen gas laser[28] involving several Lyman bands (156.7 and 161.3 nm).

Dye lasers will probably turn out to be the most useful for nanosecond and picosecond flash photochemistry because of the wide range of laser frequencies possible. These consist usually of a fluorescent organic dye in a solvent (such as water or alcohol) which is pumped to the fluorescent population-inverted state by a flash lamp or pulsed laser (such as the N_2 gas laser or the ruby or neodymium solid-state laser). The state of this art has developed so rapidly in just the last few years so that commercial units are now readily available from several companies that use the appropriate combinations of dyes, radiant pumping sources, and frequency doubling and quadrupling devices to provide pulse radiation of virtually any wavelength from 250 nm to the infrared region.[29]

It is useful to compare typical light outputs from the various types of light sources described here and available for flash photolysis experimentation, and this is done in Table I.[1, 2, 10, 11, 18, 22, 30] It is important to note that while the total number of photons emitted by the pulsed lasers per flash is in general less than that from conventional flash units, the radiation from the pulsed lasers is highly monochromatic in contrast to the flash discharge lamps and the laser radiant power is greater (particularly with mode-locked picosecond lasers). Depending on the spectral characteristics of the systems to be studied, it is apparent that laser flash systems are now capable of studying photochemical processes taking place in a time interval of the order of a single molecular vibration.

TABLE I

Representative Light Outputs From Conventional and Laser Flash Sources

Type of lamp	Total energy discharged (J)	Flash duration (seconds)	Total light output (quanta)	Peak radiant power (W)
Early flash lamp (Norrish and Porter, 1949)[1, 2]	10,000	2×10^{-3}	10^{21}	3×10^5
Very high energy flash (Claesson and Lindqvist, 1957)[30]	33,600	1.5×10^{-4}	6×10^{21}	2×10^7
Coaxial flash (Vallotton and Wild, 1971)[18]	2,000	3×10^{-6}	8×10^{19}	1.5×10^7
"Microsecond flash" (LeBlanc et al., 1972)[22]	600	0.9×10^{-6}	$4 \times 10^{18 a}$	$3.5 \times 10^{6 a}$
Frequency-doubled Q-switched Ruby laser (347 nm) (Novak and Windsor, 1967)[10]	—	3×10^{-8}	2.6×10^{17}	5×10^6
Commercially available[b] (1974) Nd:glass laser (1060 nm)	—	1.5×10^{-8}	5×10^{20}	6×10^9
Commercially available[c] (1974) dye laser (360–650 nm)	—	1×10^{-8}	1×10^{14}	4×10^3
Mode-locked frequency-doubled Nd:glass laser (530 nm) (Rentzepis, 1968)[11]	—	2×10^{-12}	4×10^{15}	7×10^8

[a] For 200–300 nm range only.
[b] Apollo Lasers, Inc., and Holobeam, Inc.
[c] Laser Energy, Inc.

B. Physical Detection of Phototransients

Spectrophotometric and spectrographic analyses, both employing absorption spectroscopy, are the two methods most widely used in the physical detection of transients produced in flash photolysis studies. The choice of method is determined to a large extent by the system to be studied and the type of information desired, although construction of a highly reliable and sensitive photometric analysis system is considerably easier and cheaper than for a comparable spectrographic system. Certainly for detection and complete structural analysis of free radicals in the gas phase a high-resolution spectrograph is essential and has the advantage that a complete spectrum can be obtained from a single flash. Kinetic data are also obtainable by this method of detection from the analysis of a series of spectrographs taken at varying times after the flash, but this requires a photochemically stable system (or a convenient method of reproducibly changing samples) and constant flash energies. On the other hand, a complete kinetic run can be obtained spectrophotometrically from a single flash, and for most excited-state intermediates in condensed media the absorption bands are so broad that satisfactory absorption spectra can be obtained through point-by-point wavelength analysis, each at a specific time after discharge of the activating flash. However, this technique also requires a laborious series of flash experiments and hence a photochemically stable system and flash discharge energies. Furthermore, spectra obtained in this manner may be erroneous if several absorbing species of varying decay times are present, unless the decay rates and orders are considered in the analysis.

1. Kinetic Spectrophotometric Analysis

A block diagram of a simple kinetic flash spectrophotometer is given in Fig. 2. Continuum light from a steady source (such as a tungsten or quartz–iodine lamp, or a dc-stabilized xenon or mercury arc) is collimated through the absorption cell to the monochromator or filter system and is monitored as a function of time with a photomultiplier or fast-response photodiode detector in conjunction with a fast recording device such as an oscilloscope

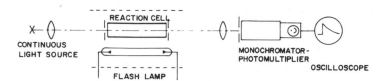

FIG. 2. Block diagram of flash spectrophotometer.

(as shown) or a computer of averaged transients (CAT), or by direct analog-to-digital interfacing to a mini- or a midicomputer.

In general, the monochromator is placed between the absorption cell and the photodetector to minimize deleterious effects due to scattered flash light; it may also be necessary to include an appropriate filter in front of the cell to prevent photolysis by the continuous analyzing light, since in general many more light quanta are delivered to the cell by the analyzing light than by the flash light integrated over the time required for a complete experiment involving a series of flashes (although of course the flash intensity is many orders of magnitude greater than that of the monitoring light).

If a conventional gas-discharge flash lamp is used, its light input to the cell can be limited to a wavelength band by appropriate filters interposed between the flash lamp and the reaction cell, but in general extensive filtering to produce even approximately monochromatic light will decrease the flash intensity to a point where this major feature of the flash technique is defeated. If monochromatic activating light is important for an adequate interpretation of the kinetic or spectral behavior of the chemical system, then a more satisfactory approach is to use an appropriate laser flash, even if the very short flash time is not required (i.e., for long-lived transients).

A limiting sensitivity factor in many flash photolysis systems is the photomultiplier tube shot noise, which results primarily from the random emission or statistical fluctuation of electrons from the photocathode (and, to a lesser extent, from the multiplying dynodes). This effect becomes serious as the time response of the apparatus is decreased, and even with conventional flash lamps special techniques may be required to reduce the shot noise-to-signal ratio (which is proportional to the square root of the detector photocathode current) to an acceptable level in the microsecond response region. These techniques include superimposing a voltage pulse on a continuous xenon arc to increase its light output,[31] voltage-pulsing the photomultiplier to minimize overloading,[32] and using a very high-inductance (long-lived, and hence approximately flat discharge) flash lamp for the analyzing light source.[33] With the nanosecond laser flash system, the discharge time of most conventional flash lamps is sufficiently long for these to serve as analyzing light sources, and in fact a particularly convenient monitoring source is the flash lamp which is also used for optical-pumping the activating laser pulse.[34] Picosecond time-resolved analysis requires special "interrogating" techniques and these have been reviewed by Rentzepis.[26]

2. Flash Spectrography

Many of the features of flash spectrographic apparatus are the same as those of the kinetic spectrophotometric equipment, as shown in Fig. 3. The major differences are (1) a pulsed monitoring source in place of the continuous

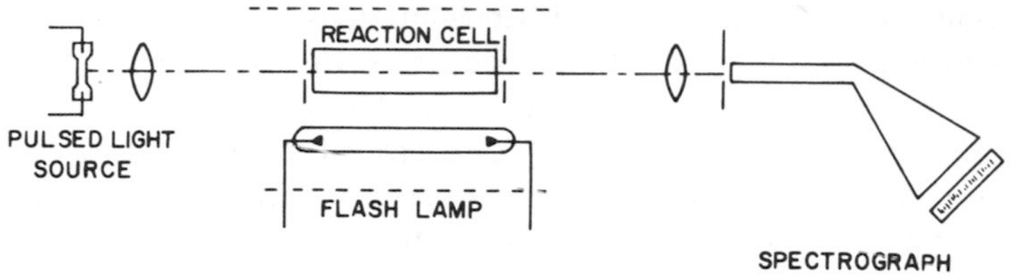

FIG. 3. Block diagram of flash spectrograph.

analyzing light, and (2) spectrographic detection rather than monochromatic photodetection.

It is essential that the monitoring pulse have a lifetime at least equal to or less than that of the activating flash (with accurate and reproducible triggering) if it is not to be the time-limiting element in the system, and hopefully the light output will be a continuum over the spectral region of interest. These conditions can generally be satisfied in a conventional flash apparatus with a synchronized second spectroflash of lower energy than the activating flash. With the nanosecond and picosecond laser flash systems, however, laser-initiated spectroflashes are required, and several methods have been developed to provide the accurate synchronization of the activating and monitoring laser flashes. For example, in experiments using the frequency-doubled ruby laser, the UV (347 nm) second harmonic is used as the activating flash and the first harmonic (694 nm) is focussed to a fine point in air (or other gases) leading to electrical breakdown of the gas and emission of continuum light that follows the time profile of the laser pulse.[10] Since this discharge originates from the same primary source as the activating laser flash, the two pulses are self-synchronized to within a few nanoseconds; light travels only 10 m in a 30-nsecond time interval, and therefore a convenient and accurate delay between the activating and analyzing flashes can be built into the system by simply incorporating an appropriate path-length difference between the two beams. Another technique for generating the pulsed monitoring source is to use the fluorescence emission, excited by part of the activating laser flash, from an appropriate scintillator[34]; in this case also the two sources are synchronized to approximately the fluorescence lifetime of the scintillator. Although the spectral output from the fluorescent source usually is not very broad, a variety of materials can be used to cover a wide spectral region, and in general the fluorescent bands are free from sharp emission lines that are present in spark gaps.[34] Synchronization with the photolyzing flash in the picosecond time region requires many of the same techniques as with the time-resolved method of analysis; generation of a "continuum" flash for picosecond spectrography involves such practices as Raman-shifted emission in liquids stimulated by the laser fundamental or its harmonics, etc.[26]

As already stated, flash spectrography is essential for the very high spectral resolution needed in the structural determinations of gas-phase free radicals in their ground and electronically excited states. In essence this technique uses many of the basic spectrographic principles known for some time and used extensively in the analyses of vibronic spectra of ground-state species; here these principles are applied to transient spectra of reactive intermediates in complex photochemical systems.[35] A particularly useful technique, which greatly increases the sensitivity of the method by increasing optical path length and hence optical absorbance without increasing the cell length, is the use of mirrors at the ends of the absorption cell to produce multiple traversals through the cell.[36]

Although flash photolysis does produce relatively high concentrations of transient intermediates, and the use of such techniques as multi-pass cells described above does greatly increase the sensitivity of the flash spectrographic method of detection, the very great resolution required for complete gas-phase rotational and vibrational analyses of free radicals does require in most cases a large number of exposure flashes. Atkinson, Laufer, and Kurylo have recently described a new technique[37] involving dye lasing, which greatly increases the sensitivity per flash and eliminates the disadvantages of the conventional flash spectrographic technique. Basically it consists of placing the absorbing species (or, in the case of transient radicals, the photochemical source of the radicals) inside the cavity of a broadly tuned laser. Each transient absorption following flash photodissociation then acts as a loss at a specific wavelength in the same manner as gratings, etc., when included in the cavity, so that the laser output has superimposed on it the spectrum of the transient species. This absorption arises from nonlinear processes and therefore requires internal calibration, but in principle quantitative determination of transient concentrations should also be possible by this technique.

3. Transient Optical Rotatory Dispersion

Optical rotatory dispersion (ORD)—the change with wavelength in rotation of plane-polarized light by an optically active substance—has found widespread usage in structural and stereochemical studies of stable optically active organic molecules.[38] Combination of flash excitation with spectropolarimetric analysis can provide similar structural information on electronically excited states and transient chiral intermediates. Furthermore, the rotational strength of an inherently symmetric chromophore (such as carbonyl) is a measure of the asymmetric environment adjacent to the chromophore, so that it should also be possible to obtain information on modifications of the electronic environment of a chromophore associated with electronic excitation.

In its basic form, the flash spectropolarimeter is simply the conventional kinetic flash spectrophotometer (Fig. 2), modified by placing a fixed polarizer between the xenon lamp and the absorption cell, and an analyzer mounted on a 360° angle divider between the cell and the monochromator.[39] The intensity of light transmitted by the polarizer and analyzer at an angle of rotation θ relative to each other is

$$I = I_0 K \cos^2 (\theta + \alpha)$$

where I_0 is the incident light, K is a constant for the apparatus and absorption cells, and α is the total rotation of the solution. Scattered flash light and absorbance changes also contribute to the total change in transmittancy (ΔI) of the cell following flash excitation, so that in practice two ΔI measurements are made as a function of time with the analyzer rotated at $+\theta'$ and $-\theta'$ from the null position, where $\theta' = \theta + \alpha_0$ and α_0 is the rotation of the solution resulting from ground-state optical activity. The transitory change in rotation at a specific wavelength λ and time t after initiation of the flash is approximately

$$\alpha_t(\lambda) = -\frac{\Delta I_+ - \Delta I_-}{4I' \tan |\theta'|}$$

if $\alpha_t(\lambda) \ll |\theta'|$. (In this expression, I' is the steady transmitted intensity at $\pm \theta'$.) The transient ORD curve is constructed from these $\alpha_t(\lambda)$ measurements taken point-by-point as a function of wavelength.

An example of the type of curves obtained by this procedure is given in Fig. 4 for d- and l-benzoin.[39] Although there is much scatter in the data, it appears that the excited-state ORD curves are roughly mirror images of each other, exhibiting multiple Cotton effects in the region of triplet–triplet absorption similar to the effects shown by the ground state in the ultraviolet (singlet–singlet) absorption region. (Apparently the basic symmetry and not the specific type of electronic transition determines the gross ORD effect for a system of this type.[39])

Results from this flash spectropolarimetric technique are limited so far, but they do show that it has the potential to become a powerful tool for determinations of excited-state structures and conformation changes accompanying photoexcitation of optically active molecules. It should be pointed out that transient optical rotation has been measured on perturbed macromolecular systems[40]; in this case, however, transient rotatory changes were produced by a pulsed electric field rather than by flash photoexcitation. Similar structural information on electronically excited chiral systems has also been obtained from circularly polarized fluorescence or phosphorescence, since the same molecular quantities are involved in light absorption and emission.[41, 42]

4. Transient Luminescence

Transient luminescence and its decay properties are important parameters in characterizing electronically excited states, and detection techniques

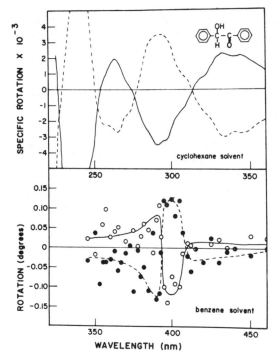

FIG. 4. Optical rotatory dispersion curves of the ground and excited state of benzoin[39]: ———, ground-state *d*-benzoin; – – – –, ground-state *l*-benzoin; ○, excited-state *d*-benzoin; ●, excited-state *l*-benzoin. Reprinted with permission from P. A. Carapellucci, H. H. Richtol, and R. L. Strong, *J. Amer. Chem. Soc.* **89**, 1742 (1967). Copyright by the American Chemical Society.

generally associated with stroboscopic or low-intensity flash methods[43] are also applicable to repetitive nanosecond and picosecond laser flashes. One of the most sensitive techniques is the single-photon method.[44] In this technique, the time of emission of a single photon is measured from a reference zero (such as the triggering of the flash lamp); this time is converted to an amplitude and the resulting pulse is stored in a multichannel pulse height analyzer, so that the number of pulses in the channel is proportional to the probability of luminescence at a given time. An obvious disadvantage of this technique is the large number of flashes required, but for photochemically stable systems this is probably one of the most sensitive and reliable methods of emission detection.

5. Other Detection Systems

In principle, any analytical device capable of fast response can be coupled to flash photoexcitation, and a wide variety of detection systems other than spectral optics have been used in special cases (dealing primarily, however, with kinetic rather than structural determinations). These include incorporation of analytical techniques such as resonance fluorescence,[45] potentiostatic chronoamperometry,[46] electron spin resonance[47] (recently reviewed by Bolton and Warden[48]), and mass spectrometry.[49] Specific details of these techniques are given in the cited references. In general, the experimental complexities are greater in these systems than for absorption spectroscopy, but they will certainly continue to be developed if the full potentials of the flash technique in structural determinations are to be realized.

III. SPECTRA OF FREE RADICALS

Much detailed structural information on gas-phase intermediates and free radicals has been obtained over the years from flame and electrical discharge emission studies, particularly on diatomic radicals. Unfortunately many polyatomic radicals cannot be followed by emission spectroscopy due to their great instability, and absorption spectroscopy holds probably the major potential for determination of molecular structure, electronic states, and dissociation energies. On the other hand, in most cases continuous illumination by conventional photochemical light sources simply does not produce high enough concentrations of reactive intermediates to give the highly resolved spectra in the gas phase necessary for complete structural characterization through detailed rotational and vibrational analyses. It was early recognized by Porter[50] and independently by Herzberg and Ramsay[3, 51] that the flash spectrography technique provides the means to produce the high concentrations of radical intermediates through photodissociation and, simultaneously, to record the spectra in a time that is short compared to the lifetimes of the intermediate radicals. Some of the first flash spectrography studies were on diatomic radicals such as ClO, SH, SD, OH, and OD,[50,51] confirming, where possible, assignments made from emission spectra.

Early extension to triatomic species led to complex spectra and clearly showed the need for high-resolution spectrographic equipment; nevertheless, a variety of radicals were spectroscopically observed and characterized, such as NH_2[52] and HCO[53] (which were shown to be bent in the ground state but linear in the excited state), NCO and NCS,[54] PH_2,[55] and HNO[56] (both ground and first excited electronic states bent). As pointed out by Thrush,[57] results on these and other relatively simple molecules have been invaluable in confirming earlier molecular orbital treatments to predict ground and excited-state structures.[58] Briefly, this approach results in the following general principles[35]: (1) the primary factor in determining if an orbital becomes more tightly bound with change of angle is whether or not it changes from being formed from a p orbital to being formed from an s orbital; (2) if the orbital is antibonding between the end atoms, the most stable configuration of the molecule or radical is linear, and vice versa; and (3) for a molecule containing n electrons, the first excited state belongs to the same symmetry class as the ground state of a similar molecule containing $(n + 1)$ or $(n + 2)$ electrons (only one electron being added for free radicals, since it is assumed that the lowest excited state results from excitation of an electron to the molecular orbital containing the single unpaired electron). These same conclusions have recently been derived by Pearson, based on the second-order Jahn-Teller effect, and applied to the structures of molecules in excited electronic states[59] and unstable free radicals.

The spectral analysis of the free PH_2 radical is a good example of the type of detailed treatment possible for triatomic transient species. The absorption spectrum of PH_2 was first observed by Ramsay[55] in 1956 following flash photolysis of gaseous phosphine at ca. 5 mm pressure (half-life of PH_2 ca. 50 μseconds). Eleven bands were detected from 360 to 550 nm, corresponding to the $^2A_1$–$^2B_1$ electronic transition. Although a detailed rotational analysis was not carried out because of the complexity of the spectrum, it was interpreted as a long progression of the bending frequency of PH_2 in its upper electronic state. Subsequently, rotational analyses have been carried out for the (000)–(000) band[60] and for the $(0v_2'0)$–(000) progression ($v_2' = 1$–8) on data obtained from a 0.075 nm/mm resolution spectrograph and a 16 m effective pathlength optical cell.[61] In all, 1000 rotational lines have been assigned, with fitting of the (000)–(000) band to an accuracy of ± 0.06 cm^{-1}. However, perturbations of up to 0.6 cm^{-1} were found in the (000) level of the excited state—perturbations which could only be caused by higher vibrational states of the ground state since no electronic states lower than the $^1A_1$ exist— and this is the first definite example of a perturbation of this type for a polyatomic species.[61] (Similar perturbations have been observed in the zeroth vibrational level of the first excited singlet state of stable molecules, but it is possible in these cases that the perturbation results from interaction with the excited triplet electronic state, which of course cannot occur with PH_2.)

One of the major driving forces in developing the flash photolysis technique was the various attempts to detect the methylene biradical, CH_2,[3, 50] a postulated reactive intermediate in a variety of organic reactions.[62] Flash photolysis of ketene (CH_2CO) gave strong absorption bands of CO, showing that decomposition actually occurred, but no Rydberg series due to CH_2 could be obtained. Absorption was ultimately attained by flash photodecomposition of diazomethane,[63, 64] but interpretation has been complicated by the fact that there are two low-lying states of methylene, the lower of the two being a triplet ($^3\Sigma_g^-$) state giving rise to vacuum UV absorption (and hence unobserved in the flash photolysis of ketene because of masking by the parent compound), and the higher one being a singlet ($^1A_1$) with absorption bands in the visible region extending from 550 to 950 nm. The detailed rotational and vibrational analysis of the vacuum UV bands led to the conclusion that the ground triplet state of CH_2 is "linear or nearly linear" (i.e., the HCH angle is greater than 150°). On the other hand, theoretical calculations[65-67] as well as electron spin resonance work in solid matrices[68, 69] strongly suggest that the ground triplet state is nonlinear, with a triatomic angle as small as 135.1°.[67] Herzberg has recently pointed out[70] that the vacuum UV spectrum that was interpreted as being a $\Sigma_u^- - \Sigma_g^-$ band of a linear molecule can also be interpreted in terms of a bent ($\angle HCH = 136°$) molecule if it arises from a 0–0 subband of an A_2-B_1 electronic transition in the radical. (It was pointed out, however, that this does require an assumption that strong predissociation occurs in the upper state so that the higher 1–1, 2–2, etc., subbands become so broadened as to be undetectable. A similar effect exists for the HCO radical.)[53] A similar analysis of the visible spectrum, which has a large number of well-defined bands, corresponding to the singlet manifold, shows that the lowest singlet state is a strongly asymmetric top ($\angle HCH = 102.4°$, C—H bond distance = 11.1 nm) but that the upper ($^1B_1$) state is "nearly" linear, with an HCH angle of 140° and a C—H distance of 10.5 nm.[71] However, sensitivity was not sufficient to arrive at an exact relationship between the singlet and triplet states through, for example, perturbations on one of the singlet states by a triplet state.[71]

A tetratomic free radical of major importance in organic reactions is the methyl radical, CH_3. Early attempts to detect this species by the Herzberg group centered on the visible spectral region, drawing analogy from the isoelectronic NH_2 radical, which does show strong absorption in the red.[72] On the other hand, if the ground state of CH_3 were symmetrical and planar, then the transition in the visible region would be forbidden. The research was then extended to the vacuum ultraviolet region using a 3 m vacuum grating spectrograph, where strong Rydberg-type transitions would be expected, and a spectrum extending to 216 nm was soon detected following flash photolysis of $Hg(CH_3)_2$[73]; further work showed the same bands from a variety of methyl-containing compounds for which much photochemical evidence exists substantiating the presence of CH_3 radicals. Subsequently it was found that a much higher radical concentration was possible from the reaction of CH_2 (produced by the flash photolysis of diazomethane) with H_2, and the resulting stronger bands have permitted a reasonably complete spectral analysis (although parent compound absorption and weakness of the continuum spectroflash in the vacuum UV region do limit resolution near the Rydberg limit).[64] The 216 nm band is quite diffuse, as shown in Fig. 5 (although considerably less so for the corresponding 214 nm band of CD_3), indicating extensive predissociation slowed down by a potential energy barrier (more so for CD_3 than for CH_3). Nevertheless, the observed intensity alternation in parallel bands clearly indicates that CH_3 and CD_3 are indeed planar or nearly so, and furthermore, the presence of only one strong band in each electronic

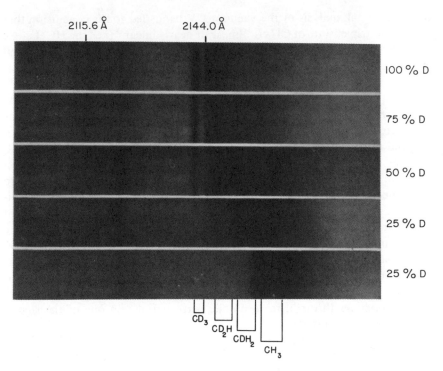

2115.6 Å   2144.0 Å

100 % D

75 % D

50 % D

25 % D

25 % D

CD$_3$

CD$_2$H

CDH$_2$

CH$_3$

FIG. 5. The 215 nm absorption band of free methyl obtained with various concentrations of deuterium in acetone. From G. Herzberg, *Proc. Chem. Soc., London* 116 (1959).

transition observed shows that there is no appreciable configurational change on excitation—i.e., the methyl radical is planar in all of the observed doublet electronic states.[64]

Another example of a flash spectrographic study involving a relatively simple tetratomic free radical is that on the HNCN radical, also produced in the flash photolysis of diazomethane.[74] The absorption band is in the ultraviolet region (344 nm), and rotational analysis of the high-resolution spectra showed that the radical in its ground state is a planar very slightly symmetric top having the structure

$$\begin{array}{c} \text{N—C—N} \\ \diagup \\ \text{H} \end{array}$$

with the HNC angle equal to 116.5°.

The flash photolysis technique is of course not limited only to simple free radicals, and absorption spectra of many polyatomic radicals have now been obtained, both in gas and liquid systems; in fact, virtually all of the spectra of polyatomic free radicals have been obtained by the flash technique. The difficulty comes in interpreting the rotational spectra, since in general even in the gas phase they are very dense and in many cases also diffuse as a result of predissociation in the excited electronic state (as, for example, the propargyl radical, CH_2—C≡CH, which is produced by flash photolysis of propargyl halides, allene, etc.[75]). Thus, although dissociation energies can often be obtained from the (0–0) vibrational transition, in general geometric determinations have not been made on radicals larger than simple 4-atom species. Nevertheless, much valuable kinetic information has been obtained through the direct observation of transient free radicals in photochemical systems, both flash spectrographically and flash spectrophotometrically, and the identification of these species has proven extremely useful in the interpretation of their roles in complex organic and biochemical mechanisms. The isoelectronic 7-π-electron benzyl, anilino, phenoxyl, and related radicals have been detected and characterized in gas[76] and liquid[77] media, and in fact it appears that almost every aromatic molecule will give a free radical spectrum under appropriate flash conditions.[78] Other examples of the types of polyatomic radicals definitely identified in flash photochemical systems are the semiquinone,[79] cyclopentadienyl[80] and tropyl,[81] and phenyl[82] radicals. This last radical is particularly interesting because of its participation in aromatic solution reactions[83] and because of early failure to detect it in absorption in spite of extensive indirect evidence for its presence. The observed band of the phenyl radical in the visible region (440–530 nm) following flash photolysis of benzene and the halobenzenes is a $^2A_1 \rightarrow {}^2B_1$ transition resulting from $\pi \rightarrow n$ excitation.[82]

Most of the free radicals studied by flash photolysis are by nature quite unstable, so that conventional analytical techniques are not applicable, and a problem may exist in determining if the observed transient spectrum is due in fact to a radical or to an excited state of the parent molecule. Confirmation of the presence of a free radical is sometimes possible through detection of its electron spin resonance spectrum (and response times of the order of 1 μsecond are possible[84]), but usually the ESR signal is so weak that many flashes and computer-averaging of the transient signal is necessary, limiting this method to very photochemically stable systems. It is also possible in many cases to distinguish between a free radical and an excited-state absorption by the effect of oxygen on the decay kinetics; oxygen is a very efficient quencher of excited states (see the following section), whereas in general radicals react much more slowly with molecular oxygen.[85]

Radicals formed from a related or homologous series of molecules may be identified if the spectrum of one of the series is known, since in general similar spectra should appear for radicals having similar electronic structures. Such an effect is clearly shown in Fig. 6 for the phenoxyl and alkyl-substituted phenoxyl free radicals.[86]

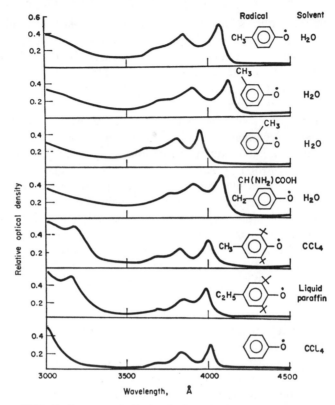

FIG. 6. Spectra of alkyl-substituted phenoxyl radicals.[86]

IV. ELECTRONIC STRUCTURE OF EXCITED STATES

A. Triplet States

One of the outstanding successes of the flash photolysis technique has been the direct study of excited electronic states in fluid or gaseous media. Many organic molecules phosphoresce in rigid or very viscous media, and it is generally accepted now that this emission is from an excited (usually the lowest) triplet state of the molecule,[87] where the radiative transition to the ground singlet state is partially forbidden by the spin selection rule. Interest in triplet states has expanded greatly in the past few years due to recognition of their importance as intermediates in photochemical reactions involving organic and biological systems.

Triplet states can be observed in absorption by continuous illumination in rigid glasses at very low temperatures (where the observed lifetimes approach their phosphorescence radiative lifetimes), but direct observations in fluid media by the same techniques are not possible because of competing fast nonradiative processes. Indirect photochemical evidence suggested, however, that triplet lifetimes in liquids should be in the range open to conventional flash photolysis methods, and in 1953 Porter and Windsor reported their first studies on transient absorption spectra of a large number of aromatic molecules at room temperature using the flash spectrographic technique.[88] In many cases similar spectral behavior was obtained, as exemplified in Fig. 7

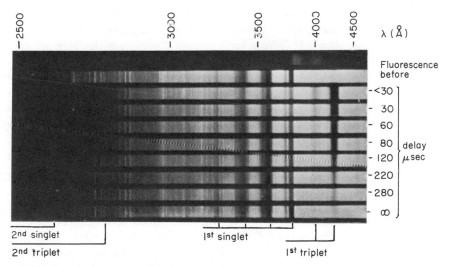

FIG. 7. Flash spectra of anthracene in hexane. Reprinted with permission from G. Porter and M. W. Windsor, *Discuss. Faraday Soc.* **17**, 178 (1954).

for anthracene; the visible spectrum consisted of several bands that were assigned to transitions to different vibrational levels of an excited triplet state, and the UV region displayed a strong structureless continuum that was attributed to a spin-allowed transition to a second electronically excited triplet state. All species had lifetimes of the order of 100 μseconds in benzene or ethanol and approximately 1 msecond in glycerol; in some cases as high as 50% conversion to the triplet state was produced by a single flash. These results showed that the formation of triplet states is a rather general occurrence in fluid solvents, at least with aromatic hydrocarbons (only with single-ring compounds were triplet spectra not observed[89]); following papers demonstrated similar formation of triplets for aromatics in the vapor phase[89] and for a wide variety of organic molecules in fluid media.[90] The higher percentage of conversions to the triplet state allowed estimations of absolute extinction coefficients and oscillator strengths in many cases. The general applicability of the flash photolysis technique—spectrographic and spectrophotometric—to triplet studies (lifetimes, triplet–triplet spectra, conversion efficiencies, quenching and energy transfer, etc.) is spectacularly shown by the large number of triplet states studied over the past two decades.

Although each type of molecular species would be expected to have a characteristic spectrum, which is the basis for the spectral identification and determination of concentration, there are similarities among most of the triplet states studied. For the most part, triplet–triplet absorption occurs in the visible and/or near-ultraviolet regions, and at longer wavelengths (lower energies) than the corresponding singlet–singlet absorption occurs from the ground electronic state. The triplet–triplet peaks may show a few vibrational bands, but since they result from transitions between two electronically excited states they are generally quite diffuse (even under high gas-phase resolution), and therefore so far they have not been very useful for configurational and conformational structure determinations of excited triplet states. Such spectra are, of course, invaluable in the characterization of electronic structure.

Just as the spin selection rule partially prevents phosphorescence emission of radiation from the excited triplet state, so also is the direct population of the triplet by photoexcitation from the ground singlet state a forbidden transition. Direct excitation is to a higher vibrationally excited singlet state, which for the most part undergoes rapid deactivation in fluid media to the lowest vibrational level followed by either radiative (fluorescence) or nonradiative [internal conversion (IC)] decay to the ground singlet state, chemical reaction, or radiationless conversion to the triplet level [intersystem crossing (ISC)]. (It should be noted that radiative $S_0 \leftrightarrow T_1$ processes are greatly enhanced by the heavy-atom effect, and this phenomenon has also been used in flash studies of triplet states.[91])

The properties of excited electronic states and interstate conversions are quite dependent on the electronic origin of the states. Aromatic hydrocarbons, enes, dienes, etc., can exist only in (π, π^*) states (at least in the readily accessible photochemical spectral region), whereas molecules such as carbonyls, enones, dienones, amines, and nitrogen heterocyclics contain both (π, π^*) and (n, π^*) excited electronic states.

El-Sayed has reviewed in detail the factors contributing to the efficiencies of interstate conversions.[92] Radiationless transition to the lowest excited singlet state from higher states of the same multiplicity is very rapid ($\tau \cong 10^{-12}$ seconds), whereas intersystem crossing is considerably slower and is governed by the selection rules

$$S_{\pi,\pi^*} \xrightarrow{\quad\times\quad} T_{\pi,\pi^*} \qquad\qquad S_{n,\pi^*} \xrightarrow{\quad\quad} T_{\pi,\pi^*}$$
$$S_{\pi,\pi^*} \xrightarrow{\quad\quad} T_{n,\pi^*} \qquad\qquad S_{n,\pi^*} \xrightarrow{\quad\times\quad} T_{n,\pi^*}$$

where transitions indicated by $\nrightarrow$ are "forbidden" and should be slower by a factor of ca. 10^3 than the allowed ($\rightarrow$) transitions. These selection rules are based, however, on the Born-Oppenheimer approximation for mixing the $S_{n,\,\pi*}$ and $S_{\pi,\,\pi*}$ states,[93] and failure of this approximation would have the effect of relaxing these rules; it is interesting to note in this regard that the intersystem crossing rate constants for 9,10-diazaphenanthrene ($S_{n,\,\pi*} \rightarrow T_{\pi,\,\pi*}$) and phenanthrene ($S_{\pi,\,\pi*} \rightarrow T_{\pi,\,\pi*}$) are reported to differ by only a factor of 3.[94]

The early flash photolysis studies of triplet states in fluid media soon established that the decay of the triplet state is kinetically first-order, usually independent of the parent molecule concentration but dependent on the solvent viscosity, and is considerably greater than the natural radiative decay determined from solid-state phosphorescence measurements.[90] The dependence on solvent viscosity (determined in subsequent flash work carried out over a wide range of viscosities and temperatures[95]) shows that the process probably involves radiationless quenching to the ground singlet state by energy transfer to some solvent impurity or impurities (oxygen, unsaturated molecules, aldehydes, etc.), which requires diffusional transport through the solution. Only in cases such as when the viscosity of the solution is very high,[95] the solvents are extensively purified,[96] or the radiationless internal conversion is enhanced by heavy-atom induced spin–orbit interaction (such as with 9,10-dibromoanthracene)[97] does the lifetime of the triplet become roughly independent of the solvent viscosity and hence indicative of the radiative or radiationless lifetime of the triplet.

Physical quenching of the triplet to the ground singlet state by collisional energy transfer in the situation where a quenching species is deliberately added (rather than present as an impurity, as above) has also been extensively studied by conventional flash photolysis, and leads to information on the relative energetics of the interacting electronically excited triplet states, which is often difficult to obtain in other ways. Conservation of energy and spin dictates that the added quencher molecule Q (if a ground-state singlet) must have an excited triplet energy less than that of the donor molecule D. If the donor concentration is sufficiently low so as to exclude a second-order triplet–triplet annihilation ($D^T + D^T \rightarrow D^{S*} + D^S$) process[98] and radiative (phosphorescence) decay is neglected, the decay of the donor triplet is given by the mechanism

$$D^T \xrightarrow{\ k^T\ } D^S \tag{5}$$

$$D^T + Q^S \xrightarrow{\ k_q{}^T\ } D^S + Q^T \tag{6}$$

where k^T is the rate constant for intersystem crossing to the ground singlet in the absence of quencher (impurity quenching or true unimolecular ISC), and k_q^T is the quenching rate constant. The observed first-order triplet decay is

$$-\frac{d[D^T]}{dt} = k_{obs}^T[D^T] = k^T[D^T] + k_q^T[D^T][Q] \tag{7}$$

so that k_q^T is obtained from the slope of a modified Stern-Volmer plot of k_{obs} vs quencher concentration [Q] according to the relation

$$k_{obs}^T = k^T + k_q^T[Q] \tag{8}$$

Some typical results obtained by flash photolysis for k_q^T in hexane solvent are given in Table II as a function of the energy gap between the donor and acceptor triplet energies ($\Delta E^T = E_D^T - E_Q^T$). The quenching rate constants are all approximately the same and independent of the energy gap ΔE^T, indicating a process governed only by the rate of diffusion of the energy-transfer particles. The results are, however, significantly lower than the value for the encounter frequency predicted by the simple Debye equation for spherical particles based on a continuum model for solvent structure and infinite coefficient of sliding friction[99]:

$$k_{q,\,diff} = \frac{8RT}{3000\eta} \tag{9}$$

This equals 11×10^9 liter/mole second for benzene at 25°; if the coefficient of sliding friction is zero, thus allowing "slip" between diffusing species and the solvent medium, the modified Debye equation becomes[100]

$$k_{q,\,diff} = \frac{8RT}{2000\eta} \tag{10}$$

TABLE II

Some Triplet Quenching Rate Constants in Benzene Solvent

Donor	Quencher	ΔE^T (kJ)	$k_q^T \times 10^{-9}$ (liter/mole second)
Triphenylene	trans-Stilbene	69	7.6[a]
Triphenylene	1,3-Cyclohexadiene	55	1.3[a]
Phenanthrene	trans-Stilbene	50	3.6[a]
2-Acetonaphthone	trans-Stilbene	39	2.3[a]
Triphenylene	2,4-Hexadiene	> 33	4.6[a]
Benzophenone	Naphthalene	32	1.2[b]
Biacetyl	1,2-Benzanthracene	32	3.0[b]
Triphenylene	Isoprene	27	6.4[a]

[a] S. L. Murov, "Handbook of Photochemistry." Dekker, New York, 1973.
[b] G. Porter and F. Wilkinson, *Proc. Roy. Soc., Ser.* A **264**, 1 (1961).

and the difference between experimental and calculated diffusion-controlled quenching constants is even greater. This effect appears to be rather general for solvents of low viscosity[101, 102] (<3 cp); the reasons are not fully understood, but may involve spin statistical factors or geometrical and solvent interrelations. These results do show, however, that care should be exercised in assuming that Eq. (9) or (10) holds for all quenching processes, even though energetically favorable.

If the two triplet energy levels are close ($\Delta E^T < 12$ kJ), then back energy transfer may occur (a temperature-dependent behavior) thereby also decreasing k_q.[103] The extent to which this occurs is sensitive to actual decay rates of the donor and quencher triplet states, and under appropriate conditions of fluorescent, phosphorescent, and nonradiative rates there may even be "negative" quenching; i.e., adding a quencher Q will increase the apparent lifetime of the donor molecule.[104]

Oxygen is undoubtedly an impurity in many room-temperature solutions even though rather rigorous outgassing procedures are followed.[96] The oxygen molecule is a triplet ground state ($^3\Sigma_g^-$) but has two low-lying singlet states—the $^1\Delta_g$ state at 94.5 kJ, and the $^1\Sigma_g^+$ state at 157 kJ—that may contribu to the importance of oxygen in photosensitization and photooxidation cesses in a variety of biological and environmental systems. Spin conserv is allowed by the energy-transfer mechanism

$$D^T + O_2(^3\Sigma_g^-) \longrightarrow D^S + O_2(^1\Delta_g \text{ or } ^1\Sigma_g^+)$$

and the availability of low-energy excited levels in oxygen make favorable process. For some time it has been assumed that oxygen is diffusion controlled in solution for all energeticall counters.[90] Recent investigations on oxygen quenching of carbons—in fluid solvents[105] or in a solid polystyrene matr triplet state carbonyl compounds in fluid solvents[107] (th not by flash excitation) have shown, however, that ox cases is appreciably less than solely diffusion limited. aromatic molecules with low triplet energies (but s state of oxygen) is approximately one-ninth di suggests[105] that the most probable quenching transfer process proceeding through an interme

$$D^T + O_2^T \rightleftharpoons (D^T \cdot O_2^T)$$

The one-ninth factor then results from multiplicities of the complex—singlet, t with a statistical weight of one-nint molecules. [With the carbonyl triple most cases an (n, π^*) state rather

molecules, the oxygen quenching rate is even lower, and may indicate that nonbonding orbitals are less accessible to the oxygen.[107]] As the aromatic triplet energy is increased, however, thereby increasing the energy gap ΔE^T between the aromatic triplet and singlet oxygen, the rate of quenching by oxygen becomes roughly inversely proportional to the aromatic triplet energy (Fig. 8); this is shown to be consistent with a nonradiative transition within the $D^T \cdot O_2^T$ collision complex.[108]

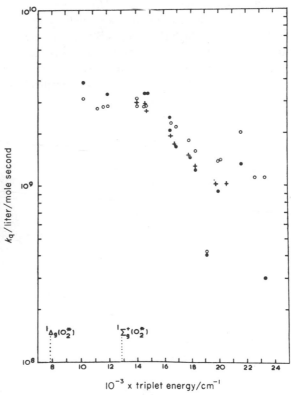

G. 8. Rate constants for oxygen quenching of triplet aromatics as a function plet energy.[105] ○, benzene; ●, hexane; +, cyclohexane.

reactivity of the excited triplet state is strongly dependent on its elec-
structure, and this is an area that has been extensively studied by the
ional flash photolysis technique. A specific example is that of liquid-
drogen abstraction by electronically excited aromatic carbonyl com-
for these it has been shown to be the lowest triplet level that is
e for H-atom abstraction. With benzophenone

in solution the lowest-lying triplet state is an (n, π^*) state, in which an n electron has been promoted from a nonbonding orbital associated with the oxygen atom to an antibonding π^* state conjugated with the aromatic rings and the carbonyl chromophore. The oxygen is therefore electron deficient in this excited state, and can readily undergo hydrogen abstraction from a hydrogen atom-donating solvent or added substrate (primary or secondary alcohols, hydrocarbons, etc.), forming the ketyl radical which dimerizes to benzopinacol[109]; the transient absorption spectrum of the ketyl radical is shown in Fig. 9. Also shown in Fig. 9 is the transient spectrum following flash

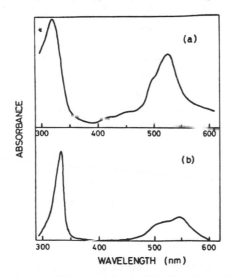

FIG. 9. Flash spectra of benzophenone[110]: (a) triplet; (b) ketyl radical.

excitation of benzophenone in trichlorotrifluoroethane[110]; since this solvent has no abstractable hydrogen atom and therefore cannot lead to ketyl radical formation, this spectrum is concluded to be that of the benzophenone triplet. The marked similarity between the two spectra suggests that the electronic structures of the two species are similar, i.e., the ketyl radical presumably has a π-electron configuration like that of the lowest benzophenone triplet, with sp^2 rather than sp^3 hybridization on the ketyl radical central carbon atom.[110]

Substituting amino, dimethylamino, or hydroxy groups on benzophenone at the meta or para ring positions (thereby excluding intramolecular reactions) drastically changes the quantum yield of ketyl radical formation, reducing it to zero from ca. unity in benzophenone itself, whereas halogen or carboxy substitution has very little effect.[111] Similarly, reactivity is strongly affected by changes in solvent polarity. Accompanying these changes in reactivity are major changes in the absorption spectrum of the lowest-lying triplet state, and these changes observable by flash spectroscopy give the best indication of the electronic structure of the excited state. The results of an extensive study of these effects are summarized as follows[8]:

1. The most reactive state in aromatic carbonyl compounds is the (n, π^*) state.

2. Lying energetically near the (n, π^*) state is a charge-transfer (CT) state, corresponding to partial transfer of charge from the ring structure to the carbonyl group, and since the transfer of charge is in the opposite direction to that occurring in the (n, π^*) state, the CT triplet state is unreactive toward hydrogen atom abstraction.

3. Abstraction by the triplet state is therefore strongly dependent on which state is the lower in energy, so that small substituent and solvent changes that reverse the relative positions of the two states greatly change the reactivity.

4. Substitution of a strong electron donor group (such as NH_2) in the ring lowers the energy of the CT state in alcohol below that of the (n, π^*) state, making the molecule unreactive even though the solvent contains readily abstractable hydrogen atoms. However, as the polarity of the solvent molecules is decreased (as for example going to liquid paraffins), the energy of the (n, π^*) triplet is lowered relative to the CT state until it again is the lowest triplet state, and reactivity (i.e., hydrogen abstraction) returns in contrast to the usual predicted behavior.

5. Between the (n, π^*) and CT states in reactivity is the (π, π^*) triplet state which, however, lies energetically above the (n, π^*) and CT states in most aromatic carbonyl compounds.

These results clearly show the importance of electronic structure in excited-state chemistry, data on which in many cases are best supplied through direct observation by flash photolysis of the excited states in fluid media.

An interesting application of flash photolysis to excited-state structure is the determination of triplet protonation pK values (acidity constants) whereby the molecule in the flash-pumped excited state serves as the indicator in the "titration." This technique was first used by Jackson and Porter,[112] and has been reviewed by Vander Donckt.[113] Essentially the same procedure is followed as in the ground-state determination of protonation equilibria, and this procedure is based on the principle that the spectra of the acid and base forms are sufficiently separated to allow the independent absorbance measurement of one of the two forms, and that excited-state protonation to equilibrium is very fast compared to the lifetime of the excited state. The absorbance is then measured as a function of pH, and the protonation pK value is the equivalence point where the absorbance has decreased to one-half its limiting value. A "titration" curve for 1-anthroic acid is shown in Fig. 10, and typical values are given in Table III and compared with those for the ground state and first excited singlet state (determined by fluorescence yields).

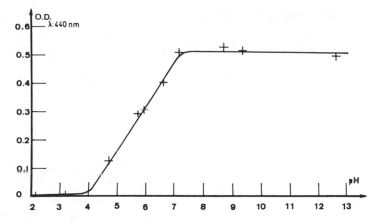

FIG. 10. Titration curve for 1-anthroic acid (optical density of transient vs pH).[113]

The trends suggest the generalization that charge transfer, which contributes to stabilization of the ground state, is even greater in the triplet state and greatest in the first excited singlet state.[34] Assuming that charge-transfer contribution to the various states is primarily a function of the separation between the CT state and the specified state, the separation is thus greatest for the S_1 state and smallest for the S_0 state.

TABLE III

Representative Values of Acidity Constants in the Ground Singlet (S_0), Excited Singlet (S_1), and Excited Triplet (T_1) Electronic States[111]

Compound	pK (S_0)	pK (S_1)	pK (T_1)
2-Naphthol	9.5	3.0	8.1
1-Naphthoic acid	3.7	7.7	3.8
2-Naphthoic acid	4.2	6.6	4.0
1-Anthroic acid	3.7	6.9	5.6
2-Anthroic acid	4.2	6.6	6.0
9-Anthroic acid	3.0	6.5	4.2
Quinoline	4.9	—	6.0
Acridine	5.5	10.6	5.6
2-Naphthylamine	4.1	−2	3.3
N,N-Dimethylaniline	4.9	—	2.7
2-Aminoanthracene	3.4	−4.4	3.3

B. Nanosecond and Picosecond Species

So far we have been concerned with systems directly observable in the microsecond time range possible with the traditional or conventional flash photolysis apparatus. On the other hand, most primary photophysical events, including those involved in photosynthesis, presumably occur in times at least on the order of 10^{-9} second. The direct study by flash photochemical means of species with nanosecond and picosecond lifetimes has become possible with the development of Q-switched and mode-locked pulsed lasers, described in Section II. As a result, it is now feasible to observe directly by absorption spectroscopy the complete primary photochemical mechanism following initial photon absorption and excitation to a specific vibrational level of an excited electronic state, involving radiative and nonradiative relaxation processes as well as formation and reaction of the transient intermediates responsible for the overall photochemistry. The very high light intensity from laser emission also greatly increases the probability of multiphoton absorption and thus photochemistry with long-wavelength radiation not "absorbed" in the traditional sense.[114]

1. Short-Lived Triplet States

Most triplet states of molecules in gas or liquid media have lifetimes greater than a few microseconds because of the partially forbidden character of the excited triplet–ground singlet radiative decay process. In some instances, however (as, for example, when the triplet states are very reactive chemically), the triplet lifetimes are in the submicrosecond region, and some of the first applications of the pulsed laser flash technique have been on systems of this type. Examples are the triplet lifetime studies of phthalocyanine,[5] acridine,[6] and chloranil[115] and other benzoquinone derivatives.[34]

Aromatic carbonyl compounds such as benzophenone and its derivatives contain both conjugated π electrons and nonbonding n electrons, and the photoreactivity of this class of compounds (showing various substituent and solvent effects) as it relates to specific electronic structures has already been discussed (Section IV,A). Again, benzophenone has been one of the molecules most thoroughly studied by laser flash excitation because of its fortuitous absorption at 347 nm, the wavelength of the frequency-doubled ruby laser. Initial conventional flash photolysis studies on benzophenone in liquid paraffins assigned the 1-msecond lifetime transient absorbance in the visible region to the triplet state[90]; subsequent flash experiments have shown, however, that this spectrum is actually that of the ketyl radical formed by H-atom abstraction: the confusion resulted from the close similarity between the spectra of the two species[110, 116] (note Fig. 9). In the presence of competing abstraction reactions with labile hydrogen atom solvents, the lifetime of the lowest triplet [the (n, π^*) state] of benzophenone would be expected to be in the nanosecond range[117]; this has indeed been shown to be the case,[34] a typical value being 46 nseconds in isopropanol as compared to ca. 20 μseconds in the hydrogen-free solvent trichlorotrifluorethane.[110]

Similarly, the triplet and ketyl radical absorption spectra for a series of alkyl phenyl ketones have been observed by laser flash spectrophotometry, with triplet decay and radical formation correlating well with hydrogen-donating efficiencies of the solvents.[118, 119] The radical spectra are very similar to those of the corresponding (n, π^*) triplet (as with the benzophenones) but significantly different from spectra when the (π, π^*) configuration is the lowest triplet state, indicating that triplet absorption spectra may be useful in determining the extent of (n, π^*) and (π, π^*) contribution to the lowest triplet state as influenced by different solvent and substituent environments.[119]

The lowest triplet state is energetically below the first excited singlet state, so that the radiationless intersystem crossing (ISC) from the singlet has to be to a vibrationally excited level of the triplet state, even if relaxation to the lowest vibrational level of the singlet state has occurred. In the liquid or solid phases or in gases at sufficiently high pressures, the pressure dependence of vibrational relaxation within the triplet manifold is very rapid, and inter-system crossing is essentially irreversible unless the singlet and triplet levels are energetically close, in which case there is a temperature-dependent reverse ISC (leading to E-type delayed fluorescence[98]). However, at sufficiently low pressures the molecule can be considered "isolated" (i.e., the effective collision time is large compared to the radiative lifetime, so that only intra-molecular pathways are possible), and reverse ISC should compete with triplet vibrational relaxation, leading to a decrease in the overall intersystem crossing efficiency. This has indeed been shown to be the case for several aromatic vapors, using conventional flash photolysis,[120] the effect becoming measurable for anthracene below 16 Torr and following a reciprocal–reciprocal relation. The detailed reaction scheme[120] required, however, that the specific vibrational level to which the molecules are excited should contri-bute to the overall intersystem crossing efficiency, an effect that could not be measured with the polychromatic light and microsecond decay times inherent in the conventional flash technique. Using a nanosecond frequency-doubled (347.1 nm) Q-switched ruby laser flash apparatus, however, it is possible to excite to a specific singlet vibrational state and hence to form a vibrationally hot triplet state at sufficiently low pressures. Recent results obtained by this method[121] have essentially confirmed the mechanism for the pressure-de-pendent intersystem crossing yield involving reverse ISC; they have further shown directly that vibrational deactivation by collision within the lowest triplet state manifold is a multistep process, but that rather than being a "stepladder" (equal amounts of energy) cascade to the lowest level there is a distribution of collisional probabilities and hence varying amounts of vibra-tional energy removed per collision.[121]

2. Excited Singlet States

The lifetimes of the lowest excited singlet states of most molecules are too short to be detected directly by conventional flash spectrography or spectrophotometry, and energetics of higher excited singlet states are in general not known because of the symmetry-forbidden nature of these transitions from the ground state.[34] Much of the work over the past several years since introduction of the Q-switched pulsed laser to flash photochemistry has therefore involved measuring the optical absorption spectra and kinetic behavior of these species for a variety of compounds,[122] results of which have been useful in determining efficiencies of the various radiative and nonradiative relaxation modes and in elucidating theories of excited-state energy levels. As with triplet–triplet absorption, these transitions between excited singlet states lead to spectra that are quite diffuse even under very high resolution, and therefore complete rotational analyses of the spectra are not possible as would be required for reasonable determinations of configurational properties of the higher excited electronic states.

Singlet–singlet absorption bands in the polynuclear aromatic hydrocarbons were among the first to be detected by laser flash excitation.[8, 10, 123, 124] Typical in behavior is coronene,[10] which has a relatively long fluorescence lifetime; in acetonitrile (10^{-3} M) at room temperature, excited singlet–singlet absorption occurs in a strong broad band centered at 520 nm and a weaker band at 380 nm, in contrast to triplet–triplet bands centered at 390, 480, and 640 nm. The 520 band and fluorescence both decay at the same first-order rate ($\tau \simeq 300$ nseconds), which is also the rate of "growing in" of the 480 and 390 nm triplet bands by intersystem crossing.[10]

The quenching of excited triplet states by ground-state triplet molecular oxygen has been discussed in Section IV,A in terms of a one-ninth spin statistical factor for triplet–triplet annihilation, although the encounter rate was considered to be essentially diffusion controlled. Such a factor should not be important for the analogous quenching of the lowest excited singlet state, since singlet quenching can occur even with no concomitant electronic excitation of the molecular oxygen[125] (although singlet oxygen will still be produced in most cases if energetically allowed), and therefore should occur with unit efficiency in a true diffusion-controlled process. The decay in the presence of O_2 of the excited singlet state for nine aromatic hydrocarbons has indeed been shown to be diffusion controlled,[126] with $k_q \simeq 2.7 \times 10^{10}$ liter/mole second (cyclohexane solvent), in agreement with Eq. (10). It is apparent that this highly efficient quenching process for singlet states (and, to a lesser extent, for triplets) can lead to the very reactive singlet molecular oxygen in a variety of environmental and biological photosystems.

As the Q-switched laser technology extends the time range of transient detection by approximately three orders of magnitude over the conventional flash technique and thereby opens a new area of transient study (such as detection and characterization of singlet states, described above), so also does the application of picosecond spectroscopy with the mode-locked flash laser lead to the direct experimental determination of molecular relaxation and radiationless transition processes in excited singlet states. Progress in this rapidly developing field has been reviewed to 1973 by Rentzepis.[26]

Two molecules representing quite different degrees of coupling of excited states have been extensively studied by this picosecond technique. Azulene, the first to be studied,[127-129] is an example of a molecule in which the excited singlet state is strongly coupled to a densely populated (essentially a "quasi continuum") vibrational manifold of another state—in this case the ground singlet state where the energy gap between the two states is large (see Fig. 11). The statistical limit for this situation then predicts similar and very rapid radiationless decay rates in solution and for an isolated molecule[130]; in this sense it displays anomalous fluorescence behavior in that it fluoresces only very weakly from the first singlet state,[128] the major fluorescence being from the second excited singlet.[127] Azulene was also a convenient choice because the transition to the lowest vibrational level of the first excited singlet state corresponds to the fundamental emission from the ruby laser (694 nm), while

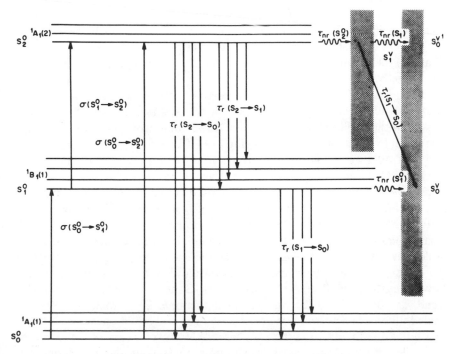

FIG. 11. Energy level diagram for azulene.[129]

excitation to a higher ($v \simeq 5$) vibrational level is possible with a frequency-doubled neodymium laser (530 nm) (both lasers in mode-locked operation). With either method of excitation relaxation was radiationless (fluorescence quantum yield ca. 10^{-6}) to the highly vibrationally excited ground singlet state (internal conversion) rather than to the lowest triplet level by intersystem crossing; this decay was very fast (ca. 7 pseconds lifetime), in good agreement with the behavior expected on the statistical limit model for coupling to a continuum.[129]

Benzophenone is an example of a molecule in which the energy gap between the excited singlet state and some lower state (in this case the lowest triplet state) is relatively small (about 2000 cm^{-1}), and therefore is intermediate between coupling to a densely populated state and to a sparse number of levels.[26] It also is a convenient molecule for studies with available mode-locked lasers, and radiationless decays have been determined[131] from two different vibronic levels of the lowest excited singlet state ($v = 2$ and $v = 0$ levels of the C—O mode) using the ruby frequency-doubled (347 nm) and stimulated Raman red-shifted (ca. 400 nm) laser lines. Unlike azulene, the lowest excited singlet state of benzophenone is strongly coupled to the "intermediately dense" triplet manifold,[131] and therefore electronic relaxation should not occur in the gas phase. In solution, however, collisional pressure broadening of the triplet vibrational bands and strong spin–orbit coupling will greatly enhance the intersystem crossing probability, and, experimentally, phosphorescence is much greater than fluorescence in the liquid phase in the absence of triplet intermolecular quenching, in contrast to the gas-phase behavior.

Time- and frequency-resolved picosecond spectra of benzophenone in benzene solution confirm this coupling model, and further show that relaxation from the higher vibrational level is slower than from the lower level, the first-order exponential decay rates being 2×10^{11} seconds^{-1} ($\tau = 5$ pseconds) and 5×10^{10} ($\tau = 20$ pseconds) respectively from the $v = 0$ and $v = 2$ levels.[131] This trend is opposite to the expected behavior simply on the basis of relative densities of states, which would predict that the intersystem crossing rate should increase with increasing vibrational energy within the singlet state manifold, but the trend is in the direction expected for radiative decay in an isolated molecule.[132] Theoretical calculations have shown[133] that for molecules with large energy gaps between the excited singlet and triplet states the crossing rate is determined by the relative density of vibrational levels, but that as the gap is decreased other Franck-Condon factors may reverse this trend, as apparently is the case of benzophenone. Confirmation of the expected difference between decay rates in the gas and liquid phases was shown by similar measurements in the gas phase.[134] However, decay of the "isolated molecule" in the excited singlet state is anomalously long and nonexponential (the strong coupling component having a lifetime of approximately 10

μseconds, greater than the 1-μsecond lifetime predicted solely from the measured oscillator strength), from which it is inferred that strong coupling exists between a singlet-state vibrational level and ca. ten vibrational levels with the triplet manifold.[134]

The above examples serve to indicate the types of detailed electronic structural parameters and modes of radiative and nonradiative relaxation directly determinable from the various available flash photoexcitation techniques described here, and show the experimental feasibility of complete characterization of the various processes and photochemical dynamics following the initial act of photoexcitation.

Until recently no clear-cut data were available showing true radiationless internal conversion between the lowest excited singlet state and the ground state (without chemical change).[8] That this type of internal conversion can occur even in a dense-media environment has now been established by indirect means for several systems such as the cyanine dyes,[135] benzene,[136] and other simple aromatic molecules,[137] as well as by the direct determinations described above. Recent calculations have shown[138] in fact that this pathway for radiationless decay of the lowest excited singlet state may become increasingly important for high vibrational levels, in agreement with the observed activation energy for internal conversion.[136, 137] Further picosecond flash studies will aid in the understanding of the coupling mechanisms leading to the stabilities of the various electronic states.

V. SOME APPLICATIONS TO PHOTOBIOLOGICAL SYSTEMS

Solar radiation is essential for life on earth, and therefore it is to be expected that research on photosensitive biological systems constitutes a large proportion of activity in current photochemical studies. It was soon recognized after the development of the flash photolysis method that this technique had wide applicability to photobiology. Extensive reviews have been written in recent years,[139, 140] and brief reviews are in the current literature (e.g., *Important Events in Photosynthesis*[141] and *Excited States of Biomolecules*[142]). It is the purpose of this section to summarize two of the research areas that have been active over the past several years in utilizing the newer flash techniques, particularly as these studies lead to new models and clarification of existing concepts of electronic structure. (These two are by no means the only ones; conventional and laser flash photolysis studies involving systems such as quinones, flavins, and carotenoids are contributing to our understanding of primary processes and mechanisms of photobiological reactions. Recent flash kinetic studies on the carbon monoxide-stabilized protoheme–imidazole complex portion of hemoglobin are also aiding in the clarification of oxygen binding structure and transport under physiological conditions.[143])

Most biological systems of course involve large molecules, so that optical absorption spectra are in general broad, structureless bands and therefore do not contribute significantly to molecular structure determinations; other analytical techniques, in particular electron spin resonance,[144] have been useful in identifying the nature of photobiological intermediates, and have contributed in some instances to the elucidation of specific molecular configurations.

A. Chlorophylls

The importance of chlorophyll in photosynthesis is of course well recognized, although its full role in light energy absorption, storage, and transfer *in vivo* is still not completely understood. It is not surprising that it was one of the early species studied by conventional flash photolysis. The visible absorption spectrum of chlorophyll a or chlorophyll b consists of two prominent bands: a long-wavelength (640–700 nm) band corresponding in polar solvents to (π, π^*) excitation to the lowest excited singlet state, and the Soret band (400–600 nm) resulting from excitation to a higher singlet. The absorbing chromophore is the conjugated tetrapyrrole ring complexed to a magnesium atom. In 1953, Livingston[145] reported flash spectrophotometric experiments (using a bank of three photographer's flash lamps fired simultaneously) on dilute solutions of chlorophyll. With chlorophyll b (and, to a lesser extent, with chlorophyll a) partial transient bleaching of the solution occurred in the visible regions of ground-state absorption ($t_{1/2} \cong 0.5$ m-second), and a new transient absorbance appeared at 524.5 nm which was assigned to the lowest triplet state. In addition, a longer-lived radical was formed by reaction with the solvent (methanol) and this is the species assumed responsible for the steady-state reversible bleaching observed in continuously illuminated systems. These results were confirmed in subsequent more sophisticated flash studies in several laboratories[146–148] using other solvents, and showed that decay of the triplet in a solvent such as benzene consists of both first- and second-order kinetics, consistent with a mechanism[146] involving self-triplet–triplet annihilation, a radiationless first-order decay, and self-quenching by a parent molecule (the last two processes possibly impurity quenching).

Although (π, π^*) triplet–triplet absorption is well established in dilute solutions of chlorophyll in many different solvents, the primary photosynthetic process in normal plants takes place in the highly organized chloroplast involving many chlorophyll molecules, and the triplet–triplet spectrum is not observed under these conditions. In a detailed series of experiments summarized in 1967, Witt[149] has studied transient intermediates from bulk chlorophyll *in vivo*—disarranged however by separating the molecules with cationic detergents or by removing chlorophyll through mutation or through nitrate deficiency—by conventional flash and (repetitive) laser flash excitation, covering a time range from 10^{-1} to 10^{-8} second.[150] The (π, π^*) triplet of chlorophyll with a spectrum as that in water is then observed under these conditions (using conventional flash), the difference apparently being that the energy migration in the chloroplasts is interrupted by the disarrangement and therefore excitation energy remains as the (π, π^*) triplet. Similarly, using algae containing chlorophyll a, a transient intermediate ($\lambda_{max} = 520$ nm, $t_{1/2} = 3 \times 10^{-6}$ second as determined by the repetitive flash method) was observed and tentatively assigned as the (n, π^*) triplet state of chlorophyll a and also as the precursor to the one-electron transfer reaction (the primary act of photosynthesis) that converts light energy into electrical energy in the photosynthetic unit.[149]

Porter and Strauss also observed triplet–triplet absorption for chlorophyll a and b similar to that detected in fluid solutions in a number of dilute solid solvents and under aggregate (high concentration) conditions more analogous to actual photosynthesis conditions.[151, 152] A special microbeam flash apparatus was developed in which the light from a conventional flash lamp was focussed onto a microscopic slide containing the sample encompassed in the solid solvent matrix and positioned on a microscope stage, with this technique, samples of from five to several hundred μm thick could be flashed and spectrophotometrically analyzed. Under chloroplast (high concentration) conditions, the triplet state yield was very low (limited by self-quenching and energy transfer from chlorophyll b to chloropyll a) and therefore incapable of contributing significantly to the overall energy accumulation; a more likely mechanism involving singlet energy transfer was proposed for light energy collection.[152] It is clear, however, that observations on triplet yields and energy transfer mechanisms are very sensitive to the physical structure of the chlorophyll, and very careful flash experiments on well-oriented structures are now needed to clarify the energy interactions of chlorophyll molecules in natural photosynthetic environments.

Light-induced one-electron transfer reactions in chlorophyll lead to free radical intermediates detectable by electron spin resonance (ESR). However, accurate comparison with optical changes is virtually impossible because of uncontrollable biological variability under varying experimental conditions. To circumvent this limitation, an apparatus has been constructed for simultaneous optical and ESR detection following dye laser flash excitation.[153] The result for spinach chloroplasts are shown in Fig. 12, where the kinetic behavior of optical bleaching at 703 nm is identical to that for the ESR signal. This correlation clearly suggests that the two signals are associated with the same species—probably the π-cation radical of chlorophyll a.[153, 154]

Photosynthetic bacteria contain a pigment (bacteriochlorophyll) that also serves as a light energy collector in a manner similar to chlorophyll in green plants, although oxygen is not evolved in the process. In an experiment

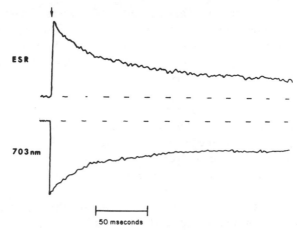

FIG. 12. Simultaneous ESR and optical absorption (703 nm) delay kinetics for spinach subchloroplast particles. Reprinted with permission from J. T. Warden and J. R. Bolton, *J. Amer. Chem. Soc.* **94**, 4351 (1972). Copyright by the American Chemical Society.

described as the first application of picosecond flash spectroscopy to a biologically important molecule, Netzel, Rentzepis, and Leigh[155] have measured the rate of bleaching of the 865 nm absorption band of bacteriochlorophyll in the protein *Rhodopseudomonas spheroides* strain R-26. Photoexcitation was in the 535 nm pheophytin band using the frequency-doubled neodymium:glass (530 nm) laser, with the continuum interrogating light produced by Stokes and anti-Stokes Raman scattering from water. Bleaching of the 865 nm band occurred within 7 ± 2 pseconds, indicating that photooxidation of the reaction center requires a finite (few picoseconds) time to transfer the energy to the 865 nm band; these results do indicate, however, that strong interactions exist between the bacteriochlorophyll and the bacteriopheophytins.

B. Retinenes and Rhodopsin

The major pigment responsible for vision in most vertebrates is rhodopsin ($\lambda_{max} = 498$ nm), which is a compound formed between retinal and the protein opsin. The photoactive isomer of retinal is the 11-cis form; it is converted to all-trans in the visual cycle and back to 11-cis either by light or by an enzyme (isomerase). The all-trans form is unstable in the bound form with opsin, and therefore detaches during the cycle. A transient species in the process is prelumirhodopsin ($\lambda_{max} = 543$ nm), which is the all-*trans*-opsin complex.

Initial flash studies on free all-*trans*-retinal (using conventional flash spectrophotometry) gave bleaching at 370 nm (λ_{max} for retinal) and transient absorption at 450 nm, with return to the initial ground state in ca. 10 μseconds (methylcyclohexane solvent).[156] Subsequent work[157] confirmed this absorption band and assigned it to the lowest (π, π^*) triplet; initial excitation is to the (π, π^*) singlet, which then rapidly goes by internal conversion to the (n, π^*) singlet followed by intersystem crossing to the observed triplet (ca. 11% efficiency). On the other hand, a recent paper[158] describes results using a frequency-quadrupled (265 nm) neodymium laser flash spectrophotometer, which confirmed the triplet character of the 450 nm band but led to an intersystem crossing quantum efficiency of 0.6 ± 0.1. [It was pointed out[158] that this higher ISC quantum yield is more compatible with the quantum efficiency of rhodopsin photobleaching (0.67)[159] and suggests a more important role for the triplet state of the retinal in the cis-trans photoisomerization process and hence in the vision cycle.] The higher quantum yield is confirmed in another report of additional flash laser studies[160] (with a pulsed nitrogen laser at 337.1 nm) on all-*trans*-retinal (0.7 ± 0.1) and on 11-*cis*-retinal (0.6 ± 0.1).

Although the triplet state of all-*trans*-retinol (vitamin A) can be produced by photosensitization or by pulse radiolysis, until recently it had not been observed by direct excitation and intersystem crossing. This has been cited in support of the rapid intersystem crossing in retinal, occurring from the (n, π^*) singlet state, which cannot exist in retinol, the $^1(\pi, \pi^*) \rightarrow {}^3(\pi, \pi^*)$ being a much slower process[92] than $^1(n, \pi^*) \rightarrow {}^3(\pi, \pi^*)$. The triplet state has now been produced directly by laser flash photoexcitation to the singlet state (using the 337.1 nm line from a pulsed nitrogen laser),[161] with $\lambda_{max} = 405$ nm. In addition, two other transient species are observed spectrally: a short-lived $(\tau < 10$ nseconds) transient absorbing at 435 nm, and assigned to the lowest excited singlet state; and in polar solvents a species absorbing at approximately 590 nm that decays in microseconds. This 590 nm species was assumed to be the retinylic cation produced by ionic photodissociation

$$ROH \xrightarrow{h\nu} R^+ + OH^-$$

Intersystem crossing is enhanced by dissolved O_2, presumably through charge-transfer interactions (which also could exist between retinol and acceptor or donor sites on a protein). These results further suggest[161] that the triplet state and possibly ionic photodissociation may be important pathways in the cis-trans photoisomerization and hence in the visual process, although as with photosynthesis the process is undoubtedly strongly dependent on the physical environment, which is quite different for retinal or retinol in dilute solution and in rhodopsin under physiological conditions.

In a further extension of the picosecond time-resolved flash spectroscopy technique, the transient species produced from bovine rhodopsin by flash excitation with a mode-locked frequency-doubled neodymium laser have been studied.[162] The only species observed is prelumirhodopsin, which absorbs at 561 nm and is produced within the lifetime of the flash (< 6 pseconds); its decay is exponential ($k = 3.7 \times 10^7$ seconds^{-1} at 22.5°) with a 25 kJ/mole activation energy. The very rapid rise-time suggests that decay of the electronically excited rhodopsin is directly to prelumirhodopsin, and hence this is the primary photochemical event. Results obtained by varying the relative polarizations of the exciting and monitoring laser pulses showed that no geometrical changes occurred on excitation that were great enough to destroy the linear nature of the retinene chromophore. The rate of formation of prelumirhodopsin ($k > 1.6 \times 10^{11}$ seconds^{-1}) is comparable to that for electronically excited azulene going to its ground state by internal conversion and benzophenone going to its triplet from the $v = 0$ level of the excited singlet state (see Section IV,2); both of these processes involve strong vibrational coupling between states, and therefore a comparable process (i.e., direct electronic relaxation) may be involved in the rhodopsin → prelumirhodopsin reaction.

VI. SUMMARY

The flash photolysis technique as developed over the past two decades has become a powerful tool for the direct study of transient and unstable intermediates in kinetically complex reacting systems. One of the more important and highly successful aspects of these studies, based on the time resolution of the technique (to 10^{-12} second), is the determination of the various kinetic parameters associated with the detailed reaction mechanisms, but this has not been an appropriate subject for this presentation.

The spectra of many gas-phase free radicals have been measured by conventional flash spectrography under conditions of very high resolution, so that accurate evaluations of the ground- and excited-state structural configurations are in principle possible through complete analyses of the rotational bands; however, just as with stable polyatomic molecules, the complexity of these spectra increases with increasing number of atoms, and so far thorough treatments have been limited to relatively simple free radical species. Similarly, the optical absorption spectra of a very large number of excited triplet states with microsecond or longer lifetimes have been measured in both gas and liquid phases, but in general they consist of broad, diffuse bands even under high resolution, which preclude accurate physical structure analyses; probably the conventional flash spectrophotometric method of detection is just as accurate spectrally in most cases and is much more convenient and accurate for kinetic analysis. Extension to times as short as 10^{-12} second is now possible with flash Q-switched and mode-locked lasers for the direct time-resolved study of vibrationally and electronically excited states formed in the initial or primary act of light absorption (usually a singlet state), so that a complete picture of radiative and nonradiative relaxation processes is being developed on the basis of specific electronic and vibrational structural parameters of molecules.

As pointed out by Professors Eigen[163] and Porter[164] in the Summarizing Lectures of the 1967 Fifth Nobel Symposium on Fast Reactions and Primary Processes in Chemical Kinetics, one begins to experience diminishing returns as the time boundary of detection is pushed back, and it is doubtful that extension beyond the picosecond region will contribute significantly to organic structure and reaction mechanism determinations. The combination of more sophisticated analytical means with conventional and laser flash photoexcitation—optical rotatory dispersion, X-ray diffraction, nuclear magnetic resonance, etc.—will undoubtedly lead to more complete molecular structure determinations in the future; these combinations are necessary extensions of the flash technique, but so far the complexities and time-limited sensitivities of such assemblies greatly restrict their applicabilities to the study of very short-lived intermediate species.

References

1 R. G. W. Norrish and G. Porter, *Nature (London)* **164**, 658 (1949).
2 G. Porter, *Proc. Roy. Soc., Ser. A* **200**, 284 (1950).
3 G. Herzberg and D. A. Ramsay, *Discuss. Faraday Soc.* **9**, 80 (1950).
4 N. Davidson, R. Marshall, A. E. Larsh, and T. Carrington, *J. Chem. Phys.* **19**, 1311 (1951).
5 W. F. Kosonocky, S. E. Harrison, and R. Stander, *J. Chem. Phys.* **43**, 831 (1965).
6 L. Lindqvist, *C. R. Acad. Sci., Ser. C* **263**, 852 (1966).
7 D. de Vault and B. Chance, *Biophys. J.* **6**, 825 (1966).
8 G. Porter, in "Fast Reactions and Primary Processes in Chemical Kinetics" (S. Claesson, ed.), p. 141. Almqvist & Wiksell, Stockholm, 1967.
9 R. M. Danziger, K. H. Bar-Eli, and K. Weiss, *J. Phys. Chem.* **71**, 2633 (1967).
10 J. R. Novak and M. W. Windsor, *J. Chem. Phys.* **47**, 3075 (1967).
11 P. M. Rentzepis, *Photochem. Photobiol.* **8**, 579 (1968).
12 G. Porter, in "Investigations of Rates and Mechanisms of Reactions" (G. D. Hammes, ed.), 3rd ed., Part II, p. 367. Wiley, New York, 1974.
13 See, for example, H. Eyring, ed., *Annu. Rev. Phys. Chem.* Annual Reviews, Inc., Palo Alto, California; D. Bryce-Smith, Senior Reporter, *Photochemistry.* Chemical Society, London.
14 F. W. Willets, *Progr. React. Kinet.* **6**, 51 (1971).
15 W. T. Ham, Jr., H. A. Mueller, A. I. Goldman, B. E. Newman, L. M. Holland, and T. Kuwabara, *Science* **185**, 362 (1974).
16 S. Claesson, L. Lindqvist, and R. L. Strong, *Ark. Kemi* **22**, 245 (1964).
17 L. Lindqvist, *Rev. Sci. Instrum.* **35**, 993 (1964).
18 M. Vallotton and U. P. Wild, *J. Phys. E.* **4**, 417 (1971).
19 T. A. Gover and G. Porter, *Proc. Roy. Soc., Ser. A* **262**, 476 (1961).
20 R. L. Strong and J. E. Willard, *J. Amer. Chem. Soc.* **79**, 2098 (1957).
21 S. Claesson and G. Wettermark, *Ark. Kemi* **11**, 561 (1957).
22 J. C. LeBlanc, M. A. Herbert, D. W. Willans, and H. E. Johns, *Rev. Sci. Instrum.* **43**, 1814 (1972).
23 E. E. Daby, J. S. Hitt, and G. J. Mains, *J. Phys. Chem.* **74**, 4204 (1970).

[24] F. J. McClung and R. W. Hellworth, *J. Appl. Phys.* **33**, 828 (1962).

[25] A. J. DeMaria, D. A. Stetser, and W. H. Glenn, Jr., *Science* **156**, 1557 (1967).

[26] P. M. Rentzepis, *Advan. Chem. Phys.* **23**, 189 (1973); T. L. Netzel, W. S. Struve, and P. M. Rentzepis, *Annu. Rev. Phys. Chem.* **24**, 473 (1973).

[27] M. M. Malley, *in* "Creation and Detection of the Excited State" (W. R. Ware, ed.), Vol. 2, p. 99. Dekker, New York, 1974.

[28] R. W. Wanant, J. D. Shipman, R. C. Elton, and A. W. Ali, *Appl. Phys. Lett.* **17**, 383 (1970).

[29] A. Dienes, C. V. Shank, and A. M. Trozzolo, *in* "Creation and Detection of the Excited State" (W. R. Ware, ed.), Vol. 2, p. 149. Dekker, New York, 1974.

[30] S. Claesson and L. Lindqvist, *Ark. Kemi* **11**, 535 (1957).

[31] H. C. Christensen, G. Nilsson, P. Pagsberg, and S. O. Nielsen, *Rev. Sci. Instrum.* **40**, 786 (1969).

[32] L. Lindqvist, *Ark. Kemi* **16**, 79 (1961).

[33] R. W. Yip, *Rev. Sci. Instrum.* **40**, 1035 (1969).

[34] G. Porter and M. R. Topp, *Proc. Roy. Soc., Ser. A* **315**, 163 (1970).

[35] D. A. Ramsay, *in* "Determination of Organic Structures by Physical Methods" (F. C. Nachod and W. D. Phillips, ed.), Vol. 2, p. 245. Academic Press, New York, 1961.

[36] J. U. White, *J. Opt. Soc. Amer.* **32**, 285 (1942).

[37] G. H. Atkinson, A. H. Laufer, and M. J. Kurylo, *J. Chem. Phys.* **59**, 350 (1973).

[38] P. Crabbé, *in* "Determination of Organic Structures by Physical Methods" (F. C. Nachod and J. J. Zuckerman, ed.), Vol. 3, p. 133. Academic Press, New York, 1971.

[39] P. A. Carapellucci, H. H. Richtol, and R. L. Strong, *J. Amer. Chem. Soc.* **89**, 1742 (1967); R. L. Strong and H. H. Richtol, *in* "Fast Reactions and Primary Processes in Chemical Kinetics" (S. Claesson, ed.), p. 71. Almqvist & Wiksell, Stockholm, 1967.

[40] B. R. Jennings and E. D. Baily, *Nature (London)* **228**, 1309 (1970).

[41] C. A. Emeis and L. J. Oosterhoff, *Chem. Phys. Lett.* **1**, 129 (1967).

[42] A. Gafni and I. Z. Steinberg, *Photochem. Photobiol.* **15**, 93 (1972).

[43] W. R. Ware, *in* "Creation and Detection of the Excited State" (A. A. Lamola, ed.), Vol. 1, Part A, p. 213. Dekker, New York, 1971.

[44] L. M. Bollinger and G. E. Thomas, *Rev. Sci. Instrum.* **32**, 1044 (1961).

[45] W. Braun and M. Lenzi, *Discuss. Faraday Soc.* **44**, 252 (1967).

[46] S. P. Perone and J. R. Birk, *Anal. Chem.* **38**, 1589 (1966).

[47] T. J. Bennett, R. C. Smith, and T. H. Wilmshurst, *Chem. Commun.* 513 (1967).

[48] J. R. Bolton and J. T. Warden, *in* "Creation and Detection of the Excited State" (W. R. Ware, ed.), Vol. 2, p. 63. Dekker, New York, 1974.

[49] G. B. Kistiakowsky and P. H. Kydd, *J. Amer. Chem. Soc.* **79**, 4825 (1957).

[50] G. Porter, *Discuss. Faraday Soc.* **9**, 60 (1950).

[51] D. A. Ramsay, *J. Chem. Phys.* **20**, 1920 (1952).

[52] D. A. Ramsay, *J. Chem. Phys.* **25**, 188 (1956).

[53] G. Herzberg and D. A. Ramsay, *Proc. Roy. Soc., Ser. A* **233**, 34 (1955).

[54] R. Holland, D. W. G. Style, R. N. Dixon, and D. A. Ramsay, *Nature (London)* **182**, 336 (1958).

[55] D. A. Ramsay, *Nature (London)* **178**, 374 (1956).

[56] F. W. Dalby, *Can. J. Phys.* **36**, 1336 (1958).

[57] B. A. Thrush, *in* "Photochemistry and Reaction Kinetics" (P. G. Ashmore, F. S. Dainton, and T. M. Sugden, eds.), p. 112. Cambridge Univ. Press, London and New York, 1967.

[58] A. D. Walsh, *J. Chem. Soc., London* 2260 (1953).

[59] R. G. Pearson, *Chem. Phys. Lett.* **10**, 31 (1971).

[60] R. N. Dixon, G. Duxbury, and D. A. Ramsay, *Proc. Roy. Soc., Ser. A* **296**, 137 (1967).

[61] J. M. Berthou, B. Pascat, H. Gruenebaut, and D. A. Ramsay, *Can. J. Phys.* **50**, 2265 (1972).

[62] H. M. Frey, *Progr. React. Kinet.* **2**, 131 (1964).

[63] G. Herzberg and J. Shoosmith, *Nature (London)* **183**, 1801 (1959).

[64] G. Herzberg, *Proc. Roy. Soc., Ser. A* **262**, 291 (1961).

[65] J. M. Foster and S. F. Boys, *Rev. Mod. Phys.* **32**, 305 (1960).

[66] J. F. Harrison and L. C. Allen, *J. Amer. Chem. Soc.* **91**, 807 (1969).

[67] C. F. Bender and H. F. Shaeffer, III, *J. Amer. Chem. Soc.* **92**, 4984 (1970).

[68] F. A. Berheim, H. W. Bernard, P. S. Wang, L. S. Wood, and P. S. Skell, *J. Chem. Phys.* **53**, 1280 (1970).

[69] E. Wasserman, W. A. Yager, and V. J. Kuck, *Chem. Phys. Lett.* **7**, 409 (1970).

[70] G. Herzberg and J. W. C. Johns, *J. Chem. Phys.* **54**, 2276 (1971).

[71] G. Herzberg and J. W. C. Johns, *Proc. Roy. Soc., Ser. A* **295**, 107 (1966).

[72] G. Herzberg and D. A. Ramsay, *Discuss. Faraday Soc.* **14**, 11 (1953).

[73] G. Herzberg and J. Shoosmith, *Can. J. Phys.* **34**, 523 (1956).

[74] G. Herzberg and P. A. Warsop, *Can. J. Phys.* **41**, 286 (1963).

[75] D. A. Ramsay and P. Thistlethwaite, *Can. J. Phys.* **44**, 1381 (1966).

[76] G. Porter and F. J. Wright, *Trans. Faraday Soc.* **51**, 1469 (1955).

[77] E. J. Land and G. Porter, *Trans. Faraday Soc.* **57**, 1885 (1961).

[78] G. Porter, *Science* **160**, 1299 (1968).

[79] D. R. Kemp and G. Porter, *Proc. Roy. Soc., Ser. A* **326**, 117 (1971).

[80] B. A. Thrush, *Nature (London)* **178**, 155 (1956).

[81] B. A. Thrush and J. J. Zwolenik, *Discuss. Faraday Soc.* **35**, 196 (1963).

[82] G. Porter and B. Ward, *Proc. Roy. Soc., Ser. A* **287**, 457 (1965).

[83] G. H. Williams, "Homolytic Aromatic Substitution." Pergamon, Oxford, 1960.

[84] P. W. Atkins, K. A. McLaughlin, and A. F. Simpson, *J. Phys. E.* **3**, 547 (1970).

[85] E. J. Land and G. Porter, *Trans. Faraday Soc.* **59**, 2016 (1963).

[86] E. J. Land, *Progr. React. Kinet.* **3**, 369 (1965).

[87] G. N. Lewis and M. Kasha, *J. Amer. Chem. Soc.* **67**, 994 (1945).

[88] G. Porter and M. W. Windsor, *J. Chem. Phys.* **21**, 2088 (1953); *Discuss. Faraday Soc.* **17**, 178 (1954).

[89] G. Porter and F. J. Wright, *Trans. Faraday Soc.* **51**, 1205 (1955).

[90] G. Porter and M. W. Windsor, *Proc. Roy. Soc., Ser. A* **245**, 238 (1958).

[91] G. G. Giachino and D. R. Kearns, *J. Chem. Phys.* **52**, 2964 (1970).

[92] M. A. El-Sayed, *Accounts Chem. Res.* **1**, 8 (1968).

[93] M. A. El-Sayed, *J. Chem. Phys.* **38**, 2834 (1963).

[94] H. Dewey and S. G. Hadley, *Chem. Phys. Lett.* **12**, 57 (1971).

[95] J. W. Halpern, G. Porter, and L. J. Stief, *Proc. Roy. Soc., Ser. A* **277**, 437 (1964).

[96] G. Jackson, R. Livingston, and A. C. Pugh, *Trans. Faraday Soc.* **56**, 1635 (1960).

[97] M. Z. Hoffman and G. Porter, *Proc. Roy. Soc., Ser. A* **268**, 46 (1962).

[98] C. A. Parker, *Advan. Photochem.* **2**, 305 (1964).

[99] P. Debye, *Trans. Electrochem. Soc.* **82**, 265 (1942).

[100] A. D. Osborne and G. Porter, *Proc. Roy. Soc., Ser. A* **284**, 9 (1965).

[101] P. J. Wagner and I. Kochevar, *J. Amer. Chem. Soc.* **90**, 2232 (1968).

[102] T. Takemura, H. Baba, and M. Fujita, *Bull. Chem. Soc. Jap.* **46**, 2625 (1973).

[103] K. Sandros and H. L. Bäckström, *Acta Chem. Scand.* **16**, 958 (1962); **18**, 2355 (1964).

[104] S. Nordin and R. L. Strong, *Chem. Phys. Lett.* **2**, 429 (1968).

[105] O. L. J. Gijzeman, F. Kaufman, and G. Porter, *J. Chem. Soc., Faraday Trans. 2* **69**, 708 (1973).

[106] R. Benson and N. E. Geacintov, *J. Chem. Phys.* **59**, 4428 (1973).

[107] P. M. Merkel and D. R. Kearns, *J. Chem. Phys.* **58**, 398 (1973).

[108] O. L. J. Gijzeman and F. Kaufman, *J. Chem. Soc., Faraday Trans. 2* **69**, 721 (1973).

[109] G. Porter and F. Wilkinson, *Trans. Faraday Soc.* **57**, 1686 (1961).

[110] H. Tsubomura, N. Yamamoto, and S. Tanaka, *Chem. Phys. Lett.* **1**, 309 (1967).

[111] G. Porter and P. Suppan, *Trans. Faraday Soc.* **61**, 1664 (1965).

[112] G. Jackson and G. Porter, *Proc. Roy. Soc., Ser. A* **260**, 13 (1961).

[113] E. Vander Donckt, *Prog. React. Kinet.* **5**, 273 (1970).

[114] M. Göppert-Mayer, *Ann. Phys. (Leipzig)* [5] **9**, 273 (1931).

[115] D. R. Kemp and G. Porter, *J. Chem. Soc., D* 1029 (1969).

[116] J. A. Bell and H. Linschitz, *J. Amer. Chem. Soc.* **85**, 528 (1963).

[117] A. Beckett and G. Porter, *Trans. Faraday Soc.* **59**, 2038 (1963).

[118] H. Lutz and L. Lindqvist, *J. Chem. Soc., D* 493 (1971).

[119] H. Lutz, E. Breheret, and L. Lindqvist, *J. Phys. Chem.* **77**, 1758 (1973).

[120] C. W. Ashpole, S. J. Formosinho, and G. Porter, *Proc. Roy. Soc., Ser. A* **323**, 11 (1971).

[121] S. J. Formosinho, G. Porter, and M. A. West, *Proc. Roy. Soc., Ser. A* **333**, 289 (1973).

[122] See, for example, C. B. Moore, *Annu. Rev. Phys. Chem.* **22**, 387 (1971).

[123] R. Bonneau, J. Faure, and J. Joussot-Dubien, *Chem. Phys. Lett.* **2**, 65 (1968).

[124] G. Porter and M. R. Topp, *Nature (London)* **220**, 1228 (1968).

[125] C. S. Parmenter and J. D. Rau, *J. Chem. Phys.* **51**, 2242 (1969).

[126] L. K. Patterson, G. Porter, and M. R. Topp, *Chem. Phys. Lett.* **7**, 612 (1970).

[127] P. M. Rentzepis, *Chem. Phys. Lett.* **2**, 117 (1968).

[128] P. M. Rentzepis, *Chem. Phys. Lett.* **3**, 717 (1969).

[129] D. Huppert, J. Jortner, and P. M. Rentzepis, *J. Chem. Phys.* **56**, 4826 (1972).

[130] M. Bixon and J. Jortner, *J. Chem. Phys.* **48**, 715 (1968).

[131] P. M. Rentzepis and G. E. Busch, *Mol. Photochem.* **4**, 353 (1972).

[132] G. R. Fleming, O. L. J. Gijzeman, and S. H. Lin, *Chem. Phys. Lett.* **21**, 527 (1973).

[133] A. Nitzan, J. Jortner, and P. M. Rentzepis, *Chem. Phys. Lett.* **8**, 445 (1971).

[134] G. E. Busch, P. M. Rentzepis, and J. Jortner, *J. Chem. Phys.* **56**, 361 (1972).

[135] A. V. Beuttner, *J. Chem. Phys.* **46**, 1398 (1967).

[136] R. B. Cundall and D. A. Robinson, *J. Chem. Soc., Faraday Trans. 2* **68**, 1145 (1972).

[137] R. B. Cundall and L. C. Pereira, *J. Chem. Soc., Faraday Trans. 2* **68**, 1152 (1972).

[138] G. S. Beddard, G. R. Fleming, O. L. J. Gijzeman, and G. Porter, *Chem. Phys. Lett.* **18**, 481 (1973).

[139] L. I. Grossweiner, *Advan. Radiat. Biol.* **2**, 83 (1966).

[140] L. I. Grossweiner, *Photophysiology* **5**, 1 (1970).

[141] L. P. Vernon, *Photochem. Photobiol.* **18**, 529 (1973).

[142] P. S. Song, *Photochem. Photobiol.* **18**, 531 (1973).

[143] T. G. Traylor, C. K. Chang, and J. Geibel, *168th Nat. Meet. Amer. Chem. Soc. Abstr.* no. INOR 30 (1974).

[144] J. R. Bolton, D. C. Borg, and H. M. Swartz, *in* "Biological Applications of Electron Spin Resonance" (H. M. Swartz, J. R. Bolton, and D. C. Borg, eds.), p. 82. Wiley, New York, 1972.

[145] R. Livingston and V. A. Ryan, *J. Amer. Chem. Soc.* **75**, 2176 (1953).

[146] R. Livingston, *J. Amer. Chem. Soc.* **77**, 2179 (1955).

[147] H. Linschitz and K. Sarkanen, *J. Amer. Chem. Soc.* **80**, 4826 (1958).

[148] S. Claesson, L. Lindqvist, and B. Hölmström, *Nature (London)* **183**, 661 (1959).

[149] H. T. Witt, *in* "Fast Reactions and Primary Processes in Chemical Kinetics" (S. Claesson, ed.), p. 261. Almqvist & Wiksell, Stockholm, 1967.

[150] H. T. Witt, *in* "Fast Reactions and Primary Processes in Chemical Kinetics" (S. Claesson, ed.), p. 81. Almqvist & Wiksell, Stockholm, 1967.

[151] G. Porter and G. Strauss, *Proc. Roy. Soc., Ser. A* **295**, 1 (1966).

[152] A. R. Kelly and G. Porter, *Proc. Roy. Soc., Ser. A* **315**, 149 (1970).

[153] J. T. Warden and J. R. Bolton, *J. Amer. Chem. Soc.* **94**, 4351 (1972).

[154] D. C. Borg, J. Fajer, R. H. Felton, and D. Dolphin, *Proc. Nat. Acad. Sci. U.S.* **67**, 813 (1970).

[155] T. L. Netzel, P. M. Rentzepis, and J. Leigh, *Science* **182**, 238 (1973).

[156] E. W. Abrahamson, R. Adams, and V. J. Wulff, *J. Phys. Chem.* **63**, 441 (1959).

[157] W. Dawson and E. W. Abrahamson, *J. Phys. Chem.* **66**, 2542 (1962).

[158] R. Bensasson, E. J. Land, and T. G. Truscott, *Chem. Phys. Lett.* **17**, 53 (1973).

[159] H. J. Dartnall, *Vision Res.* **8**, 339 (1968).

[160] T. Rosenfeld, A. Alchalel, and M. Ottolenghi, *J. Phys. Chem.* **78**, 336 (1974).

[161] T. Rosenfeld, A. Alchalel, and M. Ottolenghi, *Chem. Phys. Lett.* **20**, 29 (1973).

[162] G. E. Busch, M. L. Applebury, A. A. Lamola, and P. M. Rentzepis, *Proc. Nat. Acad. Sci. U.S.* **69**, 2802 (1972).

[163] M. Eigen, *in* "Fast Reactions and Primary Processes in Chemical Kinetics" (S. Claesson, ed.), p. 477. Almqvist & Wiksell, Stockholm, 1967.

[164] G. Porter, *in* "Fast Reactions and Primary Processes in Chemical Kinetics" (S. Claesson, ed.), p. 469. Almqvist & Wiksell, Stockholm, 1967.

²⁹Si Nuclear Magnetic Resonance

4

J. SCHRAML and J. M. BELLAMA

I. INTRODUCTION

A. Historical Development of ²⁹Si NMR

Many papers have demonstrated that nuclear magnetic resonance (NMR) is a most valuable method for determination of the structure of organosilicon compounds.[1-3] Though most of these papers have been concerned with proton resonance, the features associated with the NMR-active nucleus of silicon (^{29}Si) have been studied with increasing interest and at an increasing rate since 1954 when ^{29}Si–^{19}F coupling was observed for the first time.[4] Until recently, it was in particular ^{29}Si coupling that was frequently reported because such coupling to other nuclei could be easily measured from, e.g., the ^{1}H and ^{19}F spectra that were readily available. Only recently have commercial spectrometers become available on which ^{29}Si spectra can be routinely examined. Previously, modifications of spectrometers were necessary in order to measure these parameters.

In 1956 and 1962 the fundamental papers by Lauterbur and co-workers[5, 6] not only established the basic trends in silicon shielding but also described how some of the practical difficulties encountered in measuring ^{29}Si NMR spectra could be overcome. Apparently, the detailed account of the difficulties and the relatively large experimental errors, combined with the fact that some of the most fundamental compounds of silicon had now been measured, served to deter other workers from exploring this field. This situation lasted until 1968, except for a few indirect measurements of ^{29}Si chemical shifts.[7–12]

A paper by Hunter and Reeves[13] appeared in 1968 and contained new information about both ^{29}Si chemical shifts and relaxation times. This paper triggered an explosion of the field, and some 30 papers were published during the years 1970–1973, as compared to only five papers published on this subject during the years 1963–1968. This rapid increase in interest has occurred to some extent because the analytical potential of ^{29}Si NMR spectroscopy has now been realized,[2, 14–17] but probably to a greater extent because new spectrometers, which were constructed in the laboratories of Lippmaa[18] and of Maciel,[19] permitted more precise data to be obtained less tediously. These new spectrometers employed spin stabilization and accumulation of ^{29}Si NMR spectra that could be proton decoupled. Considerable saving of time was achieved, and solutions of lower concentration could be measured relative to an internal reference when Fourier transform pulsed spectrometers were introduced. Currently, multinuclear spectrometers that have all of these features are on the market, and ^{29}Si measurements are no longer limited to a few exclusive NMR laboratories. Spectra now can be and are being obtained in many laboratories.

Early work in ^{29}Si NMR was reviewed in previous volumes of this series.[6, 20] A later review of Marsmann[21] concentrated on shielding aspects. This review approaches the subject in a broader sense. It covers papers that have appeared recently, which have contributed to an understanding of structural correlations, and which have broadened analytical applications of ^{29}Si NMR. The original concept of providing a comprehensive review had to be reconsidered when the list of pertinent references went well past 400. Although it was possible to maintain the comprehensive concept in the section on ^{29}Si chemical shifts, a representative and illustrative approach had to be adopted in the section on ^{29}Si coupling constants.

B. Properties of the ^{29}Si Nucleus

Of the three stable isotopes of silicon that occur naturally (^{28}Si, ^{29}Si, ^{30}Si), only ^{29}Si has a nuclear spin greater than zero.[4] The value of $\frac{1}{2}$ for the nuclear spin quantum number was established for the ^{29}Si isotope (natural abundance =4.7%) by a ^{19}F NMR study[4] of isotopically enriched SiF$_4$. Several obvious advantages for NMR studies follow from this spin $\frac{1}{2}$ value: (1) The nucleus can exist in only two nuclear spin states, and thus coupling with other nuclei leads to simple two-line satellites. (The combined intensity ratio of the two satellites to the intensity of the main resonance is the ratio of the isotopic content of the sample, i.e., 4.7:95.3 in nonenriched samples.) (2) The ^{29}Si nucleus does not have a nuclear quadrupole moment that would provide an additional relaxation mechanism and broaden the ^{29}Si NMR signals.

On the other hand, a disadvantage results from the negative magnetogyric ratio of this isotope ($\gamma_{Si} = -5.314 \times 10^{-7}$ rad T^{-1} second^{-1}, calculated from the data of Varian Associates[22]). As common nuclei present in NMR samples have positive magnetogyric ratios, a negative nuclear Overhauser effect is observed. (See Section IV of this chapter and also Chapter 5.)

The above factor and the relatively long relaxation time (T_1) of ^{29}Si nuclei make it difficult to estimate or to compare the relative sensitivity of direct ^{29}Si NMR experiments with those of other nuclei. In favorable cases ^{29}Si NMR spectroscopy should be about twice* as sensitive as ^{13}C NMR, or 3.7×10^{-4} as sensitive as ^{1}H NMR.[23] Less favorable cases were briefly discussed by Levy and Cargioli[3]; the sensitivity achieved depends on the experimental method employed. According to the value of the magnetogyric ratio, the ^{29}Si resonance is observed at a frequency of 11.92 or 19.87 MHz in a field of 1.41 or 2.35 T, in which a proton resonates at a frequency of 60 or 100 MHz, respectively.

C. Characteristics of ^{29}Si NMR

The following types of spectral information related to the ^{29}Si nucleus can be obtained from NMR spectra: (1) ^{29}Si chemical shifts, (2) relative intensities, (3) spin–spin coupling constants, (4) spin relaxation times, (5) nuclear Overhauser effect enhancement (NOE), and (6) isotope effects on the shifts of other nuclei.†

---

* It is interesting to note how a typographical error in a text[6] that gives the relative sensitivity as twenty can propagate through reviews[1,20] even though the data in the tables that accompanied the original text gave correct values.

† Isotope effects on coupling between pairs of other nuclei have not yet been reported. The isotope effect of ^{29}Si on coupling constants of silicon cannot be observed since silicon has no other NMR-active isotope. Isotope effects from isotopes of other elements on ^{29}Si shielding have not yet been reported, although effects on ^{29}Si coupling have been noted.[8,12,24-28,85]

The first four of these parameters can be derived both from direct ^{29}Si NMR experiments and also from NMR studies of nuclei which are spin–spin coupled to silicon. The nuclear Overhauser effect is accessible only through direct experiments, while the last parameter can be measured only through an NMR study of the other nucleus.

The isotope effect, which has been comparatively poorly studied and which is not yet sufficiently explained,[29] will not be considered in this discussion since it has no bearing upon structure determination of organic compounds. Similarly, the knowledge of silicon shifts in solids[30–33] will not be discussed. The interesting work on ^{29}Si shielding anisotropy[34] is also beyond the scope of this chapter.

For convenience, we shall discuss chemical shifts (and relative intensity), coupling constants, and relaxation and NOE separately, although the most complete structural information can obviously be obtained by a joint consideration of all parameters. In general, the relative importance of these parameters in structure elucidation cannot at present be estimated, since previous studies have been mainly concerned with relatively simple molecules or with those of known structure. However, the basic and fundamental trends have now been established, and ^{29}Si NMR can be applied to organosilicon compounds of unknown structure.

II. ^{29}Si CHEMICAL SHIFTS

A. General

If the most extreme ^{29}Si chemical shifts are considered, the range of the silicon shifts spans some 400 ppm. However, it is clear from the chemical shift chart (Fig. 1) that most of the known silicon shifts are clustered in a considerably narrower range of about 200 ppm. This range is appreciably smaller than the corresponding ranges of the shifts of other group IV B elements. The original suggestion[35] that the smaller ^{29}Si range might be due to limited information is correct only in the sense that this smaller range obviously does not include the silicon counterparts of multiple-bonded carbon or of carbonium ions (which are responsible for the low-field extremes of ^{13}C chemical shifts). The high-field extremes for both ^{13}C and ^{29}Si chemical shifts are the tetraiodo derivatives.[36] The considerably larger ranges of ^{117}Sn, ^{119}Sn, and ^{207}Pb chemical shifts[20, 35] are a consequence of the Z (atomic number) dependence of chemical shifts.[35] Despite the different ranges, interesting correlations can be drawn[20, 37–39] between the shifts in analogous compounds of the group IV elements or between isoelectronic[36, 40] and other[41] compounds. However, the limited scope of this review permits a discussion of only those correlations that justify the theory of ^{29}Si chemical shifts in comparison with ^{13}C chemical shifts.

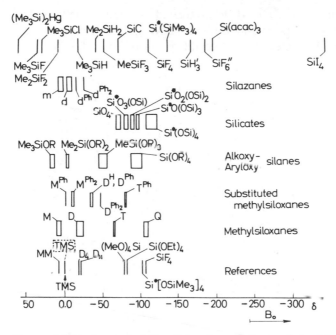

FIG. 1. Chart of ^{29}Si chemical shifts. (Meaning of the symbols: Me = CH$_3$, Et = C$_2$H$_5$, acac = acetyl acetonate, R = alkyl, aryl. Other symbols are explained in the text. In compounds which contain more than one silicon atom, the chemical shift belongs to the silicon atom with the asterisk.)

B. Experimental Techniques

1. Direct Observations of ^{29}Si Resonance

Progress in ^{29}Si NMR instrumentation has been marked by the spectrometers and techniques used by Lauterbur et al.,[5, 6] Hunter and Reeves,[13] Lippmaa et al.,[18] and Maciel et al.[19] Since Fourier transform (FT) NMR multinuclear spectrometers are now available, current experimentalists usually prefer the FT NMR spectral technique to continuous-wave (CW) NMR. (No ^{29}Si NMR spectra measured on high-field superconducting magnets have yet been reported.) The reasons for this preference are that by using FT, considerable saving of time is achieved, and it is also possible to measure at comparable resolution solutions of lower concentration with an internal reference. The power of this currently available technique is well illustrated by a part of the FT spectrum[42] of [(CH$_3$)$_3$Si]$_4$Si, which clearly shows (Fig. 2) not only satellites due to ^{29}Si–^{29}Si coupling (such coupling is present in only 0.9% of the molecules), but also ^{13}C–^{29}Si satellites (0.6% of the molecules).

The principles of FT NMR are treated in detail in the book by Farrar and Becker.[43] Also, Levy and Cargioli[3] have given an extensive account with illustrative examples of the special features of ^{29}Si FT NMR.

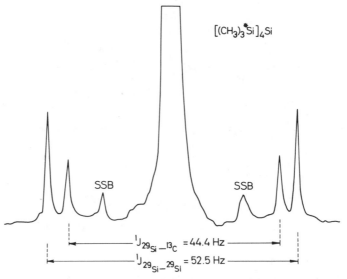

$[(CH_3)_3\overset{*}{S}i]_4Si$

FIG. 2. The $(CH_3)^*_3Si$- portion of the ^{29}Si FT NMR spectrum of tetrakis-(trimethylsilyl)silane. The figure also shows the satellites due to ^{29}Si–^{29}Si and ^{29}Si–^{13}C coupling. Spinning side bands are labeled SSB. (This spectrum is reproduced through the courtesy of J. D. Cargioli and E. A. Williams[42] of the General Electric Company.)

2. Indirect Observation of ^{29}Si Resonance

NMR experiments on nuclei (^{1}H and ^{19}F in particular) that have higher sensitivity to NMR detection than ^{29}Si can yield not only ^{29}Si coupling constants but also ^{29}Si chemical shifts. (Additionally, the relative intensities of satellites can be discerned, and the isotope effect can be measured from the asymmetry of the satellites.) This indirect observation is achieved through heteronuclear double resonance experiments. Methods that can be employed for this purpose have been reviewed by McFarlane.[44, 45] Also, an elementary account with examples drawn from the spectra of silicon-containing compounds was given by Johannesen and Coyle.[11] The applicability of the internuclear double resonance (INDOR) technique to organosilicon compounds was first demonstrated by Baker[7] and later used by others.[46–48] ^{19}F–{^{29}Si} "spin tickling" experiments were described by Johannesen et al.,[10] and ^{1}H–{^{29}Si} experiments by McFarlane and Seaby.[49]

3. Decoupling Effects

Heteronuclear double (or multiple) resonance techniques can be used not only for indirect observation of resonance but also to facilitate direct observations and the assignment of the spectral lines.

While off-resonance decoupling is helpful in assigning the lines, as it is in NMR spectroscopy of other nuclei, the outcome of decoupling (both coherent and incoherent) experiments depends on the magnitude of the nuclear Overhauser effect (NOE).

In NMR spectroscopy of nuclei with positive magnetogyric ratios, irradiation of the proton (which has also a positive magnetogyric ratio γ_H) NMR transition bands leads to an increase in the signal to noise ratio (S/N) by collapsing multiplets due to spin–spin coupling and by nuclear Overhauser enhancement. Generally, under favorable conditions the intensity of a resonance line of nucleus A is enhanced by the NOE by a factor of $[1 + (\gamma_H/2\gamma_A)]$ if proton transitions are saturated. In the case of the ^{29}Si nucleus the maximum possible NOE is -1.5; that is, by means of decoupling an inversion of a ^{29}Si signal is achieved and the inverted signal is 50% stronger than that observed without the coupling. Under less favorable conditions the enhancement would be smaller, and NOE can null the signal. In FT NMR the instrumental technique of interrupted or pulse-modulated decoupling can be used to suppress NOE.[3, 50]

If samples are doped with paramagnetic relaxation reagents (which is usually done in order to shorten relaxation time and to permit a higher pulse repetition rate), the relaxation reagent quenches the NOE.[51] Since the relaxation reagent reduces all spin–lattice relaxation times to approximately the same value, it allows quantitative intensity measurement even in ^{29}Si FT NMR.[3, 15] Acetyl acetonates of chromium and iron can be used for this purpose. Oxygen (dissolved under a pressure of 1 atm) does not suppress the NOE completely,[3, 52] but it is easier to displace from the sample.

A difference between ^{29}Si chemical shifts proton decoupled and not decoupled was reported,[14] but since sufficient experimental details have not yet been given, the origin of the effect is not clear.

C. References, Scales, Precision, and Solvent Effects

The chemical shift data reported in the literature were obtained by both direct and indirect methods at different levels of instrumentation. Therefore, the precision of the reported data varies considerably through the literature.

No ^{29}Si chemical shift scale has yet been universally accepted. The one point on which all authors seem to agree is rather trivial; they all express chemical shifts in ppm units. However, several reference compounds (polydimethylsiloxane,[5, 6] tetramethoxysilane,[13, 38, 53] tetrafluorosilane,[8, 10] octamethyltetrasiloxane,[54] and tetramethylsilane[14–17, 21, 28, 34, 36, 48, 49, 55–68]) have all been used in combination with both possible sign conventions. Some consensus on the sign convention has now apparently been reached, since in most of the more recent work, shifts to low field from the reference line (higher frequency, paramagnetic shift) are given as positive. This practice is in keeping with the tendencies and recommendations for NMR of other nuclei,[23] especially with the δ scale of proton NMR. The δ scale based on tetramethylsilane (TMS) as a reference compound has been internationally accepted for proton NMR[69] and is by now also well established in ^{13}C NMR.[70, 71] It has been suggested[44] that through heteronuclear double resonance TMS could serve as a universal reference compound for all nuclei.

It has been noted,[72] however, that some interim period of time will be necessary before an officially sanctioned [29]Si shift reference (and scale) will be accepted. Since in principle any of the common silicon compounds[3] could be used as a reference standard, the choice should be dictated by practical considerations. These practical aspects differ according to the experimental and theoretical needs. Therefore, a single primary [29]Si standard that would best suit each and every requirement is unlikely to be found and secondary references will have to be used.

It is in the interest of all workers active in the field that this interim period be as short as possible. In accord with what appears to be favored by most spectroscopists, we propose that an analogy of the TMS-based δ scale be adopted for ^{29}Si NMR.

It should be noted, however, that objections to TMS as a [29]Si standard were recently published. The principal objection[72] was that the "TMS resonance lies in the middle of the range of common [29]Si chemical shifts." This is true, and the objection is even more serious in view of the fact that the δ scale owes its acceptance in ^{1}H and ^{13}C NMR spectroscopy to the extreme position of its signals in the two spectra. However, organosilicon chemists who use ^{1}H NMR (especially those using spectrometers with internal locking systems) have become accustomed to the proximity of CH_3Si or $Si-CH_2-Si$ proton signals to that of TMS and also to the use of secondary references when necessary.

Advantages (especially the chemical inertness) of TMS over other reference compounds used in ^{1}H NMR have been cited by Tiers[73] and recently reviewed.[1] These advantages apply equally to the use of TMS in ^{29}Si NMR. In addition, if this reference compound is accepted for ^{29}Si NMR, then it would serve as a NMR reference for three important nuclei (^{1}H, ^{13}C, and [29]Si), and all three spectra could then be measured from one solution. Also, comparison with NMR of other heavier nuclei would be facilitated, since the analogous permethylated compounds are quite stable (in contrast to analogues of the suggested tetraethoxysilane). Concentration requirements,[72] relaxation times,[50, 72] and the resulting necessity to add relaxation reagents[72] also argue against the use of tetraethoxysilane as a primary reference (for its chemical shift, see Fig. 1), but the fundamental objection against the use of tetra-alkoxysilanes as reference compounds is based on their reactivity and ability to exchange alkoxy groups. (Cases in which reactivity has prevented the use of alkoxysilanes and siloxanes as reference compounds have been reported.[53, 74] This property might be particularly undesirable in ^{29}Si NMR applications to structure determination of natural products. These substances

TABLE I

Chemical Shifts of ^{29}Si Reference Materials

Compound	Symbol	^{29}Si chemical shift, δ scale		Solvent[b]
		(external reference)[a]	(internal reference)	
$(CH_3)_4Si$	TMS	0.0[c]	0.00[c]	—
$(C_2H_5O)_4Si$	Si(OEt)$_4$	−83.2[d]	−82.59[e]	CCl$_4$
$(CH_3O)_4Si$	TMOS	−79.5[f]	−79.22[g]	CCl$_4$
		−78.50[h]	−78.48[i]	CDCl$_3$
$[(CH_3)_2SiO]_4$	D$_4$	−20.0[j]	−19.51[g]	CCl$_4$
			−19.51[k]	(CD$_3$)$_2$CO
$[(CH_3)_2SiO]_x$	D$_x$	−22.0[l]	−22.22[g, m]	CCl$_4$
$[(CH_3)_2Si]_2O$	MM	6.3[j]	6.87[g]	CCl$_4$
			6.79[k]	(CD$_3$)$_2$CO
SiF$_4$	—	−113.6[n]	—	—
		−109.0[o]	—	—
$[(CH_3)_3SiO]_4^*Si^p$	—	−105.2[j]	—	—

[a] Usually measured in neat liquids.

[b] The solvent refers to δ (internal reference) values.

[c] Primary reference.

[d] Data from Mägi and Schraml,[75] ±0.3 ppm: values between −81.0 and −83.5 ppm reported.[5, 12, 13]

[e] Data from Levy et al.,[72] ±0.07 ppm.

[f] Data from Hunter and Reeves,[13] ±1 ppm; this value should be used for conversion of data.

[g] Value converted from data of Levy et al.,[72] ±0.14 ppm.

[h] Data from Scholl et al.,[38] ±0.05 ppm; this value should be used for conversion of data.

[i] Data from Ernst et al.[53]; this value should be used for conversion of data.

[j] Data from Engelhardt et al.,[16] ±0.3 ppm.

[k] Data from Levy and Cargioli,[3] ±0.05 ppm.

[l] Data from Lauterbur[6] for silicone "DC 200"; used for reference signal in Lauterbur et al.[5, 6]

[m] Data for $[(CH_3)_3SiO(CH_3)_2SiO]_xSi(CH_3)_3$ with $x = 50$.[72]

[n] Value for gas, 30 atm. Calculated on the bases of data from Johannesen et al.[10] and from Hunter and Reeves,[13] ±1.1 ppm.

[o] Data from Marsmann and Horn.[36]

[p] Chemical shift given is for the atom with asterisk; reference compound suggested by Lauterbur.[6]

are frequently isolated as silylated derivatives of unknown compounds, the chemical properties of which are uncertain.)

In accord with the proposed δ scale, all chemical shift data given here are reported in ppm units relative to the TMS signal. Positive chemical shifts correspond to shifts to lower field. Preference in the citation of numerical values has been given to those obtained with an internal reference.

To facilitate conversion of published data to the δ scale, chemical shifts of the references used were compiled in Table I.[3, 5, 6, 10, 12, 13, 16, 36, 38, 53, 72, 75] In converting chemical shifts, care should be taken to choose the shift of a reference determined under comparable conditions.

Since many data have been published in a preliminary form that does not give enough experimental detail, it is not always possible to estimate the errors involved, to find which solution was used, or to know what method of referencing was employed. To make the situation even worse, there is too little known about solvent and concentration effects on ^{29}Si chemical shifts.

Solvent effects have been extensively studied only on the ^{29}Si NMR line of TMS[38, 76, 77] and on mixtures of TMS, hexamethyldisiloxane (MM), and tetraethoxysilane.[72] It was found that solvent shifts of the ^{29}Si resonance covered a range of 0.64 ppm in 15 different solvents, which included halo-benzenes and halocyclohexanes. Hexafluorobenzene, which has been fre-quently employed as a solvent in ^{29}Si NMR studies, causes an upfield shift of 0.38 ppm relative to pure TMS; cyclohexane produces a downfield shift of 0.05 ppm.

No effect of paramagnetic relaxation reagents [Cr(acac)$_3$ and Fe(acac)$_3$; acac = acetyl acetonate] on the shifts of TMS and MM relative to internal tetraethoxysilane in carbon tetrachloride was detected,[72] but when an external lock was used, it was found necessary[15] to correct for an upfield shift of about 0.01 ppm caused by Cr(acac)$_3$. Solvent effects of dichloromethane have also been reported.[15, 49]

D. Effects of Substituents Directly Bonded to Silicon

1. The Effects of Multiple α Substitution

Our knowledge of ^{29}Si shielding is most complete for the common $(CH_3)_{4-n}SiX_n$ compounds, which were studied by early workers[5, 6, 13] and still attract attention.[10, 16, 36, 39, 47–49, 65, 66] Some of the available values are summarized in Table II. (It is clear from the legend to Table II that it would be most desirable for these compounds to be remeasured under identical con-ditions, i.e., on a modern spectrometer, in diluted solutions, and with an internal reference.) The dependence of δ on n, the number of substituents, is shown in Fig. 3. The "sagging" pattern in this relationship is characteristic for $(CH_3)_{4-n}SiX_n$ compounds with X = F, Cl, Br, O, and N. The reasons why

TABLE II

^{29}Si Chemical Shifts in Methylsilanes, $(CH_3)_{4-n}SiX_n$

X	n			
	1	2	3	4
F	35.4[a]	8.8[a]	−51.8[a]	−109.0[b, c]
Cl	30.21[d]	32.17[d]	12.47[d]	−18.5[e]
Br	26.41[d]	19.86[d]	−18.18[d]	−93.6[b]
I	8.72[d]	−33.68[d]	−17.96[d]	−346.2[b]
H	−18.5[f]	−41.5[f]	−65.2[p]	−93.1[p]
OCH$_3$	17.2[g]	−2.5[g]	−41.4[g]	−79.22[h]
OC$_2$H$_5$	13.5[i]	−6.1[i]	−44.5[i]	−82.59[j]
OC$_6$H$_5$	17.2[e]	−6.1[e]	−54.0[e]	−101.1[e]
OC(O)CH$_3$	22.3[k]	4.4[k]	−42.7[k]	−74.5[f]
OSi(CH$_3$)$_3$	6.87[h]	−21.5[i]	−65.0[g]	−105.2[g]
N(CH$_3$)$_2$	5.9[e]	−1.7[e]	−17.5[e]	−28.1[e]
C$_2$H$_5$	1.0[m]	4.6[m]	6.5[m]	8.4[p]
C$_6$H$_5$	−5.1[n]	−9.4[n]	−11.9[n]	—
CH=CH$_2$	−6.80[o]	−13.67[o]	−20.55[o]	−22.5[f]

[a] Data from Johannesen et al.,[10] relative error ±0.06 ppm; converted into δ scale using the value for SiF$_4$ given in Table I; error estimated to be ±1 ppm.
[b] Data from Marsmann and Horn,[36] error not given.
[c] Value of −117.4 in Marsmann[56] is in error.
[d] Data from van den Berghe and van der Kelen,[48] ±0.06 ppm.
[e] Data from Engelhardt et al.,[39] ±0.3 ppm.
[f] Data from Hunter and Reeves[13] converted into δ scale, error estimated ±2 ppm.
[g] Data from Engelhardt et al.,[16] ±0.3 ppm.
[h] Data derived from values of Levy et al.[72]; estimated error ±0.14 ppm.
[i] Data from Schraml et al.,[66] ±0.3 ppm.
[j] Data from Levy et al.,[72] ±0.07 ppm.
[k] Data from McFarlane and Seaby.[49]
[l] Data from Levy and Cargioli,[3] ±0.05 ppm.
[m] Data from Schraml et al.,[78] ±0.3 ppm.
[n] Data from Engelhardt et al.,[17] ±0.3 ppm.
[o] Data from Scholl et al.,[38] converted into δ scale, error estimated ±0.1 ppm.
[p] Data from Marsmann.[40]

these dependencies are not monotonous (as they are for ^{13}C chemical shifts in analogous carbon compounds) is a main concern of the theory of ^{29}Si shielding.

In view of the similarity of the relationships shown in Fig. 3 for different substituents, it is not surprising that the dependencies can be brought together into one general dependency of δ either on the electronic charge on the silicon atom (as found by Engelhardt et al.[39]) or, in an equivalent fashion, on

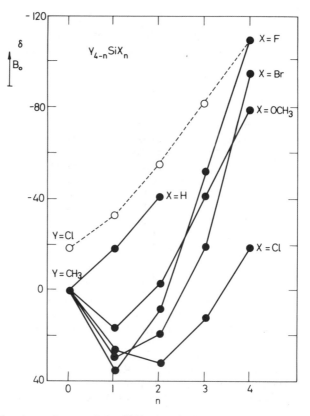

FIG. 3. The dependence of the [29]Si chemical shift on multiple substitution. (Based on a plot of Engelhardt et al.[39] in *Org. Magn. Resonance*, published by Heyden & Son, Ltd.)

the sum of electronegativities ($\sum \chi$) of the substituents R in a somewhat more general class of $R_1R_2R_3R_4Si$ compounds (as reported later by Ernst et al.[53]). These united dependencies show considerable scatter, of course, as can be seen in Fig. 4, but since the electronegativity values are readily available, the relationship shown in Fig. 4 could be used to predict chemical shifts. The least-squares fit of the data of Fig. 4 gave the following equation for [29]Si chemical shifts (δ scale)

$$\delta = 6.06 \times 10^3 - 1.82 \times 10^3(\sum \chi) + 1.95 \times 10^2(\sum \chi)^2$$
$$- 9.17(\sum \chi)^3 + 0.161(\sum \chi)^4 \tag{1}$$

The chemical shifts predicted from this expression agree to within ± 10 ppm of those found experimentally.[53]

Numerous individual data on more general classes of compounds (e.g., $Y_{4-n}SiX_n$) can be found in the literature,[6, 10, 13, 21, 39, 53, 78] but data are com-

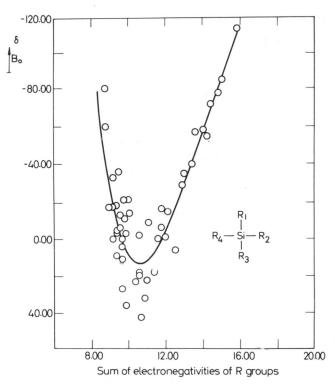

FIG. 4. The dependence of the ²⁹Si chemical shift on electronegativity. (Based
on a plot of Ernst *et al.*[53] in *J. Amer. Chem. Soc.*, published by the American
Chemical Society.)

plete for only a few of the series. Some of these data are assembled in Table
III. An informative collection of chemical shift data on compounds with
various substituents is presented in Table IV.[39, 40, 53, 55, 56, 60, 65, 66, 67, 80]

2. The Effect of the Nature of the Substituent

The linear dependence on the substituent electronegativity χ_X of the substituent in $(CH_3)_3SiX$ compounds was noted quite early.[6] This linear dependence of the shift on the electronegativity of X includes compounds with
substituent X = Cl, Br, I, and Si.[13] Compounds with X = N, O, and F do
not fall on the line, supposedly due to an additional shielding mechanism.[13]
These latter compounds together with those with X = C_6H_5 and C_2H_3 fall on
a second line.[20] The different shielding effects of sp^2- and sp^3-hybridized α
carbon atoms are readily apparent from Table V. In any case, vinyl and
phenyl silanes are shielded more than the corresponding alkylsilanes.

TABLE III

^{29}Si Chemical Shifts for $Y_{4-n}SiX_n$ Compounds

				n		
X	Y	0	1	2	3	4
F	Cl	-18.7^a	-32.1^b	-55.0^b	-81.7^b	-109.0^a
F	Br	-93.6^a	-67.0^b	-67.4^b	-82.4^b	-109.0^a
F	C_6H_5	^{c}	-4.7^d	-29.1^e	-72.73^f	-109.0^a
H	C_6H_5	^{c}	-21.1^a	-34.5^g	-61.5^g	^{h}

^{a} Data from Marsmann and Horn,[36] ±0.5 ppm.[21]
^{b} Data from Johannesen et al.,[10] converted into δ scale using the value for SiF_4 from Table I, error estimated ±1 ppm.
^{c} Not measured because of low solubility.
^{d} Data from Engelhardt et al.,[39] ±0.3 ppm.
^{e} Data from Marsmann,[56] ±0.5 ppm.[21]
^{f} Data from Ernst et al.[53]
^{g} Data from Hunter and Reeves,[13] ±1 ppm.
^{h} Not measured.

TABLE IV

^{29}Si Chemical Shifts in Miscellaneous Compounds^a

Compound	δ	Ref.	Compound	δ	Ref.
$(AcO)_3SiCH_2OAc$	-82.5	66	$(Me_3Si)_3P$	0.4	56
F_3SiCH_2Cl	-71.3	67	$(Me_3Si)_3N$	0.5	40
$MeSi(OC_2H_4)_3N$	-66.6	40	$Me_2Si(N_3)_2$	0.6	56
$PhCH_2SiF_3$	-64.2	65	$(Me_3Si)_2PH$	1.8	56
$PhSi(OEt)_3$	-59.4	65	$(Me_3Si)_2NH$	2.2	60
$PhSi(OMe)_3$	-57.5	53	$(Me_3Si)_2NPNSi*Me_3$	2.2	80
$Ph_2Si*(NHSiMe_3)_2$	-24.2	60	Me_3SiNCS	5.4	56
$(EtO)_2MeSiH$	-16.1	53	Me_3SiNCO	7.4	56
$PhMeSi*(NHSiMe_3)_2$	-15.0	60	$(Me_3Si)_2S$	12.8	55
$(i-C_4H_9)_3SiH$	-14.7	39	Me_3SiN_3	15.2	56
$(Me_3Si*)_2NPNSiMe_3$	-8.2	80	$PhMe_2SiCl$	19.9	65
$Ph_2Si(N_3)_2$	-6.5	56	$(Me_2SiS)_3$	21.1	40
$Cl_3SiCHCl_2$	-6.5	53	Me_2FSiCH_2Cl	24.6	39
$PhSiHCl_2$	-2.2	53	$(n-C_4H_9)_3SiF$	28.8	39
$(Me_3Si)_2NCN$	-0.8	56	$(Me_3Si)_2Hg$	60.2	40

^{a} Abbreviations: $Me = CH_3$, $Et = C_2H_5$, $Ph = C_6H_5$, $OAc = OC(O)CH_3$. An asterisk denotes the silicon atom for which the shift is given.

In these cases in which the pK_a values of the acids HX were known, the pK_a values and the silicon shifts in $(CH_3)_3SiX$ compounds could be linearly correlated, as was independently shown by McFarlane and Seaby[49] and by Marsmann and Horn.[55] Actually, the correlations reported by the former were more concerned with the effects of remote substituents since they involved only carboxylic acids with the Si—OC(O)— group kept constant through the series (therefore, the variation of pK_a values was smaller than in the latter study), but because the pK_a value is related to the whole HX molecule, the work is mentioned in this section. The relationship of the shifts in $(CH_3)_3SiX$ compounds was such that the more electronegative (or more electron-withdrawing) substituents caused shifts to lower field.

With one exception, the effects of the nature of substituents (X) on the shielding in $(CH_3)_2SiX_2$, $(CH_3)SiX_3$, and SiX_4 compounds have not been discussed in the literature, but the reader can visualize the trends from the data in Table II. The one exception is the above-mentioned work of McFarlane and Seaby.[49] They found separate linear correlations between pK_a and the silicon shifts for $(CH_3)_{3-n}SiX_n$ ($n = 1, 2, 3$) carboxylates. The slopes of the corresponding lines all have the same sign (decreasing δ with increasing pK_a), but the slopes have a ratio of 1.00:0.88:0.42 (instead of a 1:2:3 ratio which might be expected).

TABLE V

^{29}Si Chemical Shifts in $(CH_3)_{3-n}X_nSiR$ Compounds

X	n	R						
		CH_3	C_2H_5	$CH{=}CH_2{}^a$	$C_6H_5{}^b$	$n\text{-}C_3H_7{}^a$	$CH_2CH{\overset{\|}{CH_2}}{}^a$	$CH_2C_6H_5{}^b$
Cl	0	0.00	1.6^a	−7.6	−5.1	0.7	−0.4	0.4
	1	30.21^c	—	16.7	19.9	—	27.2	26.6
	2	32.17^c	34.0^a	16.5	17.9	—	26.8	26.9
	3	12.47^c	14.6^a	−3.5	−0.8	—	8.0	7.2
OC_2H_5	1	13.5^d	15.8^d	2.7	5.1	14.8	—	11.7
	2	−6.1^d	−6.7^d	—	−20.2	−7.6	—	−11.9
	3	−44.5^d	−45.9^d	−60.3	−59.4	−47.0	−51.6	−52.7
C_2H_5	1	1.6^a	4.6^a	−4.4	—	—	—	—
	2	4.6^a	6.5^a	−2.3	—	—	—	—
	3	6.5^a	8.4^e	−1.7	—	—	—	—

^{a} Data from Mägi and Schraml,[75] ±0.3 ppm.
^{b} Data from Schraml et al.,[65] ±0.3 ppm.
^{c} Data from van den Berghe and van der Kelen,[48] ±0.06 ppm.
^{d} Data from Schraml et al.,[66] ±0.3 ppm.
^{e} Data from Marsmann.[21]

E. Effects of Substituents Not Directly Bonded to Silicon

Though considerable data on compounds with substituents in the β or in other positions relative to the silicon atom can be found in the literature,[13, 16, 81–85] systematic studies of the effects of such substituents have just begun to appear.[49, 53, 57, 63, 65–67, 86–90] These reports have thus far been concerned with the inductive influence of these substituents on silicon shielding; the effects of nonbonded interactions on ^{29}Si shielding have not yet been observed.

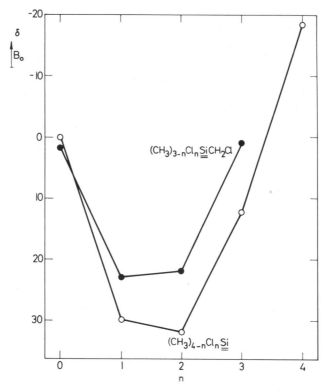

FIG. 5. Comparison of ^{29}Si chemical shift dependence on multiple substitution in series with and without a chlorine atom in the β position to the silicon atom. (Based on data of Schraml *et al.*[66])

As would be expected, the influence of substituents that are not directly bonded to the silicon atom is smaller than that of substituents directly attached. This dominant influence of the α substituent can be seen in Fig. 5 by comparing the dependencies of the shifts on n, the number of α chlorine atoms in two series of compounds, $(CH_3)_{3-n}Cl_nSiCH_2Cl$ and $(CH_3)_{4-n}Cl_nSi$. Clearly, the (chloromethyl) compounds follow the pattern described earlier for methylchlorosilanes.

Two other consequences of the dominant role of the α substituents were recently noted: (1) it was found that chemical shifts in $(CH_3)_{3-n}X_nSiCH_2Y$ and in $(CH_3)_{4-n}X_nSi$ compounds can be linearly correlated,[78] and (2) that an excellent linear correlation exists between the shifts in substituted phenyl and benzylsilanes, $(CH_3)_{3-n}X_nSi(CH_2)_mC_6H_5$ ($m = 0$ and 1).[62, 65]

The minor long-range (over two or more bonds) effects of the substituent Y in $(CH_3)_{3-n}X_nSi\ldots Y$ are conveniently discussed in terms of the substituent chemical shift (SCS) $\varDelta\delta$, which is defined as the difference between the chemical shift in the substituted compound, $\delta[(CH_3)_{3-n}X_nSi\ldots Y]$, and in the parent compound in which a hydrogen atom replaces the substituent Y, $\delta[(CH_3)_{3-n}X_nSi\ldots H]$. That is,

$$\varDelta\delta = \delta[(CH_3)_{3-n}X_nSi\ldots Y] - \delta[(CH_3)_{3-n}X_nSi\ldots H] \qquad (2)$$

(The SCS value is thus the vertical distance in ppm between the corresponding points on the two lines in plots analogous to those in Fig. 5.)

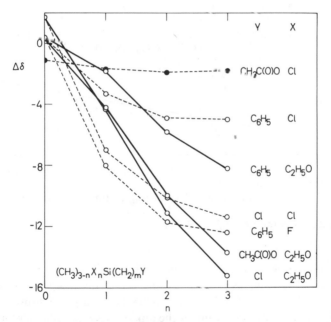

FIG. 6. ^{29}Si substituent chemical shift ($\varDelta\delta$) dependence on multiple substitution on the silicon atom. Open circles refer to series with $m = 1$; solid circles to series with $m = 2$. (Based on data of Schraml et al.[78])

Results of several studies in which these effects were investigated in the following series of $(CH_3)_{3-n}X_nSi(CH_2)_mY$ [X = OC_2H_5, Cl, F, $OC(O)CH_3$ and Y = C_6H_5, C_2H_3, Cl, $OC(O)CH_3$, NH_2],[37, 62, 65–67, 78, 86–88, 91] $(CH_3)_3SiOY$ [Y = $(CH_2)_mCl$, $(CH_2)_mSi(CH_3)_3$],[88] and $(CH_3)_{3-n}X_nSiC_6H_4Y$ ($n = 0, 3$; various X and Y)[37, 38, 53, 63, 89] compounds are summarized below. (It should be emphasized that the conclusions given here cannot necessarily

TABLE VI

^{29}Si Chemical Shifts and Substituent Chemical Shifts in $(CH_3)_{3-n}X_nSiCH_2Y$ Compounds[a]

Substituent					Substituent				
Y	X	n	δ	Δδ[b]	Y	X	n	δ	Δδ[b]
CH$_3$	Cl	0	1.6	1.6	Cl	Cl	0	2.79	2.79[c]
		1	31.3	1.7			1	22.9	−7.0
		2	34.0	1.8			2	21.7	−10.1
		3	14.6	2.1			3	0.8	−11.4
	OC$_2$H$_5$	1	15.8	2.3		OC$_2$H$_5$	1	8.9	−4.6
		2	−6.7	−0.6			2	−17.2	−11.1
		3	−45.9	−1.4			3	−59.7	−15.2
C$_6$H$_5$	Cl	0	0.4	0.4		F	1	24.83[c]	−10.57
		1	26.6	−3.3			2	−9.03[c]	−17.83
		2	26.9	−4.9			3	−71.34[c]	−19.54
		3	7.2	−5.0	OC(O)CH$_3$	OC$_2$H$_5$	0	0.3	0.3
	OC$_2$H$_5$	1	11.7	−1.8			1	9.0	−4.5
		2	−11.9	−5.8			2	−16.1	−10.0
		3	−52.7	−8.2			3	−58.2	−13.7
NH$_2$	OC$_2$H$_5$	0	0.5	0.5		OC(O)CH$_3$	1	13.8	−8.5
		1	13.2	−0.3			2	−18.1	−22.5
		2	−9.9	−3.8			3	−82.5	−39.8
		3	−50.5	−6.0					

[a] Data from Lippmaa et al.[57, 87, 88] and Schraml et al.,[65, 66, 78, 86] maximum error for δ ±0.3 ppm, for Δδ ±1 ppm.

[b] SCS of substituent Y, calculated according to Eq. (2) using values given here and in Table II.

[c] Data from Schraml et al.,[67] ±0.1 ppm.

be extended to other series of compounds yet to be investigated. In ^{29}Si NMR the danger of erroneous extrapolations from limited data is considerable.[57, 63, 66] The sign of SCS values depends for a given substituent Y on essentially two factors: the nature of the link between the silicon atom and the substituent Y, and on the other substituents that are directly attached to the silicon.

Two cases must be distinguished when discussing the effect of substituents in $(CH_3)_{3-n}X_nSi\ldots Y$ compounds. The first case occurs when $n = 0$. In such a case, electronegative (electron-withdrawing) substituents cause either (a) a downfield shift (positive SCS value) if the substituent Y is attached to the silicon atom through a carbon atom (i.e., a SiC...Y fragment), or (b) an upfield shift (negative SCS value) if the substituent Y is attached to the silicon atom through a Si—O bond (i.e., a SiO...Y fragment). (An exception was reported[99] for $(CH_3)_3SiOCH_2Cl$.)

TABLE VII

^{29}Si Chemical Shifts and Substituent Chemical Shifts in $(CH_3)_{3-n}(C_2H_5O)_nSi(CH_2)_mY$ Compounds

Y	n	m = 1		m = 2		m = 3		m = 4	
		δ	$\varDelta\delta^a$	δ	$\varDelta\delta^a$	δ	$\varDelta\delta^a$	δ	$\varDelta\delta^a$
NH$_2^b$	0	0.5	0.5	−0.2	−1.8	1.6	0.9	1.4	0.8
	1	13.2	−0.3	—	—	15.8	1.0	—	—
	2	−9.9	−3.8	—	—	−5.7	1.9	−5.9	1.9
	3	−50.5	−6.0	−46.5	−0.6	−45.3	1.7	−45.6	1.8
O(CO)CH$_3^c$	0	0.3	0.3	−1.0	−2.6	0.9	0.2	1.3	0.7
	1	9.0	−4.5	12.7	−3.1	—	—	—	—
	2	−16.1	−10.0	—	—	−8.4	−0.8	—	—
	3	−58.2	−13.7	−50.7	−4.8	—	—	—	—
Cl^c	0	1.7	1.7	−0.4	−2.0	1.5	0.8	—	—
	1	8.9	−4.6	—	—	14.8	0.0	—	—
	2	−17.2	−11.1	−11.5	−4.8	−7.1	0.5	—	—
	3	−59.9	−15.2	−52.3	−6.4	−47.0	0.0	—	—
H^c	0	0.0	—	1.6	—	0.7	—	0.6	—
	1	13.5	—	15.8	—	14.8	—	14.1	—
	2	−6.1	—	−6.7	—	−7.6	—	−7.8	—
	3	−44.5	—	−45.9	—	−47.0	—	−47.4	—

^{a} SCS of Substituent Y, calculated according to Eq. (2).
^{b} Data from Schraml et al.,[86] error for δ ±0.3 ppm, for $\varDelta\delta$ ±1 ppm.
^{c} Data from Mägi and Schraml,[75] error as in footnote b.

The second case occurs when n is greater than zero. In that case, (a) upfield shifts were observed in cases with $SiC(sp^3)$...Y fragments. In compounds with a SiC_6H_4Y link, upfield shifts were also observed for electronegative X substituents (X = F, Cl, and OC_2H_5) but only when $n = 3$.[53] In cases with intermediate values of n, the shift is probably independent of the substituent Y.[37] When X = H and $n = 3$, the shift varies in the same direction as in trimethylsilyl compounds. (b) Only a few compounds with SiO...Y fragments have been investigated thus far. Pehk et al.[92] studied aminoalkoxysilanes, a few of which would fall in this category.

Corollary data to the above conclusions are given in Tables VI and VII and in Fig. 6. They illustrate cases 1(a) and 2(a) and also show the dependence of the shifts on the length of a Si...Y link. The dependencies of ^{29}Si shieldings on Hammett constants (reported by Ernst et al.[53]) for phenyl substituted silanes are shown in Fig. 7 to further illustrate case 2(a). A few examples of case 1(b) are shown in Table VIII.

As is indicated by these data, the magnitudes of SCS values are not constant since they depend on the electronegativities of substituents X and Y, on the number (n) of X substituents, and, of course, on the ability of the SiC...Y

222   J. Schraml and J. M. Bellama

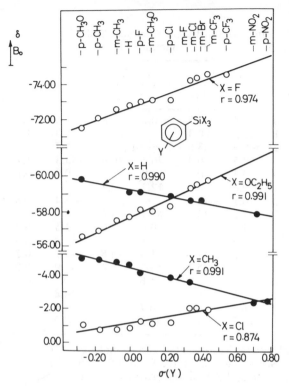

FIG. 7. ^{29}Si chemical shift dependence on Hammett σ constant of substituent Y in substituted phenylsilanes. (Based on a plot of Ernst et al.[53] in J. Amer. Chem. Soc., published by the American Chemical Society.)

link to transmit the influence of substituent Y. The more electronegative the substituents X and Y, the higher the value of n, and the shorter the SiC...Y link, the more the shielding of the silicon nucleus is increased by substituent Y (for additional comment on these upfield shifts, see Section II,H on ^{29}Si shielding theory).

TABLE VIII

^{29}Si Substituent Chemical Shift of Substituent Y[a]

Molecules compared		SCS
SiO...H	SiO...Y	(ppm)
$(CH_3)_3SiO(CH_2)_2H$	$(CH_3)_3SiO(CH_2)_2Cl$	+4.8
$(CH_3)_3SiO(CH_2)_3H$	$(CH_3)_3SiO(CH_2)_3Cl$	+2.9
$(CH_3)_3SiO(CH_2)_3H$	$(CH_3)_3SiO(CH_2)_3Si(CH_3)_3$	−0.4

[a] Data from Lippmaa et al.,[88] SCS values ±0.6 ppm.

F. Miscellaneous Compounds

Silicon-containing compounds can be treated systematically, but many of the compounds fall into several categories and some repetition cannot be avoided. We will be concerned in this section only with silicon hydrides, polysilanes, and silacyclic compounds, which are of current research interest. Other important types of compounds (siloxanes, silicates, and silazanes), which represent the important analytical applications, will be discussed in Section II,G. Data on several other types of compounds (especially the substituted tetraorganosilanes) have already been discussed in connection with substituent effects.

1. Silicon Hydrides and Polysilanes

Papers giving chemical shift data for silicon hydrides are relatively sparse and are scattered through the literature.[13, 38, 39, 53, 56, 65, 93] The situation in polysilanes is similar,[10] although both types of compounds are now being intensively studied.[94, 95] Some of the available data have already been cited in Table II, and it was seen in Fig. 3 that methylsilanes do not show the typical "sagging" pattern. Additional data on hydrides and polysilanes, arranged in the order of decreasing shielding, can be found in Tables IX and X. The chemical shifts in monosilane derivatives cover a δ range from -90 to $+10$ ppm. The shift is dependent both on the number of hydrogen atoms

TABLE IX

^{29}Si Chemical Shifts for Some Silicon Hydrides^a

Compound	δ	Ref.	Compound	δ	Ref.
H_3Si hydrides			HSi hydrides		
$SiH_2(CH_2Si*H_3)_2$	-69.04	54	$Si*H(Si(CH_3)_3)_3$	-117.40	40
$SiCl_2(CH_2Si*H_3)_2$	-64.74	54	$Si*H(CH_3)(OSi(CH_3)_3)_3$	-38.70	14
$C_6H_5SiH_3$	-59.9	65	$SiH(C_6H_5)_3$	-21.1	56
$C_6H_5CH_2SiH_3$	-56.0	65	$SiH(CH_3)(C_6H_5)_2$	-19.5	13
$N(SiH_3)_3$	-39.92	93	$SiH(CH_3)_2(C_6H_5)$	-17.15	38
$CCl_2(SiH_3)_2$	-39.25	54	$SiH(CH_3)(OC_2H_5)_2$	-16.1	53
$SiH_2(CCl_2Si*H_3)_2$	-38.89	54	$SiH(i\text{-}C_4H_9)_3$	-14.7	39
			$SiH(CH_3)_2CH_2Cl$	-12.29	38
H_2Si hydrides			$SiH(n\text{-}C_3H_7)_3$	-8.5	39
$SiH_2(CH_3)(C_6H_5)$	-36.80	6	$SiH(n\text{-}C_4H_9)_3$	-6.7	39
$SiH_2(C_6H_5)_2$	-34.5	13	$SiH(C_6H_5)Cl_2$	-2.2	53
$Si*H_2(CH_2SiH_3)_2$	-30.44	54			
$SiHBr(CH_2Si*H_2Br)_2$	-21.54	54			
$Si*H_2(CCl_2SiH_3)_2$	-5.09	54			

^{a} For data on other hydrides see also Tables II and XIII; the asterisk denotes the silicon atom for which the shift is indicated.

TABLE X

^{29}Si Chemical Shifts for Some Polysilanes^a

Compound	δ	Ref.	Compound	δ	Ref.
Si*(Si(CH$_3$)$_3$)$_4$	-135.2	40	Si(CH$_3$)$_2$[Si*(CH$_3$)$_3$]$_2$	-15.9	38
(F$_3$Si*)$_2$SiF$_2$	-75.5	10	(F$_3$Si)$_2$Si*F$_2$	-13.4	10
Si$_2$F$_6$	-73.5	10	[(CH$_3$)$_3$Si*]$_3$SiH	-13.4	40
Si$_2$(OCH$_3$)$_6$	-52.5	13	[(CH$_3$)$_3$Si*]$_4$Si	-8.7	40
Si*(CH$_3$)$_2$(Si(CH$_3$)$_3$)$_2$	-48.45	38	Si$_2$Cl$_6$	-8.0	40
Si$_2$(CH$_3$)$_6$	-19.58	38	[(CH$_3$O)$_2$(CH$_3$)Si]$_2$	-7.5	13
Si*FCl$_2$SiCl$_3$	-18.7	10	FSiCl$_2$Si*Cl$_3$	-2.5	10

^{a} For data on other polysilanes see Table IX; the asterisk denotes the silicon atom for which the shift is indicated.

bonded to the silicon atom and also on the nature of the other substituents. Nevertheless, the hydrides are easily identified in ^{29}Si NMR spectra by means of ^{29}Si–^{1}H coupling.

The resonance in the SiH$_3$$^-$ anion occurs[28] at considerably higher field ($\delta = -165$), which is well above the range for silicon hydrides but is within the range where other anionic species are found (see Fig. 1).

2. Silacyclic Compounds

Some regular trends emerge from the few chemical shifts available for silacyclic compounds.[6, 38, 54, 81, 82, 84] Lauterbur[6] commented on the effect of strain and substitution on the spectra of cyclic siloxanes (see Section II,G). The strain seemed to decrease the shielding in 1,3-disila-2-oxacyclopentanes as compared to linear siloxanes.[6] The data listed in Table XI indicate a similar tendency for silacycloalkanes of different ring size. (Silacyclopropanes showed, however, signals at $\delta = -52$ ppm,[96] which is rather anomalous.)

In contrast to patterns observed in ^{13}C shielding in similar compounds,[38] the incorporation of a dimethylsilyl fragment into a six-membered ring increased the shielding of the silicon by about 4 ppm (compared to TMS). In five-membered rings the silicon is deshielded by some 16 ppm, and in four-membered rings the shielding is even 2 ppm less.[38] Obviously, the effect of ring size is of comparable magnitude to the effects of α substituents.

Though there are some data on 1,3-disila-, 1,3,5-trisila-, and 1,3,5,7,9-pentasilacyclic compounds,[54, 81, 82, 84] it is difficult to find regularity for ^{29}Si chemical shifts in a Si—C—Si chain. Certainly, the shift of the silicon atom in a C—Si(CH$_3$)$_2$—C fragment undergoes little variation with ring size. In ten-membered rings all of the silicon signals are found around $\delta = 0 \pm 1.5$ ppm (but different stereochemical positions can be distinguished in ^{29}Si NMR

TABLE XI

^{29}Si Chemical Shifts in Some Cyclic Compounds[a]

Compound	δ[b]	Ref.
6-membered rings		
$(CH_2)_5Si(CH_3)(CH_2)_2OH$	-4.11	38
$(CH_2)_5Si(CH_3)CH_2Cl$	-1.31	38
$(CH_2)_5Si(CH_3)CH_2COOC_2H_5$	-1.25	38
$[(CH_3)_2Si-CH_2]_3$	0.3	82
$[H_2Si-CH_2]_3$	-34.1	82
$[H_2Si-CCl_2]_3$	-18.4	82
$[Cl_2Si-CCl_2]_3$	-1.0	82
$[Cl_2Si-CH_2]_3$	19.5	82
5-membered rings		
$(CH_2)_4Si(CH_3)(CH_2)_2OH$[c]	16.33	38
$(CH_2)_4Si(CH_3)_2$[c]	16.77	38
$SiCl_2(a)-CH_2$	(a) 15.1	54
	16.4	54
$SiCl_2(b)-CCl_2$	(b) 7.3	54
$SiCl_2(a)-CCl_2$	(a) 12.8	54
$SiCl_2(b)-CHCl$	(b) 18.1	54
4-membered rings		
$(CH_2)_3Si(CH_3)_2$[c]	18.90	38
$[(CH_3)_2Si-CH_2]_2$	2.73	38

[a] For data on cyclic siloxanes, see Table XIV.
[b] Values converted from source literature using data of Table I.
[c] Typographical error in source literature corrected.

spectra); six-membered rings give lines that occur arround $\delta = 0$ ppm and five-membered rings give lines occurring at $\delta = 2.7$ ppm.

Inclusion of the silicon atom in the ring apparently does not alter its response to substitution. The effects of α substituents, as demonstrated by the data in Table XI on 1,3,5-trisilacyclohexane derivatives, are similar to those discussed earlier for silane derivatives (Section II,D). Similarly, the effects of β substituents exhibit the same trends as described earlier (Section II,E). In particular, the data show that the ^{29}Si SCS values of β substituents depend on the nature of the α substituents. Thus, the SCS value of the two β chlorine atoms is -20.5 ppm on the basis of a comparison of $(Cl_2Si-CCl_2)_3$ and

$(Cl_2Si—CH_2)_3$, but it is $+15.7$ ppm from the comparison of $(H_2Si—CH_2)_3$, and $(H_2Si—CH_2)_3$.

G. Analytical Applications

1. General

Although a comparison of the overall ranges of ^{29}Si and ^{13}C chemical shifts is unfavorable to ^{29}Si NMR applications, there are cases (e.g., siloxanes) when the spectral dispersion of ^{29}Si shifts exceeds that of ^{13}C (and, of course, ^{1}H) chemical shifts in a given class of compounds. The nonmonotonous behavior of ^{29}Si chemical shifts shown in Figs. 3 and 4 could appear as a severe problem in the analytical exploitation of ^{29}Si chemical shifts. The situation is not too serious, however, since for many of the substituents the dependencies are monotonous for all cases in which $n > 0$, and the case of $n = 0$ is not analytically important for such compounds. Nevertheless, it is not possible to construct such simple direct additive schemes for ^{29}Si chemical shifts as are known in ^{13}C NMR spectroscopy (see Levy and Nelson,[70] p. 38) and which would hold for all values of n.

In making ^{29}Si chemical shift spectral assignments by analogy, the utmost care should be taken that the closest possible model is chosen for comparison. In ^{29}Si NMR, because of the nonmonotonous behavior described above, the danger of making erroneous assignments or conclusions is significant.[57, 63]

2. Siloxanes

Since the initial work[6] had very clearly shown that valuable information can be obtained from the ^{29}Si NMR spectra of siloxanes, the spectra of mono-

TABLE XII

Ranges of Chemical Shifts of ^{1}H, ^{13}C, and ^{29}Si Nuclei in Unsubstituted Methyl-siloxanes[a]

Unit[b]	$\delta(^1H)^c$	$\delta(^{13}C)^c$	$\delta(^{29}Si)^c$
M	0.1 to 0.05	1.6 to 1.2	8.0 to 6.1
D	0.14 to 0.00	0.7 to 0.3	-17.8 to -23.0^d
T	0.00 to -0.03	2.1	-65.0 to -66.2
Q	—	—	-105.2 to -110.2^e

[a] Ranges taken from Engelhardt *et al.*[16]
[b] For designation of units see text.
[c] δ Scale in ppm relative to TMS; positive values indicate shifts to low field.
[d] Cyclic compound D_3 not in this range; $\delta = -9.2$ ppm.[16]
[e] Limit of upper field corrected according to Jancke *et al.*[59] Silicate ions occur at higher field; see Fig. 1.

and oligomeric siloxanes have been studied in many laboratories with the goal of developing analytical methods applicable to the industrially important silicone polymers. (For a brief discussion of the advantages of ^{29}Si NMR over other analytical methods in solving structural and analytical problems of polysiloxanes, see Horn and Marsmann.[14] Also, illustrative comparisons with ^{13}C NMR spectra are given by Levy and Cargioli[3] and by the data in Table XII.)

As one would expect from the trend in the shielding of the silicon in $(CH_3)_{4-n}Si(OR)_n$ compounds (see Section I,D,1), the silicon chemical shifts in a number of M, D, T, and Q units in dimethylsiloxanes fall into four separate and nonoverlapping spectral regions,[16,17] whereas the proton and carbon-13 shieldings in these compounds are all very similar. This information is given in Table XIII and depicted in Fig. 1. [The symbols M, D, T, and Q stand for $(CH_3)_3SiO_{0.5}$, $(CH_3)_2Si(O_{0.5})_2$, $CH_3Si(O_{0.5})_3$, and $Si(O_{0.5})_4$ siloxane units, respectively. A substitution for the methyl group is indicated by a superscript; e.g., M^H and D^{CH_2Cl} stand for the $(CH_3)_2SiHO_{0.5}$ and $CH_3(ClCH_2)Si(O_{0.5})_2$ units, respectively.] The occurrence of these four distinct regions is what makes the application of ^{29}Si NMR techniques such a promising tool for structural analyses of silicone polymers.[14,59] Even in

TABLE XIII

^{29}Si Chemical Shifts for Some Polysiloxanes[a,b,c]

Compound	M	^{1}D	^{2}D	^{3}D	^{4}D
MM	6.79	—	—	—	—
MDM	6.70	−21.5	—	—	—
MD$_2$M	6.80	−22.0	—	—	—
MD$_3$M	6.90	−21.8	−22.6	—	—
MD$_4$M	7.0	−21.8	−23.4	—	—
MD$_5$M	7.0	−21.8	−22.4	−22.3	—
MD$_6$M	7.0	−21.8	−22.3	−22.2	—
MD$_7$M	7.0	−21.89	−22.49	−22.33	−22.29
MD$_8$M	6.93	−21.86	−22.45	−22.30	−22.20
MD^{H}M	8.72	−36.86	—	—	—
MD$^H_2$M	9.32	−36.41	—	—	—
MD$^H_3$M	9.52	−36.00	−35.15	—	—
MD$^H_4$M	9.57	−35.90	−35.55	—	—
MD$^H_5$M	9.63	−35.90	−35.40	−35.09	—
MD$^H_6$M	9.63	−35.85	−35.40	−34.99	—

[a] Data for MD$_m$M compounds were obtained in acetone-d_6 solutions relative to internal TMS with an error less than 0.05 ppm, data are taken from Levy and Cargioli.[3]

[b] Data for MD^H_nM compounds were obtained in neat liquids relative to external TMS, estimated error ±0.05 ppm, data taken from Harris and Kimber.[15]

[c] For designation of M, D, and D^H units see the text. D or D^H units are numbered from the ends of the compounds, e.g. M^{1}D^{2}D^{1}DM.

siloxanes that are substituted by chlorine atoms, or by hydroxyl or alkoxy groups, the terminal and middle groups differ considerably in their ^{29}Si shifts.[3] (The only ambiguity reported so far was the assignment of a line in a T^{Cl} unit that occurred at -121.8 ppm[14]; that is, in the region of Q groups.)

With better resolving power, several D units in linear dimethylsiloxanes can be differentiated by ^{29}Si NMR, as is illustrated by the data of Table XIII. If such resolution is required, it is evident that an appropriate solvent must be chosen. For example, Levy and Cargioli[3] could distinguish D^2, D^3, and D^4 groups when such groups were measured in acetone-d_6, but Harris and Kimber[15] could not make such distinctions on a comparable instrument in neat liquids. (The latter authors could, however, differentiate D^H units under the same conditions.)

Open-chain and ten-membered ring compounds can be clearly distinguished from strained cyclic systems by the ^{29}Si spectra, as noted by Lauterbur.[6] In cyclic poly(dimethylsiloxanes) the silicon atoms appeared equivalent[3, 16] and with the exception of a D_3 siloxane (in which $\delta = -9.12$ ppm[3, 59]) the shifts fell in the narrow range of $\delta = -19.5$ to -23.0 ppm. A trend analogous to that in silacycloalkanes is apparent from Table XIV. The shielding increased with increasing ring size.

TABLE XIV

^{29}Si Chemical Shifts in Cyclic Dimethylsiloxanes[a]

Siloxane	D_3	D_4	D_5	D_6
δ	-9.2	-20.0	-22.8	-23.0

[a] Data from Engelhardt et al.,[16] ± 0.3 ppm.

The equivalence of silicon atoms in a D_4 cyclosiloxane is removed if one or two hydrogen atoms in one of the eight methyl groups are replaced by chlorine atoms[16] (D^{CH_2Cl} or D^{CHCl_2}). A substitution by one chlorine atom shifts the resonance of the silicon to which the methyl group was attached upfield by 10 ppm and the other silicon atoms downfield by approximately 1 ppm (see Section II.E).

The effect of a similar substitution in the end M group in a linear polysiloxane is analogous, as the data of Table XV demonstrate. (Data for larger molecules can also be found in the literature.[14, 16] Also available are data on branched and cross-linked siloxanes, some of which have hydroxyl or chlorine atoms as substituents.) The substitution by a phenyl group (Ph) to form a M^{Ph}, M^{Ph_2}, D^{Ph}, D^{Ph_2}, or even a T^{Ph} unit results in a high-field shift that is similar to that found for a D^H group silicon atom when compared to that in a D group (see Fig. 1). The shielding of silicon in M and D groups is sensitive

TABLE XV

^{29}Si Chemical Shifts for Some Substituted Dimethylsiloxanes[a, b]

Compound	M			D		
	X = CH$_3$[c]	X = CH$_2$Cl	X = OCH$_3$	X = CH$_3$[c]	X = CH$_2$Cl	X = OCH$_3$
M$_2^X$	6.3	3.5	−12.0	—	—	—
M$_2^X$D	6.1	2.5	−12.5	−22.1	−19.6	−22.0
M$_2^X$D$_2$	6.7	2.1	−12.0	−22.3	−20.7	−20.4

[a] Data taken from Engelhardt *et al.*[10]
[b] Values relative to external TMS; neat liquids; error ±0.3 ppm.
[c] M^X = M, D^X = D.

not only to the substitution in this unit, but also to the arrangement of the neighboring units. Thus, for example, the spectrum of a mixed polymer with D and D^H units shows in the M region 8 lines, which correspond to all possible combinations of the three neighboring D units (MD^{H}D^{H}D^H, MD^{H}D^{H}D, MD^{H}DD^H, MD^{H}DD, MDD^{H}D^H, MDD^{H}D, MDDD^H, and MDDD).[15] Therefore, the silicon nucleus in M units is susceptible to changes in substitution at distances up to six bonds.[15] Similarly, the silicon in D units responds to changes four bonds away and in D^H units to changes three bonds away.[15] Recently, tacticity effects were reported in the ^{29}Si spectra of silicones that contain D^H groups.[97]

On the basis of this knowledge the spectrum of a polymer can indicate the minimum length of the units repeated in the skeleton (diads, triads, etc.) and which of the possible units are actually present in the given polymer. Such an approach was first applied by Engelhardt *et al.*[17] when a methylphenylsiloxane was shown to have a skeleton consisting essentially of D^{Ph}—D—D^{Ph} triads with D^{Ph}—D^{Ph}—D^{Ph} triads randomly mixed.

3. Silicates

In silicates that have four oxygen atoms bonded to the silicon, the shift depends on the position of the silicon in the silicate chain.[58, 98, 98a] In orthosilicate anions (without any Si—O—Si grouping) the ^{29}Si resonance occurs at δ = −68.7 to −72.3 ppm. It is found at −77.4 to −81.0 ppm for a silicon in the end position (one Si—O—Si group), at −85.3 to −89.2 ppm for a silicon in the middle (two Si—O—Si groups), at −92.6 to −96.6 ppm for branching silicons (three Si—O—Si groups), and at −107 to −120 ppm for crosslinked silicon atoms (four Si—O—Si groups),[58, 98] as illustrated in Fig. 1.

4. Silazanes

The ^{29}Si chemical shifts in silazanes cover a much smaller range than they do in the analogous siloxanes. This is illustrated in Fig. 1, where the abbreviations m, d, d^{Ph}, and d^{Ph_2} were used for groups analogous to M, D, D^{Ph}, and D^{Ph_2}, respectively, and in which an oxygen was replaced by a nitrogen atom. These compounds have been studied much less than siloxanes.[60] (For model dimethylaminosilanes, see Engelhardt et al.[39]) It was shown that despite the smaller range, ^{29}Si NMR can be applied in exactly the same way to silazanes as it is to siloxanes and that measurements of NOE enhancement factors provide additional important information, which is useful for spectral assignments.[60]

5. Miscellaneous

^{29}Si NMR has been used very efficiently for structural determinations in a series of papers on carbosilanes (especially cyclic ones) from the laboratory of Fritz.[54, 81, 82, 84] The proof of the structure of the first silacyclopropane was aided by ^{29}Si NMR[96]; the signal that occurred at $\delta = -52$ ppm in a substituted dimethylsilacyclopropane is shifted upfield from the resonance in other tetraalkylsilanes and was taken as indicative of an unusual SiC_2 cyclic system. The work of Mägi et al.[61, 68] was aided by ^{29}Si NMR in a study of exchange reactions.

In the future, ^{29}Si NMR will be used for the elucidation of the structure of various organic materials that are frequently silylated in synthetic or separation procedures. The potential of this method to indicate the number of different hydroxy groups present in the molecule has been demonstrated by Schraml et al.[99] who showed that the ^{29}Si NMR spectrum of 1,6-anhydro-2,3,4-tri-O-trimethylsilyl-β-D-glucopyranose consists of three distinct lines.

H. ^{29}Si Shielding Theory

At present, knowledge of the current trends in ^{29}Si shielding theory is not a prerequisite for successful application of ^{29}Si NMR to structural problems. However, since discussion of shielding theory is a pervasive theme in papers on ^{29}Si NMR, it seems desirable to present here a brief account of these theories. Such an account not only helps in reading the literature, but also, when correlations between ^{29}Si chemical shifts and structural parameters are better established (and their limitations understood), a knowledge of theory could help in solving structural problems.

Generally, the chemical shift δ^A observed for nucleus A in a sample is (within some precision limits[100]) equal to the difference of the shielding constants σ_{obs}^R and σ_{obs}^A in the reference and in sample (A).

$$\delta^A = \sigma_{obs}^R - \sigma_{obs}^A \tag{3}$$

Both σ_{obs}^R and σ_{obs}^A can be written as the sums of the shielding constants of the isolated molecules (σ) and the contributions originating outside the molecule: the overall bulk susceptibility effect (σ_b); and contributions from anisotropy in the molecular susceptibility (σ_a), from van der Waals forces between the molecules (σ_w), from the electric field of neighboring molecules (σ_e), and from specific interactions (σ_s).[101] Thus,

$$\delta^A = -\sigma^A + \sigma^R + \Delta(\sigma_b + \sigma_a + \sigma_w + \sigma_e + \sigma_s) \tag{4}$$

where the symbol Δ denotes the difference between the corresponding terms of reference (R) and sample (A).

In principle, $\Delta\sigma_b$ could be eliminated from Eq. (4) either by employing an internal reference or by application of a bulk susceptibility correction. Other Δ terms of Eq. (4) can be evaluated or eliminated through gas-phase or solvent and concentration effect studies, but such studies have thus far been concerned very little with ^{29}Si NMR, and there is no reliable estimate of the total magnitude of the Δ terms for this nucleus. Usually, it is estimated that in a series of related compounds the variation in the sum of the Δ terms does not exceed 1-2 ppm and the shielding constant is identified with the chemical shift.

Interpretation of variations in shielding constants has been largely based on models in which σ is split into various contributions. The shielding constant σ^A may be written[102, 103] as

$$\sigma^A = \sigma_d^{AA} + \sigma_p^{AA} + \sum_{B \neq A} \sigma^{AB} + \sigma^{A, \text{ring}} \tag{5}$$

The meaning of the terms in Eq. (5) is as follows:

(a) σ_d^{AA} is the diamagnetic contribution arising from electronic circulations centered on atom A; it is the diamagnetic part of the Ramsey expression for shielding[104]

$$\sigma_d^{AA} = \frac{e^2}{3mc^2} \sum_k \langle r_k^{-1} \rangle \tag{6}$$

in which the mean inverse distances of electron k from the nucleus A in the molecular ground state are summed over the electrons on atom A. (Recent calculations,[105] however, have demonstrated that nonlocal contributions to σ_d can contribute as much as 35% of the total of σ_d.)

(b) σ_d^{AA} is a paramagnetic contribution due to restricted currents on atom A.[104] In the usual approximation of average excitation energy, ΔE_{av}, this term can be expressed[35] as

$$\sigma_p^{AA} = -\frac{2e^2h^2}{3m^2c^2}\frac{1}{\Delta E_{av}}\{\langle r_p^{-3}\rangle \cdot P^A + \langle r_d^{-3}\rangle \cdot D^A\} \tag{7}$$

where the average inverse cubes of the distances of the valence p and d electrons from nucleus A are multiplied by P^A and D^A, respectively, which are complex expressions[35] representing the "unbalance" of the valence p and d electrons.

(c) σ^{AB} represents the effects from currents on other atoms in the molecule ("neighbor anisotropy" effects).[103]

(d) $\sigma^{A, \text{ring}}$ is the ring current contribution arising from nonlocalized circulations within the molecule.[103]

The last two contributions are believed by many authors to be of smaller magnitude than the first two terms, at least for atoms other than hydrogen.[103, 104]

Since Saika and Slichter[102] demonstrated that the σ_p^{AA} term alone can account for the main aspects of fluorine chemical shifts, it has been generally accepted that the changes in the shifts of heavier nuclei are due to the changes in this paramagnetic term. It was recently claimed,[106] however, that σ_d^{AA} is much more important than had been previously recognized.

Another possibility, in addition to that given by Eq. (5), is to split up σ^A into contributions arising from electric fields created by polar groups (σ_{electric}) and contributions from electronic charge densities ($\sigma_{\text{electronic}}$)[107]:

$$\sigma^A = \sigma_{\text{electric}}^A + \sigma_{\text{electronic}}^A \qquad (8)$$

The theoretical approaches to ^{29}Si shielding will be discussed in reverse chronological order, since the two most recent interpretations[39, 53] are the most rigorous despite the approximations which are required.

Both approaches neglect all other terms in Eq. (5) except σ_p^{AA}, which is used as given in Eq. (7). Thus, both make use of the average excitation energy approximation; neither considers the d orbitals of the Si atom. In further stages of development the two theories proceed differently but arrive at similar conclusions which agree with experimental results.

Ernst et al.[53] simply adopted the theory of ^{31}P shielding, which was developed by Letcher and Van Wazer.[108, 109] In addition to the above approximations, this theory considers $\langle r_p^{-3} \rangle / \Delta E_{av}$ to be a constant for a given coordination number. Assuming that σ molecular orbitals use only s and p orbitals from silicon, considerations of normalization and the electronegativity requirements for molecules of the $YSiX_3$ type result in the following expression for ^{29}Si shielding.[109]

$$\sigma = B \cdot [0.5h_X^2 + h_Y - h_Y h_X + 0.5(h_X - h_Y^2 - h_Y h_X - h_Y)/\sin^2(\theta/2)] \qquad (9)$$

where B is a negative constant, θ is the X—Si—X bond angle, and the quantities h_A are related to the electronegativities χ_A of atoms A (A = X or Y) and silicon through the relationship

$$h_A = 1.0 + 0.16(\chi_{Si} - \chi_A) + 0.035(\chi_{Si} - \chi_A)^2 \qquad (10)$$

The primary concern of this theory was to explain the different signs of the slopes observed in Hammett-type plots of ^{29}Si chemical shifts in substituted phenylsilanes (Fig. 7). Indeed, depending on the electronegativity of the X substituents, Eqs. (9) and (10) predict (for a constant angle) both positive and negative slopes of the ^{29}Si shift dependence on the electronegativity of sub-stituent Y.[53] (There is, however, some discrepancy between the theoretically calculated and experimentally found χ_Y values at which the reversal in slope occurs.[53]) In addition, Eqs. (9) and (10) suggest that the dependence of σ on χ_X (for a given substituent Y and a fixed bond angle) exhibits a sagging pattern. Such a pattern was found experimentally for $YSiX_3$ compounds.[53] The more general dependence of $R^1R^2R^3R^4Si$ compounds on $\sum \chi$ [Fig. 4, Eq.(1)] appears as a natural extension of this finding, although it had not been con-sidered theoretically.[53] The authors have not shown how this theory can ex-plain the difference between the trends in ^{29}Si and ^{13}C shieldings.

The theory of Engelhardt et al.[39, 89, 110] was developed earlier than the one discussed above. Its goal was to account specifically for the large effects observed in the shielding of $R^1R^2R^3R^4Si$ compounds, and also to account for the difference between ^{29}Si and ^{13}C shifts. It was shown, however,[37, 89] that this theory can also account for much smaller effects, e.g., those observed by Ernst et al.[53, 63, 90] Engelhardt's theory, on the one hand is more general than Ernst's theory since it also considers variations in the $\langle r_p^{-3} \rangle$ term. On the other hand, it assumes tetrahedral bond angles on silicon. Instead of calcu-lating the shielding constant σ, Engelhardt[39] calculates a "relative shielding constant" σ^*, which is the shielding constant given as a fraction of the shielding constant σ° in a completely nonpolar molecule. Thus, within the discussed approximations,

$$\sigma^* = \frac{\sigma}{\sigma^{\circ}} = \frac{\langle r^{-3} \rangle_p}{\langle r^{-3} \rangle_{p^{\circ}}} \cdot \frac{P}{P^0} \tag{11}$$

If Slater orbitals are used, this expression can be rewritten as

$$\sigma^* = (1 + fkq)^3 \cdot \frac{P}{P^0} \tag{12}$$

where $q = 4 - (4 - n)h_Y + nh_X$ and $k = 0.0843$. The factor f was introduced to correct for the known inadequacies of Slater orbitals. Its value is found empirically. The P/P^0 ratio must be calculated for various types of $Y_{4-n}SiX_n$ compounds from h_A values [Eq. (10)] according to the symmetry of the compound.

With the similar behavior of σ^* and the observed shifts (and a scaling factor between the two), this theory has been remarkably successful in pre-dicting trends in ^{29}Si shieldings in $(CH_3)_{4-n}SiX_n$ compounds and in accounting for the different trends in the carbon analogues. See, for example, Figs. 3 and 8. Apparently, the theory can also accommodate the effects of substituents not directly bonded to the silicon atom except when the compound contains a SiO...Y link. In the latter case the theory is apparently not adequate to account for the subtle changes in the shielding.[37]

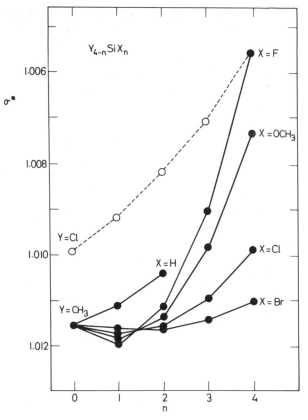

FIG. 8. ^{29}Si calculated relative shielding constant dependence on multiple substitution. (Compare with the experimental dependence in Fig. 3. Based on a plot of Engelhardt *et al.*[39] in *Org. Magn. Resonance* published by Heyden & Son, Ltd.)

The success of the theory of Engelhardt *et al.*[39] is especially interesting in view of another theoretical approach to ^{29}Si shielding. Lyubimov and Ionov[111] have shown that in a series of $Y_{4-n}SiX_n$ compounds (X and Y = halogens) the variations in the average excitation energy ΔE_{av} (the variations are due to changes in symmetry with substitution) are such that they alone can account for the observed trends in the shielding.

Van den Berghe and van der Kelen[48] claim that it is more likely that the major factors influencing the upfield shifts that occur with increasing halogen substitution on silicon arise from σ^{AB} and $\sigma^{A}_{electric}$, and also that σ_d must be taken into account.

The first model to account for ^{29}Si shielding was suggested in the early work of Lauterbur.[6] It stemmed from the observation of the "sagging pattern," and from the highly shielded silicon atoms encountered in those compounds in which silicon $3d$ orbital participation in bonding was considered to be related to certain other properties (for a review, see Bažant et al.,[1] Chapter 3). More recently, however, the concept of hyperconjugation[112] as an alternative interpretation of these unusual properties has gained popularity. Since the discussions in the original papers are given in the terms of $(p \rightarrow d)\pi$ bonding, that concept is utilized in this chapter.

Some details of this model were elaborated by Hunter and Reeves,[13] who used the $(p \rightarrow d)\pi$ donation concept to interpret the sagging pattern and the additional shielding in $(CH_3)_3SiX$ compounds. Correlations of δ with the results of the quantum chemical calculations of Nagy et al.[113, 114] were also interpreted as consistent with $(p \rightarrow d)\pi$ bonding in organosilicon compounds.

In brief, this model would at present view ^{29}Si shielding as a result of two (usually opposing) effects. The inductive effect is supposed to operate on silicon in essentially the same way that it does on carbon. The difference in the trends in shielding would then be due to the other effect, $(p \rightarrow d)\pi$ interaction, which is not encountered in carbon analogues. This interaction, or back-bonding, increases the shielding of silicon by increasing the electron density on the silicon atom. The different upfield shifts are then easily interpreted by the differing ability of substituents to interact. In compounds with XSiC...Y groupings, the substituent Y also acts inductively, but with the principal result of enhancing $(p \rightarrow d)\pi$ interaction, X $\rightarrow$ Si, if the substituent X is capable of such interaction. If not, a simple inductive shift to low field is observed for an electronegative substituent Y. (SCS would then be positive, as is found in trimethyl derivatives). In compounds with SiO...Y groupings, the main effect is the weakening of the Si—O $(p \rightarrow d)\pi$ interaction which results in shifts to low field.[37]

In summary, this model gives a good qualitative account of ^{29}Si shielding trends even though it is based on rather naive oversimplifications, which neglect all theory of shielding. On the other hand, the description provided thus helps to visualize trends in shielding in terms used to explain other properties of organosilicon compounds.

III. ^{29}Si SPIN–SPIN COUPLING CONSTANTS

A. General

The theory of spin–spin coupling is treated extensively elsewhere[115–117] and will be mentioned here only when necessary to consider calculation of coupling constants or correlations of coupling constants with structure. This section will instead be primarily concerned with the magnitudes, signs, and semiempirical calculations of coupling constants between the ^{29}Si nucleus and other nuclei, and with correlations of these coupling constants with structural features of the molecule.

The ease of examining ^{29}Si coupling has had the effect of making this parameter and its associated information easily accessible to organic chemists who routinely measure proton spectra, and it has already provided voluminous data in the literature. Since it is not feasible to list comprehensively all coupling constant data in this section, individual papers are cited only when the information contained can give an indication of the typical magnitude, sign, or calculation of ^{29}Si coupling with other nuclei, or when structural correlations are apparent.

B. Calculation of Spin–Spin Coupling Constants

It was shown by Ramsey[118] that of the interactions leading to nuclear spin–spin coupling (a Fermi contact mechanism, a spin–orbital mechanism, and a spin-polarization or dipole–dipole mechanism), the contact mechanism provides the major contribution to the coupling. McConnell[119] used LCAO molecular orbitals to extend this work to larger molecules, but his calculations used an "average excitation energy approximation" that resulted in the same relative sign for all directly bonded coupling constants. Pople and Santry[120] avoided the average excitation energy approximation and calculated signs and magnitudes of coupling constants of directly bonded atoms as heavy as fluorine, but the problem of excitation energy continued to plague later workers who attempted to calculate and correlate ^{29}Si coupling constants. Reeves[121, 122] suggested that ΔE_{av} is relatively independent of the atomic number (Z) of the nucleus, while other workers stated that ΔE_{av} would become relatively less important as the atomic number increased.[35] Dreeskamp[123] assumed ΔE_{av} to be constant when coupling between directly bonded nuclei is considered, and Smith[124] also concluded that an assumption of a constant ΔE_{av} permits a better calculation of $^{2}J_{^{29}Si-C-^{1}H}$ than calculations that sum the bond energies.

Most authors agree, however, that the Fermi contact is the most important term in determining the magnitude of the coupling. Recent theoretical studies[125] have shown, however, that in some cases of $^{13}C-^{19}F$ coupling, orbital contributions must also be considered, and the same may also be true in coupling involving the silicon nucleus. Certainly, as the H—Si—H angle deviates from tetrahedral, the orbital contribution would become more important.[126]

Much consideration has been devoted to $^{13}C-^{1}H$ coupling,[127, 128] and as frequently is the case in organosilicon chemistry, it was natural to attempt to explain ^{29}Si$-^{1}H$ coupling by similar arguments. However, there are several problems in evaluating the factors that determine carbon and silicon coupling constants: (1) the evaluation of the above-mentioned average excitation energies from singlet to triplet states; (2) the effective electron densities at the nuclei of elements found in the second and third rows, respectively, of the periodic table; and (3) bond polarities, which obviously differ considerably. For example, the sp^3 ($-$)C—H($+$) bond has a dipole moment of 0.4 D, while the ($+$)Si—H($-$) bond has the opposite polarity and a surprisingly large bond dipole moment of 1.0 D.[129]

Malinowski had observed that the variation of $^1J_{^{13}C-^1H}$ in substituted methanes of the type CHXYZ could be correlated by a linear additivity rule.[130] This additivity rule was applied to silicon but it was found that with electronegative substituents, deviations occurred.[126,131]

Specifically, Juan and Gutowsky[131] noticed that halosilanes deviated rather systematically from additivity: (1) deviations were always positive, i.e., $|J_{obs}| = |J_{add}| + \varDelta$; (2) the more electronegative halogens produced the greater deviation, i.e. $\varDelta$ values decrease in the order F > OCH$_3$ > OSiH$_3$ > Cl > Br > N(CH$_3$)$_2$ > I for SiH$_2$X$_2$ compounds; (3) for a given halogen, the deviation increased according to the number of halogen atoms bonded to silicon, i.e., $\varDelta$(SiHX$_3$) > $\varDelta$(SiH$_2$X$_2$). Deviations can be considerable. For example, Ebsworth[132] has noted that the calculated $^1J_{^{29}Si-^1H}$ for SiHF$_3$ is 26% low.

Thus, it appeared that coupling constants were affected by the electronegativity of the substituent, and that the greater polarizability of silicon (compared to carbon) made difficult an extrapolation of ^{13}C–^{1}H coupling considerations to the analogous silicon systems.

Jensen and co-workers used the Pople-Santry method of calculating coupling constants in a study of substituent effects on couplings of directly bonded ^{29}Si–^{1}H (and ^{13}C–^{1}H) nuclei,[133–135] and both Malinowski and Jensen found that the use of pairwise interaction terms gave considerable improvement.[134,136] For SiHXYZ compounds, the ^{29}Si–^{1}H coupling constant is given by the expression

$$^1J_{X,Y,Z} = J'_{XY} + J'_{YZ} + J'_{ZX} \tag{13}$$

in which J' represents a constant for each pair of substituents. The parameters[134] for some substituent pairs are given in Table XVI.

TABLE XVI

Coupling Constant Parameters for Pairwise Additivity[a]

J'_{H-H}	67.6						
J'_{H-Cl}	84.7	J'_{Cl-Cl}	120.5				
J'_{H-CH_3}	67.5	J'_{Cl-CH_3}	79.9	$J'_{CH_3-CH_3}$	63.1		
$J'_{H-C_6H_5}$	66.0	$J'_{Cl-C_6H_5}$	84.6	$J'_{CH_3-C_6H_5}$	63.4	$J'_{C_6H_5-C_6H_5}$	66.1

[a] Data from Bishop and Jensen,[134] for use in Eq. (13).

In general, $^1J_{^{29}Si-^1H}$ varies over a range of some 200 Hz as the number of substituents for F- or Cl-containing species increases from 0 to 3, varies up to 100 Hz for O- and N-containing species, and remains essentially constant or decreases slightly with an increasing number of H, C_6H_5, and CH_3 substituents.[134]

Although pairwise interactions can give satisfactory results, much succeeding work has been based on the use of a quadratic equation proposed by Jensen[135]

$$J(MHXYZ) = J(MH_4) + A(\alpha_X + \alpha_Y + \alpha_Z) + B(\alpha_X^2 + \alpha_Y^2 + \alpha_Z^2)$$
$$- C(\alpha_X\alpha_Y + \alpha_Y\alpha_Z + \alpha_Z\alpha_X) \tag{14}$$

in which M can be either C or Si and α_X is the Coulomb integral or "effective electronegativity" of the substituent X. In the simplest case, α_X can be calculated from $J(SiH_3X)$, in which case Jensen's equation simplifies to

$$J(SiH_3X) = J(SiH_4) + A\alpha_X + B\alpha_X^2 \tag{15}$$

with the derived values of α_X falling in the range of -4 to $+1$ eV. Since slightly different values of α_X are found for each X-substituted silane, a least-squares fit of α_X is desirable. By definition, α_H is zero. Jensen found a standard deviation in his study of 2.3%, which he considered quite satisfactory for a method that considers only the Fermi contact contribution to coupling.[135]

The effect on $^1J_{^{29}Si-^1H}$ of a β substituent (one atom removed from the M—H fragment in which the coupling is measured) is obviously less than the effect of an analogous α substituent. The effect of a β substituent can be calculated from the expression

$$\Delta(\beta - X) = K\alpha_X \tag{16}$$

in which $K = 3.9$ Hz/eV. The calculated value of $\Delta(\beta - X)$ is added to the value of the constant calculated without considering the β substituent.[135]

Other workers have suggested additional modifications of the additivity relationship proposed by Jensen. For example, Ebsworth and co-workers[132] have noted that the difference between J_{obs} and a J_A calculated from the additivity relationship

$$J_A(^{29}SiH_2XY) = J(^{29}SiH_3X) + J(^{29}SiH_3Y) - J(^{29}SiH_4) \tag{17}$$

is linearly proportional to the square of the electronegativity difference between the substituent and hydrogen. They propose a modification such that

$$J_B = J_A - 7.5(2\chi_H - \chi_X - \chi_Y)^2 \tag{18}$$

and find that values of J_B are within experimental error of the observed coupling constants. The authors comment that this successful calculation involved only one empirical parameter,[137] whereas the quadratic additivity scheme suggested[133-135] for halogenosilanes is in error by 10 Hz for SiH_2ICl despite a larger ratio of parameters to observations.

Yoshioka and MacDiarmid[138] have proposed for compounds of the general formula $M'H_3MH_2X$ (M and M' may be the same or different Group IV elements) that Eq. (14) simplifies to the form

$$J_{MH} = J_{MH}(MH_3M'H_3) + [J_{MH}(MH_3X) - J_{MH}(MH_4)] \qquad (19)$$

This equation is satisfactory for C_2H_5X, CH_3SiH_2X, and SiH_3CH_2X compounds, but it does not hold for Si_2H_5X compounds. For the latter case, Eq. (20) was proposed.

$$J_{^{29}Si-^1H}(SiH_3SiH_2X) = J_{^{29}Si-^1H}(Si_2H_6) + \tfrac{3}{4}J_{^{29}Si-^1H}(SiH_3X) - J_{^{29}Si-^1H}(SiH_4) \qquad (20)$$

Drake has also based correlations of disilanyl derivatives on Jensen's calculations. Using Jensen's equation[135] [Eq. (14)], and the values $A = 18.26$ Hz/eV, $B = 1.78$ Hz/eV, and $C = -1.28$ Hz/eV, the typical literature values of α, and also $\alpha_{SiH_3} = 0.57$, $\alpha_{SiSiH_3} = -1.37$, and $\alpha_{N-SiH_3} = -0.72$ (positive values denote substituents less electronegative than hydrogen), Drake found good agreement between predicted and experimental values for disilanyl species.[139, 140] He suggested an additional modification for a silicon atom that is contiguous with a substituted silicon [Eq. (21)].

$$J_{^{29}Si-^1H} = J_{^{29}Si-^1H}(Si_2H_6) + 2.25\alpha \qquad (21)$$

Jensen's work suggests again that ^{29}Si coupling observations can be explained on the basis of the Fermi contact term alone, and that important perturbations are related to the electronegativity of the substituents.[133] Juan and Gutowsky[131] have shown that if the contact mechanism is dominant, then the magnitude of the coupling constant is a function of s character utilized by the silicon in its bonding orbital to the coupled nucleus. The degree of hybridization of an orbital can be calculated by the method of Juan and Gutowsky.[131] For a discussion of the limitations of the method as applied to ^1H coupling constants, see Gil and Geraldes.[128]

Rastelli and Pozzoli[141] correlated $J_{^{29}Si-^1H}$ with the s character of the hybrid orbitals (calculated by the Del Re method) and found that, apart from additive contributions, the simple relationship $J_{^{29}Si-^1H} = 810\,\alpha_H^2$ is obeyed, where α_H^2 represents the s character of the silicon orbital directed toward hydrogen. When the substituents were phenyl or methyl groups, extra contributions to the coupling constants appear to be present. Juan and Gutowsky[131] concluded from their study on halosilanes that owing to the high polarizability of the electron cloud of Si, a substituent more electronegative than hydrogen should cause a significant positive deviation from additivity as a result of its $-I$ effect. Rastelli and Pozzoli[141] found this result to hold for phenyl groups, whereas methyl groups show a smaller effect in the opposite direction in agreement with the respective values of σ^I inductive substituent constants for the phenyl and methyl groups.

The coupling appears to be sensitive to so many details of molecular structure that the general nature of the coupling mechanism can be obscured.[142] It was noted quite early that $^1J_{M-H}$ of the Group IV hydrides increased monotonically with the nuclear charge Z_M, a trend compatible with the Fermi contact term dominating the coupling mechanism.[35]

Smith[124] has related the M–C–H geminal coupling in $(CH_3)_4M$ compounds of Group IV to the H–C–H geminal coupling in methane, but the problem of choosing a form for the orbitals of the central atom is more difficult for the heavier elements. Smith adopted an empirical approach and calculated effective nuclear charges (Z^*) for s-like orbitals that gave the best fit with observed coupling constants.

However, coupling constants for different pairs of nuclei can be meaningfully compared only if their values are "corrected" for the different signs and magnitudes of the magnetogyric ratios of the nuclei. For this purpose the "reduced" coupling constant (K) is favored. According to the accepted theory,[23] the reduced coupling constant, as proposed by Pople and Santry,[120] is independent of nuclear spins and moments and therefore eliminates the trivial dependence of J_{M-X} on the magnitude of the dipole moments of the coupled nuclei. The reduced coupling constant is given by the expression

$$K_{M-X} = (2\pi/\hbar\gamma_M\gamma_X)J_{M-X} \tag{22}$$

It was shown that $|^2K_{M-C-H}|$ from $(CH_3)_4M$ compounds could be correlated with Z_M,[122, 124, 143, 144] and the determination of $|^1K_{M-H}|$ in the MH_4 hydrides established a similar dependence on the atomic number of the Group IV element.[123]

Reeves and co-workers[121, 122] noted a linear relationship between Z_M^2 and $^2K_{M-C-H}$ in $(CH_3)_4M$ compounds. They also found that a linear equation is obeyed for other Group IV molecules[145] and proposed that a general expression [Eq. (23)] when it has been established for a given vertical family in

$$(J_{M-X}/\gamma_M\gamma_X)^{1/2} = AZ_M + B \tag{23}$$

the periodic table, is a sensitive test for hybridization at atom M.[146] This relationship correlates geminal M–H coupling constants quite well, but the fit for one-bond C–M coupling constants is less satisfactory. McFarlane[79] suggested that hybridization effects should be considered in such correlations, and Dreeskamp[147] noted that $\ln(K_{M-X})$ gives good correlations with Z_M. Dalling and Gutowsky[148] have also studied this relationship.

Reeves[121, 145] has also calculated a related quantity which he terms a "deviation parameter," D_{M-H}, which is given as

$$D_{M-H} = [(J_{MH_4}/\gamma_M\gamma_H)^{1/2} - (J_{MH_nX_{4-n}}/\gamma_M\gamma_H)^{1/2}] \tag{24}$$

The variation of D_{M-H} is much greater in silicon than in carbon compounds, which may be due to the greater polarization of silicon. It has been suggested[132] that D/n (i.e., deviation per bond) might represent the relative polarizing power of a substituent.

Thus far, no comment has been made about the sign of the coupling constant. Values were initially reported only as magnitudes, but with the use of various methods of determining relative signs and with the determination of the absolute sign of a few coupling constants, signs are now well known. Jameson and Gutowsky[142] have listed a number of methods for determining relative signs, such as spectral analysis, double resonance and heteronuclear multiple resonance, double quantum transitions, and nuclear Overhauser effects. Absolute signs have been variously determined from high-resolution NMR in those cases in which cross terms between different relaxation mechanisms are important, from spectra of partially oriented molecules obtained by application of a strong electric field to a sample of polar molecules, or by dissolving the molecule of interest in a liquid crystal matrix, by a molecular beam technique, and by line-shape analysis of broad-line spectra in solids. In long-range coupling (over more than two intervening bonds), the determination of sign is of particular interest because of the possibility of cancellation of contributing terms.[149] Initial efforts to calculate signs were not without difficulty. Cowley and co-workers[150, 151] found that the most satisfactory calculation of J resulted from the extended Hückel method (which included overlap but excluded electron–electron interactions). This method, however, and also self-consistent field (SCF) calculations in which differential overlap was neglected, both failed to reproduce the sign of $^1J_{^{29}Si-^{19}F}$. The trouble was not rectified by inclusion of spin–orbital or spin–dipolar contributions. A LCAO/SCF method was then tried. The best results were obtained from a parameterized SCF method that included overlap in the diagonalization of the secular equation. The results of the calculations gave the proper signs and a close fit of the magnitude of $^1J_{^{29}Si-^{19}F}$ in SiF_4, although the calculated magnitudes of CH_3Y ($Y = NH_2$, PH_2, and SiH_3) couplings were much too low.

SiF_4 proved a troublesome compound to other workers. Pople and Santry[120] suggested that the reduced coupling constant depends not only on the s character of the orbitals of the central element, but also on a term that is a linear function of the ionicity of the bond. In MF_4 compounds, this ionicity is important and hence such compounds tend to deviate from empirical relationships.

When the magnitudes of reduced coupling constants are similar over a number of compounds, the signs are expected to be identical. The evidence suggests that $^1K_{^{29}Si-^1H}$ is positive,[93] and Pople and Santry[120] suggested that K should be positive in all cases except those involving coupling to fluorine. Since silicon has a negative magnetogyric ratio (Section I,B), J and K are opposite in sign. References to studies involving the determination of signs of (nonreduced) coupling constants involving the ^{29}Si nucleus are given in Table XVII.[8, 142, 149, 150, 152–155]

TABLE XVII

The Signs of ^{29}Si Coupling Constants

Nuclei	Sign	Ref.
$^1J_{^{29}Si-^1H}$	−	8, 142, 150, 152–154
$^1J_{^{29}Si-^{19}F}$	+	8, 142, 150, 152
$^1J_{^{29}Si-^{13}C}$	−	142, 150, 153, 155
$^2J_{^{29}Si-C-^1H}$	+	153
$^3J_{^{29}Si-C-C-^1H}$	−	149
$^3J_{^{29}Si...^1H}$	+	154

A description of a typical experimental technique used in a $^1H-\{^{29}Si\}$ "tickling" double resonance experiment on $(CH_3)_3SiF$ was given by Danyluk.[152] Since $^1J_{^{29}Si-^{19}F} \gg {}^2J_{^{29}Si-C-^1H}$ it is possible to irradiate near one set of ^{29}Si multiplets without perturbing transitions of the other set. The relative signs of $^1J_{^{29}Si-^{19}F}$ and $^3J_{^1H-C-Si-^{19}F}$ can therefore be determined by irradiating one of the ^{29}Si multiplets and observing which pair of satellite lines collapses. For $(CH_3)_3SiF$, irradiation of the lowest ^{29}Si frequency collapsed the low-field satellite. Increasing the ^{29}Si frequency by 280 Hz ($^1J_{^{29}Si-^{19}F}$) collapsed the high-field satellite. Therefore, $^1J_{^{29}Si-^{19}F}$ and $^3J_{^1H-C-Si-^{19}F}$ have the same sign (note that γ for Si is negative).

C. One-Bond Coupling

1. $^{29}Si-^1H$ Coupling Constants

Coupling between directly bonded $^{29}Si-^1H$ nuclei in SiH_4, the parent hydride, amounts to -202.5 Hz. In higher members of the series, Si_nH_{2n+2}, $J(SiH_3)$ is usually -198 to -200 Hz, while $J(SiH_2)$ is slightly smaller, about -192 to -194 Hz.[156] Isotope effects are very small. The substitution of one deuterium for protium in SiH_3I reduces $|^1J_{^{29}Si-^1H}|$ by only 0.3 Hz, and the same substitution in $SiH_2(CH_3)_2$ lowers J by 0.2 Hz.[144]

The primary cause of larger $|^1J_{^{29}Si-^1H}|$ values in many substituted compounds can be attributed to the presence of the electronegative substituents. The effect of halogen substitution is given in Table XVIII.[126, 157]

It can be seen that in any given series $SiH_{4-n}X_n$ ($n = 1–3$), the substitution of a more electronegative halogen for another halogen has an irregular effect with the overall magnitude within the series not varying widely. As the halogen substitution increases from 0 to 3, however, the absolute value of the coupling

TABLE XVIII

Coupling Constants of Halosilanes[a,b]

Compound	F	Cl	Br	I
SiH₃X	−229.0	−238.1	−240.5	−240.1
SiH₂X₂	−282	−288.0	−289.0	−280.5
SiHX₃	−381.7	−362.9	−360[c]	−325[c]

[a] In Hz.
[b] Data from Ebsworth and Turner.[126]
[c] Data from Hengge and Höfler.[157]

increases approximately 40 Hz from silane to monohalogenosilanes, approximately 50 Hz from mono- to disubstituted silanes, and nearly 50 Hz from di- to trihalogenated silanes.

In Table XIX are given the coupling constants for two series of mixed trihalogenated silanes, $HSiX_nY_{3-n}$, in which X and Y are different halogens. In these cases a regular variation is readily apparent.

TABLE XIX

$^1J_{^{29}Si-^1H}$ Coupling Constants of Mixed Trihalogenosilanes[a]

Compound	Coupling constant[b]	Compound	Coupling constant[b]
HSiCl₃	−364	HSiBr₃	−360
HSiCl₂I	−353	HSiBr₂I	−349
HSiClI₂	−341	HSiBrI₂	−338
HSiI₃	−325	HSiI₃	−325

[a] Data from Hengge and Höfler.[157]
[b] In Hz.

It is generally considered that as electronegative substituents withdraw electron density from silicon, shielding of the protons is similarly reduced and the effective nuclear charge seen by an s-electron is increased. The Fermi contact contribution is directly proportional to the product of the electron densities of the two bonding orbitals at the respective nuclei. When hybridized orbitals are present, this contact contribution is proportional to the percentage of s character used in forming the bond.[158] It is interesting to note that $^1J_{^{29}Si-^1H}$ has been reported to vary linearly with the Si–H force constant in X₃SiH compounds and with the Si–Si force constant in X₃Si—SiX₃ compounds,[159] although the correlation is only fair.

Several compounds with rather large $^1J_{^{29}Si-^1H}$ coupling constants are listed in Table XX. It can be seen that four of the five compounds are similar in

TABLE XX

Compounds with Large $^1J_{^{29}Si-^1H}$ Values[a]

Compound	Group	Coupling constant	Ref.
$H_2Si[Mn(CO)_5]_2 \cdot 4py$		-420	14
$HSiF_3$		-383.9	206
$HSi(OOCCF_3)_3$		-382	190
	Middle	-381	195
$\{SiHClO\}_n$[b]	End	-360	195
	Cross-linked	-331	195
$HSi(OOCCH_3)_3$		-344	190
$(HSiCl_2)_2NC_2H_3$		-328	81

[a] In Hz.

[b] Although the formula was given in Horn and Marsmann[14] as $\{SiCl_2O\}_n$ it must certainly be $\{SiHClO\}_n$, which was also discussed in that paper.

type to those in Table XIX, i.e., of the formula $HSiX_3$, where X is an electronegative substituent. The markedly high value for $SiH_2[Mn(CO)_5]_2 \cdot 4py$ can only be accounted for by attributing this large coupling to the presence of four coordinated pyridine molecules. It can be seen from Table XXI[160, 161] that $|^1J_{^{29}Si-^1H}|$ in silyl transition metal carbonyls, under normal circumstances, is reduced below those of unsubstituted silanes. This is not unexpected when it

TABLE XXI

$^1J_{^{29}Si-^1H}$ in Silylmetal Carbonyl Systems[a]

Unsubstituted silyl group			Substituted silyl group		
Compound	$^1J_{^{29}Si-^1H}$	Ref.	Compound	$^1J_{^{29}Si-^1H}$	Ref.
$SiH_2[Re(CO)_5]_2$	-168	160	$SiH_2ClMn(CO)_5$	-210	161
$SiH_2[Mn(CO)_5]_2$	-173	160	$SiH_2ClCo(CO)_4$	-239	161
$SiH_3Mn(CO)_5$	-195.6	160	$SiHCl_2Co(CO)_4$	-280	161
$SiH_2[Co(CO)_4]_2$	-203	161			

Pyridine coordinated		
Compound	$^1J_{^{29}Si-^1H}$	Ref.
$SiH_3Mn(CO)_5 \cdot 2py$	-273	160
$SiH_2[Mn(CO)_5]_2 \cdot 4py$	-420	160

[a] In Hz.

is considered that the transition metals are not electronegative substituents; in addition, it is possible that interactions between the $M_x(CO)_y$ group and the silicon d orbitals can actually result in considerable electron donation to the silicon by the metal carbonyl moiety. Thus, in unsubstituted systems, $^1J^{29}{}_{Si^{-1}H}$ has a value ranging from -168 to -203 Hz, values that might be expected and indeed are similar to those in alkyl substituted compounds (*vide infra*). In the second column of Table XXI can be seen the effect of mono- and dichloro substitution. As expected, the coupling constant is increased about 40 Hz by the substitution of one chloro substituent. In column 3 it is apparent that the dicoordinated species has a coupling constant similar to that of the dichloro compound and that the tetracoordinated species has the largest $|^1J^{29}{}_{Si^{-1}H}|$ value yet reported, by virtue of having four electronegative substituents in an expanded coordination sphere of silicon.

Increasing methyl substitution in carbosilanes produces a lessening of the coupling constant.[162] Some typical values for alkyl- and phenyl-substituted compounds are given in Table XXII.[141, 163-167]

TABLE XXII

$^1J^{29}{}_{Si^{-1}H}$ Values in Alkyl- and Phenyl-silanes[a]

Compound	$^1J^{29}{}_{Si^{-1}H}$	Ref.
$(C_6H_5)_3SiH$	-198.0	163
$(C_6H_5)_2(CH_3)SiH$	-187	141
$(CH_3)_3SiH$	-184.0	164
$[(C_6H_5)_2P]_3SiH$	-182.5	165
$(C_2H_5)_3SiH$	-172.2	163
$(CH_3)_5Si_2H$	-173	166
$(C_6H_5)_5Si_5H_5$	-169	167
$[(CH_3)_3Si]_3SiH$	-155	164
$[(C_2H_5)_3Si]_3SiH$	-147	163

[a] In Hz.

It was suggested that the relatively small coupling in the cyclic $(C_6H_5)_5Si_5H_5$ represents a weakening of the Si—H bonds because of increased electron density in the Si_5 ring.[167] Similarly, it would appear from the low $^1J^{29}{}_{Si^{-1}H}$ in $[(CH_3)_3Si]_3SiH$ that the central silicon is relatively negative.[164] Thus, it is apparent that the presence of electron-donating substituents on silicon results in a decrease of $^1J^{29}{}_{Si^{-1}H}$ values.

A study of $CH_3OSi(H)_n(CH_3)_{3-n}$ $(n = 0-3)$ compounds and the corresponding disiloxanes found that $|^1J^{29}{}_{Si^{-1}H}|$ lessens with an increasing number of methyl groups.

In silyl amines, the incorporation of electron-donating substituents were associated not only with smaller $^1J_{^{29}Si^{-1}H}$ coupling constants but also with short Si—N bonds.[168] No correlation was found with the Si—O bond length in silyl ethers,[169] however, and since the electronic effects might be expected to be similar in silyl amines and ethers, it suggests that the bond-length observations in the case of the amines may be fortuitous.

The concept that $^1J_{^{29}Si^{-1}H}$ is related to the electronegativity of the substituent, however, has been questioned by Drake.[170] In a series of $RSiH_2SeCH_3$ (R = H, CH_3, and $SeCH_3$) compounds, $^1J_{^{29}Si^{-1}H}$ was found to be -216.4, -208.5, and -233.7 Hz, respectively. Drake suggested that since these coupling constants are close to the values of other Group VI species (oxygen and sulfur) of similar electron configurations but different electronegativities, but different from analogous Group V and Group VII derivatives which have similar electronegativities, the s character in the Si—H bond in an E—Si—H system is apparently more strongly related to the electron configuration of element E rather than its size or electronegativity.

The effect of β substituents on $^1J_{^{29}Si^{-1}H}$ is relatively unimportant, as demonstrated in several series of compounds. In SiH_3SiH_2X, the coupling constant of the SiH_3 group[171] is essentially that of SiH_4. For the SiH_3SiHF_2 [$J(SiH_3) = -206.2$ Hz] and SiH_3SiF_3 [$J(SiH_3) = -212.0$ Hz] compounds,[172] and for the $SiH_3SiH[N(CH_3)_2]_2$ ($J = -184.5$ Hz) species,[139] the presence of multiple β-fluorine or β-nitrogen substitution has little effect on $^1J_{^{29}Si^{-1}H}$. Similarly, it is not surprising that HCH_2SiH_3 and $GeH_3CH_2SiH_3$ have the same $^1J_{^{29}Si^{-1}H}$ (-198.7 Hz) coupling constant,[173] and although the authors note an increase in $|^1J_{^{29}Si^{-1}H}|$ in SiH_3SeX as X varies from SiH_3 to CF_3, the increase is only 4 Hz.[174]

It has been reported, however, that insertion of CO_2, CS_2, and CSe_2 into SiH_3NR_2 to form the corresponding carbamates and analogues increases $|^1J_{^{29}Si^{-1}H}|$ by some 30 Hz.[168]

Of particular interest is any work which relates $^1J_{^{29}Si^{-1}H}$ to other fundamental parameters. In this respect it is noteworthy that Nagai et al,[175] found a good correlation between $^1J_{^{29}Si^{-1}H}$ and Taft's polar substituent constants (σ^* values) for substituted phenyl compounds $R_1R_2R_3SiH$ (R = alkyl, phenyl, and hydrogen) such that

$$^1J_{^{29}Si^{-1}H} = -10.21 \sum \sigma^* - 182.9 \tag{25}$$

It is thus possible to evaluate the polar substituent constant by fitting its NMR parameter to the above equation.[176] This procedure has recently been utilized for the suggested purpose of determining substituent constants of various silyl groups.[177] It has also been shown that $^1J_{^{29}Si^{-1}H}$ for an individual series of aryl hydrosilanes correlates well with Hammett's σ constants.[175] In such compounds $J_{^{29}Si^{-1}H}$ appears more sensitive to a change in structure than is

δ_{Si}.[175] The presence of an electron-attracting group leads to an increase in the absolute magnitude of the coupling constant, whereas an electron-donating group decreases it. Nagai's trends are consistent with the concept that in a given series of compounds, $^1J_{^{29}Si-^1H}$ is proportional to s character.[175]

Nagai and co-workers have also found that the relative reactivities of Si—H bonds toward the $\cdot CCl_3$ radical are linearly correlated with $^1J_{^{29}Si-^1H}$ and hence reflect the s character in the Si—H bond.[178, 179] They also suggest that correlations between $^1J_{^{29}Si-^1H}$ and $\log k_{rel}$ might imply a change in the s character of the silicon to hydrogen bond, and thus reflect a change in the electron density around hydrogen and the consequent ease of removal of hydrogen by an abstracting radical.[179]

2. $^{29}Si-^{19}F$ Coupling Constants

The range of reported values for $^1J_{^{29}Si-^{19}F}$ in tetracovalent silicon compounds varies from $+167$ Hz in SiF_3SiF_2H[180] to $+488$ Hz in $\underline{CH_2CH_2SiF_2SiF_2}$.[181]

SiF_4, the parent compound of the fluorosilanes, has been measured to have a $^1J_{^{29}Si-^{19}F}$ of $+169.00$ Hz (gas at 30 atm), $+169.84$ Hz (gas at 110 atm), and $+169.97$ Hz (liquid at $-52°$),[182] and several higher values ranging from $+169.3$ Hz[26] to $+178$ Hz[183] in solutions. In Si_2F_6 the $^1J_{^{29}Si-^{19}F}$ coupling constant is $+321.8$ Hz,[12] and in Si_3F_8 it is $+356.6$ Hz for the $^{29}SiF_2$ group and $+344.4$ Hz for the $^{29}SiF_3$ group.[8]

As with silicon–proton coupling, the magnitude of the coupling constant in anions is smaller. In SiF_6^{-2} the $^1J_{^{29}Si-^{19}F}$ coupling constant has been reported as $+110$ Hz[183] and $+108$ Hz.[26]

The isotope effect is somewhat larger than in silicon–proton coupling, but it is still small. The coupling constant decreases from $+276.6$ Hz in $HSiF_3$ to $+274.6$ Hz in $DSiF_3$. It was noted that this effect is in accord with the concept that X—D bonds have more s character than X—H bonds (bonds to deuterium are shorter). Alternatively, it was suggested that a faster relaxation time of 0.04 seconds for the central (silicon) nucleus would produce a lowering of $^1J_{^{29}Si-^{19}F}$ by 2.0 Hz. However, this explanation is judged unlikely because there should be a concomitant broadening of the linewidth to some 25 Hz, and the observed lines in $DSiF_3$ are narrower than 1 Hz.[184]

It has been suggested that $^1J_{^{29}Si-^{19}F}$ should decrease in the order $J_{SiF} > J_{SiF_2} > J_{SiF_3}$.[27] However, it has been observed that in $F_2Si(NR_2)_2$ compounds the coupling is $+218–219$ Hz, while in F_3SiNR_2 compounds the coupling is $+202–206$ Hz.[185] In the series F_nSiH_{4-n} ($n = 1–3$), the $^1J_{^{29}Si-^{19}F}$ values are $+281$ Hz, $+297.8$ Hz, and $+274.8$ Hz, respectively.[126] Such data do not fit any monotonous function. It has also been reported that GeH_3SiF_2H (344 Hz) and the (suspected) GeH_3SiF_3 species (343 Hz) have virtually identical $^1J_{^{29}Si-^{19}F}$ values,[186] and that in the $H(SiF_2)_nSiF_3$ series J_{SiF_3} is $+340–350$ Hz while J_{SiF_2} is lower at $+323$ Hz.[187]

It has been suggested that there is an "inverse" relationship[188] between $^1J_{^{29}Si-^{19}F}$ and substituent electronegativity. The coupling constants of the compounds shown below would appear to be consistent with that trend, although more supportive evidence is needed.

Compound	J
$SiFCl_2Br$	332
$SiFClBr_2$	351
$SiFBr_3$	365
$SiFClBrI$	380
$SiFClI_2$	401

Other suggestions of this inverse coupling and substituent electronegativity relationship have been offered.[10] It was observed that $^1J_{^{29}Si-^{19}F}$ is similar to SiF_3CF_3 and $SiF_3C_2H_5$, but that the value for $SiF_2IC_2F_5$ is larger,[189] as are the coupling constants of SiF_3I[190] and SiF_2ICF_3.[191] With SiF_3X species, there is a rough correlation between $^1J_{^{29}Si-^{19}F}$ and substituent electronegativity, but the I value is larger than "expected."[189] It therefore appears that the question of the effect of substituent electronegativity on $^1J_{^{29}Si-^{19}F}$ is in need of further examination. Coupling constants of a few SiF_3X compounds in which X is relatively electronegative are given in Table XXIII,[192–195] while compounds in which X is comparatively less electronegative are given in Table XXIV.[11, 24, 56, 159, 163, 172, 190, 191, 196, 197]

TABLE XXIII

$^1J_{^{29}Si-^{19}F}$ for SiF_3X (X = O, N) Species[a]

Compound	$^1J_{^{29}Si-^{19}F}$	Ref.	Compound	$^1J_{^{29}Si-^{19}F}$	Ref.
SiF_3OCH_3	+181	192	SiF_3NCO	+181	195
$SiF_3OSi(CH_3)_3$	+184.2	193	$SiF_3NCNSi(CH_3)_3$	+187.6	193
SiF_3OCOCF_3	+194	194	$SiF_3N(CH_3)_2$	+201.4	193

[a] In Hz.

3. Coupling between ^{29}Si and Nuclei Other than ^1H and ^{19}F

Spin–spin coupling constants between ^{29}Si and nuclei other than ^1H and ^{19}F are less frequently reported. Several values have been reported for coupling to ^{13}C and to ^{31}P, and there are also reports of coupling to ^{15}N, ^{29}Si, ^{77}Se, ^{119}Sn, and ^{195}Pt. The correct way to compare spin–spin coupling constants among different nuclei, of course, is by the use of reduced coupling constants (K_{M-X}; cf., Section III,B). However, coupling constants are typically reported in the literature in the nonreduced form (J_{M-X}, i.e., the form in which the magnetogyric ratio plays a role). If reduced coupling constants are used, however, it can be seen that $^1J_{^{29}Si-^{31}P}$ is very similar to $^1J_{^{31}P-^{31}P}$.[198] On the other hand, $^1J_{^{29}Si-^{195}Pt}$ in *trans*-Pt(Cl)(SiH$_2$Cl)[P(C$_2$H$_5$)$_3$]$_2$ is $+3120 \times 10^{19}$ N A^{-2} m^{-3}, which is considerably higher than the value of 2330×10^{19} N A^{-2} m^{-3} found for *cis*-PtCl$_2$[P(C$_2$H$_5$)$_3$]$_2$ where the ligand trans to the phosphorus is also chlorine.

TABLE XXIV

$^1J_{^{29}Si-^{19}F}$ for SiF$_n$X$_{4-n}$ (X = Group III, IV, V) Species[a]

n = 3			n = 2		
Compound	$^1J_{^{29}Si-^{19}F}$	Ref.	Compound	$^1J_{^{29}Si-^{19}F}$	Ref.
SiF$_3$SiH$_3$	+356	172	SiF$_2$I$_2$	375	190
SiF$_3$(SiF$_2$)$_2$BF$_2$	+355	196	SiF$_2$ICF$_3$	340	191
SiF$_3$SiF$_2$BF$_2$	+351	196	SiF$_2$(C$_6$H$_5$)$_2$	302.7	56
SiF$_3$I	+296	190			
SiF$_3$R	+268–284	197			
SiF$_3$(C$_6$H$_5$)	264.5	56			

n = 1		
Compound	$^1J_{^{29}Si-^{19}F}$	Ref.
SiFCl$_2$SiCl$_3$	384.9	11
SiFBr$_3$	368.7	159
SiF[Si(CH$_3$)$_3$]$_3$	335	163
SiF[Si(C$_2$H$_5$)$_3$]$_3$	326	163
SiF(C$_2$H$_5$)$_3$	288	163
SiFH$_3$	281	163
SiFH$_2$(CH$_3$)	279.8	24
SiF(CH$_3$)$_3$	274	163

[a] In Hz.

Again, the Fermi contact mechanism (and therefore the amount of s character in the orbital of the bond) plays a predominant role in determining the magnitudes of the one-bond coupling constants. The values of $^1J_{^{29}Si-^{13}C}$ range upwards from a low of 44 Hz in [(CH$_3$)$_3$Si]$_4$Si (see Fig. 2) which is somewhat low; an sp^3-hybridized system is usually ≥ 50 Hz ($K = +86.6 \times 10^{19}$ N A^{-2} m^{-3}).[142] This particular value of 44 Hz was also confirmed from the ^{13}C spectrum.[42] In general, it can be seen from the values listed in Table XXV[2, 3, 42, 80, 93, 96, 141, 154, 198–202, 204, 205] that $^1J_{^{29}Si-^{13}C}$ values are plateaued according to (1) the hybridization of the carbon atom (with sp^3 values around 50 Hz, sp^2 values equal to 65–70 Hz, and sp-hybridized carbons > 80 Hz), and (2) substituent electronegativity factors.

It is also interesting to note that $^1K_{^{29}Si-^{31}P}$ has been reported to be -43.7×10^{19} N A^{-2} m^{-3}.[204] Such values of reduced coupling constants over one bond

TABLE XXV

$^1J_{29_{\text{Si}-M}}$ between ^{29}Si and Nuclei Other than ^{1}H and ^{19}F

Nucleus M	Compound	$^1J_{29_{\text{Si}-M}}$	Ref.
^{13}C	[(CH$_3$)$_3$Si]$_4$Si	-44.4	42
^{13}C	(CH$_3$)$_4$Si	-50.3	199
^{13}C	(C$_2$H$_5$)$_4$Si	-50.2	199
^{13}C	(CH$_3$)$_3$SiCl	-57.4	200
^{13}C	(CH$_3$)$_2$Si—C⟨(CH$_3$)$_2$ / C⟨H$_2$ / H$_2$ / (CH$_3$)$_2$	-58	96
^{13}C	(CH$_2$=CH)$_2$Si(CH$_3$)$_2$	-66	3
^{13}C	C$_6$H$_5$Si(CH$_3$)$_3$	-66	3
^{13}C	(CH$_3$)$_2$SiCl$_2$	-68.3	200
^{13}C	(CH$_2$=CH)$_4$Si	-70	200
^{13}C	(CH$_2$=CH)(CH$_3$)SiCl$_2$	(Me) -71	200
		(Vi) -92	200
^{13}C	(C$_6$H$_5$)C≡CSi(CH$_3$)$_3$	-83.6	199
^{13}C	CH$_3$SiCl$_3$	-86.6	200
^{13}C	ClCH$_2$SiCl$_3$	-97.6	200
^{13}C	(CH$_2$=CH)SiCl$_3$	-113	200
^{31}P	(C$_6$H$_5$)$_2$PSi(CH$_3$)$_3$	$+6.6$	201
^{31}P	[(CH$_3$)$_3$Si]$_2$N—P[=NSi(CH$_3$)$_3$]$_2$	(amino) $+1.2$	202
		(imino) $+16.6$	
^{31}P	[(CH$_3$)$_3$Si]$_2$N—P=NSi(CH$_3$)$_3$	(amino) $+9.1$	80
		(imino) $+26.8$	
^{31}P	(SiH$_3$)$_3$P	$+42.1$	198
			204
^{15}N	(SiH$_3$)$_3$N	-4.2	93
^{29}Si	[(CH$_3$)$_3$Si]$_4$Si	52.5	42
^{77}Se	(SiH$_3$)$_2$Se	$+110.6$	154
^{119}Sn	(C$_6$H$_5$)$_3$SiSn(CH$_3$)$_3$	-1.4	201
^{195}Pt	trans-Pt(Cl)(SiH$_2$Cl)[P(C$_2$H$_5$)$_3$]$_2$	-1600	205

have been associated with considerable s electron character in the lone pair,[206] and thus this value of K was judged to be consistent with the phosphorus valence angle of 96.5°, which was found by an electron diffraction study,[207] but not consistent with an earlier suggestion that the vibrational spectrum indicated a planar molecule.[203] In general, coupling to phosphorus occurs over a wide range. Particularly interesting are the amino and imino couplings in bis(trimethylsilyl)aminobis(trimethylsilyl)iminophosphorane, a phosphorus(V) derivative with a coordination number of 3. The influence of s character on the coupling constant is readily apparent in comparing the amino and imino derivatives.

D. Two- (or More) Bond Coupling

1. $^{29}Si-C-^{1}H$ Coupling Constants

The range of $^2J^{29}{}_{Si-C-^1H}$ coupling constants is most frequently regarded[2] to extend from about 6 to over 9 Hz, the larger values occurring with electronegative ligands. For example, in $(CH_3)_nCl_{3-n}SiOC_2H_5$ the $^2J^{29}{}_{Si-C-^1H}$ value increases from 6.3 to 7.6 to 9.2 Hz as n increases from 1 to 3, respectively,[208] and from 6.6 to 7.1 to 7.8 to 9.2 Hz as n increases from 0 to 3 in $(CH_3)_{4-n}SiCl_n$.[208] Although such values are typical, the range extends much more widely.

It was observed that $^2J^{29}{}_{Si-C-^1H}$ increases in the series $(CH_3)_0SiX$ as X is varied from CH_3 to F to Cl to Br to I, and also increases in the series $[(CH_3)_3Si]_2Y$ in which $Y = O$, S, and Se. Since these trends run counter to the expected relationships based on substituent electronegativity, it was originally suggested that π bonding could contribute to this relationship.[209] Alternate explanations[210] involve an increasing amount of s character in the silicon orbital to carbon, and possible rehybridization with angle changes as the substituent is varied; however, this explanation would not be consistent with Bent's rules of s character. The most satisfactory explanation would seem to be that it is not only the s character of the silicon orbital that must be considered, but also the s character of the orbitals of the intervening carbon atom, which may be even more important.[200]

Other relationships that are similar to those observed for one-bond coupling include a decrease in coupling constant (somewhat less than in the case of $^1J^{29}{}_{Si-^1H}$) with increasing methyl substitution in carbosilanes.[210] Also, the internal coupling constants in permethylpolysilanes are 0.2–0.3 Hz lower for

TABLE XXVI

Some Species in Which $^2J^{29}{}_{Si-C-^1H}$ Is Less than 6 Hz

Compound	$^2J^{29}{}_{Si-C-^1H}$	Ref.
$(CH_3)_3SiCH_2Cl^a$		
SiCH$_2$	3.6	200
SiCH$_2$	3.7	212
$[(CH_3)_3Si]_3HgLi \cdot 3DME^b$	5.2	213
$[(CH_3)_3Si]_4HgLi_2 \cdot 4DME^b$	4.8	213
$(CH_3)_3SiLi$	3.5^c	213
	2.8^d	213

^{a} The ^2J value for the SiCH$_3$ group has been reported as 6.8 Hz[200] and 6.6 Hz.[212]

^{b} The ^2J value for $[(CH_3)_3Si]_2Hg$ is 6.7 Hz.[213]

^{c} In cyclopentane.

^{d} In DME (1,2-dimethoxyethane).

internal as opposed to terminal positions.[211] In carbosilanes, the values of $^2J_{^{29}Si-C-^1H}$ are 2–3 Hz smaller for $SiCH_3$ groups than for $SiCH_2Si$ groups. A relationship was observed between a decreasing ^2J and decreasing ring size.[210]

A very large number of $^2J_{^{29}Si-C-^1H}$ coupling constants have been measured, a not surprising outcome of (1) the extensive studies of compounds containing the $Si(CH_3)_3$ group, and (2) the usual occurence of the NMR signal of this group as a narrow singlet, which makes possible the determination of the $^{29}Si-C-^1H$ satellites. Only those values which seem remarkable in some way are discussed here. For example, although the minimum $^2J_{^{29}Si-C-^1H}$ value is often considered to be about 6 Hz, several lower values have been reported and are listed in Table XXVI.[200, 212, 213]

In the mixed mercury–lithium systems listed in Table XXVI, the authors noted that some $[(CH_3)_3Si]_2Hg$ was always present and exchanged with the lithium species so that no ^{199}Hg satellites were seen in the ^1H spectra. $^2J_{^{29}Si-C-^1H}$ satellites were always present, however, which indicates that exchange of methyl groups between silicon atoms is slow.[213]

Values for $^2J_{^{29}Si-C-^1H}$, which are larger than the commonly regarded 9 Hz maximum, have also been reported and are listed in Table XXVII.[54, 81, 200, 210, 214–219] For $(C_2H_5)_3SiI$ it is assumed that this value represents $^2J_{^{29}Si-C-^1H}$[210] and not $^3J_{^{29}Si-C-C-^1H}$ as originally reported.[214]

TABLE XXVII

Some Species in Which $^2J_{^{29}Si-C-^1H}$ Is Larger than 9 Hz

Compound		$^2J_{^{29}Si-C-^1H}$	Ref.
$(C_2H_5)_3SiI$		9.0	210, 214
$[Cl_2(CH_3)SiN]_2S$		9.1	215
CH_3SiCl_3		9.3	200
	(SiCH)	9.0	216
	(SiCH$_2$)	10.0	216
$(CH_3)_3SiCH_2Li$			
(SiCH$_3$)		6.1	217
(SiCH$_2$)		10.8	217
$Cl_3Si—CBr_2—SiCl_2—CH_2—SiCl_3$		12.4	54
$Cl_3Si—CCl_2—SiCl_2—CH_2—SiCl_3$		12.6	54, 218
$Cl_3Si—CCl(CH_3)—Si(CH_3)_2—C≡C—Si(CH_3)_3$		12.6	54
$Cl_3Si—N—CH=CH—CH(SiCl_3)—CH=CH$		13	54

4. ^{29}Si Nuclear Magnetic Resonance   253

2. Other Two-Bond Coupling

Typical values are given in Table XXVIII.[8, 11, 12, 154, 163, 187, 199, 205, 220–224] The linewidth of the signal in $[(CH_3)_3Si]_3SiF$ was given as 3.5 Hz.[221]

TABLE XXVIII

Two-Bond Coupling Involving ^{29}Si (Other than $^{2}J_{^{29}Si-C-^1H}$)

Nuclei	Compound		Coupling constant	Ref.
^{29}Si–Si–^{19}F	Si_2F_6		−90.5	12
	Si_3F_8		−66.1	220
	Si_3F_8	(SiF_2)	−64.4	8
		(SiF_3)	−50.1	
	$HSiF_2SiF_3$	(SiF_2)	−59	187
		(SiF_3)	−63	
	$H_2Si(SiF_3)_2$		−38	187
	$HSi(SiF_3)_3$		11	187
	$Si(SiF_3)_4$		−32	187
	$SiCl_3SiCl_2F$		−64.8	11
	$[(CH_3)_3Si]_3SiF$		−16.8	221
	$[(C_2H_5)_3Si]_3SiF$		−14.5	163
^{29}Si–C–^{13}C	$(CH_3)_3SiC{\equiv}C(C_6H_5)$		16.1	199
^{29}Si–O–^{29}Si	$(SiH_3)_2O$		+1.0	154
^{29}Si–Fe–^{1}H	$(Cl_3Si)_2FeH(CO)(C_5H_5)$		20	222
^{29}Si–Fe–^{13}C	cis-$[(CH_3)_3Si]_2Fe(CO)_4$		5.3	223
^{29}Si–Pt–^{31}P	$trans$-$Pt(Cl)(SiH_2Cl)[P(C_2H_5)_3]_2$		+18	205
^{29}Si–N–^{119}Sn	$\{[(CH_3)_3Si]_2N\}_2Sn$		16.2	224

3. Three-Bond Coupling

A surprisingly wide variety of three-bond coupling constants have been determined and are listed in Table XXIX.[8, 12, 42, 54, 82, 149, 154, 198, 201, 204, 208, 221, 225–235] The usual trends appear. Increasing the number of chlorine substituents on silicon raises the $^{3}J_{^{29}Si-O-C-^1H}$ coupling constants in $Cl_n(CH_3)_{4-n}SiOCH_3$ ($n = 0$–3) compounds,[208] and a similar effect is noted in the $(CH_3)_{4-n}Si(OCH_3)_n$ ($n = 1$–4) species.[208] In $(SiH_3)_3Y$ (Y = ^{14}N, ^{15}N, P, As, and Sb) compounds, a strong correlation was observed between $^{3}J_{^{29}Si-Y-Si-^1H}$ and $^{4}J_{^1H\ldots^1H}$ coupling constants.[198]

The presence or absence of satellites from long-range coupling has also been used to examine chemical behavior. The $^{3}J_{^{29}Si-N-C-^1H}$ satellites were used by Fukui et al.[228] to explain the nonequivalent trimethylsilyl resonances in

TABLE XXIX

Three-Bond Coupling Involving ^{29}Si

Nuclei	Compound		Coupling constant	Ref.
^{29}Si–Si–Si–^{19}F	Si_3F_8		-15.7	8
^{29}Si–O–Si–^{19}F	$(SiF_3)_2O$		2.5	12
^{29}Si–C–C–^{1}H	$(C_6H_5)_2SiHCl$		0.9	42
	$(C_6H_5)_3SiC{\equiv}CH$		4.3	225
	$(CH_3)_3SiC(Cl){=}CH_2$	cis-	4.4	149
	$(CH_3)_3SiC(Cl){=}CH_2$	trans-	-10.5	
	$Cl_3SiC(Cl){=}CH_2$	cis-	7.5	149
	$Cl_3SiC(Cl){=}CH_2$	trans-	-21.2	
	$(CH_2{=}CH)_4Si$	cis-	8.8	226
	$(CH_2{=}CH)_4Si$	trans-	-16.5	
^{29}Si–Si–C–^{1}H	$Si_5(CH_3)_{10}$		3.2	227
	$Si_6(CH_3)_{12}$		3.4	227
	$Si_7(CH_3)_{14}$		3.2	227
	$[(CH_3)_3Si]_3SiH$		2.3	227
	$[(CH_3)_3Si]_3SiF$		2.73	221
	$[(CH_3)_3Si]_3SiCl$		3.20	221
	$[(CH_3)_3Si]_3SiBr$		3.15	221
	$[(CH_3)_3Si]_3SiI$		3.20	221
	$[(CH_3)_3Si]_4Si$		2.2	231
	$(CH_3)_3SiSi(C_6H_5)_3$		-2.8	201
	$(CH_3)_3SiSi[P(C_2H_5)_2]_3$		2.3	232
	$(CH_3)_2HSiSi[P(C_2H_5)_2]_3$		2.6	232
^{29}Si–C–Si–^{1}H	$(SiH_3CCl_2)_2SiH_2$		1.75	54
	$(SiH_3CH_2)_2SiH_2$		4.6	233
	$(BrSiH_2CH_2)_2SiHBr$		3.4	54
	$(SiH_2CCl_2)_3$		2.8	82, 233
	$SiH_2CH_2SiH_2CCl_2SiH_2CCl_2$		0.9	82
^{29}Si–N–C–^{1}H	$[(CH_3)_3Si]_2NCH_3$		4.7	228
	$(CH_3)_3SiN(CH_3)_2$		3.7	228
	$(CH_3)_3SiN(CH_3)COC_6H_5$		2.9	228
	$(CH_3)_3SiN(CH_3)C(CH_3)_3$		3.5	229
	$(R_3Si)_2NCH_3$		4.8	230
	$(CH_3)_3SiN(CH_3)Ge(CH_3)_3$		4.9	230
	$(SiH_3)_3N$		3.7	198
	$(SiH_3)_3P$		5.4	198, 204
	$(SiH_3)_3As$		-5.3	198
	$(SiH_3)_3Sb$		-4.5	198, 204
^{29}Si–O–C–^{1}H	$(CH_3O)_3SiH$		4.2	42
	$(SiH_3)_2O$		-2.4	154
	$(CH_3)_3SiOCH_3$		3.9	208
	$Cl(CH_3)_2SiOCH_3$		4.7	208
	$Cl_2(CH_3)SiOCH_3$		5.3	208
	Cl_3SiOCH_3		6.0	208

TABLE XXIX (*Continued*)

Nuclei	Compound	Coupling constant	Ref.
	$(CH_3)_2Si(OCH_3)_2$	3.9	208
	$CH_3Si(OCH_3)_3$	3.6	208
	$Si(OCH_3)_4$	3.2	208
^{29}Si–S–Si–^1H	$(SiH_3)_2S$	−4.9	154
		0.70	198
	$SiH_3SSi(CH_3)_0$	2.5	234
^{29}Si–Se–Si–^1H	$(SiH_3)_2Se$	−4.6	154
	$SiH_3SeSi(CH_3)_3$	2.1	234
^{29}Si–Se–C–^1H	$(CH_3)_3SiSeCH_3$	3.6	235
	$(CH_3)_2Si(SeCH_3)_2$	4.2	235
	$CH_3Si(SeCH_3)_3$	5.4	235
	$Si(SeCH_3)_4$	5.7	235

$C_6H_5CON(CH_3)Si(CH_3)_3$ in terms of a 1,3 migration of trimethylsilyl groups between the nitrogen and oxygen atoms.

Thus, coupling was observed in form (A) but was not seen in form (B).

It is also interesting to note that Ebsworth et al.[236] observed sharp quartets in $Li[YSiH_3]$ (Y = S and Se) compounds, i.e., no $^3J_{^{29}Si-Y-Si-^1H}$ coupling; however, in $Li[Z(SiH_3)_2]$ (Z = P and As) species, such coupling was implied by the "quartet of quartets" appearance of the spectrum.

E. Solvent Effects

Coyle et al.[132] demonstrated a substantial change in $J_{^{29}Si-^{19}F}$ of SiF_4 as a function of solvent and suggested that specific interactions between solute and solvent might be responsible. The usual solvents were examined, and also two series of related solvents of formula CX_nF_{4-n} and SiX_nF_{4-n}. In these series of fluorinated solvents, the coupling constant increased monotonically as n increased. There is no evidence of any correlation with the dielectric constant of the solvent since the smallest increase in coupling occurs with the most highly fluorinated solvent. It was suggested by Raynes[237] that these shifts were additive in terms of a characteristic shift per substituent on carbon or silicon. The order found is given in Table XXX and has been rationalized on the basis of potential functions of SiF_4.[238]

The shifts in C_6H_6 (7.98 Hz) and C_6F_6 (4.44 Hz) were larger than with the series of fluorinated solvents. Since SiF_4 is nonpolar, the data obtained afforded an opportunity to see whether intermolecular dispersion interactions affect coupling. Calculations of such interactions require the availability of the polarizability, ionization potentials, absorption frequencies, and intermolecular distances of approach for solute–solvent systems.[239] Most of this information is not available for the solvents employed in the cited study. Assuming a rough correlation of dispersion free energy and heat of vaporization, Hutton, Bock, and Schaefer[239] postulated that the variation in $^1J_{^{29}Si-^{19}F}$ can be attributed mainly to time-varying electric fields, i.e., dispersion interactions, which affect the electronic states of the SiF_4 molecule. It was suggested that as these intermolecular dispersion interactions increase, so does the magnitude of $^1J_{^{29}Si-^{19}F}$. In terms of the Pople-Santry theory, an increase in dispersion interactions would produce a lower excitation energy between the highest bonding and the lowest antibonding molecular orbitals in the Si—F bond. That is, dispersion terms would produce an expansion of the electron clouds within the molecule.

TABLE XXX

Changes in $^1J_{^{29}Si-^{19}F}$ with Variation in Solvent Substituent in the Series MX_nF_{4-n}

Solvent substituent	Change in $^1J_{^{29}Si-^{19}F}$ (Hz)	Solvent substituent	Change in $^1J_{^{29}Si-^{19}F}$ (Hz)
Si–F	0	C–CN	1.66
C–F	0	C–Cl	2.0
C–H	1.2	Si–Br	2.5
Si–CH$_3$	1.5	Si–C$_2$H$_5$	3.0

In the study cited above, Coyle et al.[182] found little variation (0.18 Hz) in $^1J_{^{29}Si-^{19}F}$ at 30 atm and 110 atm, but Jameson and Reger[240] found the gas-phase coupling constant to be density dependent in SiF_4 but not in SiH_4. These authors commented that although one might be tempted to explain solvent effects in coupling on the basis of some property of the solvent, no significant dependence on density in the gas phase should occur. If such a shift did occur, it might reflect instead the presence of gas-phase (liquid-phase) shifts.

The solvent dependence of SiF_4 coupling has speculatively been attributed by Barfield and Johnston[241] to dispersion interactions of a nonpolar solute and orbital (Fermi contact) contributions.

F. Applications

1. Exchange Phenomena

The stereochemistry and "rigidity" of trigonal bipyramidal molecules has received much attention in recent years, and various mechanisms have been cited to explain the rapid exchange of axial and equatorial fluorines in silicon compounds (and phosphorus and sulfur fluorides as well). Since ^{29}Si has a spin of $\frac{1}{2}$, NMR techniques may be used to distinguish between intermolecular and intramolecular fluorine exchange. Intermolecular exchange involves bond breaking and loss of ^{29}Si-^{19}F coupling, whereas intramolecular exchange retains bonding and coupling.

In an NMR study by Gibson and Janzen,[242] the diethylamine-catalyzed hydrolysis of $(CH_3)_3SiF$ established that fluorine exchange occurs via an intermolecular mechanism with the loss of ^{29}Si-^{19}F coupling information.

$$(CH_3)_3SiF + H_2O \leftrightarrows (CH_3)_3SiOH + HF$$

On the basis of the kinetics and rate expression, it appeared that one or two H_2O molecules may coordinate to $(CH_3)_3SiF$ in the transition state prior to Si—F bond breaking. Therefore, depending on the symmetry of the molecule, a rapid preequilibrium involving water coordinated to the central atom may result in an observable intramolecular averaging of fluorine environments. It was suggested that if intramolecular exchange involving H_2O is too rapid to be seen on the NMR time scale, then perhaps other Lewis bases such as amines or ethers might slow intramolecular exchange. Similarly, the appearance of satellites in aqueous solutions of $(NH_4)_2SiF_6$ suggests that exchange of fluorine atoms must occur at a rate less than 10^{-3} second^{-1}.[183] In $(CH_3)_3SiCN$, however, the lack of ^{29}Si-^{13}C coupling implies that exchange processes here are relatively rapid.[243]

It was found by Klanberg and Muetterties[244] that SiF_5^-, $RSiF_4^-$, and $R_2SiF_3^-$ showed spectroscopic equivalence of fluorine atoms at room temperature, while $(C_6H_5)_2SiF_3^-$ showed axial and equatorial fluorines at lower temperatures. It was suggested that the dominant exchange process for the equilibration of the fluorine atom environment in SiF_5^- is probably the intramolecular Berry pseudorotation. In addition, the loss of ^{29}Si-^{19}F coupling above $-60°$ suggested an intermolecular mechanism, which could be either a dissociative or an associative (fluorine bridging) process. At the same time, the influence of impurities on the NMR spectra was noted, and interestingly, the existence of $CH_3SiF_4(H_2O)^-$ was proposed.

Conclusive evidence that intermolecular fluorine exchange in SiF_5^- is due to hydrolysis was suggested by Janzen and co-workers[245] from the ^{19}F NMR spectra at $38°$ of $(n-C_3H_7)_4N^+SiF_5^-$ in CH_2Cl_2 solvent. Since no ^{29}Si-^{19}F coupling was observed in the untreated solution, SiF_5^- must have been undergoing rapid intermolecular fluorine exchange. After addition of

$[(CH_3)_3Si]_2NH$ the $^{29}Si-^{19}F$ coupling (140 Hz) became clearly evident; hence, intermolecular exchange was slowed. Introduction of H_2O resulted once more in loss of $^{29}Si-^{19}F$ coupling. Such results are entirely consistent with hydrolysis as the mechanism of intermolecular fluorine exchange.

$$SiF_5^- + H_2O \leftrightarrows SiF_4OH^- + HF$$

While the evidence based on $^{29}Si-^{19}F$ coupling shows that intermolecular exchange may be slowed, the fact that all five fluorine atoms give a single peak at 38° (and also at $-90°$) is proof that a lower energy intramolecular mechanism of exchange was still equilibrating axial and equatorial fluorines.

Adducts of SiF_5^- are also known. Infrared studies by Clark et al.[246] suggested that a variety of Lewis bases form adducts of the type $SiF_5:base^-$, in which the base may be $(C_2H_5)_2NH$, $(C_2H_5)_3N$, $(n\text{-}C_3H_7)_3N$, C_5H_5N, and $(C_2H_5)_3P$. The variable temperature NMR spectra show a single sharp peak that broadens on cooling and eventually splits into two components, a low-field peak assigned to the four basal fluorines, and a peak at 136.6 ppm assigned to the apical fluorine, F^a.

The spectrum was reversible upon warming the sample, and the results were thus consistent with the rapid equilibrium

$$SiF_5^- + NH_3 \leftrightarrows [SiF_5:NH_3]^-$$

This equilibrium would involve intramolecular fluorine exchange and require retention of $^{29}Si-^{19}F$ coupling. Coupling could be observed at 24° ($^1J_{^{29}Si-^{19}F} = 120 \pm 5$ Hz) and at $-83°$ ($^1J_{^{29}Si-^{19}F^a} = 120 \pm 10$ Hz and $^1J_{^{29}Si-^{19}F^b} = 125 \pm 5$ Hz).

2. π Bonding

The traditional question of d orbital participation in organosilicon chemistry, or $(p \rightarrow d)\pi$ bonding as it is often termed, is not solved by consideration of spin–spin coupling involving the ^{29}Si nucleus. Certain experimental results have been cited by various authors as being consistent with such interactions, and a brief indication of such suggestion follows.

An early interpretation of $^2J_{^{29}Si-C-^1H}$ in compounds such as $(CH_3)_3SiX$ was both suggested[209] and later questioned by Schmidbaur[247]; it was also suggested by Brune[208, 248] that in compounds such as the $Cl_n(CH_3)_{3-n}SiOR$,

$(CH_3)_{4-n}Si(OR)_n$, and $R_{3-n}HSi(OR)_n$ series, it would be possible to involve $(p \rightarrow d)\pi$ bonding to help explain coupling constants including $^2J_{^{29}Si-C-^1H}$.

Newton and Rochow[158] suggested that $(p \rightarrow d)\pi$ bonding played a role in $(RO)_3SiH$ compounds, but that π donation was outweighed by σ withdrawal from the electronegative substituents. It was noted, however, that the decrease in coupling constants observed could result from a flattening of the $-Si(OR)_3$ group due to π bonding, which would result in the silicon atom moving closer to the plane of the oxygen atoms. Thus, the O—Si—O angles would increase to a value greater than the tetrahedral value, cause rehybridization at silicon, and produce greater s character in the Si—O bonds and less s character in the Si—H bonds.

In the trialkoxysilanes, J decreases in the order MeO > i-BuO > EtO > n-PrO = n-BuO > i-PrO > s-BuO > t-BuO, a trend opposed to that expected from inductive considerations. This reversed trend was explained by assuming an increased inductive effort which strengthened multiple Si—O bonding by increasing the electron density on the oxygen, thereby increasing π donation and further decreasing the s character of the Si—H bond. The authors concluded that steric interactions would be similar to the inductive effect but were not important.[158]

It was questioned whether deviations of Si_2H_5X compounds from an equation to fit coupling constants of $M'H_3MH_2X$ species might be consistent with an X $\rightarrow$ Si interaction, which could occur either through space or through the intervening silicon atom.[249] Alternatively, such effects might merely reflect the greater polarizability of silicon compared to carbon.[131] It was also stated that $^3J_{^{29}Si-O-C-^1H}$ could be understood on the basis of $(p \rightarrow d)\pi$ bonding, which is consistent with a decrease in the s character exhibited by silicon towards oxygen, thus suggesting a rise in the "effective electronegativity" of oxygen and increased back donation.[47]

Ebsworth and Turner noted in early work that when $^1J_{^1H-^1H}$ was plotted against $^1J_{^{29}Si-^1H}$, some effect appeared to increase $^1J_{^1H-^1H}$ while, if anything, reducing $^1J_{^{29}Si-^1H}$ when silicon was bound to a first row element of Groups V, VI, or VII. They commented that it was tempting to associate this factor with the d orbitals of silicon, although Ebsworth[250] later warned against "easy" interpretations of $(p \rightarrow d)\pi$ bonding to explain coupling constants.

Extensive studies by other workers have failed to discover any correlations which would need to invoke π interactions for satisfactory interpretation. For example, the work by Nagai et al.,[175, 176] in which the $^1J_{^{29}Si-^1H}$ values gave good correlations with Hammett σ constants in phenyl- and benzylsilanes, found no evidence that would be consistent with $(p \rightarrow d)\pi$ interactions.

Cartledge and Riedel[251] found no irregularities in a disilane series for substituents such as p-OCH$_3$ and p-Cl which might be involved in $(p \rightarrow d)\pi$ interactions between the ring and silicon. The $^1J_{^{29}Si-^1H}$ values show the same sensitivity in substituents in the two other series with no enhanced transmission which could be due to d-orbital utilization or polarizability differences. They also noted that $^1J_{^{29}Si-^1H}$ for a Si—H bond is appreciably more sensitive to substituents than in $^1J_{^{13}C-^1H}$ for either σ or π C—H bonds. The authors suggested that perhaps $^1J_{^{29}Si-^1H}$ is simply inherently more sensitive to substituent effects; alternatively, it has been proposed that there may be some interaction between an aromatic ring and a β silicon atom which allows an effective transmission of substituent effects.[113]

Finally, Nagai et al.[175] noted that the failure of three series of aryl hydrosilanes to provide a better correlation with σ^* constants provides little support for the $(p \to d)\pi$ back bonding proposed[252] to explain $^1J_{^{13}C-^1H}$ values for a series of substituted phenyltrimethylsilanes.

IV. RELAXATION AND THE NUCLEAR OVERHAUSER EFFECT

These two related phenomena, which are so important in ^{29}Si FT NMR,[3] are considered in detail elsewhere in this volume by Saunders (Chapter 5). The present knowledge of ^{29}Si relaxation and the nuclear Overhauser effect is almost entirely due to work done in the laboratory of Levy and Cargioli[3, 50–52, 253, 254] who have recently reviewed this field.[3] Data have also been reported by other authors.[12, 38, 255]

Since the above review, only three papers have appeared in which quantitative data are given. One paper[256] describes experiments by which the silicon T_1 relaxation time (and its temperature dependence) can be determined from ^{1}H NMR experiments on compounds having Si—H bonds. The experiments involved measurements of proton T_1 relaxation time of the central (^{28}SiH) line and measurements of NOE on one ^{29}Si satellite when the other is irradiated. The other two papers deal with NOE in phenylsilanes[255] and methyl- and methylphenylsiloxanes.[257] A possible analytical utilization of the regular trends in enhancement factors in the latter compounds was also discussed. The observed trends of analytical importance are (1) the lowest enhancements are observed for permethylated silicon atoms (e.g., TMS, M units in permethylated siloxanes); (2) the NOE in siloxanes increases for D, T, and Q units with increasing length of the chain and distance from the end of the chain; (3) the introduction of a phenyl group increases the NOE in siloxanes; and (4) the NOE in silazanes is larger for D units than for M units.[60] A theoretical justification of these trends follows from the classification of relaxation mechanisms in organosilicon compounds as detailed by Levy and Cargioli.[3]

Since a more complete account of relaxation would necessarily result in a duplication of the above reviews, the interested reader is referred to them for details of these processes. In general, the uses that can be expected from measurements of T_1 and NOE include the characterization of molecular motion[3] and also aid in spectral assignments and structure determination, particularly in compounds containing several silicon atoms.[257]

Acknowledgments

This review was written while Dr. J. M. Bellama was at the Institute of Chemical Process Fundamentals of the Czechoslovak Academy of Sciences under a scientific exchange program between the National Academy of Science (U.S.A.) and the Czechoslovak Academy. We appreciate the encouragement of Dr. V. Chvalovský, head of the Department of Homogeneous Reactions of the Institute. We would also like to thank those who have made available unpublished data, especially Prof. Dr. G. Fritz, Dr. H. C. Marsmann, and Dr. J. D. Cargioli. In addition, a helpful discussion with Dr. G. Engelhardt is gratefully acknowledged.

References

[1] V. Bažant, M. Horák, V. Chvalovský, and J. Schraml, "Organosilicon Compounds," Vol. 3, Chapter 1. Institute of Chemical Process Fundamentals, ČSAV (Czech. Acad. Sci.), Prague, 1973.

[2] D. W. Williams, in "Analysis of Silicones" (A. L. Smith, ed.), Chapter 11. Wiley (Interscience), New York, 1974.

[3] G. C. Levy and J. D. Cargioli, in "Nuclear Magnetic Resonance Spectroscopy of Nuclei Other than Protons" (T. Axenrod and G. A. Webb, eds.), Chapter 17. Wiley (Interscience), New York, 1974.

[4] G. A. Williams, D. W. McCall, and H. S. Gutowsky, Phys. Rev. 93, 1428 (1954).

[5] G. R. Holzman, P. C. Lauterbur, J. H. Anderson, and W. Koth, J. Chem. Phys. 25, 172 (1956).

[6] P. C. Lauterbur, in "Determination of Organic Structure by Physical Methods" (F. C. Nachod and W. D. Phillips, eds.), Vol. 2, p. 465. Academic Press, New York, 1962.

[7] E. B. Baker, J. Chem. Phys. 37, 911 (1962).

[8] R. B. Johannesen, J. Chem. Phys. 47, 955 (1967).

[9] R. B. Johannesen, J. Chem. Phys. 47, 3088 (1967).

[10] R. B. Johannesen, F. E. Brinckman, and T. D. Coyle, J. Phys. Chem. 72, 660 (1968).

[11] R. B. Johannesen and T. D. Coyle, Endeavour 31, 10 (1972).

[12] R. B. Johannesen, T. C. Farrar, F. E. Brinckman, and T. D. Coyle, J. Chem. Phys. 44, 962 (1966).

[13] B. K. Hunter and L. W. Reeves, Can. J. Chem. 46, 1399 (1968).

[14] H. G. Horn and H. C. Marsmann, Makromol. Chem. 162, 255 (1972).

[15] R. K. Harris and B. J. Kimber, J. Organometal. Chem. 70, 43 (1974).

[16] G. Engelhardt, H. Jancke, M. Mägi, T. Pehk, and E. Lippmaa, J. Organometal. Chem. 28, 293 (1971).

[17] G. Engelhardt, M. Mägi, and E. Lippmaa, J. Organometal. Chem. 54, 115 (1973).

[18] E. Lippmaa, T. Pehk, and J. Past, Eesti NSV Tead. Akad. Toim., Fuus.-Mat. Tehnikatead. Seer. 16, 345 (1967).

[19] V. J. Bartuska, T. T. Nakashima, and G. E. Maciel, Rev. Sci. Instrum. 41, 1458 (1970).

262   J. Schraml and J. M. Bellama

20 P. R. Wells, *in* "Determination of Organic Structures by Physical Methods" (F. C. Nachod and J. J. Zuckerman, eds.), Vol. 4, p. 233. Academic Press, New York, 1971.
21 H. C. Marsmann, *Chem. Ztg.* **97**, 128 (1973).
22 Varian Associates, "NMR Table," 5th ed. Varian, Palo Alto, 1965.
23 E. D. Becker, *in* "Nuclear Magnetic Resonance of Nuclei Other than Protons" (T. Axenrod and G. A. Webb, eds.), p. 1. Wiley (Interscience), New York, 1974.
24 E. A. V. Ebsworth and S. G. Frankiss, *Trans. Faraday Soc.* **59**, 1518 (1963).
25 E. A. V. Ebsworth and J. J. Turner, *Trans. Faraday Soc.* **60**, 256 (1964).
26 R. J. Gillespie and W. Quail, *J. Chem. Phys.* **39**, 2555 (1963).
27 S. G. Frankiss, *J. Phys. Chem.* **71**, 3418 (1967).
28 H. Bürger, R. Eujen, and H. C. Marsmann, *Z. Naturforsch. B* **29**, 149 (1974).
29 H. Batiz-Hernandez and R. A. Bernheim, *in* "Progress in Nuclear Magnetic Resonance Spectroscopy" (J. W. Emsley, J. Feeney, and L. H. Sutcliffe, eds.), Vol. III, p. 63. Pergamon, Oxford, 1967.
30 H. T. Weaver, R. C. Knauer, R. K. Quinn, and R. J. Baughman, *Solid State Commun.* **11**, 453 (1972).
31 H. T. Weaver, R. K. Quinn, R. J. Baughman, and R. C. Knauer, *J. Chem. Phys.* **59**, 4961 (1973).
32 H. Saji, T. Yamadaya, and M. Asanuma, *Phys. Lett. A* **45**, 109 (1973).
33 W. Sasaki, S. Ikehata, and S. Kobayashi, *Phys. Lett. A* **42**, 429 (1973).
34 M. G. Gibby, A. Pines, and J. S. Waugh, *J. Amer. Chem. Soc.* **94**, 6231 (1972).
35 C. J. Jameson and H. S. Gutowsky, *J. Chem. Phys.* **40**, 1714 (1964).
36 H. C. Marsmann and H. G. Horn, *Chem. Ztg.* **96**, 456 (1972).
37 J. Schraml, V. Chvalovský, M. Mägi, and E. Lippmaa, *Collect. Czech. Chem. Commun.* **40**, 897 (1975).
38 R. L. Scholl, G. E. Maciel, and W. K. Musker, *J. Amer. Chem. Soc.* **94**, 6376 (1972).
39 G. Engelhardt, R. Radeglia, H. Jancke, E. Lippmaa, and M. Mägi, *Org. Magn. Resonance* **5**, 561 (1973).
40 H. C. Marsmann, unpublished results (1974).
41 Yu. A. Buslaev, V. D. Kopanev, and V. P. Tarasov, *Chem. Commun.* p. 1175 (1971).
42 J. D. Cargioli and E. A. Williams, unpublished results (1974).
43 T. C. Farrar and E. D. Becker, "Pulse and Fourier Transform NMR," Academic Press, New York, 1971.
44 W. McFarlane, *Ann. Rev. NMR (Nucl. Magn. Resonance) Spectrosc.* **1**, 135 (1968).
45 W. McFarlane, *in* "Nuclear Magnetic Resonance Spectroscopy of Nuclei Other than Protons" (T. Axenrod and G. A. Webb, eds.), p. 31. Wiley (Interscience), New York, 1974.
46 E. V. van den Berghe and G. P. van der Kelen, *Abstr., Int. 1971 Congr. Organometal. Chem., 5th*, Vol. 2, p. 614 (1971).
47 E. V. van den Berghe and G. P. van der Kelen, *Int. Conf. Organometal. Chem., 6th, 1973* Abstract No. 166 (1973).
48 E. V. van den Berghe and G. P. van der Kelen, *J. Organometal. Chem.* **59**, 175 (1973).
49 W. McFarlane and J. M. Seaby, *J. Chem. Soc., Perkin Trans. 2* p. 1561 (1972).
50 G. C. Levy, *J. Amer. Chem. Soc.* **94**, 4793 (1972).
51 G. C. Levy and J. D. Cargioli, *J. Magn. Resonance* **10**, 231 (1973).

[52] G. C. Levy, J. D. Cargioli, P. C. Juliano, and T. D. Mitchell, *J. Magn. Resonance* **8**, 399 (1972).

[53] C. R. Ernst, L. Spialter, G. R. Buell, and D. L. Wilhite, *J. Amer. Chem. Soc.* **96**, 5375 (1974).

[54] G. Fritz, private communication (1974).

[55] H. C. Marsmann and H. G. Horn, *Z. Naturforsch.* **B 27**, 1448 (1972).

[56] H. C. Marsmann, *Chem. Ztg.* **96**, 288 (1972).

[57] E. Lippmaa, M. Mägi, J. Schraml, and V. Chvalovský, *Collect. Czech. Chem. Commun.* **39**, 1041 (1974).

[58] G. Engelhardt, H. Jancke, D. Hoebbel, and W. Wieker, *Z. Chem.* **14**, 109 (1974).

[59] H. Jancke, G. Engelhardt, M. Mägi, and E. Lippmaa, *Z. Chem.* **13**, 392 (1973).

[60] H. Jancke, G. Engelhardt, M. Mägi, and E. Lippmaa, *Z. Chem.* **13**, 435 (1973).

[61] M. Ya. Mägi, E. T. Lippmaa, S. L. Ioffe, V. A. Tartakovskii, A. S. Shashkov, B. N. Khasapov, and L. M. Makarenkova, *Izv. Akad. Nauk SSSR, Ser. Khim.* p. 1431 (1973); *Bull. Acad. Sci. USSR, Div. Chem. Sci.* **22**, 1401 (1973).

[62] Nguyen-Duc-Chuy, V. Chvalovsky, J. Schraml, M. Mägi, and E. Lippmaa, *Collect. Czech. Chem. Commun.* **40**, 875 (1975).

[63] C. R. Ernst, L. Spialter, G. R. Buell, and D. L. Wilhite, *J. Organometal. Chem.* **59**, C-13 (1973).

[64] H. C. Marsmann and R. Löwer, *Chem. Ztg.* **97**, 660 (1973).

[65] J. Schraml, Nguyen-Duc-Chuy, V. Chvalovský, M. Mägi, and E. Lippmaa, *J. Organometal. Chem.* **51**, C-5 (1973).

[66] J. Schraml, J. Pola, V. Chvalovský, M. Mägi, and E. Lippmaa, *J. Organometal. Chem.* **49**, C-19 (1973).

[67] J. Schraml, J. Včelák, and V. Chvalovský, *Collect. Czech. Chem. Commun.* **39**, 267 (1974).

[68] S. L. Ioffe, V. M. Shitkin, B. N. Khasapov, M. V. Kashutina, V. A. Tartakovskii, M. Ya. Mägi, and E. T. Lippmaa, *Izv. Akad. Nauk SSSR, Ser. Khim.* p. 2146 (1973); *Bull. Acad. Sci. USSR, Div. Chem. Sci.* **22**, 2100 (1973).

[69] IUPAC, "Recommendations for the Presentation of NMR Data," Inform. Bull. No. 4. IUPAC Secretariat, Oxford, 1970.

[70] G. C. Levy and G. L. Nelson, "Carbon-13 Nuclear Magnetic Resonance for Organic Chemists," p. 22. Wiley (Interscience), New York, 1972.

[71] J. B. Stothes, "Carbon-13 NMR Spectroscopy," p. 49. Academic Press, New York, 1972.

[72] G. C. Levy, J. D. Cargioli, G. E. Maciel, J. J. Natterstad, E. B. Whipple, and M. Ruta, *J. Magn. Resonance* **11**, 352 (1973).

[73] G. V. D. Tiers, *J. Phys. Chem.* **62**, 1151 (1958).

[74] D. Kummer, V. Chvalovský, and J. Schraml, unpublished results.

[75] M. Mägi and J. Schraml, unpublished results.

[76] M. R. Bacon and G. E. Maciel, *J. Amer. Chem. Soc.* **95**, 2413 (1973).

[77] M. Bacon, G. E. Maciel, W. K. Musker, and R. Scholl, *J. Amer. Chem. Soc.* **93**, 2537 (1971).

[78] J. Schraml, Nguyen-Duc-Chuy, J. Pola, P. Novák, V. Chvalovský, M. Mägi, and E. Lippmaa, unpublished results (1974).

[79] W. McFarlane, *Mol. Phys.* **13**, 587 (1967).

[80] E. Niecke and W. Flick, *Angew. Chem.* **85**, 586 (1973); *Int. Ed. in Engl.* **12**, 585 (1973).

[81] G. Fritz and P. Böttinger, *Z. Anorg. Allg. Chem.* **395**, 159 (1973).

[82] G. Fritz and N. Braunagel, *Z. Anorg. Allg. Chem.* **399**, 280 (1973).

[83] G. Fritz, H. Fröhlich, and D. Kummer, *Z. Anorg. Allg. Chem.* **353**, 34 (1967).

[84] G. Fritz and M. Hähnke, *Z. Anorg. Allg. Chem.* **390**, 137 (1972).

[85] J. Dyer and J. Lee, *Spectrochim. Acta A* **26**, 1045 (1970).

[86] J. Schraml, Nguyen-Duc-Chuy, V. Chvalovský, M. Mägi, and E. Lippmaa, *J. Org. Magn. Resonance* **7**, 379 (1975).

[87] E. Lippmaa, M. Mägi, V. Chvalovský, and J. Schraml, unpublished results (1974).

[88] E. Lippmaa, M. Mägi, J. Schraml, and V. Chvalovský, *Eur. Congr. Mol. Spectrosc.*, *11th 1973* Abstract No. 146 (1973).

[89] R. Radeglia and G. Engelhardt, *J. Organometal. Chem.* **67**, C-45 (1974).

[90] C. R. Ernst, G. R. Buell, and L. Spialter, *Organosilicon Symp.*, *1973* Abstract No. 9, p. 17 (1973).

[91] J. Pola, J. Schraml, and V. Chvalovský, *Collect. Czech. Chem. Commun.* **38**, 3158 (1973).

[92] T. Pehk, E. Lippmaa, E. Lukevits, and L. I. Simchenko, *Zh. Obshch. Khim.* **46**, 602 (1976).

[93] D. W. W. Anderson, J. E. Bentham, and D. W. H. Rankin, *J. Chem. Soc.*, *Dalton Trans.* p. 1215 (1973).

[94] E. A. V. Ebsworth, private communication (1974).

[95] R. West, Lecture (Institute of Chemical Process Fundamentals, Czech. Acad. Sci.), Prague, 1973.

[96] R. L. Lambert, Jr. and D. Seyferth, *J. Amer. Chem. Soc.* **94**, 9246 (1972).

[97] R. K. Harris and B. J. Kimber, *J. Chem. Soc.*, *Chem. Commun.* p. 559 (1974).

[98] H. C. Marsmann, *Z. Naturforsch. B* **29**, 495 (1974).

[98a] R. O. Gould, B. M. Lowe, and N. A. MacGilp, *J. Chem. Soc.*, *Chem. Commun.* p. 720 (1974).

[99] J. Schraml, J. Pola, H. Jancke, G. Engelhardt, M. Černý, and V. Chvalovský, *Collect. Czech. Chem. Commun.*, in press.

[100] F. H. A. Rummens, *Org. Magn. Resonance* **2**, 209 (1970).

[101] A. D. Buckingham, T. Schaefer, and W. G. Schneider, *J. Chem. Phys.* **32**, 1227 (1960).

[102] A. Saika and C. P. Slichter, *J. Chem. Phys.* **22**, 26 (1954).

[103] J. A. Pople, *Discuss. Faraday Soc.* **34**, 7 (1962).

[104] N. F. Ramsey, *Phys. Rev.* **78**, 699 (1950).

[105] M. M. Gofman, E. L. Rozenberg, and M. E. Dyatkina, *Dokl. Akad. Nauk SSSR* **199**, 635 (1971); *Proc. Acad. Sci. (USSR), Phys. Chem. Section* **199**, 635 (1971).

[106] J. Mason, *J. Chem. Soc.*, *A* p. 1038 (1971).

[107] J. Feeney, L. H. Sutcliffe, and S. M. Walker, *Mol. Phys.* **11**, 117, 129, 137, and 145 (1966).

[108] J. H. Letcher and J. R. Van Wazer, *J. Chem. Phys.* **44**, 815 (1966); **45**, 2916 and 2926 (1966).

[109] J. H. Letcher and J. R. Van Wazer, *Top. Phosphorus Chem.* **5**, Chapter 2 (1967).

[110] G. Engelhardt, H. Jancke, E. Lippmaa, and M. Mägi, *Eur. Congr. Mol. Spectrosc.*, *11th, 1973* Abstract No. 147 (1973).

[111] V. S. Lyubimov and S. P. Ionov, *Zh. Fiz. Khim.* **46**, 838 (1972); *Russ. J. Phys. Chem.* **46**, 486 (1972).

[112] C. G. Pitt, *J. Organometal. Chem.* **61**, 49 (1973).

[113] J. Nagy and J. Réffy, *Symp. Int. Chim. Composes Org. Silicium*, 2nd, 1968 p. 143 (1968).

[114] J. Réffy, G. Pongor, and J. Nagy, *Period. Polytech., Chem. Eng.* **15**, 185 (1971).

[115] J. N. Murrell, *in* "Progress in Nuclear Magnetic Resonance Spectroscopy" (J. W. Emsley, J. Feeney, and L. H. Sutcliffe, eds.), Vol. VI, p. 1. Pergamon, Oxford, 1971.

[116] M. Barfield and M. D. Johnston, Jr., *Chem. Rev.* **73**, 53 (1973).

[117] J. D. Mcmory, "Quantum Theory of Magnetic Resonance Parameters," McGraw-Hill, New York, 1968.

[118] N. F. Ramsey, *Phys. Rev.* **91**, 303 (1953).

[119] H. M. McConnell, *J. Chem. Phys.* **24**, 460 (1956).

[120] J. A. Pople and D. P. Santry, *Mol. Phys.* **8**, 1 (1964).

[121] L. W. Reeves, *J. Chem. Phys.* **40**, 2128, 2132, and 2423 (1964).

[122] L. W. Reeves and F. J. Wells, *Can. J. Chem.* **41**, 2698 (1963).

[123] H. Dreeskamp, *Z. Naturforsch. A* **19**, 139 (1964).

[124] G. W. Smith, *J. Chem. Phys.* **39**, 2031 (1963); **40**, 2037 (1964); **42**, 435 (1965).

[125] A. C. Blizzard and D. P. Santry, *J. Chem. Phys.* **55**, 950 (1971).

[126] E. A. V. Ebsworth and J. J. Turner, *J. Chem. Phys.* **36**, 2628 (1962).

[127] J. H. Goldstein, V. S. Watts, and L. S. Rattet, *in* "Progress in Nuclear Magnetic Resonance Spectroscopy," (J. W. Emsley, J. Feeney, and L. H. Sutcliffe, eds.), Vol. VIII, p. 103–104. Pergamon, Oxford, 1971.

[128] V. M. S. Gil and C. F. G. C. Geraldos, *in* "Nuclear Magnetic Resonance Spectroscopy of Nuclei Other than Protons" (T. Axenrod and G. A. Webb, eds.), p. 219. Wiley (Interscience), New York, 1974.

[129] J. M. Bellama, R. S. Evans, and J. E. Huheey, *J. Amer. Chem. Soc.* **95**, 7242 (1973).

[130] E. R. Malinowski, *J. Amer. Chem. Soc.* **83**, 4479 (1961).

[131] C. Juan and H. S. Gutowsky, *J. Chem. Phys.* **37**, 2198 (1962).

[132] J. H. Campbell-Ferguson, E. A. V. Ebsworth, A. G. MacDiarmid, and T. Yoshioka, *J. Phys. Chem.* **71**, 723 (1967).

[133] R. Ditchfield, M. A. Jensen, and J. N. Murrell, *J. Chem. Soc., A* p. 1674 (1967).

[134] E. O. Bishop and M. A. Jensen, *Chem. Commun.* p. 922 (1966).

[135] M. A. Jensen, *J. Organometal. Chem.* **11**, 423 (1968).

[136] E. R. Malinowski and T. Vladimiroff, *J. Amer. Chem. Soc.* **86**, 3575 (1964).

[137] E. A. V. Ebsworth, A. G. Lee, and G. M. Sheldrick, *J. Chem. Soc., A* p. 2294 (1968).

[138] T. Yoshioka and A. G. MacDiarmid, *J. Mol. Spectrosc.* **21**, 103 (1966).

[139] J. Schraml, V. Chvalovský, M. Mägi, E. Lippmaa, *Collect. Czech. Chem. Commun.* **40**, 897 (1975).

[140] J. E. Drake and N. Goddard, *J. Chem. Soc., A* p. 2587 (1970).

[141] A. Rastelli and S. A. Pozzoli, *J. Mol. Struct.* **18**, 463 (1973).

[142] C. J. Jameson and H. S. Gutowsky, *J. Chem. Phys.* **51**, 2790 (1969).

[143] H. Dreeskamp, *Z. Phys. Chem. (Frankfurt am Main)* [N.S.] **38**, 121 (1963).

[144] C. Schumann and H. Dreeskamp, *J. Magn. Resonance* **3**, 204 (1970).

[145] P. T. Inglefield and L. W. Reeves, *J. Chem. Phys.* **40**, 2424 (1964).

[146] E. J. Wells and L. W. Reeves, *J. Chem. Phys.* **40**, 2036 (1964).

[147] H. Dreeskamp and G. Stegmeier, *Z. Naturforsch. A* **22**, 1458 (1967).

[148] D. K. Dalling and H. S. Gutowsky, *J. Chem. Phys.* **55**, 4959 (1971).

[149] S. S. Danyluk, *J. Amer. Chem. Soc.* **87**, 2300 (1965).

[150] A. H. Cowley and W. D. White, *J. Amer. Chem. Soc.* **91**, 1913 and 1917 (1969).

151 A. H. Cowley, W. D. White, and S. L. Manatt, *J. Amer. Chem. Soc.* **89**, 6433 (1967).
152 S. S. Danyluk, *J. Amer. Chem. Soc.* **86**, 4504 (1964).
153 W. McFarlane, *J. Chem. Soc., A* p. 1275 (1967).
154 G. Pfisterer and H. Dreeskamp, *Ber. Bunsenges. Phys. Chem.* **73**, 654 (1969).
155 R. R. Dean and W. McFarlane, *Mol. Phys.* **12**, 289 and 364 (1967).
156 F. Fehér, P. Hädicke, and H. Frings, *Inorg. Nucl. Chem. Lett.* **9**, 931 (1973).
157 E. Hengge and F. Höfler, *Z. Naturforsch. A* **26**, 768 (1971).
158 W. E. Newton and E. G. Rochow, *Inorg. Chim. Acta* **4**, 133 (1970).
159 E. Hengge, *Monatsh. Chem.* **102**, 734 (1971).
160 B. J. Aylett, unpublished results (1974).
161 K. M. Abraham and G. Urry, *Inorg. Chem.* **12**, 2850 (1973).
162 G. Fritz, G. Grobe, and D. Kummer, *Advan. Inorg. Chem. Radiochem.* **7**, 400 (1965).
163 H. Bürger and W. Kilian, *J. Organometal. Chem.* **26**, 47 (1971).
164 H. Bürger and W. Kilian, *J. Organometal. Chem.* **18**, 299 (1969).
165 G. Fritz, G. Becker, and D. Kummer, *Z. Anorg. Allg. Chem.* **372**, 171 (1970).
166 J. V. Urenovitch and R. West, *J. Organometal. Chem.* **3**, 138 (1965).
167 E. Hengge and H. Marketz, *Monatsh. Chem.* **101**, 528 (1970).
168 C. Glidewell and D. W. H. Rankin, *J. Chem. Soc., A* p. 279 (1970).
169 C. Glidewell, *J. Chem. Soc., A* p. 823 (1971).
170 G. K. Barker, J. E. Drake, and R. T. Hemmings, *J. Chem. Soc., Dalton Trans.* p. 450 (1974).
171 C. H. Van Dyke and A. G. MacDiarmid, *Inorg. Chem.* **3**, 1071 (1964).
172 J. E. Drake and N. P. C. Westwood, *J. Chem. Soc., A* p. 3300 (1971).
173 C. H. Van Dyke, E. W. Kiefer, and G. A. Gibbon, *Inorg. Chem.* **11**, 408 (1972).
174 E. A. V. Ebsworth, H. J. Emeléus, and N. Welcman, *J. Chem. Soc., London* p. 2290 (1962).
175 Y. Nagai, M. Ohtsuki, T. Nakano, and H. Watanabe, *J. Organometal. Chem.* **35**, 81 (1972).
176 Y. Nagai, H. Matsumoto, T. Nakano, and H. Watanabe, *Bull. Chem. Soc. Jap.* **45**, 2560 (1972).
177 J. Schraml and V. Chvalovský, to be published.
178 Y. Nagai, S. Inaka, H. Matsumoto, and H. Watanabe, *Bull. Chem. Soc. Jap.* **45**, 3224 (1972).
179 Y. Nagai, H. Matsumoto, M. Hayashi, E. Tajima, and H. Watanabe, *Bull. Chem. Soc. Jap.* **44**, 3113 (1971).
180 J. F. Bald, Jr., K. G. Sharp, and A. G. MacDiarmid, *J. Fluorine Chem.* **3**, 433 (1973).
181 J. C. Thompson and J. L. Margrave, *Chem. Commun.* p. 566 (1966).
182 T. D. Coyle, R. B. Johannesen, F. E. Brinckman, and T. C. Farrar, *J. Phys. Chem.* **70**, 1682 (1966).
183 E. L. Muetterties and W. D. Phillips, *J. Amer. Chem. Soc.* **81**, 1084 (1959).
184 M. Murray, *J. Magn. Resonance* **9**, 326 (1973).
185 W. Airey, G. M. Sheldrick, B. J. Aylett, and I. A. Ellis, *Spectrochim. Acta, Part A* **27**, 1505 (1971).
186 D. Solan and P. L. Timms, *Inorg. Chem.* **7**, 2157 (1968).
187 D. Solan and A. B. Burg, *Inorg. Chem.* **11**, 1253 (1972).
188 F. Höfler and W. Veigl, *Angew. Chem.* **83**, 977 (1971); *Int. Ed. in Engl.* **10**, 919 (1971).
189 K. G. Sharp and T. D. Coyle, *Inorg. Chem.* **11**, 1259 (1972).

[190] J. L. Margrave, K. G. Sharp, and P. W. Wilson, *J. Inorg. Nucl. Chem.* **32**, 1813 (1970).
[191] J. L. Margrave, K. G. Sharp, and P. W. Wilson, *J. Inorg. Nucl. Chem.* **32**, 1817 (1970).
[192] W. Airey and G. M. Sheldrick, *J. Inorg. Nucl. Chem.* **32**, 1827 (1970).
[193] J. J. Moscony and A. G. MacDiarmid, *Chem. Commun.* p. 307 (1965).
[194] W. Airey and G. M. Sheldrick, *J. Chem. Soc.*, A p. 1222 (1970).
[195] W. Airey and G. M. Sheldrick, *J. Chem. Soc.*, A p. 2865 (1969).
[196] P. L. Timms, T. C. Ehlert, J. L. Margrave, F. E. Brinckman, T. C. Farrar, and T. D. Coyle, *J. Amer. Chem. Soc.* **87**, 3819 (1965).
[107] J. Dyer and J. Lee, *Spectrochim. Acta, Part A* **26**, 1045 (1970).
[198] E. A. V. Ebsworth and G. M. Sheldrick, *Trans. Faraday Soc.* **62**, 3282 (1966).
[199] D. M. White and G. C. Levy, *Macromolecules* **5**, 526 (1972).
[200] H. Dreeskamp and K. Hildenbrand, *Justus Liebig's Ann. Chem.* p. 712 (1975).
[201] H. Elser and H. Dreeskamp, *Ber. Bunsenges. Phys. Chem.* **73**, 619 (1969).
[202] E. Niecke and W. Flick, *Angew. Chem.* **86**, 128 (1974); *Int. Ed. in Engl.* **13**, 134 (1974).
[203] G. Davidson, E. A. V. Ebsworth, G. M. Sheldrick, and L. A. Woodward, *Spectrochim. Acta* **22**, 67 (1966).
[204] K. D. Crosbie and G. M. Sheldrick, *Mol. Phys.* **20**, 317 (1971).
[205] D. W. W. Anderson, E. A. V. Ebsworth, and D. W. H. Rankin, *J. Chem. Soc., Dalton Trans.* p. 2370 (1973).
[206] W. McFarlane, *Quart. Rev., Chem. Soc.* **23**, 187 (1969).
[207] B. Beagley, A. G. Robiette, and G. M. Sheldrick, *J. Chem. Soc.*, A p. 3002 (1968).
[208] H. A. Brune, *Chem. Ber.* **97**, 2848 (1964).
[209] H. Schmidbaur, *J. Amer. Chem. Soc.* **85**, 2336 (1963).
[210] K. M. Mackay, A. E. Watt, and R. Watt, *J. Organometal. Chem.* **12**, 49 (1968).
[211] R. West, F. A. Kramer, E. Carberry, M. Kumada, and M. Ishikawa, *J. Organometal. Chem.* **8**, 79 (1967).
[212] D. H. O'Brien and C. M. Harbordt, *J. Organometal. Chem.* **21**, 321 (1970).
[213] T. F. Schaaf and J. P. Oliver, *J. Amer. Chem. Soc.* **91**, 4327 (1969).
[214] H. Schmidbaur and F. Schindler, *J. Organometal. Chem.* **2**, 466 (1964).
[215] W. Wolfsberger and H. H. Pickel, *Z. Anorg. Allg. Chem.* **384**, 131 (1971).
[216] G. Fritz, H. J. Dunnappel, and E. Matern, *Z. Anorg. Allg. Chem.* **399**, 263 (1973).
[217] G. E. Hartwell and T. L. Brown, *J. Amer. Chem. Soc.* **88**, 4625 (1966).
[218] G. Fritz and E. Bosch, *Z. Anorg. Allg. Chem.* **404**, 103 (1974).
[219] S. Cawley and S. S. Danyluk, *J. Phys. Chem.* **68**, 1240 (1964).
[220] P. L. Timms, R. A. Kent, T. C. Ehlert, and J. L. Margrave, *J. Amer. Chem. Soc.* **87**, 2824 (1965).
[221] H. Bürger, W. Kilian, and K. Burczyk, *J. Organometal. Chem.* **21**, 291 (1970).
[222] W. Jetz and W. A. G. Graham, *Inorg. Chem.* **10**, 1159 (1971).
[223] W. A. G. Graham, *Organosilicon Symp., 1974* (1974).
[224] C. D. Schaeffer, Jr. and J. J. Zuckerman, *J. Amer. Chem. Soc.* **96**, 7160 (1974).
[225] M. P. Simonnin, *J. Organometal. Chem.* **5**, 155 (1966).
[226] P. Krebs and H. Dreeskamp, *Spectrochim. Acta, Part A* **25**, 1399 (1969).
[227] E. Carberry, R. West, and G. E. Glass, *J. Amer. Chem. Soc.* **91**, 5446 (1969).
[228] M. Fukui, K. Itoh, and Y. Ishii, *J. Chem. Soc., Perkin Trans. 2*, p. 1043 (1972).
[229] H. Schumann, I. Schumann-Ruidisch, and S. Ronecker, *Z. Naturforsch. B* **25**, 565 (1970).

[230] I. Schumann-Ruidisch and B. Jutzi-Mebert, *J. Organometal. Chem.* **11**, 77 (1968).
[231] H. Bürger and U. Goetze, *Angew. Chem.* **80**, 192 (1968); *Int. Ed. in Engl.* **7**, 212 (1968).
[232] G. Fritz and G. Becker, *Z. Anorg. Allg. Chem.* **372**, 180 (1970).
[233] G. Fritz, H. Fröhlich, and D. Kummer, *Z. Anorg. Allg. Chem.* **353**, 34 (1967).
[234] S. Cradock, E. A. V. Ebsworth, and H. F. Jessep, *J. Chem. Soc., Dalton Trans.* p. 359 (1972).
[235] J. W. Anderson, G. K. Barker, J. E. Drake, and M. Rodger, *J. Chem. Soc., Dalton Trans.* p. 1716 (1973).
[236] S. Cradock, E. A. V. Ebsworth, H. Moretto, D. W. H. Rankin, and W. J. Savage, *Angew. Chem.* **85**, 344 (1973); *Int. Ed. in Engl.* **12**, 317 (1973).
[237] W. T. Raynes, *Mol. Phys.* **15**, 435 (1968).
[238] A. K. Jameson, C. J. Jameson, and H. S. Gutowsky, *J. Chem. Phys.* **53**, 2310 (1970).
[239] H. M. Hutton, E. Bock, and T. Schaefer, *Can. J. Chem.* **44**, 2772 (1966).
[240] A. K. Jameson and J. P. Reger, *J. Phys. Chem.* **75**, 437 (1971).
[241] M. Barfield and M. D. Johnston, Jr., *Chem. Rev.* **73**, 53 (1973).
[242] J. A. Gibson and A. F. Janzen, *Can. J. Chem.* **50**, 3087 (1972).
[243] M. R. Booth and S. G. Frankiss, *Spectrochim. Acta, Part A* **26**, 859 (1970).
[244] F. Klanberg and E. L. Muetterties, *Inorg. Chem.* **7**, 155 (1968).
[245] J. A. Gibson, D. G. Ibbott, and A. F. Janzen, *Can. J. Chem.* **51**, 3203 (1973).
[246] H. C. Clark, K. R. Dixon, and J. G. Nicolson, *Inorg. Chem.* **8**, 450 (1970).
[247] H. Schmidbaur, *Chem. Ber.* **97**, 842 (1964).
[248] H. A. Brune, *Tetrahedron* **24**, 79 (1968).
[249] C. H. Van Dyke and A. G. MacDiarmid, *Inorg. Chem.* **3**, 1071 (1964).
[250] E. A. V. Ebsworth, *in* "Organometallic Compounds of the Group IV Elements" (A. G. MacDiarmid, ed.), Vol. I, Part 1, pp. 70–76. Dekker, New York, 1968.
[251] F. K. Cartledge and K. H. Riedel, *J. Organometal. Chem.* **34**, 11 (1972).
[252] M. E. Freeburger and L. Spialter, *J. Amer. Chem. Soc.* **93**, 1894 (1971).
[253] G. C. Levy, *Organosilicon Symp., 1973* Abstract No. 8, p. 16 (1973).
[254] G. C. Levy, J. D. Cargioli, P. C. Juliano, and T. D. Mitchell, *J. Amer. Chem. Soc.* **95**, 3445 (1973).
[255] R. K. Harris and B. J. Kimber, *J. Chem. Soc., Chem. Commun.* p. 255 (1973).
[256] A. Briguet and A. Erbeia, *J. Phys. C* **5**, L-58 (1972).
[257] G. Engelhardt and H. Jancke, *Z. Chem.* **14**, 206 (1974).

Supplementary Readings

Theory of ²⁹Si Shielding

R. Radeglia, *Z. Phys. Chem.* (*Leipzig*) **256**, 453 (1975).
R. Wolff and R. Radeglia, *Z. Phys. Chem.* (*Leipzig*) (in press) (1976).

²⁹Si Shifts and Relaxation in Silicates

G. Engelhardt, D. Zeigan, H. Jancke, D. Hoebbel, and W. Wieker, *Z. Anorg. Allg. Chem.*, **418**, 17 (1975).
G. Engelhardt, *Z. Chem.* (in press) (1976).

Silicon Hydrides

R. Löwer, M. Vongehr, and H. C. Marsmann, *Chem. Ztg.* **99**, 33 (1975).

E. A. V. Ebsworth, *Organosilicon Award Symp.*, *9th*, p. 15. Case Western Reserve Univ., Cleveland, Ohio, 1975.

Solvent Effects on ^{29}Si Shielding
R. W. LaRochelle, J. D. Cargioli, and E. A. Williams, *Organosilicon Award Symp.*, *9th*, p. 19, Case Western Reserve Univ., Cleveland, Ohio, 1975.

^{29}Si Study of Trimethylsilyl Compounds
R. K. Harris and B. J. Kimber, *J. Magn. Resonance* **17**, 174 (1975).

Silicon Halogenides
U. Niemann and H. C. Marsmann, *Z. Naturforsch. B* **30**, 202 (1975).

Experimental Technique
S. A. Linde, H. J. Jakobsen, and B. J. Kimber, *J. Amer. Chem. Soc.* **97**, 3219 (1975).

The Nuclear Overhauser Effect

5

JOHN K. SAUNDERS AND JOHN W. EASTON

I. INTRODUCTION

The nuclear Overhauser effect (NOE) is defined as the change in area of the absorption of a nuclear spin when the resonance of another spin is saturated. The first NOE reported was by Solomon and Bloembergen[1] during their studies on HF. Several years later, Anet and Bourn[2] demonstrated the potential use of the technique as an aid in structural and conformational analysis in organic chemistry. Since that time, many articles[3–5] have appeared in which the technique has been put to good advantage, including several[6, 7] in which information on internuclear distances was obtained from observation of the NOE. In order to observe a NOE between two nuclear spins, there must be a time-dependent coupling between them, and thus to understand the NOE we must also understand the basics of spin–lattice relaxation.

In the following section, we present the relevant theory from which information on internuclear distances can be obtained, including a brief discussion on spin–lattice relaxation. In Section III, the various experimental methods are outlined, including steady-state measurements, transient methods, as well as the use of the Fourier transform (FT) technique. The latter sections are concerned with the application of NOE measurements to structural and stereochemical problems. It is not intended that these sections be an exhaustive literature survey, but rather that they represent a wide array of problems that can be solved, at least in part, by use of the NOE.

II. THEORY

Bloch[8] defined the spin–lattice relaxation time T_1 for a nuclear spin placed in a magnetic field in the z direction by Eq. (1)

$$\frac{dM_z}{dt} = -\frac{1}{T_1}(M_z - M_0) \tag{1}$$

where M_z is the instantaneous magnetization in the z direction and M_0 is the equilibrium value in this direction. Thus to determine values of T_1, the magnetization must be altered in a specific way and the return to equilibrium monitored. This can be achieved by a number of methods such as (1) taking the sample outside the magnetic field and observing its attainment of equilibrium; (2) saturating the resonance of the nucleus in question and monitoring its return to equilibrium; and (3) utilization of various pulse sequences, e.g., $180° - \tau - 90° - T\,(T > 5T_1)$, and $90° - \tau - 90° - T\,(\tau$, the time between pulses, being varied stepwise).

With the advent of FT techniques the last is probably the predominant method utilized by organic chemists. Its advantage is that the T_1's of a number of nuclei can be determined simultaneously.

Bloch defined the transverse relaxation processes by

$$\frac{dM_x}{dt} = \frac{-M_x}{T_2} \quad \text{and} \quad \frac{dM_y}{dt} = \frac{-M_y}{T_2}$$

The constant T_2, is called the spin–spin or transverse relaxation time, and for nuclei with $I = \frac{1}{2}$ it is much more difficult to obtain than T_1. One common method is to measure the linewidth ($\Delta\nu$) at half-peak height to give $T_2{}^* = 1/2\Delta\nu$, where $T_2{}^*$ is the "effective T_2" and includes factors, such as field inhomogeneity, which contribute to the linewidth, as well as the true T_2. For the purpose of this chapter, T_2 is of secondary importance.

Equation (1) is for isolated spins, a situation generally not observed in practice when effects of other spins on the observed ones must be considered. Solomon[9] modified the Bloch equations to account for two spins A and B as shown in Eqs. (2) and (3).

$$\frac{dM_z{}^A}{dt} = -\rho_A(M_z{}^A - M_0{}^A) - \sigma_{AB}(M_z{}^B - M_0{}^B) \tag{2}$$

$$\frac{dM_z{}^B}{dt} = -\rho_B(M_z{}^B - M_0{}^B) - \sigma_{BA}(M_z{}^A - M_0{}^A) \tag{3}$$

The definitions of the relaxation rates ρ and σ will be the same as in other NOE articles.[3,4] In Solomon's terminology, $\rho_A \equiv 1/T_{AA}$ and $\sigma_{AB} \equiv 1/T_{AB}$. The term ρ_A is the total relaxation rate of nucleus A, whereas σ_{AB} describes the mutual interaction between A and B, and is called the cross-relaxation term. For the σ terms to be finite there must be a relaxation mechanism that couples the two nuclei. There are three such possibilities:

1. Dipole–dipole coupling.
2. Scalar coupling between A and B, which is modulated by chemical exchange or by internal motion, that is scalar coupling which is time dependent.
3. Scalar coupling between the two nuclei which is modulated by rapid relaxation of one of the nuclei.

For the systems that we shall study, the first mechanism is the most important, the other two being significant only in a few specific examples.

Before we proceed further, a brief but more general consideration of nuclear relaxation mechanisms is pertinent. For a given nucleus, a number of relaxation processes is possible, with the total relaxation rate being given by $R = \sum_n \rho_n$, where the summation is for all mechanisms n. The possible mechanisms in addition to those above are spin–rotation interactions, magnetic anisotropy interactions, quadrupole interactions, and finally relaxation by a paramagnetic material. The dominant mechanism for our discussion is the mutual dipole–dipole interaction, and in the limit of extreme narrowing ($\omega\tau_c \ll 1$) this can be expressed for two nuclei, A and B, each with $I = \frac{1}{2}$, separated by a distance r, by

$$\rho_{AB}(DD) = \gamma_A{}^2\gamma_B{}^2\hbar^2\tau_c/r_{AB}{}^6 \tag{4}$$
$$\sigma_{AB}(DD) = \tfrac{1}{2}\gamma_A{}^2\gamma_B{}^2\hbar^2\tau_c/r_{AB}{}^6 \tag{5}$$

where τ_c is the isotropic rotational correlation time.

Relaxation by a time-dependent scalar interaction can be of two types. First, where one, B, of the coupled pair of nuclei is undergoing chemical exchange, and second, where the scalar coupling, J, between the two nuclei is modulated by a very rapid relaxation of one of them. For the former, Abragam[10] gave for two spins of $I = \frac{1}{2}$ and a rate of exchange τ_{ex},

$$\rho_{AB} = 2\pi^2 J^2 \frac{\tau_{ex}}{1 + (\omega_A - \omega_B)^2\tau_{ex}}$$
$$\sigma_{AB} = -2\pi^2 J^2 \frac{\tau_{ex}}{1 + (\omega_A - \omega_B)^2\tau_{ex}} \tag{6}$$

The final mechanism of interest is the interaction between the nucleus and one or more unpaired electrons. The relaxation rate depends on the magnetic moment of the species causing the relaxation and since the magnetic moment, μ, of an electron is approximately 1600 times that of a proton, an unpaired electron will generally cause rapid nuclear relaxation. The relaxation rate is[10]

$$\rho_e = \frac{16\pi^2}{15} N \langle\mu\rangle \frac{\gamma_I{}^2\eta}{kt} \tag{7}$$

where N is the concentration of the paramagnetic material and η is the viscosity. Thus since the internuclear NOE is a result of the much smaller nuclear–nuclear interaction, it is essential that all paramagnetic materials be absent when attempts are made to measure internuclear NOE's.

For a multinuclear system, the Bloch equations can be written[3] as:

$$\frac{dM_z^A}{dt} = -R_A(M_z^A - M_0^A) - \sigma_{AB}(M_z^B - M_0^B)$$
$$- \sum_{d \neq A, B} \sigma_{Ad}(M_z^d - M_0^d) \qquad (8)$$
$$R_A = \sum_{d \neq A} \rho_{Ad} + \rho_A{}^*$$

The terms ρ_{Ad} and σ_{Ad} for the dipolar interaction are defined in Eqs. (4) and (5) with ρ^* being the contribution from all other relaxation mechanisms. The Overhauser enhancement for a nucleus A when nucleus B is saturated is defined as

$$f_A(B) = \frac{M_z^A - M_0^A}{M_0^A}$$

Equation (8) can be evaluated for the steady-state situation where $dM_z^A/dt = 0$ and $M_z^B = 0$ to give

$$f_A(B) = \frac{M_z^A - M_0^A}{M_0^A} = \frac{\sigma_{AB}}{R_A} \cdot \frac{M_0^B}{M_0^A} - \sum_d \left(\frac{M_z^d - M_0^d}{M_0^A} \frac{\sigma_{Ad}}{R_A} \right)$$

However

$$M_0^i \propto I_i(I_i + 1)\gamma_i$$

thus for spin-$\frac{1}{2}$ nuclei

$$f_A(B) = \frac{\sigma_{AB}}{R_A} \cdot \frac{\gamma_B}{\gamma_A} - \sum_d \frac{M_z^d - M_0^d}{M_0^d} \cdot \frac{\gamma_d}{\gamma_A} \cdot \frac{\sigma_{Ad}}{R_A}$$

and since

$$\frac{M_z^d - M_0^d}{M_0^d} = f_d(B)$$

then

$$f_A(B) = \frac{\gamma_B}{\gamma_A} \frac{\sigma_{AB}}{R_A} - \sum_d \frac{\gamma_d}{\gamma_A} \cdot \frac{\sigma_{Ad}}{R_A} f_d(B) \qquad (9)$$

This equation is the general solution for the nuclear Overhauser enhancement in a multispin system where all spins have $I = \frac{1}{2}$. In the case where the dipolar coupling is the sole interaction between nuclei A and B, Eq. (9) becomes

$$f_A(B) = \frac{\gamma_B}{\gamma_A} \cdot \frac{\rho_{AB}}{2R_A} - \sum_d \frac{\gamma_d}{\gamma_A} \cdot f_d(B) \cdot \frac{\rho_{Ad}}{2R_A} \qquad (10)$$

since

$$\sigma_{id} = \frac{\rho_{id}}{2} \cdot \frac{I_i(I_i + 1)}{I_d(I_d + 1)}$$

This equation is valid for all spin numbers provided the only time-dependent coupling between the nuclei is the dipolar interaction.

The term σ_{id} depends on the internuclear distance between the two nuclei, and thus information on molecular geometry is attainable by appropriate substitution in Eq. (9). This is done below for a number of spin systems.

(a) Two-spin system, AB:

$$f_A(B) = \frac{\gamma_B}{\gamma_A} \cdot \frac{\sigma_{AB}}{R_A} = \frac{\gamma_B}{\gamma_A} \left(\frac{\sigma_{AB}}{\rho_{AB} + \rho^*} \right)$$

If ρ^* is negligible, then

$$f_A(B) = \frac{\gamma_B}{2\gamma_A}$$

This is the maximum NOE possible.

(b) Three-spin system, AMX:

$$f_A(M) = \frac{\gamma_M}{\gamma_A} \cdot \frac{\sigma_{AM}}{R_A} - \frac{\gamma_X}{\gamma_A} \cdot \frac{\sigma_{AX}}{R_A} f_X(M) \tag{11}$$

with

$$R_A = \rho_{AM} + \rho_{AX} + \rho^*$$

and also

$$f_A(X) = \frac{\gamma_X}{\gamma_A} \cdot \frac{\sigma_{AX}}{R_A} - \frac{\gamma_M}{\gamma_A} \cdot \frac{\sigma_{AM}}{R_A} f_M(X) \tag{12}$$

When combined with Eq. (11), this gives

$$\frac{\sigma_{AM}}{R_A} = \left(\frac{\gamma_A}{\gamma_M} \right) \left[\frac{f_A(M) + f_A(X) f_X(M)}{1 - f_M(X) f_X(M)} \right] \tag{13}$$

or better, if we divide Eq. (12) by Eq. (11) we have on rearrangement

$$\frac{\sigma_{AX}}{\sigma_{AM}} = \frac{\gamma_M}{\gamma_X} \left[\frac{f_A(X) + f_A(M) f_M(X)}{f_A(M) + f_A(X) f_X(M)} \right]$$

with σ_{AX} and σ_{AM} being given by Eq. (5). Provided $\tau_c(AX) = \tau_c(AM)$, a reasonable assumption, then

$$\left(\frac{r_{AM}}{r_{AX}} \right)^6 = \left(\frac{\gamma_M}{\gamma_X} \right)^3 \left[\frac{f_A(X) + f_A(M) f_M(X)}{f_A(M) + f_A(X) f_X(M)} \right] \tag{14}$$

Thus if all NOE's can be measured, the various relative intermolecular distances can be obtained. In order to obtain the absolute internuclear distances, τ_c and R_A must be determined. A value of τ_c can be calculated using $\tau_c = \frac{4}{3}\pi\alpha^3\eta$ where α is the effective radius of the molecule and η is the solution viscosity. But if, for example, we are studying a proton–proton interaction, a better method is to determine τ_c experimentally from the T_1 data for ^{13}C. Provided the molecule is undergoing isotropic motion and is in the limit of extreme narrowing, we may use

$$\frac{1}{T_1} = \frac{N\gamma_H{}^2\gamma_c{}^2\hbar^2\,\tau_c}{r_{C-H}{}^6}$$

where N is the number of hydrogens attached to the carbon and r_{C-H} is the C—H bond length. The proton relaxation rate, R_A, can then be measured and the internuclear distance calculated using Eqs. (13) and (5).

(c) Three-spin system AX_2:

For the observation of A when X is saturated,

$$f_A(X) = \frac{2\gamma_X}{\gamma_A} \cdot \frac{\sigma_{AX}}{(2\rho_{AX} + \rho_A{}^*)}$$

For the dipole–dipole interaction this gives

$$f_A(X) = \frac{2\gamma_A}{2\gamma_A} \cdot \frac{\rho_{AX}}{(2\rho_{AX} + \rho^*)}$$

which has a maximum of $\gamma_X/2\gamma_A$.

In the reverse experiment, $R_X = \rho_{XX} + \rho_{AX} + \rho_A{}^*$ since each of the equivalent spins must be considered separately. The value of the enhancement when A and X are coupled only by the dipolar interaction is

$$f_X(A) = \frac{\gamma_A}{\gamma_X} \cdot \frac{\rho_{XA}}{2R_X} - \frac{\gamma_X}{\gamma_X} \cdot \frac{\rho_{XX}}{2R_X} f_X(A)$$

which gives

$$f_X(A) = \frac{\gamma_A}{\gamma_X}\left(\frac{\rho_{XA}}{R_X + \rho_{XX}}\right)$$

If we take, for example, the case where $\rho_{XA} = \rho_{XX}$ and $\rho_X{}^* = 0$, then

$$f_X(A) = \frac{\gamma_A}{2\gamma_X}\left(\frac{\rho_{XA}}{\frac{5}{2}\rho_{XA}}\right)$$

$$= \frac{\gamma_A}{5\gamma_X}$$

Comparing this with the value of $f_X(A)$ but neglecting to account for the second equivalent spin, we obtain

$$f_X(A) = \frac{\gamma_A}{4\gamma_X}$$

for each nucleus. Thus the total enhancement for this example should be $2/5 \, (\gamma_A/\gamma_X)$ whereas that estimated without consideration of the effect of the other spin would be $\gamma_A/2\gamma_X$.

For a four-spin system, equations such as Eq. (14) can be derived. However, these are extremely large and cumbersome and recourse is usually made to Eq. (9), which is readily adaptable to computer application.

Another method of obtaining relative internuclear distances from NOE measurements has been proposed by Bell and Saunders[4] and by Grant et al.[11] This involves utilization of the solution for the two-spin system and the assumption that ρ^* is constant. Bell and Saunders studied the H–H' interaction using

$$f_A(B) = \frac{\sigma_{AB}}{R_A} \left(\frac{\sigma_{AB}}{\rho_{AB} + \rho^*} \right) \tag{15}$$

In dilute solution, the assumption was made that for a two-spin system ρ^* would be only the intermolecular dipole–dipole interaction, which can be expressed in the limits of extreme narrowing by

$$\rho_A \text{ (intermolecular)} = \frac{h^2}{10\pi^2} \gamma_A{}^2 \tau_t \sum_S \gamma_S{}^2 S (S + 1) \frac{1}{b^6} + \rho^{**}$$

where τ_t is the translational correlation time, $\sum_S$ is the summation over all solvent molecules, b is the distance of closest approach of the solvent nuclei to the proton under observation, and ρ^{**} is all other intermolecular interactions that contribute to the relaxation of A.

Substitution in Eq. (15) gives

$$\frac{1}{f_A(B)} = \frac{R_A}{\sigma_{AB}} = 2 + A r_{AB}{}^6 \tag{16}$$

provided that the ratio τ_t/τ_c, b, and the interaction designated as ρ^{**} are all independent of the solute. Equation (16) is a composite of two equations which take into account the 0.5 maximum. At values of $r \leq r_{AB}{}^0$, $f_A(B) = 0.5$, whereas when $r > r_{AB}{}^0$, $f_A(B) = A^{-1} r_{AB}{}^{-6}$ with $r_{AB}{}^0$ being a specific value experimentally determined.

Equation (16) applies only to a two-spin system, a situation seldom realized in practice, and thus it should be modified to allow for the presence of other intramolecular spins. In the absence of cross-correlational effects,

$$\frac{1}{f_A(B)} = 2 + G r_{AB}{}^6 \tag{17}$$

where

$$G = \left(\frac{2}{\gamma_A^2} \sum_i \frac{\gamma_i^2}{r_{Ai}^6}\right) + H$$

For the existence of a simple correlation between $f_B(A)$ and internuclear distance G must be a constant, a condition which from a theoretical standpoint is suspect. However, examination of Eq. (14) shows that the terms at the right are in fact in $(f)^2$ and will in most cases be small. Quantitatively, we can see from Eq. (17) that the relaxation at a specific distance consists of a dipolar interaction between the two spins and other relaxation processes. The latter are normally considered to be constant and independent of the nature of the interaction involved. If other spins in the molecule significantly relax the nucleus, then the term H will be small whereas if the interaction with the other spins is negligible, H will be significant, i.e., there is a general compensation between the two terms.

In compounds where more than one spin is efficiently relaxing the observed spin A, the total NOE can be computed, provided cross terms are negligible, using

$$f_A(\text{total}) = f_A(B) + f_A(C) + \ldots$$

with a maximum $f_A(\text{total}) = 0.50$ when A, B, C are of the same species. In such examples the internuclear distances can be calculated from

$$\frac{f_A(B)}{f_A(C)} = \left(\frac{r_{AC}}{r_{AB}}\right)^6 \tag{18}$$

which is an approximation of Eq. (14).

In order to test the validity of Eq. (17) and to assess its application to structural problems, a series of relatively rigid compounds of unequivocal structure were examined[7] in which single proton–proton interaction and proton–methyl group interaction existed. In all the compounds used for the latter study, the methyl group was rotating rapidly and the protons were considered to be acting as a net dipole from a point defined by the intersection of the C-3 axis of rotation with the plane containing the 3 protons of the CH_3 group.[12] In each study a log–log plot of NOE against internuclear distance gave a straight line of slope -6 as predicted, thus showing that the original assumption that ρ^* is constant was reasonable. In order to aid the use of the observed results as a guide to internuclear distances, a least-squares plot of NOE vs $(r_{AB})^{-6}$ was obtained. It is reproduced in Figs. 1 and 2. The standard deviations for intercept and slope were each less than 5% and the line passed through the origin for each case.

Grant et al.[11] used Eq. (19) for the application of NOE's to spectral assignment in ^{13}C spectra:

$$\frac{M_z^C}{M_0^C} = 1 + \frac{\gamma_H}{2\gamma_C} \frac{\alpha}{(\alpha + u)} \tag{19}$$

where α is the $\sum \left(\dfrac{1}{r_{C-H}}\right)^6$ and u is the contribution from other relaxation mechanisms. The assumption was made that u was constant for all quaternary carbons.

The methods described for determination of internuclear distances suffer from some limitations. The assumption that ρ^*/τ_c is constant is a gross one

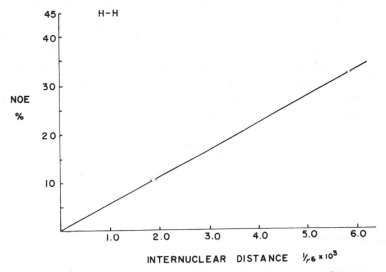

FIG. 1. NOE vs internuclear distance for H–H interactions.

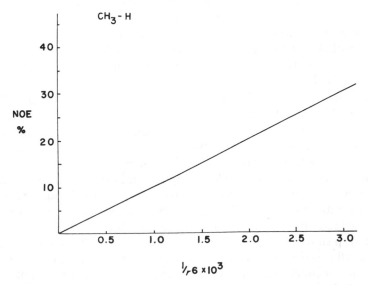

FIG. 2. NOE vs internuclear separation for CH_3–H interactions.

and receives as little theoretical justification as does the notion that a multi-spin system can be described by two spins. The method outlined by Noggle et al.[3, 5] is superior from a theoretical standpoint provided that relative distances are determined from NOE measurement and not the reverse. Equation (9) can be rearranged to estimate $f_A(B)$ from a knowledge of inter-nuclear distance assuming that ρ^* is negligible. However, this procedure can lead to large discrepancies if the values of r are greater than 2.4 Å and there-fore it is not recommended. An additional limitation to this method is that four NOE measurements must be obtained, and these data are often difficult to obtain accurately. The most serious drawback to utilization of quantitative NOE's is in fact the experimental accuracy. It has been our experience that where possible one should use Eq. (14) or at least Eq. (18), although in over 100 experiments on a large variety of compounds[13, 14] good correlation has been found with Figs. 1 and 2, except for linear systems as will be shown later.

III. EXPERIMENTAL METHODS

As stated above, the limiting problem in the determination of internuclear distances is the actual NOE measurement itself. In order to minimize errors, care must be taken in sample preparation, including the choice of solvent and removal of oxygen, as well as in the measurements themselves. The following is an outline of the precautions that should be taken.

The sample concentration should be at a minimum consistent with an adequate signal-to-noise ratio, and the solvent should not contain protons, fluorine, or other nuclei of high gyromagnetic ratio. The dipole–dipole interaction of a deuterium with a proton is small, and thus deuterated sol-vents are quite acceptable. The internal lock material also should be pref-erably without protons or fluorine. Where possible an external lock or an internal lock on the deuterium signal of the solvent should be used. However, for H–H NOE experiments we have found that sufficient tetramethylsilane (TMS) can be added for locking purposes without significant change in the recorded NOE enhancement. Paramagnetic materials efficiently relax any nucleus, and it is thus imperative that no such materials, including oxygen, be present in the solution. The best method to remove oxygen is the so-called freeze–thaw cycle, or alternative chemical methods.[15]

The observing field H_1 should be kept small in order to avoid saturation. However, a maximum H_1 is necessary in order to obtain the maximum signal-to-noise ratio. A few simple experiments can determine the approximate power conditions for saturation, and one then uses a value about 10% less. The sweep should be slow enough to avoid transient effects and thus reason-ably small sweep widths should be used. The field H_2 must saturate the proton resonance, a condition that normally requires less rf power than decoupling. In Fig. 3, the percent enhancement as a function of power is given, from which

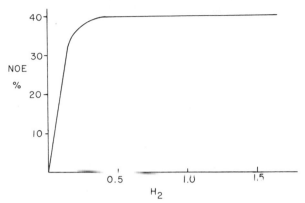

FIG. 3. Effect of decoupling field H_2 power on the observed NOE $f_A(B)$ in ochotensimine (1).

it can be seen that a plateau is reached where the observed Overhauser enhancement becomes constant. If H_2 is made too large, the value of the NOE recorded could be inaccurate due either to partial saturation of other resonances or to overloading of the phase detector of the receiver. This latter problem can be checked by measuring the intensity of the resonance with the decoupler off and then on but with a frequency removed from any resonance positions. The determination of NOE's by the saturation of resonances from protons that have J couplings requires that all transitions be saturated, and this frequently demands the use of higher H_2 power levels. Because of the possibility of detector overloading it is often advantageous to use more than one decoupling field. Thus we have found that when irradiating one proton of an AB spin system where $J_{AB} = 20$ Hz, we obtain better results by using two decoupling fields, one at each of the peaks, rather than one field located between the lines.

The Overhauser enhancement should be measured as *area* increases and not simply as peak height increases. There are three methods of integration: (1) electronic; (2) mechanical—spectra are recorded normally and integrated with a planimeter or by the cut-and-weigh method; (3) digital integration—the spectrum is recorded on a signal averager or mini computer and the integral is then taken numerically. For example, the recorded results for P–[H] NOE's were obtained in this fashion (see Section IV,B). The spectra were obtained by the FT method and then the area of a small part of each spectrum was recorded digitally with and without the presence of the H_2 field.

For each determination at least 10 area measurements should be taken at a number of frequencies close to that of the nuclear resonance being saturated. In Fig. 4 the effect of small changes of frequency is shown for a series of measurements in which 10 integrations were recorded at each frequency.

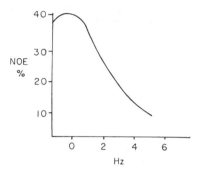

FIG. 4. Dependence on the decoupling field position on the observed NOE $f_A(B)$ in **1**.

It should be emphasized that H_2 should never be turned off in the measurement of equilibrium intensities but should simply be used in an off-resonance manner, that is, the frequency at which H_2 is applied should be moved some 50–100 Hz from any resonance lines in the spectrum.

So far the discussion has centered on continuous wave (CW) measurement but it is also possible to obtain NOE data by transient methods. If we consider the two-spin system, then

$$\frac{dM_z{}^A}{dt} = -R_A (M_z{}^A - M_0{}^A) - \sigma_{AB} (M_z{}^B - M_0{}^B)$$

$$\frac{dM_z{}^B}{dt} = -R_B (M_z{}^B - M_0{}^B) - \sigma_{BA} (M_z{}^A - M_0{}^A)$$

which have the general solutions

$$M_z{}^A = M_0{}^A + C_1 e^{-\lambda_1 t} + C_2 e^{-\lambda_2 t}$$

$$M_z{}^B = M_0{}^B + \frac{\lambda_1 - R_A}{\sigma_{BA}} C_1 e^{-\lambda_1 t} + \frac{\lambda_2 - R_A}{\sigma_{BA}} C_2 e^{-\lambda_2 t} \tag{20}$$

where

$$\lambda_{1,2} = \tfrac{1}{2} |(R_A + R_B) \pm \tfrac{1}{2} [(R_A - R_B)^2 + 4\sigma_{BA}\sigma_{AB}]^{1/2}| \tag{21}$$

Explicit solutions for a number of particular experiments can then be derived, four of which are given below.

1. The recovery of A after a 180° pulse.

At $t = 0$,
$$M_z{}^A = -M_0{}^A$$
$$M_z{}^B = M_0{}^B$$

For these conditions we obtain

$$C_1 = 2M_0{}^A \left(\frac{\lambda_2 - R_A}{\lambda_1 - \lambda_2}\right) \qquad C_2 = 2M_0{}^A \left(\frac{\lambda_1 - R_A}{\lambda_1 - \lambda_2}\right)$$

Thus the solution is a double exponential function and determination of R_A and σ_{AB} will be difficult. However, the decay could be a single exponential under certain conditions as shown below.

If

$$R_A - R_B \gg \sigma$$

then

$$\lambda_1 = R_A \qquad \lambda_2 = R_B$$

$$C_1 = \frac{R_B - R_A}{R_A - B_B} = -1 \qquad C_2 = \frac{R_A - R_A}{\lambda_1 - \lambda_2} = 0$$

and

$$\frac{M_0{}^A - M_z{}^A}{M_0{}^A} = -2e^{-R_A t}$$

Thus a plot of $\ln(M_0{}^A - M_z{}^A)$ vs t under these conditions would give a straight line with slope of $-R_A$. Generally, if $\sigma \ll R$ then a single exponential decay is observed.

2. The recovery of A from saturation while B remains saturated.

At $t = t$,

$$M_z{}^B = 0$$

and at $t = 0$,

$$M_z{}^A = 0$$

$$\frac{dM_z{}^A}{dt} = -R_A(M_z{}^A - M_0{}^A) - \sigma_{AB}(-M_0{}^B)$$

The solution is thus

$$M_z{}^A = \left[M_0{}^A + \frac{(\sigma_{AB})}{R_A} M_0{}^B\right]\left(1 - e^{-R_A t}\right)$$

For $t = \infty$,

$$M_z{}^A = M_0{}^A + \frac{\sigma_{AB}}{R_A} M_0{}^B$$

which is the steady-state Overhauser enhancement. A plot of $\ln[M_z{}^A(\infty) - M_z{}^A(t)]$ vs t will give a straight line of slope $-R_A$. Thus it is possible to determine all the dynamics of the spin system.

3. The recovery of A after a 180° pulse with B continuously saturated.

At $t = 0$,
$$M_z{}^A = -M_0{}^A - \frac{\sigma_{AB}}{R_A} M_0{}^B$$

For all values of t, $M_z{}^B = 0$, so that

$$M_z{}^A = \left(M_0{}^A + \frac{\sigma_{AB}}{R_A} M_0{}^B\right)\left(1 - 2e^{-R_A t}\right) \tag{22}$$

Again, $M_z{}^A(\infty)$ is $M_0{}^A + (\sigma_{AB}/R_A)M_0{}^B$ and a plot of $\ln[M_z{}^A(\infty) - M_z{}^A(t)]$ vs t will give the value of R_A.

4. The time dependence of A upon instantaneous saturation of B.

At $t = 0$, $\qquad\qquad\qquad M_z{}^A = M_0{}^A \qquad M_z{}^B = 0$

For all t $\qquad\qquad\qquad\qquad M_z{}^B = 0$

The solution is thus

$$M_z{}^A = M_0{}^A + \left(\frac{\sigma_{AB}}{R_A}\right) M_0{}^B(1 - e^{-R_A t})$$

Values of R_A and σ_{AB}/R_A can be deduced as described previously. This type of experiment has the additional difficulty that undesired transient effects are present in the observed peak, since the saturation of the B peak requires a finite time.

Unfortunately, most spin systems contain more than two spins. The general equation is of the form

$$\frac{dM_z{}^A}{dt} = -R_A(M_z{}^A - M_0{}^A) - \sum_d \sigma_{Ad}(M_z{}^d - M_0{}^d)$$

Various solutions similar to those described for the two-spin system can be found; for example, from a 180° pulse for A while all other nuclei are saturated, we have

$$M_z{}^A = \left(M_0{}^A + \frac{\sum \sigma_{Ad}}{R_A} M_0{}^d\right)(1 - 2e^{-R_A t}) \qquad (23)$$

It is then simple to determine the value of R_A but the individual values of σ_{Ad} must be determined by steady-state methods.

The FT technique offers wide possibilities for the determination of both relaxation time and Overhauser enhancement data. With this technique the spectrum is recorded in the time domain as a free induction decay and then is changed into the frequency domain by a Fourier transform.[16] The method has several advantages over the CW method, including the possibility of rapid spectrum accumulation and direct determination of T_1 values. The procedure involves utilization of a nonselective pulse to flip the spins toward the xy plane. The amount of flip, α, is determined by the pulse power and the pulsewidth, t_p (the time for which the pulse is on),

$$\alpha = \frac{\gamma H_1}{2\pi} t_p$$

Provided a time of $5T_1$ or greater is allowed between pulses, then the transformed signal will be a maximum for pulses of 90°, 270°, etc., and will be null for 180°, 360°, etc. Thus, the pulse width is adjusted until the peak is a maximum or a null in order to determine the pulse widths that correspond to

90° and 180° pulses, respectively. At $t = 0$ the 180° pulse is applied, the effect of which is inversion of magnetization. The spins begin to relax, and after time τ the 90° pulse is applied and the free induction decay recorded. The intensity of the peak will depend on T_1 for the nuclei and the time τ. The spectrum is recorded for a number of values of τ, and T_1 can be obtained from a plot of $\ln(M_0{}^A - M_z{}^A)$ vs τ. As can be seen from the discussion above [point 3, above], the best values of T_1 are obtained when all other spins with which the observed one interacts are saturated. This is of particular interest in the determination of ^{13}C relaxation times since the spectra are normally recorded with noise decoupling of all protons, a condition described by Eq. (23). A benefit derived from the FT technique is that the T_1's for a number of resonances can be obtained at the same time: for example, it is possible to determine the T_1 values for all carbons of a steroid simultaneously (provided accidental equivalence is not present).

The determination of Overhauser enhancements by the Fourier technique is also feasible. For *homonuclear* NOE determination various methods have been utilized depending on the type of spectrometer available. There are two such methods in which the decoupler is turned off at the actual time of sampling, in order to eliminate "beats" and also overloading of the digital-to-analog converter.

1. In the first the time between pulses is greater than $5T_1$ and the pulse angle is 90°. At $t = 0$, the sample is pulsed and the decoupler is turned off. The intensity is then $M_z{}^A = M_0{}^A + (\sigma_{AB}/R_A)$. After sampling, the system is allowed to reach equilibrium with the decoupler on and then the procedure is repeated.

2. In the second, use is made of the difference between the dwell time and the sampling time at each point. The sweep width is determined by the dwell time at each point of the computer. For example, a sweep width of 1000 Hz requires 500 μseconds at each point. However, the computer requires only 20 μseconds for sampling at each point. If the decoupler is off for each 20 μseconds of sampling but is on for the remainder of the time, this gives the same result as continuous decoupling but eliminates any interaction between sampling and decoupling fields.

For *heteronuclear* Overhauser effects, the spectrum can be taken with the decoupler on or off. However, particularly for ^{13}C and ^{31}P, for example, the coupled spectrum can be extremely complicated for a medium-sized compound and it is then impossible to gain NOE data for individual nuclei. The ideal solution then is to observe the decoupled spectrum, but with the absence of the Overhauser effect, in order to observe M_0. The time constant R_A is given as $\sum\rho_{Ad} + \rho^*$. If we make ρ^* sufficiently large that $\rho^* \gg \sum\rho_{Ad}$, then the Overhauser effect will be very close to zero. This condition can often be met by the use of the so called "shiftless" relaxation reagents such as Cr^{III}(acetyl acetonate)$_3$.[16] These paramagnetic compounds efficiently relax nuclei, thus causing ρ^* to become large. The maximum value of ρ^* induced by such a compound is of the order of 100 seconds^{-1}. Thus if T_1 is 20 seconds, the value of $M_z{}^A$ in the presence of the paramagnetic material is

$$M_z{}^A = M_0{}^A + \frac{\sum \rho_{Ad}}{2R_A} = M_0{}^A + \frac{0.05}{200} M_0{}^A \approx M_0{}^A$$

Thus, we can determine $M_0{}^A$ by taking the spectrum with sufficient paramagnetic material. However, for complex compounds such as steroids, R_A for the CH_2 and CH carbons is sufficiently large that the paramagnetic material will not satisfy the condition $\rho^* > \rho_{AB}$. A better method is to turn the decoupler on at the same instant that sampling commences and to maintain decoupling just for the time that the spectrum is recorded, and then to wait for a period of $5T_1$ with the decoupler off before recording the next accumulation. This corresponds to the condition for 4 written above. At $t = 0$, all the nuclei are decoupled with $M_z{}^A = M_0{}^A$ and thus after the application of the pulse the free induction decay then measures $M_0{}^A$.

To this point, only the benefits that accrue from the presence of the Overhauser effects have been considered. However, there are times when it is desirable to eliminate the enhancement. This is particularly true when spectra of nuclei such as ^{29}Si and ^{15}N are being recorded since these possess negative gyromagnetic ratios. Thus for a dipole–dipole interaction between ^{29}Si–[H]

$$f_{Si}(H) = \frac{\frac{1}{2}\gamma_H}{\gamma_{Si}} = \frac{-4257.7}{846.0} = -2.52$$

A negative NOE of such proportions presents no problems since the magnitude of the enhancement is > 1. A *partial* NOE, however, might result in very small signals with intensities between -1 and $+1$. In order to alleviate this problem, paramagnetic materials can be added as described above.

IV. CHEMICAL APPLICATIONS

The following sections include a number of examples in which the NOE has been applied in structural, configurational, and conformational analysis as well as in spectral assignment in a wide variety of compounds. For convenience, the NOE's will be given as percentage enhancements in the form $f_2(1) = 18\%$ or [H-1] H-2 $= 18\%$ which should be read as "saturation of the absorption corresponding to H-1 caused an enhancement of 18% in the signal of H-2." The use of the so-called "general Overhauser effect,"[17] in which the effect of saturation of one nuclear transition on other transitions is observed in order to assign transitions and determine the sign of coupling constants, is omitted as it is beyond the scope of this chapter.

A. Proton–Proton Applications

The data for the alkaloid ochotensimine (1) given in Table I provide an excellent example of the type of information that can be obtained from an

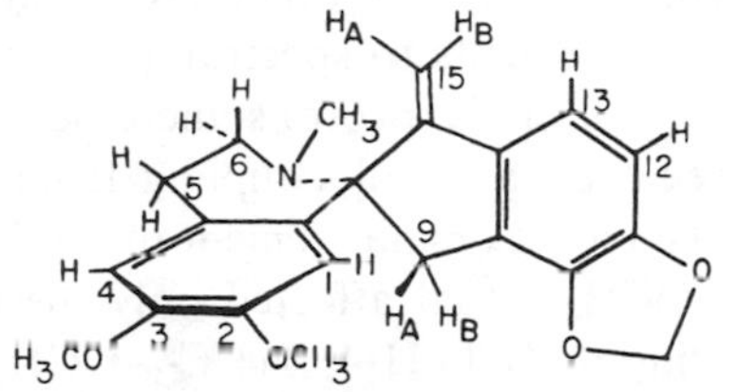

1

NOE study.[18] The four protons H-15$_A$, H-15$_B$, H-13, and H-12 form a linear four-spin system and, as such, exhibit interesting and informative NOE values. Saturation of H-15$_A$ results in a 40% increase in the area of H-15$_B$, an 8% decrease in the area of H-13, and finally a 3% increase in the area of H-12. This type of behavior is specific for linear or nearly linear systems, where the effect of saturation is transmitted to the third and fourth spin by the intervening protons. The observation of a negative NOE in an all-proton system where there is no exchange indicates the presence of a third spin between the saturated and observed ones. Saturation of H-15$_B$ causes a 40% area increase in the H-15$_A$ signal, a 24% increase in that for H-13, and a 7% decrease in the absorption of H-12. The NOE [H-13] H-15$_B$ = 8% was also recorded. Utilization of either Eq. (14) or (16) to determine the distance between H-15$_B$ and H-13 affords the value 2.4 Å, which is identical to the

TABLE I

NOE's in Ochotensimine (1)[18]

Proton irradiated	Proton observed	NOE (%)	Proton irradiated	Proton observed	NOE (%)
H-15$_A$	H-15$_B$	40	H-15$_B$[a]	—	—
	H-13	−8	H-15$_A$[a]	H-15$_B$	47
	H-12	+3	H-13[a]	—	—
H-15$_B$	H-15$_A$	40	H-5(ax)	H-4	25
	H-13	24	H-5(eq)	—	—
	H-12	−7	3-OCH$_3$	H-4	25
H-13	H-15$_B$	8	2-OCH$_3$	H-1	25
N-CH$_3$	H-15$_A$	2	H-9$_A$	H-1	14
N-CH$_3$[a]	H-15$_A$	42	N-CH$_3$	H-9$_B$	8

[a] Triple irradiation.

value measured from molecular models. The difference between [H-15$_B$] H-13, and [H-13] H-15$_B$ is a manifestation of the maximum NOE possible, 50%, and the difference in relaxation efficiency of H-15$_A$ on H-15$_B$ relative to H-12 for H-13. This large difference can be very useful in structural and conformational analysis. The use of other NOE values demonstrates the application of NOE measurements to spectral assignment. Saturation of the multiplet from the C-5 aliphatic protons increased the area of the H-4 resonance and also removed a small coupling from it. Saturation of the C-$_3$–OCH$_3$ signal caused an area enhancement of the H-4 resonance whereas saturation of the C-$_2$–OCH$_3$ signal affected only the intensity of H-1. The resonances corresponding to H-1, H-4, the C-$_2$–OCH$_3$ and the C-$_3$–OCH$_3$, could thus be assigned, often a difficult task. In other alkaloids from the same plant ring A contained an OCH$_3$ and an OH, and a similar series of experiments was used to determine the relative positions of the OH and the OCH$_3$ resonances.[19] The protons at C-9 give rise to an AB quartet with A and B being assigned from the NOE, [H-9$_A$] H-1 = 14%.

Another series of compounds in which positive and negative NOE's are observed are the three dibenzthiepin compounds 2–4 studied by Fraser and

2 R = S
3 R = SO$_2$
4 R = SO

Schuber (see Table II), in order to determine the configuration of each of the benzylic protons.[20] Inspection of molecular models indicates that in the sulfide **2**, one of each pair of diastereotopic protons lies much closer to the ortho aromatic proton than the other does and this proton (H-1) has the pro-(S) configuration in the overall (R)-biphenyl system. Thus, observation of an area increase of H-X on saturation of the low-field benzylic resonance assigns the spectral position of H-1. Saturation of the high-field resonance of the AB quartet causes an area decrease in H-X, confirming that H-1 has the "pro-(S) in (R)" configuration. By a similar process, H-1 in the sulfone **3** (low-field portion of the AB quartet) also has the "pro-(S) in (R)" configuration. The sulfoxide **4** has two benzylic AB quartets due to the presence of the asymmetric sulfoxide group (see Fig. 5). When CDCl$_3$ was used as solvent,

TABLE II

NOE's for Dibenzthiepins[20]

Proton irradiated	Proton observed	NOE (%)				
		2	3	3[a]	4	4[b]
H-X	H-1	8	10	—	—	31
H-X	H-2	−5	−1	6	—	—
H-1	H-X	23	21	—	—	33
H-1	II-2	47	48	—	—	—
H-2	H-X	−5	0	0	—	—
H-2	H-1	40	30			—
H-1	H-4	—	—	—	28	—
H-4	H-1	—	—	—	24	—

[a] $H_1 = D$
[b] $H_2, H_3, H_4 = D$

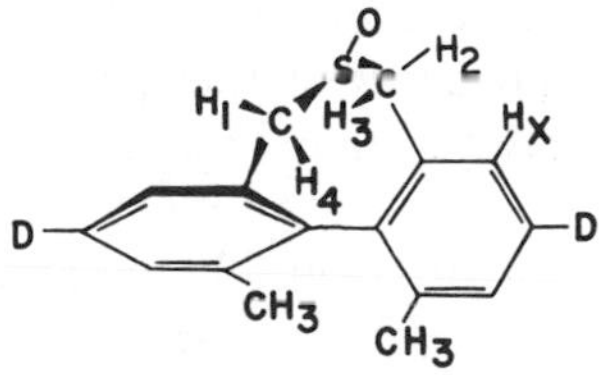

FIG. 5. Configuration of dibenzylthiepin sulfoxide (4).

no NOE was observed between H-1 and the ortho aromatic proton. However, when the solvent was changed to DMSO-d_6, this NOE was observed. The lack of an observable NOE in CDCl$_3$ was attributed to dimerization. The results obtained (Table II) establish a pro-(S) configuration for H-1 and a pro-(R) configuration for H-4 in the (R)-biphenyl. The configurations of H-2 and H-3 as pro-(S) and pro-(R), respectively, were obtained by a chemical inversion of the sulfoxide group. The final stereochemical assignment, the configuration of the sulfoxide group relative to each pair of geminal benzylic protons, was established by a consideration of chemical shift and coupling constant data. Several interesting facts emerge from these results. Comparison of the interaction between H-2 and H-X when H-1 is a proton or a deuterium resulted in NOE's of −1% and +6% respectively when compound 3 was studied. This is a direct consequence of the interdependence of the dipolar interaction when H-1 is a proton, and is a good illustration of the complexities that can occur in NOE studies.

One of the first applications of the NOE in organic chemistry was reported by Anet and Bourn[2] who studied the compounds β,β-dimethylacrylic acid,

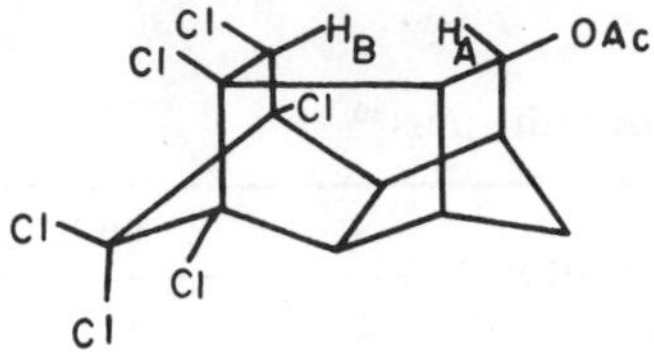

5

dimethylformamide (DMF), and the half-cage acetate **5**. The vinylic proton of β,β-dimethylacrylic acid exhibited an area increase of 17% or a slight area decrease depending upon which one of the resonances of the two methyl groups was saturated, thus confirming the spectral assignment of these two groups. In DMF the observance of an 18% and a −2% enhancement of the formyl proton on saturation of the high-field and low-field resonances of the methyl groups, respectively, unequivocally confirmed the chemical shifts of the two methyls. The two protons designated as H_A and H_B in the half-cage acetate **5** gave mutual NOE's the order of 45%, even in the presence of oxygen, illustrating the effectiveness of the dipolar interaction between the two protons forced so closely together by the cagelike configuration of the molecule.

The identification[21] of the diterpenoid ginkolides (Fig. 6) provides an excellent example of the use of NOE's as an aid in structural elucidation of natural products. Saturation of signal from the tertiary butyl group caused significant enhancements of H-7, H-8, H-10, and H-12, showing that these protons were in close proximity to the tertiary butyl group. From chemical and NMR results the partial structure shown in Fig. 6b were obtained.

TABLE III

Nuclear Overhauser Effects in Ginkolide and Ginkolide Acetates[21]

Proton irradiated	Proton observed	NOE[a] (%)			
		6	**7**	**8**	**9**
t-Butyl	H-6	0	0	12	6
t-Butyl	H-7	4	16	24	19
t-Butyl	H-8	—	6	14	13
t-Butyl	H-10	20	33	30	27
t-Butyl	H-12	20	22	0	0
H-8	H-12	0	0	7	10
H-12	H-8	0	0	20	13

[a] Data were recorded for trifluoroacetic acid solutions.

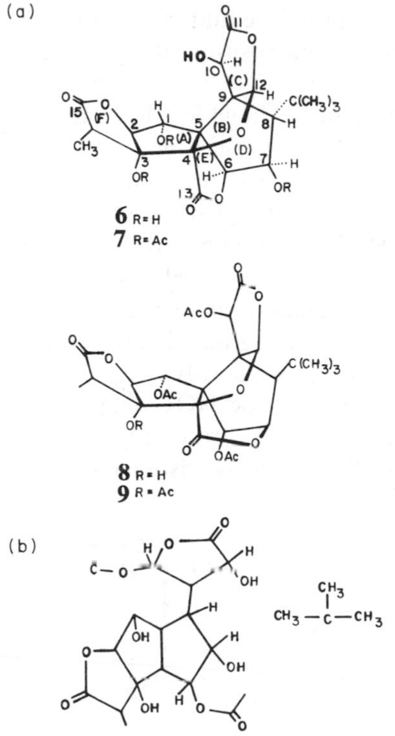

(a)

6 R = H
7 R = Ac

8 R = H
9 R = Ac

(b)

FIG. 6. a. Stereostructure of ginkolide **6** and ginkolide acetates **7**, **8**, and **9**.
b. Partial structure of ginkolides.

These rings could be fused to give eight possible structures. Since the tertiary butyl group interacts with the four protons, it cannot be attached at either C-5 or C-6 and thus four of the possible structures could be discarded. However, of the remaining four possibilities only one, namely that shown in Fig. 6, is consistent with both the NOE and NMR results.[21] The observance of an NOE between the tertiary butyl group and H-10 indicates that the C-10 hydroxyl is cis to the C-9–C-5 bond. Mild acetylation of **6** gives the 1,3,7-tri-O-acetate **7**, the NOE results of which are also recorded in Table II. More vigorous acetylation of **6** yields an isomeric triacetate **8**, which can be further acetylated to the tetraacetate **9**. The isolactones are formed by translactonization of ring E from C-6 to C-7. The lack of an NOE between the tertiary butyl and H-12 and the observance of an NOE between the tertiary butyl and H-6 are in complete agreement with this change in lactone terminus. The translactonization in addition confirms that the C-6 and C-7 oxygen functions are cis.

In order for the tertiary butyl group to be close to the protons 7, 8, 10, and 12, it must have a pseudo equatorial orientation. If the tertiary butyl group adopted a pseudo axial position, it would efficiently relax H-6 but not H-12, and also the internuclear distance between H-6 and H-12 would be small. Thus for the compounds in which the lactone ring is the same as in the natural products 6 and 7, the tertiary butyl group exists in a pseudo equatorial orientation as shown, whereas for the isomeric lactones, 8 and 9, the tertiary butyl is in a pseudo axial position since the NOE's [t-Bu], H-6, H-7, H-8, and H-10, and [H-12] H-8 are observed. It is of interest to note that these compounds were studied in trifluoroacetic acid, and because this is a solvent that should be extremely efficient at relaxing the solute protons, it must be assumed that the tertiary butyl group is protecting the protons 6, 7, 8, 10, and 12 from close approach of the solvent.

Brown et al.[22] attempted to use NOE data in order to assign the stereochemistry of 10 and 11 since the cyclobutyl methine proton has two CH_3 groups in close proximity in the syn isomer but only one CH_3 in the anti

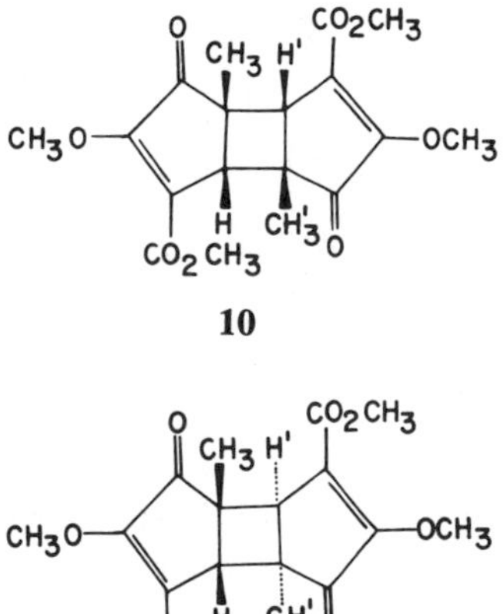

10

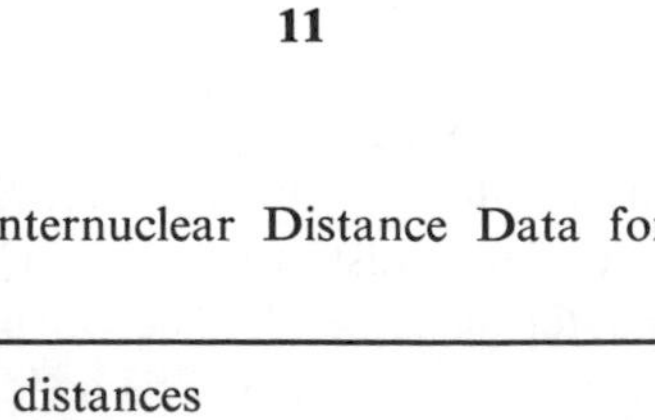

11

TABLE IV

Relaxation Time and Internuclear Distance Data for Substituted Cyclobutyl Compounds 10 and 11

Com-pound	Internuclear distances			T_1(CH)	T_1(C–CH₃)	T_1(OCH₃)	T_1(CO₂CH₃)
	H–H′	CH₃–H	CH₃′–H				
10	3.5	3.0	3.0	3.4	0.9	1.6	2.6
11	3.8	3.0	3.8	4.8	0.6	1.5	2.4

isomer (see Table IV). However, saturation of the CH_3 resonance caused an enhancement of 40% of the methine protons in each compound. This illustrates that care must be used when applying Fig. 2, since this graph was obtained from compounds in which the rate of rotation of the CH_3 group was faster than τ_c^{-1} (see Section V). In order to distinguish between the two possibilities, the proton T_1 data were recorded. They are given in Table IV. The T_1 values for the carbomethoxy and methoxy methyl protons, respectively, are approximately the same in both compounds, whereas the cyclobutyl protons exhibit significantly different values. Provided τ_c is the same in each compound, the ratio of T_1's can be expressed as

$$\frac{T_1{}^A}{T_1{}^B} = \frac{\frac{3}{2}\sum_i(r_{Ai})^{-6} + \sum_\alpha(r_{Ai})^{-6}}{\frac{3}{2}\sum_i(r_{Ai})^{-6} + \sum_\alpha(r_{Bj})^{-6}}$$

where i represents protons identical to the observed and j is for nonequivalent nuclei. Examination of the relative internuclear distances in Table IV demonstrates that T_1 for the cyclobutyl protons of the syn isomer should be significantly shorter than that of the anti, consequently 10 can be assigned as being syn and 11 as anti. The large difference between the $C-CH_3$ and the $O-CH_3$ is due to differences in rotation rates, and comparison of the ratio $T_1(C-CH_3)$: $T_1(O-CH_3)$ with those determined from ^{13}C T_1 data for enol ethers[23] suggests that the rotation rate of the $C-CH_3$ is significantly less than τ_c^{-1}.

12

The stereochemistry[24] of one of the photodimers 12 obtained from ultraviolet irradiation of 2,3-dihydro-2,6-dimethyl pyrone was substantiated as being anti from the NOE values recorded below. Saturation of the 2-CH_3 group resonance increased the signals of H-3, H-6, and H-6' by 8%, 10%, and 9%, respectively, but did not affect H-3', whereas the only effect of saturation of the 2'-CH_3 group signal was an enhancement of 8% of the H-3' absorption. These values confirmed not only the anti stereochemistry but also the relative orientations of the CH_3 groups at C-6 and C-6'. The NOE value for the interaction between the CH_3 group and the cyclobutyl methine proton recorded for 12 are significantly less than that of 10 above. A possible explanation is that the rates of rotation of the methyl groups in the two compounds are markedly different, as it has been shown that methyl group rotation rates are very dependent on their steric environment.[25] This reasoning can be substantiated only by determination of these rates, but the large NOE difference emphasizes that care must be used when CH_3 group–proton interactions are being studied.

The NOE results[26] recorded for the 9-substituted 9,10-dihydroanthracenes **13–17** (see Table V) show that in the cases where the substituent at C-9 is

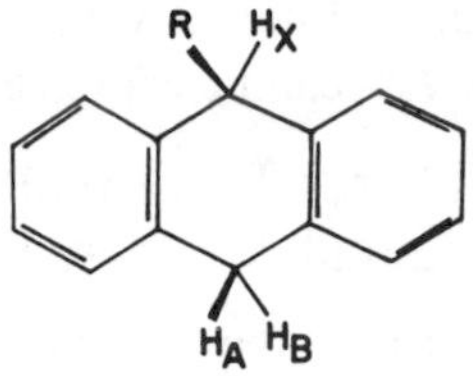

	R
13	CH_3
14	$(CH_3)_2CH$
15	$(CH_3)_3C$
16	C_6H_5
17	$1,4,9-(CH_3)_3$
18	CO_2CH_3

methyl, isopropyl, or tertiary butyl its orientation is predominantly axial whereas if the substituent is methyl or phenyl the equilibrium mixture contains more of the equatorial conformer, although this conformer is still the less populated one. It should be noted that in these examples all three benzylic protons are significantly relaxed by protons other than adjacent aryl protons, but to varying degrees. This is the reason for the differences found when

TABLE V

NOE Data for Some 9-Substituted 9,10-Dihydroanthracenes[26]

Protons irradiated	Proton observed	NOE (%)					
		13	**14**	**15**	**16**	**17**	**18**[a]
Aromatic	H–A	—	−1	−1	1	0	7
	H–B	—	5	5	10	0	11
	H–X	15	13	16	15	7	—

[a] Bell and Saunders.[27]

observing H_B and H_X, since H_A relaxes H_B more efficiently than the R substituent relaxes H_X. The results for the 9-carbomethoxy derivative[27] demonstrates that the conformer with the equatorial substituent is well populated. However, care must be exercised before quantitative comparisons can be made since the relaxation interactions of the carbomethoxy group with H_A will be negligibly small in comparison to the relaxation effects of a C-9 alkyl or aryl substituent. In a similar study, Stothers *et al.*[28] used the NOE measurements shown in Table VI in conjunction with long range couplings to ascertain the stereochemistry of 2- and 6-substituents in a series of 1,3,5,7-tetramethyl tricyclo[5.1.0.0³,⁵]octane derivatives. Although both the 2-trans and 2-cis protons show NOE's on saturation of the 1 and 3 methyl groups, the consistently larger effect shown by the cis protons is sufficient to define the stereochemistries. The observation of NOE's at both 2-trans and 2-cis protons indicates that the central ring must be flatter than is depicted in Fig. 7 and in fact the large NOE (29%) observed for both H-2 cis and H-2 trans in **17** strongly suggests that the six-membered ring here is

TABLE VI

Nuclear Overhauser Effects for Selected Tricyclooctanes[28]

Proton irradiated	Proton observed	NOE (%)			
		19	**20**	**21**	**22**
CH₃-3,1	H-2 (cis)	33	—	39	28
CH₃-3,1	H-2 (trans)	—	14	—	6
CH₃-3,1	H-B	1	13	9	15

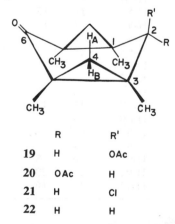

	R	R'
19	H	OAc
20	OAc	H
21	H	Cl
22	H	H

FIG. 7. Configuration and conformation of tricyclooctane derivatives **19–22**.

essentially planar. The variation in NOE's also implies that the overall geometry of the central ring is strongly influenced by the nature of the 2- and 6-substituents.

23   **24**

The NOE has been successfully applied to the structural elucidation of sesquiterpenes, particularly with respect to conformation. The NOE results for the ten-membered ring compounds **23–27** are given in Table VII. From

TABLE VII

NOE Data for a Number of Ten-Membered Ring Sesquiter-penes[29, 30]

Proton irradiated	Proton observed	NOE (%)				
		23[a]	**24**[a]	**25**[b]	**26**[b]	**27**[b]
H-5	H-1α	2	0	—	—	—
H-10	H-1β	—	—	12	20	—
CH$_3$-14	H-1β	—	—	—	—	10[c]
CH$_3$-15	H-1β	—	—	10	12	0
H-6	H-5	11	15	—	—	—
H-9	H-5	8	10	—	—	—
CH$_3$-14	H-5	2	0	—	—	—
H-5	H-6β	18	15	0	—	—
H-7	H-6β	—	—	—	—	15[c]
H-8	H-6β	—	—	—	7	—
H-9	H-6β	−4	−4	—	—	—
CH$_3$-13	H-6β	13	17	—	—	—
CH$_3$-15	H-6β	—	—	18	20	10[c]
H-7	H-8	—	—	—	—	20
CH$_3$-13	H-8	—	—	—	—	15[c]
H-5	H-9	6	8	—	—	—
CH$_3$-14	H-9	20	20	—	—	—
CH$_3$-13	H-12	19	16	—	—	—

[a] Tori *et al.*[29]
[b] Toubiana and Toubiana.[30]
[c] Approximate.

molecular models it appeared[29] that two conformations were possible for
23 and **24**, one placing H-5 and CH_3-14 syn while the alternate placed them
anti. The observance of an NOE between H-5 and H-9 conclusively demonstrates that the ten-membered rings in **23** and **24** adopt a conformation such
that H-5 and H-9 are in close proximity. For **23** the coupling constants
$J_{1\alpha, 2\alpha}$ and $J_{1\alpha, 2\beta}$ are 5.0 Hz and 12.0 Hz, respectively, and these, together with
the observation that the NOE [13-CH_3] H-6β = 17%, defines the conformation
of **24** as shown in Fig. 8. The conformation of **23** is similar. The compounds
25–**27**[30] are much more flexible. The observation that [H-8] H-6 = 7% and
[H-13] H-8 = 0% for **26** defined the stereochemistry at C-8. The NOE's
[CH_3-15] H-1 (= 12%), and [H-10] H-1 (= 20%) defined the stereochemistry
at C-1 and C-10. A conformation that follows from consideration of the
recorded NOE's of **25** and **26** is shown in Fig. 8. Reduction of the C-7–C-11
double bond caused a significant change in the conformation of the ten-
membered ring, as is particularly evident from the NOE's for **27** of [CH_3-14]
H-1 (= 10%), and [CH_3-15] H-1 (= 0%). Examination of molecular models

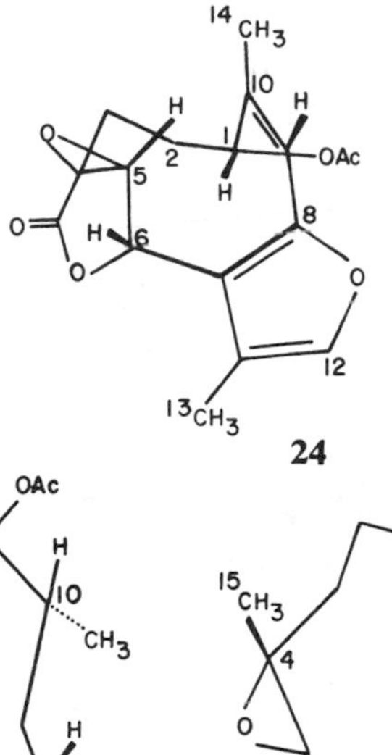

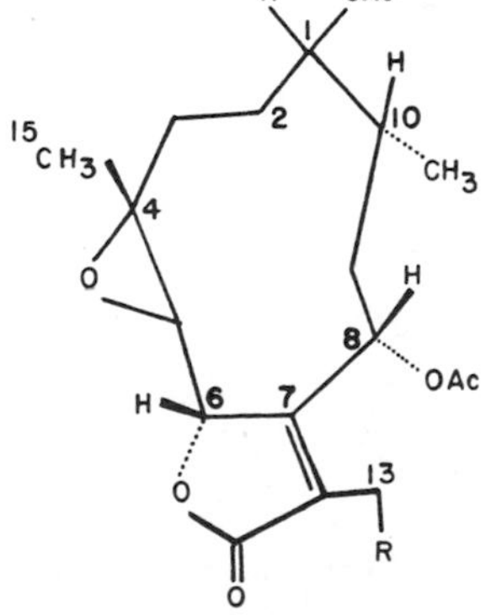

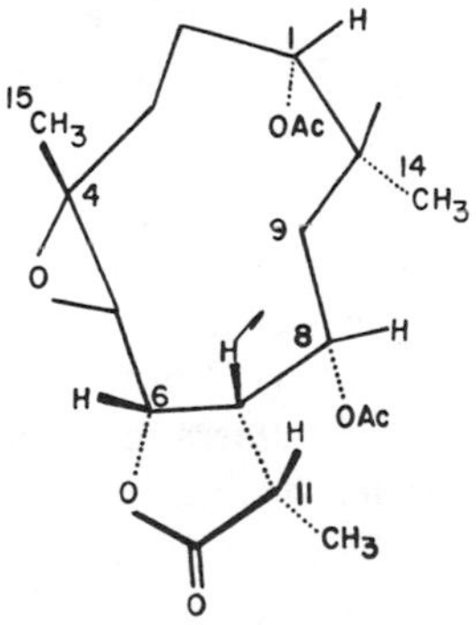

FIG. 8. Stereostructure of ten-membered ring sesquiterpenes **24–27**.

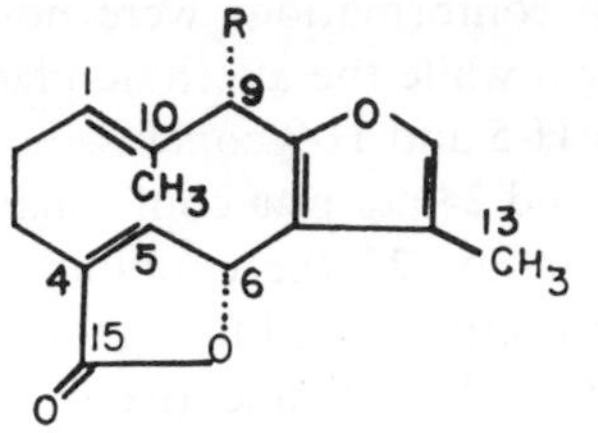

28 R = H
29 R = OAc

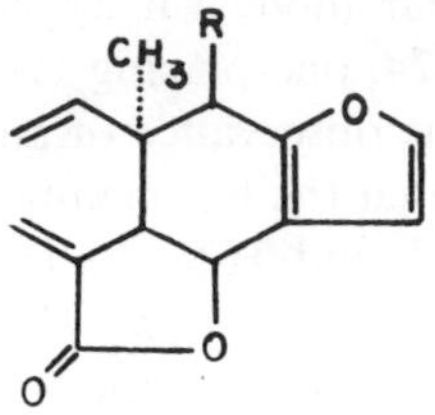

30

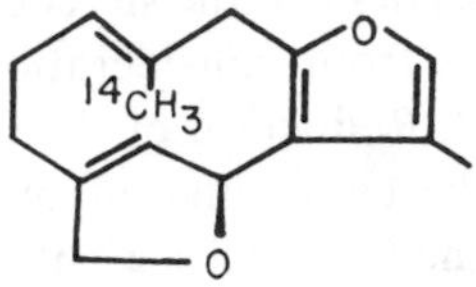

31

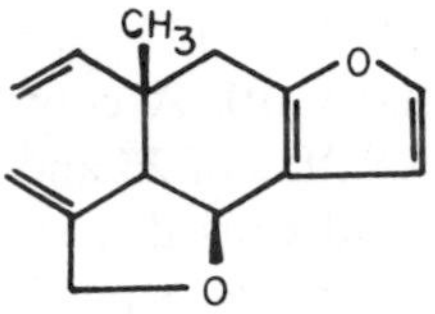

30a

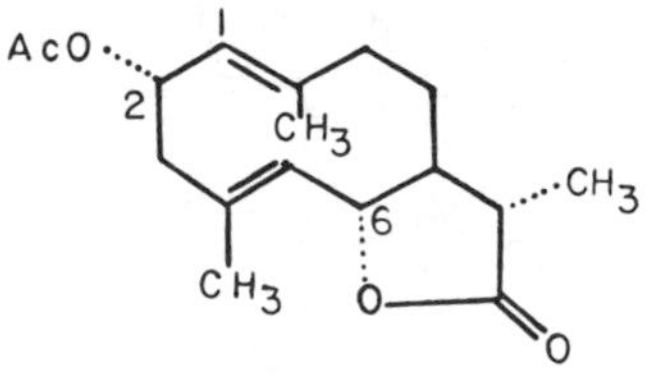

32

on the basis of the NOE results suggested that **27** has the structure represented in Fig. 8.

The Cope rearrangement of compounds **28** and **29** leads to the antipodal elemane products such as **30**, whereas the rearrangement of **31** and **32** afforded the normal elemane-type products **30a**. From a study of the NOE values (see Table VIII) for these compounds Takeda et al.[31] were able to correlate the type of rearrangement products obtained with the conformation adopted by the ten-membered ring. The conformation of **32** was defined as shown in Fig. 9 by the observation of the NOE's [CH$_3$-15] H-2 = 15%, [CH$_3$-14] H-2 = 10%, and [CH$_3$-15] H-6 = 10%.[32] The NOE data for **31**, although somewhat sparse, are best interpreted by a similar conformation.[31] The NOE values present in **28** and **29**, particularly [H-1] H-5 = 9% and [H-9β] H-1 = 11%, implied that the conformation of these compounds is as depicted in Fig. 9, in which the double bonds are cross orientated in the

TABLE VIII

NOE's for Germacrane-Type Sesquiterpenes[31, 32]

Proton saturated	Proton observed	NOE (%)			
		28[a]	**29**[a]	**31**[a]	**32**[b]
H-5	H-1	7	9	c	—
H-6	H-1	0	9	1	—
H-9α	H-1	−5	—	9	—
H-9β	H-1	11	18	0	—
CH$_3$-14	H-2	—	—	—	10
CH$_3$-15	H-2	—	—	—	10
H-6	H-5	8	9	9	—
CH$_3$-15	H-5	—	—	—	15
CH$_3$-14	H-15	—	—	3	—

[a] Takeda et al.[31]
[b] Bacca and Fischer.[32]
[c] Undeterminable.

opposite direction to that determined for **32**. A chair-shaped transition state for the [3,3]-sigmatropic shift will lead to a specific product depending on whether the double bonds are cross orientated as in **28** or **32**. The former gives antipodal elemane-type products whereas the latter gives the normal elemane-type product.

The 8β-formyl steroids are intriguing because it is known from X-ray studies[33] that 8β-methyl steroids show a distortion of their skeletal structures.

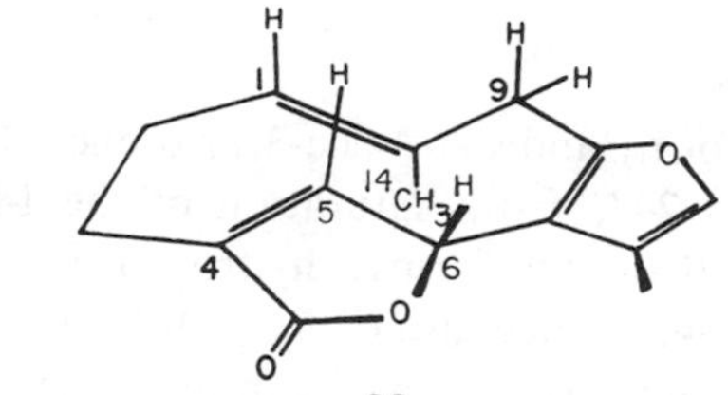

28

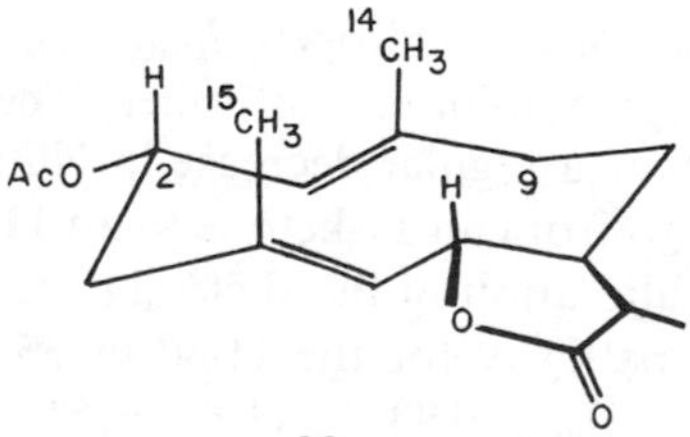

32

FIG. 9. Conformation of germacrane-type sesquiterpenes **28** and **32**.

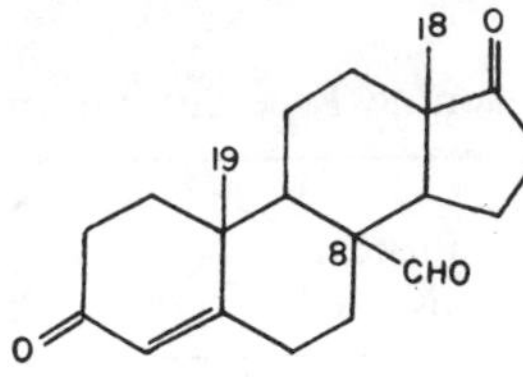

33

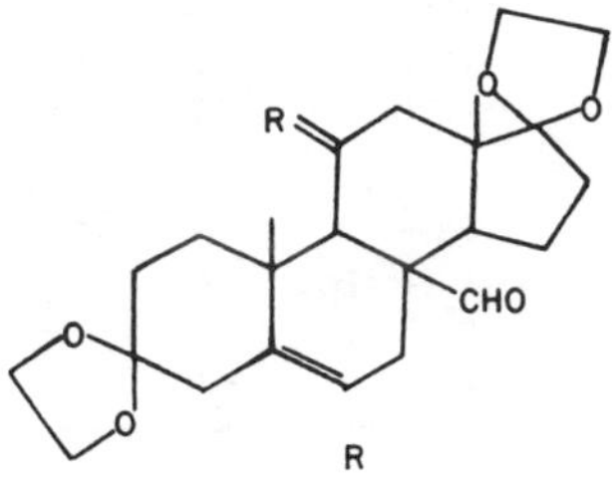

	R
34	O
35	H$_2$
36	H + OH

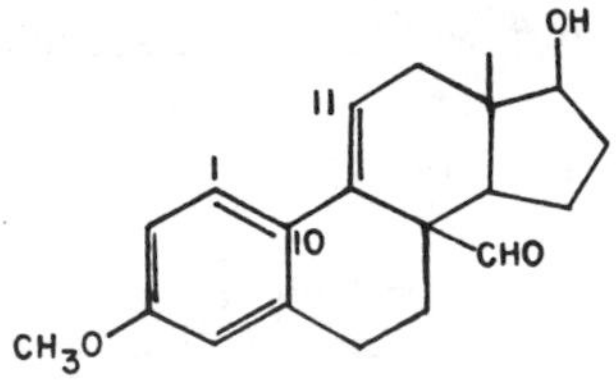

37

The compound 8β-formylandrost-4-ene-3,17-dione, **33**, has NOE's at the 8β-CHO of 33% and 24% from saturation of the 18- and 19-methyls, respectively, showing that the formyl hydrogen is effectively sandwiched between the two methyls.[34] Because these NOE's are large, we can conclude that the 8β-formyl group exists primarily in only one conformation, with the oxygen pointing away from the 18- and 19-methyls. The only long-range coupling observed for the formyl hydrogen, $J_{CHO, H-7\alpha} = 1.8$ Hz, is additional compelling evidence for a dominant conformer. The 8β-formyl androstene ketals, **34**, **35** and **36**, show a regular decrease in NOE (see Table IX) at the 8β-formyl proton as we go from an 11-ketone to an 11β-alcohol. The changes can be attributed to an alteration of the skeletal distortions and the addition of a further relaxation pathway for the H-8β in **35** and **36**. Indeed Tori[34] reports the interesting NOE, [OH-11β] H-8β = 5%, for **36c**, implying that the hydroxyl proton must have an appreciably populated conformation where it is pointing inside the C ring.

TABLE IX

NOE Values for 8β-Formyl Steroids[34]

Proton irradiated	Proton observed	NOE (%)				
		33	34	35	36	37
CH$_3$-18	H-8β	33	37	30	26	22
CH$_3$-19	H-8β	24	19	16	14	—
OH-11β	H-8β	—	—	—	5	—
H-6α	H-4	—	—	—	—	10
H-6β	H-4	—	—	—	—	3
H-1	H-11	—	—	—	—	20
H-12α ⊥ β	H-4	—	—	—	—	13
H-11	H-1	—	—	—	—	20
H-1	H-2	—	—	—	—	20

The value of the NOE, [CH$_3$-18] H-8β = 22%, for the estrol derivative **37** illustrates the loss in steric compression attendant upon removal of the 19-methyl group. The NOE of 20% between H-1 and H-11 (identical in this case in both directions) is fully consistent with the known steric interactions between H-1 and H-11 in the steroid skeleton. The remaining NOE's are included in Table IX since they show the application of NOE's to the assignment of chemical shifts in a typical estrol.

During a recent synthesis of tetracyclic intermediates in the total synthesis of pentacyclic triterpene, ApSimon et al.[35] isolated two isomeric compounds **38** and **39**. The assignment of the B–C ring junction of **38** as being cis and that of **39** as trans was made from the observation of a 13% NOE between the C-10b methyl and the C-10 aromatic proton for **38**, whereas for **39** this interaction was negligible. This assignment was later confirmed from consideration of the ^{13}C spectra for the two compounds.[36]

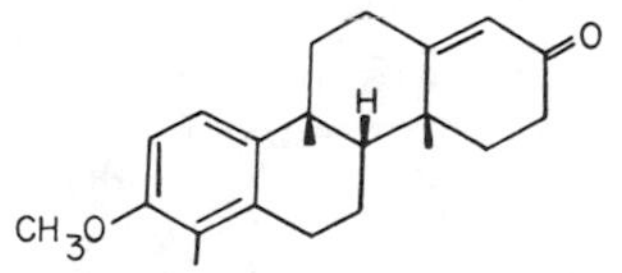

38

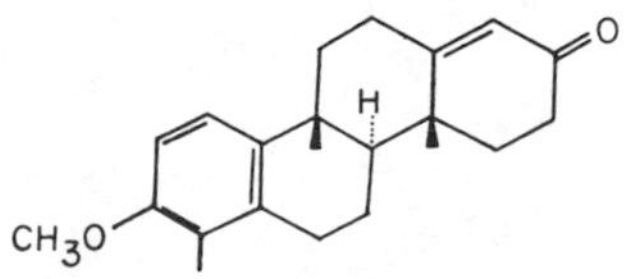

39

TABLE X

NOE's Observed in Phenoxymethyl Derivatives of Penicillin[37]

Proton irradiated	Proton observed	NOE (%)			
		40	**41**	**42**	**43**
CH_3-2β	H-3	21	26	20	22
CH_3-2β	H-5	0	0	0	0
CH_3-2α	H-3	7	0	0	0
CH_3-2α	H-5	0	14	13	11

Although the conformation of a number of penicillin derivatives in the solid state has been determined by X-ray studies, it was of interest to determine their conformation in solution.[37] These compounds are amenable to an NOE study and the results for several derivatives are given in Table X. It is immediately apparent that the thiazolidine ring undergoes a dramatic change

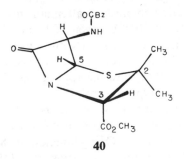

40

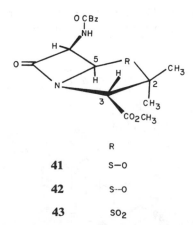

	R
41	S—O
42	S---O
43	SO_2

FIG. 10. Conformation of selected penicillin derivatives.

in shape in going from the sulfide **40** to the sulfoxides and sulfone (see Fig. 10). The 2β-methyl of the sulfide is axial (see Fig. 11) and H-3 is disposed between the two methyl groups. The larger NOE [CH$_3$-2β] H-3 suggests that H-3 is closer to CH$_3$-2β than to CH$_3$-2α, and the absence of an NOE from CH$_3$-2α to H-5 shows that H-5 must be some 3.6–3.8 Å away from CH$_3$-2α. In the (S)-sulfoxide **41** H-3 is now relaxed solely by CH$_3$$2\beta$, and CH$_3$-$2\alpha$ is sufficiently close (3.0–3.2 Å) to H-5 to affect its relaxation. In addition the NOE data show that the (R)-sulfoxide **42** and the sulfone **43** both adopt similar conformations to the (S)-sulfoxide, and this suggests that it is not the size of the oxygen atom nor hydrogen bonding with the amide hydrogen that determines the thiazolidine ring conformation, but rather the bonding at the sulfur atom. The conformations illustrated in Fig. 10 are consistent with the NOE data and with solvent-induced shifts, and are similar to the geometry determined in the solid state.[38]

A series of compounds which illustrate the use of NOE's in order to gain information on steric compressions are the terpenoid bicyclic lactones **44–47**[39] for which the interactions between the 20-methyl group and the deshielded proton of C-8 can be monitored in order to ascertain any molecular distortions. Unfortunately, protons at C-6 and C-7 which are relatively close to H-8, also efficiently relax H-8 and thus only qualitative assessments can be made. In compound **44** the NOE value, [CH$_3$-20] H-8 = 18%, shows that the maximum internuclear distance possible between these two moieties is 2.9 Å, which is significantly less than the 3.3 Å measured on molecular models. The origin of the distortion is the steric compression between H-1 and H-11 (2.30 Å apart in the normal model) which can be alleviated by a twisting of ring B such that H-8 is forced towards the 20-methyl (see Fig. 11). A similar argument applies to the cis lactone **45**, for which the NOE [20-CH$_3$] H-8 = 13%, and where an H-5–H-11α interaction further complicates the B ring distortion. The end result is again a twist that leaves H-8 closer to CH$_3$-20. The two isomeric lactones **46** and **47** are readily distinguished by the presence of the NOE [CH$_3$-20] H-8 (= 21%) in **46** and the absence of an NOE in **47**. The interesting aspect of **46** is that a molecular model shows the NOE to be far smaller than expected (the CH$_3$-20–H-8 distance is 2.55 Å) in contrast to the preceding two lactones. However, the equatorial–equatorial fusion of the lactone ring has relieved the H-1–H-11 interaction and left the CH$_3$-20–H-8 compression as the most intense in the molecule. The rings therefore distort to move H-8 further away from the 20-methyl, resulting in a lower observed NOE. Lactone **46** appears to be one of the few examples of tricyclic diterpenoids in which the H-1–H-11 compression is not the dominant one in the molecule, and it behooves one to beware the dictum that all perhydrophenanthrenes suffer strong H-1–H-11 interactions.

A publication to be recommended for a more detailed analysis of the use of Eqs. (21), (22), and (23) is that of Hoffman and Forsén,[40] who studied the

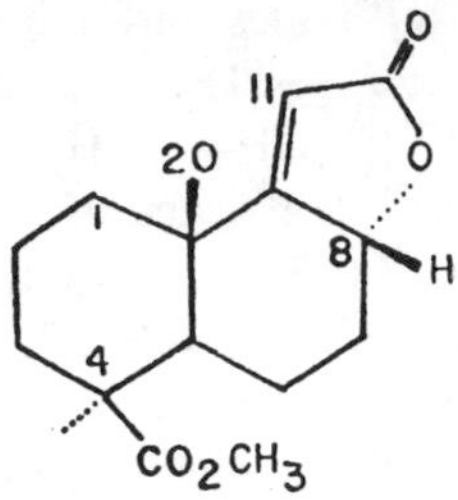

FIG. 11. a. Stereostructures of unsaturated bicyclic lactone **44** and bicyclic lactones **45–47**. b. Arrows indicate distortion caused by H-11–H-1(eq) interaction in **44**.

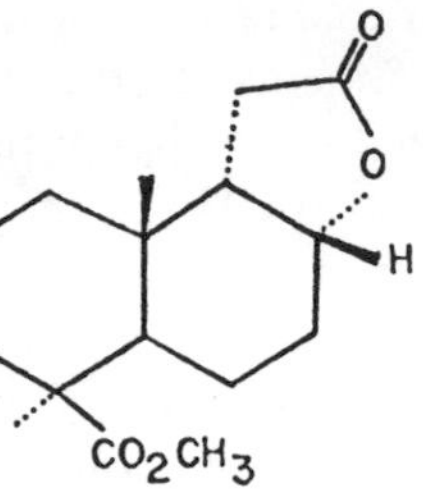

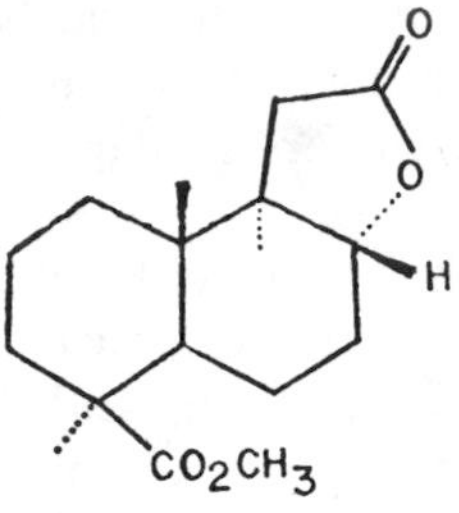

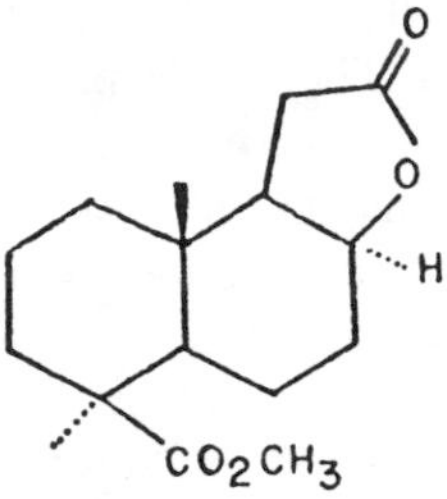

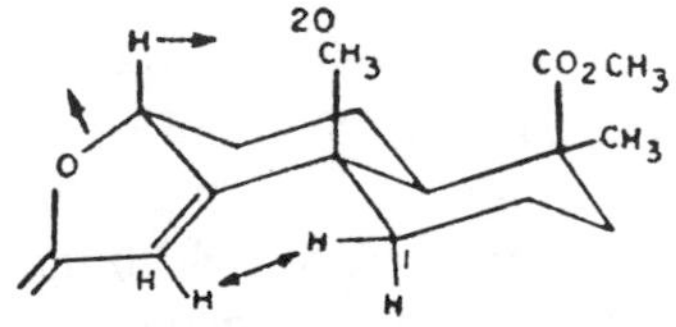

relaxation behavior of formic acid and acetaldehyde. The steady-state Overhauser experiments gave $22.5 \pm 0.5\%$ for observation of the formyl proton, $17.5 \pm 1.3\%$ when observing the acid proton in formic acid, and $23 \pm 0.5\%$ and $3.5 \pm 0.3\%$ for observation of the aldehydic proton and the methyl group, respectively for acetaldehyde. Using the method outlined in Section III above, R_A was obtained and from this σ_{AB} was calculated for each proton. A number of experiments were then performed to illustrate the nonexponential behavior of Eqs. (20) and (21). In each case the experimental curve was compared with the theoretical curve calculated using the known values of R_A and σ_{AB}. The conclusion drawn from these experiments was that the nonexponential decays obtained in the general case [Eqs. (20) and (21)] are normally unsuitable for determination of R_A and σ_{AB} but may be of some utility as cross checks of steady-state NOE data.

The first example of an intermolecular nuclear Overhauser effect was reported by Kaiser,[41] who used a mixture of chloroform and cyclohexane to demonstrate the application of the NOE in organic chemistry. The sample contained $CHCl_3$, cyclohexane, and tetramethylsilane in the ratio 1:4.5:0.5, and a 34% NOE was observed for the $CHCl_3$ proton on saturation of the cyclohexane resonance. Although the potential for intermolecular NOE's in solvation studies would appear to be great, only a limited number of inter-molecular NOE's have been reported. The main difficulty appears to be the selection of suitable substrates, since the measurement of intermolecular NOE's is practical only when the observed proton has essentially no intra-molecular relaxation pathways available to it (as in the case of $CHCl_3$).

The effect of intermolecular association can be observed in Table XI where the values of α-methyl methacrolein, 48,[27] and methyl methacrylate, 49,[42] are compared. The values for 48 in 5% DMSO-d_6 solution are signific-antly different from those obtained from a 5% CS_2 (or $CDCl_3$) solution.

48   49

TABLE XI

Comparison of the NOE Results for α-Methyl Methacrolein (48)[27] and Methyl Methacrylate (49)[42]

Proton saturated	Proton observed	NOE (%)			
		48		49	
		5% DMSO-d_0	5% CS_2[a]	5% CS_2	Pure liquid
CH_3	H_A	9	9	9	9
CH_3	H_B	0	6	0	6
H_B	H_A	39	30	42	28
H_A	H_B	42	28	48	29
H_1	H_A	0	0	—	—
H_1	H_B	7	6	—	—
OCH_3	H_A	—	—	0	5
OCH_3	H_B	—	—	0	5

[a] Similar values are also recorded in a 5% $CDCl_3$ solution.

The DMSO-d_6 results for **48** are similar to those for **49** in CS_2, whereas the CS_2 solution of **48** is similar to **49** as a pure liquid. Intermolecular interactions should play a significant role in the relaxation processes for a pure liquid, as is illustrated by the decreased NOE observed between the geminal protons, and also by the positive NOE between the CH_3 group and H_B recorded for **49** as a neat liquid. By analogy, it is reasonable to suppose that **48** in CS_2 exists as an associated species although the exact nature cannot be ascertained. These results emphasize why it is necessary when one studies intramolecular NOE's that intermolecular interactions be minimized, but conversely they demonstrate that NOE's can be useful in intermolecular interaction studies.

B. Heteronuclear NOE's

The most widely studied heteronuclear NOE has been the $[H]-^{13}C$ interaction either for the determination of the relaxation mechanism, as an aid in spectral assignment, or in order to assess correlation times. From Eq. (9) under conditions of saturation of all protons we get

$$f_C(H) = \frac{\gamma_H \sum \sigma_{CH}}{\gamma_C \left(\sum \rho_{CH} + \rho^* \right)}$$

which gives a maximum for dipolar coupling of $f_C(H) = 1.98$. However, if other mechanisms such as spin rotation are also efficiently relaxing the observed carbon, the NOE will be less than 1.98. In Table XII NOE and observed T_1 values, together with $T_1(DD)$ and $T_1(SR)$ contributions are given for two methylated benzene derivatives.[43] The protonated ring carbons each have NOE's of 2.0 ± 0.2 and thus the carbon–proton dipole–dipole interaction dominates. The methyl carbons in the two compounds show

TABLE XII

NOE and Spin–Lattice Relaxation Time Data for o-Xylene and Mesitylene[43]

	Carbon(s)	NOE	T_1 (seconds)	T_1 (DD)	T_1 (SR)
	C-1,2	0.74	38	103	60
	C-3,6	2.0	13	13	> 60
	C-4,5	2.0	13	13	> 60
	CH_3	2.0	12	12	> 60
	C-1,3,5	0.96	41	85	> 8
	C-2,4,6	2.0	9	8	> 80
	CH_3	1.0	11	23	25

significantly different NOE values although the observed T_1 values are similar. For o-xylene the maximum is observed, whereas for mesitylene, [H] $CH_3 =$ 1.0. Thus for the former, the dipolar interaction is the dominant relaxation process whereas for the latter another mechanism, probably spin rotation, is equally as important. Qualitatively, the T_1 and NOE data show that in mesitylene the CH_3 group rotation is extremely rapid resulting in a longer $T_1(DD)$ and causing $T_1(SR)$ to become relatively important, whereas for o-xylene the rotation is restricted, $T_1(DD)$ is shorter, and the contribution from spin rotation becomes negligible. The more quantitative use of these data is outside the scope of this chapter and the reader is referred to the literature.[43, 44]

The NOE in compounds such as steroids and terpenes has values of 2.0 even for quaternary carbons and thus its application to structural studies is somewhat limited. However, Grant et al.[45] used Eq. (19) in order to assign the carbon resonances of a number of polycyclic aromatic compounds. The experimental NOE results were compared with those calculated using $\eta = \sum_i (r_{CHi})^{-6}$ for all quaternary carbons for various values of μ. The value of μ was determined from a least-squares criterion of fit between experimental and calculated values as shown in Table XIII for fluoranthene, 50. From these results, the carbon spectrum of 50 was assigned. A more complete method is to use T_1 data in conjunction with the NOE values with the NOE being used to determine $\sigma_{AB}(DD)$ from which $(\sum_i r_{C-Hi}^{-6})$ can be calculated since τ_c can be deduced from T_1 for protonated carbons. This method has been used in the assignment of quaternary carbons in a number of natural products.[46]

In the derivation of Eqs. (4) and (5) and thus of the maximum possible value for the NOE, [H] C-13, of 1.98, we assumed that the correlation time was such that $\omega^2\tau_R^2 \ll 1$, i.e., the limit of extreme narrowing applied. However, in the limit of slow rotation, $\omega^2\tau_R^2 \gg 1$, Eqs. (4) and (5) must be expanded to include frequency terms and then the maximum possible value for

50

TABLE XIII

Calculated and Experimental Enhancements for Fluoranthene (50)[45]

| Carbon(s) | $\sum \alpha_i$ | Calculated | | Experimental |
		$\dfrac{M_z{}^a}{M_0}$	$\dfrac{(M_z/M_0)_i}{(M_z/M_0)_{15}}$	$\dfrac{M_z}{M_0}$
1,6	0.6933	2.82	2.52	3.0
2,5	0.7019	2.82	2.52	3.0
3,4	0.6951	2.82	2.52	3.0
7,10	0.6936	2.82	2.52	3.0
8,9	0.7023	2.82	2.52	3.0
11,12	0.0132	1.35	1.20	1.2
13,14	0.0135	1.35	1.20	1.2
15	0.0040	1.12	1.00	1.0
16	0.021	1.52	1.52	1.4

^{a} Calculated using $\mu = 0.06$.

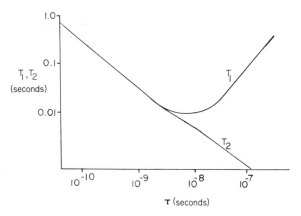

FIG. 12. Plot of T_1 and T_2 as a function of correlation time τ.

the NOE [H] C-13 is 0.153. In Figs. 12 and 13 the NOE, T_1, and T_2 as a function of correlation time are given.

Table XIV[47] contains the T_1, T_2, and NOE data of helical and random coil forms of poly(γ-benzyl)L-glutamate, **51**. The T_1 values for the α-carbons change only by about 50% in the two forms of the polymer. However, the NOE values and T_2 data indicate, in fact, that the effective correlation times of the two forms change drastically and actually lie on different sides of the T_1 curve (Fig. 12). The negligible NOE value for the α-carbon of the helical form verifies experimentally the theoretically predicted value under conditions of slow rotation and also suggests little or no contribution to the

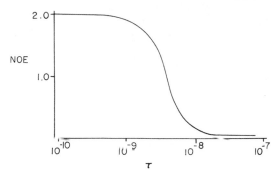

FIG. 13. Dependence of observed NOE [H] ^{13}C on the correlation time τ for the dipolar relaxation mechanism.

TABLE XIV

Carbon-13 T_1, NOE, and Linewidth Data for Helical and Random Coil Poly-(γ-benzyl)(L-Glutamate) at 40[64]

Carbon	Molecular weight	NT_1[a] (m seconds) Helix	Coil	NOE Helix	Coil	$\Delta\gamma$ Helix	Coil
α	7,000	35.5	52	0.0	1.3	28	≤ 8
	17,000	39	70	0.2	1.6	37	≤ 8
	46,000	44	64	0.1	1.2	48	≤ 12
β	7,000	65	85	0.6	1.4	—	—
	17,000	71	107	1.0	1.8	—	—
	46,000	72	91	0.9	1.7	—	—
γ	7,000	108	139	1.1	1.8	—	—
	17,000	127	159	1.4	2.1	—	—
	46,000	109	143	1.2	2.1	—	—
Benzylic	7,000	264	419	1.0	1.7	—	—
	17,000	279	480	0.9	1.5	—	—
	46,000	281	397	0.9	1.6	—	—

[a] $NT_1 = T_1$ of carbon times number of attached protons.

effective correlation time from segmental motion. The effective correlation time is given by

$$(\tau_{\text{eff}})^{-1} = (\tau_R)^{-1} + (\tau_G)^{-1}$$

where τ_G is the internal rotational correlation time for segmental motion of the backbone or for rotation of side-chain functional groups.[48]

The results, however, show that on going from the helical to the random coil conformation, NOE values increase from 0.1 to 1.5 and the linewidths

51

decrease from 28–48 Hz to less than 12 Hz. These observations are consistent with a drastic decrease in τ_{eff} as a result of rapid segmental motion occuring in the flexible random coil form which was not available to the rigid structure of the helical form. NOE studies therefore can detect transitions between the two forms. Thus, for the α-carbons of the helical form, τ_{eff} satisfies the slow rotation condition whereas for all other carbons in both forms, τ_{eff} is already close to the limit of extreme narrowing and the changes in NOE values are minimal.

The analysis of polymer materials, then, is facilitated by recourse to spin-lattice relaxation times, linewidths, and nuclear Overhauser enhancements. All carbons expected to have a correlation time satisfying the limit of slow rotation are expected to have a negligible NOE. Even in large polymers such as biomacromolecules, where spectra may be too poorly resolved when coupling is retained, it has been shown that it is possible to estimate NOE values by comparing resonance intensities of the observed and computer-simulated spectra.[49]

An interesting intermolecular [H]–^{13}C value has been observed in the system $CHCl_3$–CS_2: irradiation of the proton of the $CHCl_3$ caused a 30% *decrease* in the area of the CS_2 carbon absorption.[50] When $CHCl_3$ was replaced by $CDCl_3$, no such change was observed. Since dipolar coupling precludes a *negative* NOE in the ^{13}C–[^{1}H] case, and quadrupolar relaxation by the ^{35}Cl could only reduce the magnitude of the NOE, but not change its sign, the negative value of the NOE is suggested to result from time-dependent scalar coupling involving essentially an exchange modulation of the spin–spin coupling due to electron overlap in the collision complex. However, although a good theoretical interpretation is at present lacking, the effect itself should have application in the studies of intermolecular interactions.

The variation of NOE values for different carbon nuclei in the same molecule precludes a direct relationship between the signal intensity and the number of each type of ^{13}C nucleus. In view of the success of quantitative applications in proton magnetic resonance, suppression of the NOE in proton-decoupled ^{13}C NMR spectra would retain the spectral simplification and, at the same time, generate a quantitative utilization of ^{13}C spectral intensities. Also, in order to use ^{13}C NMR in a quantitative manner, a time of at least $4T_1$ must be allowed between pulses to prevent saturation. Thus if the carbon of interest is quaternary, much of the benefits which accrue from use of FT are lost. These problems can be circumvented by the addition of the para-magnetic material[16] such as Cr^{III}(acetyl acetonate)$_3$ which does not induce a chemical shift but does significantly increase ρ^* and thus R_A. The concomitant decrease in the NOE was discussed in Section III.

The nuclear spins of ^{15}N and ^{29}Si have the common characteristic of a negative gyromagnetic ratio. Thus the NOE will be negative and hence the signal of either nucleus, under conditions of proton decoupling when the dipolar mechanism dominates, will be inverted. Lichter et al.[51] have studied a number of amines and have found that solution preparation is extremely important for observation of good signals. With protonated nitrogens, two mechanisms of relaxation are possible, namely the dipolar interaction and the time-dependent spin–spin coupling. The former will give an inverted signal whereas the latter [compare with Eq. (6)] will give a signal in the normal direction. When 1,4-butanediamine was examined without purification, no natural abundance ^{15}N signal could be observed, while a strong absorption was easily detected for a freshly distilled sample. This is a result of the importance of exchange process in the relaxation of the undistilled compound, whereas in the purified compound the exchange rate is much slower and consequently the dipolar interaction is the dominant relaxation process. The effect of modulated scalar coupling[51a] can also be seen from the observed NOE's at different pH values for glycine. In very strong acid solution (12 N HCl), the NOE, [H] ^{15}N, approaches the maximum of -4.93; at pH = -0.3, [H] ^{15}N = 2.9; and at pH 6.4 it is -2.15. Thus the NOE decreases with increasing pH, i.e., with increasing chemical exchange. In small molecules such as NH$_4$Cl spin rotation can also make an important contribution to the relaxation processes as evidenced by a decreased NOE.

A limited number of reports[52, 53] has appeared on [H] ^{29}Si NOE values. In Table XV[53] the NOE values of TMS as a function of temperature are given. The observed signal, which is positive at 25°C, decreases in relative intensity as the temperature decreases, becoming zero at $-62.5°$. At temperatures below $-62.5°$ the signal is inverted. At 25°, $\rho^* \gg \sum \rho_{\text{Si}-\text{H}}$ with ρ^* being the contribution from spin rotation while at $-83°$, $\rho^* = 1.6 \sum \rho_{\text{Si}-\text{H}}$ since ρ(SR) and ρ(DD) have opposite temperature dependences. The maximum NOE is -2.517 and thus a ^{29}Si signal where ρ(DD) dominates will have a relative intensity of -1.517 when the proton resonances are saturated.

TABLE XV

Silicon-29 NOE Values as a Function of Temperature for Tetramethylsilane[53]

Temperature (°C)	NOE	Relative intensity[a]
25	−0.09	0.91
0	−0.205	0.795
−20	−0.35	0.65
−50	−0.495	0.505
−62.5	−1.0	0.0
−64	−1.03	−0.03
−83	−1.59	−0.59

[a] Relative to undecoupled intensity.

The silicon absorption may be positive, negative, or zero. The proton-decoupled spectrum of TMS at ambient temperature is positive, that of $[(CH_3)_3SiCH_2]_2Si(CH_3)_2$ has one inverted and one noninverted signal for the nonequivalent silicon resonances, whereas $[(CH_3)_3SiCH_2]_2Si(CH_3)CH_2-CH_2OH$ has two inverted signals.[52]

There have been no reports, to our knowledge, on the [H] ^{31}P NOE's, which is surprising since phosphorus is a relatively important nucleus with a positive gyromagnetic ratio and 100% natural abundance, and is consequently much easier to observe than either ^{29}Si or ^{15}N. The theoretical maximum for the NOE is 1.23. In our laboratory, we have studied[54] a limited number of simple compounds. In tetrabutyl phosphonium bromide, methyltriphenyl phosphonium bromide, and related compounds the maximum NOE was observed. In the methyltriphenyl phosphonium bromide, the CH_3 group protons and the aromatic protons could be saturated selectively and gave enhancements of 80% and 45%, respectively. These values cannot however be used directly in Eq. (14) as the $\tau_c(P-CH_3)$ is not necessarily equal to $\tau_c(P-\phi H)$. In order to obtain information on the internuclear distances, it is necessary to determine R_A for the phosphorus together with the relative rotation rates about the C—P bonds[54] from ^{13}C T_1 data.

Compounds such as **52** also exhibit the maximum NOE[55] when subjected to complete proton decoupling. In these examples information on inter-

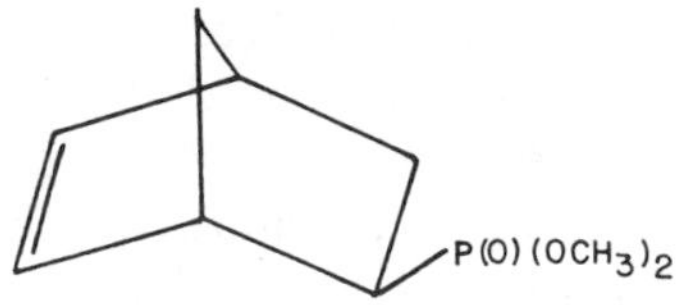

52

nuclear distances between the phosphorus and protons in the rigid part of the molecule can be obtained directly from selective saturation.

Other heteronuclear NOE's that have been observed are the [F] C,[56] [F] H, and [H] F cases.[57] The maximum value possible for the carbon–fluorine interaction is 2.87 and this was in fact observed for perfluorobenzene. The interaction between protons and fluorine was the first NOE recorded, although the original work[1] and several subsequent investigations used the technique only as an aid in relaxation studies. In small molecules the spin–rotation interaction makes a significant contribution to the relaxation processes. However, in a number of simple aromatic compounds both the proton and fluorine relaxation were shown to be almost exclusively the result of the dipolar interaction because the NOE values recorded were close to the value predicted by Eq. (14). Thus in many compounds of interest to organic chemists, the relaxation processes will be dominated by the dipolar interactions and the NOE will give information on molecular geometry.

V. THE EFFECT OF CHEMICAL EXCHANGE AND INTERNAL MOTION

Since the NOE depends on the internuclear distances, any process that makes the nuclear separation time dependent will obviously affect the magnitude of the observed enhancement. We can study the effect of time dependence in the following kinetic regions, where k is the first-order rate constant: (1) slow intramolecular exchange with the resonances for each conformer being observable; (2) slow internal motion where only an average absorption is observed; $k < R_d$; (3) intermediate rate of internal motion; $R_d < k < \tau_c^{-1}$; (4) rapid internal motion $k \geq \tau_c^{-1}$; (5) intermolecular exchange.

1. For a nuclear spin exchanging between two sites of equal population with a first-order rate constant k, the enhancement of the resonance at site A upon saturation of the signal of the second site B is given by

$$f_A(B) = \frac{\sigma_{AB}}{R_A}$$

where

$$R_A = \sum_d \rho_{Ad} + k + \rho^* = R_A' + k$$

and

$$\sigma_{AB} = -k$$

which gives

$$f_A(B) = \frac{-k}{R_A' + k}$$

Thus if the exchange is rapid, $k > R_A'$ and $f_A(B)$ will approach -1, i.e., no absorption will be observed, whereas when $R_A' > k$, the A absorption will be unaffected by saturation of B. We can thus use this method for the determination of activation parameters provided the magnitude of $f_A(B)$ is reasonable $[-0.05 < f_A(B) > -0.95]$ by recording $f_A(B)$ and R_A at a number of temperatures.

An example of this behavior was seen for annulene, **53**,[58] where the inner

53

and outer protons are exchanging. At $-60°$ the spectrum consists of 2 multiplets centered at 9.25 ppm and -4.22 ppm, respectively and at $+20°$ two broad peaks are observed, whereas above 50° a broad singlet is obtained which decreases in linewidth with increasing temperature. Irradiation of one multiplet at low temperature ($-60°$) resulted in decoupling of the other multiplet but no change in its area, whereas at 20° the same experiment caused a complete loss of signal for the observed resonance. At intermediate temperatures, intermediate values of $f_A(B)$ were observed. Several other studies using this technique have been performed,[59] including a study of[60] the inversion of cyclohexane-d_{11} where transient methods were used to determine both R_A and k as a function of temperature, from which $\Delta G\ddagger$, $\Delta H\ddagger$ and $\Delta S\ddagger$ were determined.

An extension of this can be found in the results obtained for N,N-dimethylformamide[61] (Table VXI). At 30° saturation of the resonance of the CH_3 groups B and A resulted in enhancements of the formyl proton resonance of 28% and 3%, respectively, whereas at 90° the enhancement was 28% whether A or B or both was saturated. The activation parameters can be deduced by use of $f_F(A)$ as a function of temperature as outlined below. The direct dipolar interactions $\sigma_{AB}(DD)$, $\sigma_{BA}(DD)$, and σ_{AF} are assumed to be negligible. The spin system at equilibrium with saturation of A can be expressed as:

$$\frac{dM_z^B}{dt} = 0 = -R_B(M_z^B - M_0^B) - \sigma_{AB}(M_0^A)$$

TABLE XVI

Variable Temperature NOE's in Dimethylformamide[57]

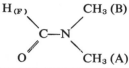

Temperature	NOE (%) at formyl proton	
(°C)	Saturation of B	Saturation of A
31	28	3
40	—	6
50	28	9
55	—	13
60	28	18
70	28	24
80	28	26
90	28	27–28

and

$$\frac{dM_z^F}{dt} = 0 = -R_F(M_z^F - M_0^F) - \sigma_{BF}(M_z^B - M_0^B) - \sigma_{AF}(-M_A^0)$$

and since σ_{AF} is negligible,

$$M_z^F - M_0^F = \frac{\sigma_{BF}}{R_F} \cdot \frac{\sigma_{AB}}{R_B}(-M_0^A)$$

But $f_C(B) = \sigma_{BF}/R_F$, $k = -\sigma_{AB}$, and $R_B = R_B' + k$; therefore

$$f_F(A) = \frac{M_z^F - M_0^F}{M_0^F} = f_F(B) \cdot \frac{k}{R_B' + k}$$

and thus

$$k = \frac{kT}{k} \exp\left(\frac{-\Delta H^{\ddagger}}{RT}\right) \exp\left(\frac{-\Delta S^{\ddagger}}{R}\right) = \left[\frac{f_F(A)}{f_F(B) - f_F(A)}\right] R_B'$$

The values of $\Delta H^{\ddagger}$, $\Delta S^{\ddagger}$, and $\Delta G^{\ddagger}$ can thus be determined from knowledge of the respective enhancements and R_B' as a function of temperature. If the R_B' data are difficult to obtain, an approximate value of $\Delta H^{\ddagger}$ can be determined for a dilute solution, provided the dominant mechanism is the dipolar interaction, by the substitution $R_B' = K\eta/T$ where η is the viscosity at temperature T of the pure solvent. Figure 14 illustrates the graph obtained from a plot of

$$\frac{1}{T} \text{ vs } \log\left[\left(\frac{f_F(A)}{f_F(B) - f_F(A)}\right)\frac{\eta}{T^2}\right]$$

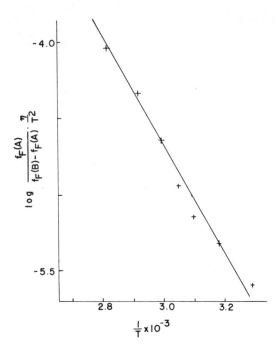

FIG. 14. Plot of the variation with temperature for dimethylformamide (see text).

where η is the viscosity of dimethyl sulfoxide at the various temperatures.[62] The value of $\Delta H^{\ddagger}$ is 17 kcal/mole, which agrees favorably with the result obtained for the same solution by lineshape analysis. Noggle and Schirmer[3] calculated the energy of activation to be 20 kcal/mole (the value quoted is in error by a factor of 2.3) by assuming that R_B' is independent of temperature. The assumption was based on the temperature independence of $f_F(B)$ which however implies only that the ratio σ_{BF}/R_B' is independent of temperature.

The examples quoted illustrate how nuclear magnetic double resonance can be of considerable aid in determination of activation parameters either directly or with the NOE.

2. In this region only the average resonance of all the possible conformations is observed although the rate of conformational change is so slow that $k < R_A$. The observed NOE is given by[3]

$$f_A(B) = \sum_i P_i f_A(B,i)$$

where P_i is the fractional population in conformation i. In the limit that the number of conformations approaches ∞ and the lifetime of each is very small, then

$$f_A(B) = \int_0^\infty f_A(B,\Omega)P(\Omega)d\Omega$$

where $P(\Omega)$ is a distribution function of the internal variable Ω.

3. In contrast to 2 above where the NOE's were averaged, if $R_A < k$ then the *internuclear distances* must be averaged because the molecule does not reside in one conformation long enough to reach a steady state. Thus for a homonuclear spin system[3]

$$f_A(B) = \frac{\langle \sigma_{AB} \rangle}{\langle R_A \rangle} - \sum_d \frac{\langle \sigma_{Ad} \rangle}{\langle R_A \rangle} f_d(A) \tag{24}$$

where

$$\langle \sigma_{AB} \rangle = \sum_i \sigma_{AB}(i)P_i \tag{25}$$

and

$$\langle R_A \rangle = \sum_i \left[\sum_d \sigma_{Ad}(i)P_i + \rho^*(i) \right]$$

For a continuous path,

$$\langle \sigma_{AB} \rangle = \int_0^\infty \sigma_{AB}(\Omega)P(\Omega)d\Omega \tag{26}$$

where $P(\Omega)$ is again the distribution function of the variable Ω.

The barrier to rotation of an aromatic aldehyde group[63] about the C-1–C-7 bond is 10 kcal/mole and thus k at ambient temperature is of the order of 10^5–10^6 seconds^{-1}. The recorded NOE's for piperonal, **54**, are $f_6(5) = 26\%$

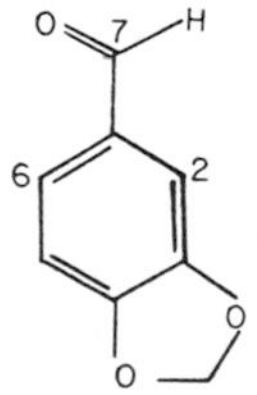

54

and $f_7(6) = 0.14\%$, with $f_7(5)$ and $f_5(7)$ being negligibly small.[13] Only the two conformations in which the carbonyl and aromatic ring are coplanar need be considered, and thus the data can be interpreted by the use of Eqs.

(24) and (25) in a form similar to Eq. (4) and the assumption that the spins 5, 6, and 7 can be considered to form a three-spin system.*

$$\frac{\langle r_{7,6}^{-6}\rangle}{\langle r_{6,5}^{-6}\rangle} = \frac{f_6(5) + f_6(7)f_7(5)}{f_6(7) + f_6(5)f_5(7)}$$

As $r_{6,5}$ is invariant and measurable, an absolute value can be calculated for $\langle r_{7,6}^{-6}\rangle$, and since

$$\langle r_{7,6}^{-6}\rangle = P_{syn}(r_{7,6}^{-6})_{syn} + (1 - P_{syn})(r_{7,6}^{-6})_{anti}$$

P_{syn} can be determined. Substitution of the recorded NOE values gives $P_{syn} = 0.49$.

For 6-bromopiperonal, $f_2(7) = 10\%$, which gives, from Fig. 1 and the same procedure described above, a value of $P_{syn} = 0.3$.

The nucleosides are a class of compounds that have undergone extensive conformational analyses by spectroscopic methods, with NOE's being no exception. The most detailed example is that of Noggle et al.[64] who utilized the technique in a conformational analysis of 2',3'-isopropylideneinosine, **55**, the experimental results of which are given in Table XVII. The problem then is to interpret these results in terms of the conformational preferences of the ribose ring and about the glycosyl bond. Equation (14) can be rearranged so

	R	R'	R"
56	PO_3	H	H
57	H	PO_3	H
58	H	H	PO_3

---

*Substitution into the corresponding equation for 4 spins[3] gives the same value of $\langle r_{7,6}^{-6}\rangle$ as $f_7(5) = f_7(5) = f_2(5) = 0$.

anti

	R	R'
59	NH$_2$	OH
60	H	NH$_2$

	R	R'	R"
55	OH	H	H
61	NH$_2$	H	H
62	NH$_2$	H	OAc
63	OH	NH$_2$	H

that for a given internuclear distance an NOE value can be calculated provided ρ^* is known. An arbitary small value was given to ρ^* and the ribose
geometry was deduced to be $r_{1,4} = 3.0$ Å, with dihedral angles H_1'–H_2' =
90°, H_2'–H_3' = 30° and H_3'–H_4' = 170°. All other reasonable ribose conformations gave similar predicted NOE values at a given glycosyl angle

TABLE XVII

Calculated and Experimental Enhancements (%) for 55[64]

Parameter	1[a]	2[a]	3[a]	4[b]	5[b]	6[c]
Y_1^0	90	180	360	355	355	359
ΔY_1^0	—	—	───	64	10	88°
Y_2^0	—	—	—	166	166	—
ΔY_2^0	───	──	—	10	64	──
P_{syn}	1.0	0	1.0	0.8	0.2	1.0

	Experimental		Calculated			
$f_8(1')$	18	0	0	36	17	0	21
$f_8(2')$	14	48	10	0	14	46	16
$f_8(3')$	4	−3	18	0	2	0	−4
$f_8(5')$	4	0	4	0	1	1	1
$f_2(1')$	0	2	6	0	1	4	0
$f_2(2')$	7	0	0	3	5	1	6
$f_2(3')$	7	0	0	10	8	4	7
$f_2(5')$	6	2	0	8	5	1	5
$f_1'(8)$	16	−2	0	17	12	12	13
$f_2'(8)$	10	40	5	0	6	6	6
$f_3'(8)$	1	−6	10	0	1	1	−1

[a] One conformation.
[b] Best fit double Gaussian.
[c] Best fit single Gaussian.

Y ($Y = -X - 120° = \phi_{CN} - 120°$[65]). Thus the major factor influencing the NOE values is the value of Y. In columns 2–4 of Table XVII, the calculated enhancements for various Y values are collected together with the experimentally determined values. Coincidence between experimental and calculated values could not be obtained and thus at least two conformations must be present. The results were then treated as both a single or a double Gaussian distribution function, the best values being given in columns 4 and 6. As can be seen, a slightly better fit was obtained using the double Gaussian function with a predominance of the molecules being in the syn conformation (syn/anti = 80/20).

In Table XVIII, the results for a number of other purine bases and nucleosides are given together with an estimation of Y, ΔY, and the syn/anti ratio. The latter were calculated using one of the methods described below.

a.[64] Using an assigned small value of ρ^*, the respective NOE values are calculated, using Eq. (26) where $P(\Omega)$ is either a single or double Gaussian distribution function, for various values of Y and ΔY. A least-squares analysis is used to find the best fit between experimental and calculated NOE's.

TABLE XVIII

NOE's Recorded for Selected Purine Base Nucleosides

Compound:	56[a]	56[a]	57[a]	57[a]	58[a]	58[a]	59[b]	60[b]	61[c]	61[d]	61[c]$_1$	62[e]	62[e]	62[e]	62[c]	62[d]	63[b]
Solvent:	D$_2$O	D$_2$O	D$_2$O	D$_2$O	D$_2$O	D$_2$O	D$_2$O	D$_2$O	Pyr[f]	Pyr[f]	DMSO/Acetone	DMSO	Acetone	CDCl$_3$	Pyr[f]	Pyr[f]	
pD:	8.4	1.3	8.9	1.4	8.7	1.2	7.2	7.2	—	—	—	—	—	—	—	—	—
$f_8(1')$:	10.5	7.5	18	15	16	16.5	7	8	20	19	17	21	24	22	27	26	4
$f_8(2')$:	18	9	—	—	18.5	8.0	1	0	7	6	8	4	8	2	5	6	11
$f_8(3')$:	9.5	14.5	—	—	9.5	4	0	—	—	—	—	2	0	—	—	—	2
$f_8(2'+3')$:	—	—	21	13	—	—	—	—	—	—	—	—	—	—	—	—	—
$f_8(5'+5')$:	3	7.5	3.5	5	2.5	4	0	0	0	0	5	—	—	—	0	0	3
$f_2(1')$:	—	—	—	—	—	—	—	—	8	—	6	—	—	—	—	—	—
$\sum f, f_8(i)$:	41	38.5	42.5	33	46.5	32.5	8	8	27	25	30	25	34	24	32	32	—
Υ_1^0:	—	—	—	—	—	—	—	—	350	350	340	300	320	300	15	17	—
$\Delta\Upsilon_1^0$:	—	—	—	—	—	—	—	—	—	170	180	90	120	120	—	—	—
Υ_2^0:	—	—	—	—	—	—	—	—	170	—	—	—	—	—	—	—	—
$\Delta\Upsilon_2^0$:	—	—	—	—	—	—	—	—	—	—	—	—	—	—	—	—	—
P_{syn}:	0.53	0.49	0.72	0.78	0.63	0.80	1.0	1.0	0.8	0.8	0.75	1.0	1.0	1.0	1.0	1.0	0.10
Syn/anti:	1.1	1.0	2.6	3.5	1.7	4.0	—	—	4	4	3	—	—	—	—	—	—

[a] Son et al.[65]
[b] Govil and Smith.[66]
[c] Bell and Saunders.[13]
[d] −50° Bell and Saunders[13] and Horner et al.[14]
[e] Hart and Davis.[67]
[f] Pyridine-d$_5$.

b.[65] The NOE and T_1 measurements are combined with calculated τ_c values to calculate $\langle r_{ij}^{-6} \rangle^{-1/6}$ using Eqs. (4), (5), and (14). The internuclear distances are measured from molecular models for three conformations, the conformations being chosen such that one of the r_{ij} values is a minimum, i.e., for compound **56** the conformations are $Y = 0$, 120, and 170 in which $r_{8,1'}$, $r_{8,2'}$, and $r_{8,3'}$, respectively, are a minimum. The proportion of each conformer is then calculated using

$$\langle r^{-6} \rangle^{-1/6} = r_{min}P^{1/6}$$

Thus, for example, for **56** at neutral pD, $\langle r_{8,1'}^{-6} \rangle^{-1/6}$ is determined to be 2.5 Å, $r_{min} = 2.3$ Å, and thus $P_{1'}$ is 0.5; $\langle r_{8,2'}^{-6} \rangle^{-1/6} = 2.2$ Å and $r_{min} = 0.18$ to give $P_{2'}$ as 0.3; and finally $\langle r_{8,3'}^{-6} \rangle^{-1/6} = 2.5$ Å and $r_{min} = 1.9$ Å, which gives $P_{3'} = 0.2$. Thus **56** is determined to be 50% syn and 50% anti.

c.[13] The average $\langle r_{ij}^{-6} \rangle$ values are deduced from the NOE data using Fig. 1 or Eq. (14). The compound is then assumed to be described in terms of two conformations of very small ΔY. Equations such as

$$\langle r_{ij}^{-6} \rangle = \sum_k P_k r_{ij}^{6'}(k)$$

and

$$\sum P_k = 1$$

are then evaluated simultaneously for all interactions by least-squares analysis for various values of Y.

As can be seen from Table XVII, $f_8(1')$ and the syn/anti ratio show a large range of values. For example, the guanosine phosphates in D_2O range from essentially a completely syn conformation for the cyclic phosphate at pD = 7.2 to approximately a 50% syn: 50% anti ratio for the 5'-phosphate at pD 1.4 and 8.4. Guéron et al.[65] also studied the effect of temperature and of concentration on the NOE values. As the temperature was increased the total NOE at H-8 for **56–58** increased, although the relative values are virtually independent of temperature. As the concentration was increased, f_8(total) decreased, T_1 decreased, and the relative enhancement $f_8(1')$ decreased for compounds **57** and **58**. These results indicate that intermolecular association is important particularly at low temperature and high concentration. The effect of aggregation is interpreted in terms of τ_c for the aggregate being long enough that the conditions of extreme narrowing are no longer operable, thus causing a decrease in the total NOE as was described above (see Section IV, B). A decrease in f_8(total) will occur even if the molecule was part of an aggregate for a small proportion of its time. Reexamination of the values for **59**, obtained from a 0.16 M solution, suggests that these results may have been observed in a solution where aggregation was present as f_8(total) = 8%. For **57**, a significant proportion of aggregation is indicated at 0.1 M.

	R	R'	R"
64	H	H	H
65	H	$C(CH_3)_2$	
66	I	$C(CH_3)_2$	
67	Br	$C(CH_3)_2$	
68	Cl	$C(CH_3)_2$	
69	F	$C(CH_3)_2$	
70	H	$P(O)OH$	
71	NH_2	H	H
72	NH_2	$C(CH_3)_2$	
73	NH_2	$P(O)OH$	

A number of pyrimidine base nucleosides have also been subjected to NOE studies, the results of which are reproduced in Table XIX.[13, 17, 67, 66-69] In these compounds the interaction between H-6 and the ribose protons is used to gauge conformer population about the glycosyl bond. Unfortunately, in a number of these compounds the resonances of H-2' and H-3' are too close to allow independent saturation. In Table XIX several of the $f^6(3')$ NOE's are recorded but in a number of these compounds f_6(total) > 50%, suggesting that in fact saturation of the signal of H-3' also caused partial saturation of the H-2' absorption. For example, in **65** a good fit for the distribution given is obtained with the exception of $f_6(3')$ where $\Delta f = 8\%$. However, $\sum_i f_6(i)$ is 59%, suggesting that the value of $f_6(3')$ could in fact be too large. Another example is for **74** where CNDO/2 calculations[69] indicated that the syn

74

TABLE XIX

Nuclear Overhauser Effects for Pyrimidine Base Nucleosides

Compound:	64^a	64^a $C_6H_6/$	65^b	65^a	65^c	65^d	65^e	66^e	66^e	67^e	67^e
Solvent: pD:	D_2O —	DMSO —	DMSO —	D_2O —	Pyr^h —	Pyr^h —	DMSO/ acetone —	$D_2O/$ DMSO —	DMSO —	D_2O —	DMSO —
$f_6(1')$:	11	7	20	14	10	10	13	7	3	7	6
$f_6(2')$:	13	—	10	4	—	—	—	27	27	24	32
$f_6(3')$:	8	—	6	5	—	—	—	7	17	17	26
$f_6(2' + 3')$:	—	11	—	—	10	9	9	—	—	—	—
$f_6(5' + 5')$:	0	3	0	0	0	—	4	0	5	6	13
$f_6(5)$:	—	—	23	—	28	26	26	—	—	—	—
$\Sigma_i f_6(i)$:	—	—	59	—	48	45	52	41	52	54	77
Y_1°:	—	—	30	—	—	—	—	—	—	—	—
ΔY_1°:	—	—	80	—	—	—	—	—	—	—	—
Y_2°:	—	—	—	—	—	—	—	100	130	130	150
ΔY_2°:	—	—	—	—	—	—	—	140	100	100	90
P_{syn}:	—	—	1.0	—	—	—	—	0	0	0	0

[a] Hart and Davis[67]
[b] Schirmer et al.[64]
[c] Bell and Saunders[13]
[d] −50°; Bell and Saunders[13]
[e] Hart and Davis[68]
[f] Govill and Smith[66]
[g] Nanda et al.[69]
[h] Pyridine-d_5.

conformation is the more stable conformer by the order of 9 kcal/mole, whereas the NOE results were interpreted in terms of a significant population of the anti conformer, largely because a sizable $f_6(3')$ was observed. Here again, $\Sigma_i f_6(i)$ is 0.56 and the most probable source of error is in $f_6(3')$. Thus it is quite possible that the CNDO/2 calculations and the NOE results are compatible. The results for the halogenated uridine derivatives must be treated with caution since in a number of examples f_6(total) is much greater than 50%. As was the case for the purine base compounds, the conformation ranges from almost exclusively anti in **66** to almost exclusively syn for **73** with a large number exhibiting significant populations of both conformations.

The uridine derivatives **75** and **76** which are (R) and (S) stereoisomers at C-1″ of the tetrahydropyranyl ether ring exhibit markedly different NMR and chromatographic behavior and were studied in order to determine the absolute stereochemistry at C-1″.[13] From the NOE data for the stereoisomers recorded in Table XX, the major difference that appears to be critical for the assignment of the stereochemistry is the NOE, [H-1″] H-1′, which is 12% for **75** but is zero for **76**. The compounds were assigned on the basis of a consideration of the most energetically favored conformations about the C(2′)–O–C(1″) bonds for both isomers. The most favored conformation for

58ᵉ D₂O/ MSO	68ᵉ DMSO	69ᵉ D₂O	69ᵉ DMSO	70' D₂O	71ᵃ Pyr/ D₂O	71ᵃ DMSO	72ᵃ DMSO	72ᵃ DMSO/ D₂O	73' D₂O	74ᵍ D₂O	74ᵍ D₂O
—	—	—	—	—	—	—	—	—	—	7.1	11.6
6	7	7	0	—	6	9	18	19	30	22	16
32	24	22	20	2	—	—	10	10	2	19	11
30	14	20	16	2	—	—	4	3	0	12	8
—	7	12	8	—	18	19	—	—	—	—	—
16	7	12	8	—	14	4	0	0	0	1	0
84	52	61	44	—	—	—	—	—	—	56	39
—	—	—	—	—	—	—	—	—	—	—	—
40	140	150	150	—	—	—	—	—	—	—	—
40	100	120	120	—	—	—	—	—	—	—	—
0	0	0	0	1.0	—	—	—	—	1.0	—	—

the (S) stereoisomer places H-1″ midway between H-2′ and H-3′ as shown in Fig. 15 for 76. The conformation that places H-1″ between H-1′ and H-2′ causes severe steric interactions between H-2′ and pyranose protons whereas the conformer that places H-1″ on the underside of the ribose ring results in serious steric interaction between H-1″ and H-1′, H-4′, or O-3′. In the preferred conformation the H-1′–H-1″ distance is 3.5 Å, which would give rise

TABLE XX

NOE Values for 2′-Tetrahydropyranyluridine Isomers 75 and 76 in Pyridine-d_5

		NOE (%)	
Irradiated	Observed	75	76
H-1′	H-6	7	8
H-2′ + H-3′	H-6	10	17
H-5′ + H-5″	H-6	0	0
H-5	H-6	28	28
H-6	H-5	32	—
H-1″	H-1″	12	0

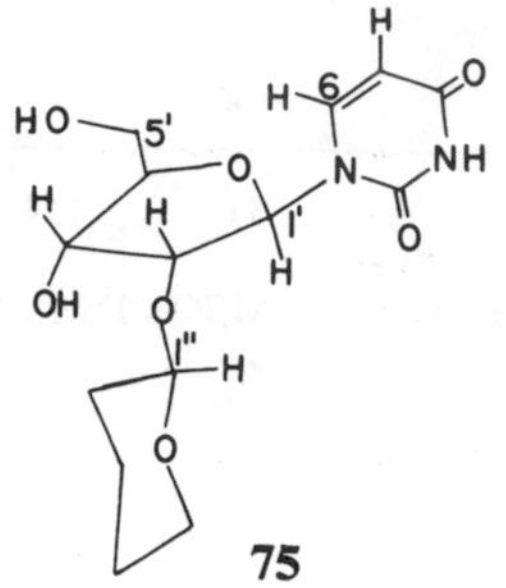

75

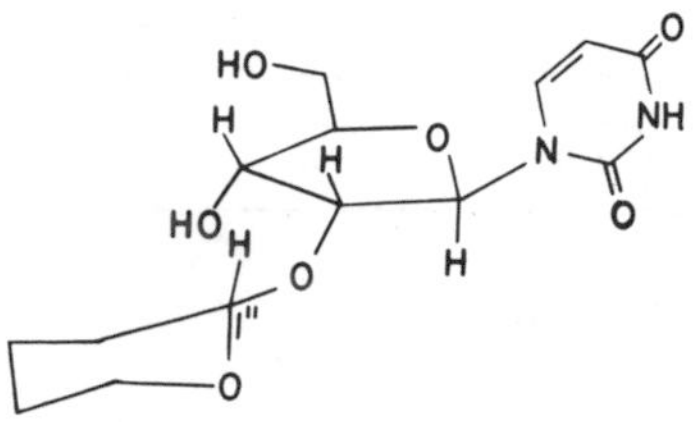

76

FIG. 15. The most favored conformation adopted by C-1″ (S)-2′-O-tetrahydro-pyranyluridine, **76**, and one of the conformations adopted by C-1″ (R)-2′-O-tetrahydropyranyluridine **75**.

to a negligible NOE. The only nonbonded interaction in the depicted conformation is between H-2′ and H-1″, which can be alleviated by forcing C-2′ into an exo position. This twisting would result in a larger value of $J_{1′, 2′}$ than in the parent compound. Thus the (S) stereochemistry is assigned to compound **76** because of the zero NOE between H-1′ and H-1″ and also because the coupling constant $J_{1′, 2′}$ for **76** is greater than in **64** and **65**. The (R) stereoisomer can exist in two conformations, namely one that is analogous to **76** where weak hydrogen bonding can occur between the C(1″)–O and the C(3′)–OH and also one in which H-1″ is midway between H-1′ and H-2′, as shown in Fig. 15. For the latter the distance between H-1′ and H-1″ is approximately 2.8 Å, which will give rise to a significant NOE. We can thus assign the (R) stereochemistry to **75**. For both isomers the NOE data that concerns the orientation about the glycosidic bond are best explained in terms of approximately equal populations for both orientations, with the larger NOE recorded between H-2′ and H-6 for **69** a consequence of having C-2′ in an exo position. Thus we can, with a fair degree of certainty, deduce **76** to be the (S) stereoisomer and **75** the (R), with rapid rotation about the glycosidic bond occurring in both isomers. An X-ray analysis[70] has subsequently confirmed these stereochemistries.

It is apparent from the above description that NOE's can play a significant role in the determination of conformation in systems undergoing internal motion. However, it is also apparent that the information so gleaned cannot always be interpreted in a quantitative fashion. The first problem is to obtain a meaningful value of ρ^*. Three methods have been described here, none of which is exact. The best method, which has recently become available, is to use ^{13}C T_1 data to determine τ_c and then determine R_A for the individual proton. However, care must be exercised as the proton relaxation data may not follow a single exponential. If possible, R_A should be determined when all other protons that have a dipolar interaction with that observed are saturated. Thus, for example, T_1 for H-8 of 55 would be measured while saturating both H-1′ and H-2′, whence the conditions of Eq. (23) would be approached. Having determined the experimental distances with the best accuracy possible, the problem remains to interpret these in terms of conformational position and population. It is inaccurate to predict that a nucleoside will have one or two conformations without giving any distribution between them. However, the problem with the Gaussian function is that another variable has been introduced, namely the width of the distribution. Even with all possible enhancements, we can often describe equally well the conformational picture in terms of one Gaussian function or two conformers of unequal population with very narrow distribution. Thus, the values of Y and the relative conformer populations quoted must be considered as approximations. We can, however, use the NOE data to determine which is the dominant conformation and approximately to what extent.

4. The theory for the region where $k \geq \tau_c^{-1}$ has not been fully developed. The graph for Fig. 2 was obtained by considering a CH_3 group as a net dipole, i.e., averaging the position of the protons as compared to averaging the NOE's as in point 2 or averaging the internuclear distances as in point 3, above. The NOE values, [H-2] H-7 = 8% and [H-7] H-2 = 9%, for 77 give an internuclear distance of 2.92 Å (Fig. 1) which corresponds to the distance X–H-2, where X is the intersection of the axis of rotation and the plane containing the three protons. In this compound $k > \tau_c^{-1}$ and thus

77

the CH_3 group can be considered as a net dipole. However, as was seen above, for the interaction between CH_3 groups and protons in substituted cyclobutanes care must be used when interpreting CH_3 group–proton interactions in terms of Fig. 2.

The effect of internal rotation of a CH_3 group on the observed NOE has recently been illustrated by Rowan et al.[70a] who studied cis- and trans-crotonaldehyde. The absolute internuclear distances for all proton–proton interactions were obtained from the NOE measurements, and the ^{13}C and ^1H nuclear spin relaxation data given in Table XXI. The effect of the methyl group on an adjacent spin was analyzed in terms of two correlation times, namely that of the molecule itself as well as that for internal rotation about the C—C bond. The methyl group was assumed to jump between three potential wells separated by 120°, with the preferred conformation having one of the CH_3 group protons eclipsing the double bond. The values of ρ_{ij} were calculated by substituting the relative NOE and ^1H relaxation time data in Eq. (10). From the ^{13}C relaxation times, the effective correlation time for molecular motion and the correlation time for internal rotation of the CH_3 group were calculated. The absolute values of the respective internuclear distances could thus be determined using Eq. (4) to give the values in Table XXI. In the case of the CH_3 group–proton interactions, an effective internuclear separation was obtained. In Table XXI, the experimentally determined values are compared with (a) distances obtained from either a microwave or theoretically optimized X-ray structure, (b) distances measured from the intersection of the threefold axis of rotation with the plane containing the three protons, and (c) distances estimated from the experimental NOE values using Figs. 1 or 2. In the case of the CH_3 group–proton interaction for (a) above, the theoretical value could not be obtained directly and hence it was calculated using the internuclear separation in the preferred conformation together with the experimentally determined correlation times. As can be observed, the experimental and theoretical values are in close agreement for protons on the rigid portion of the molecule when the protons are separated by less than 3.5 Å. Longer distances correspond to ρ_{ij}'s which are not at all well determined. The internuclear distance involving a methyl proton is reasonably well predicted (distances less than 3.5 Å) by the "centroid" model although the other method is more precise. However, internuclear distances cannot be obtained directly when using the latter method. Thus it appears that a combination of NOE and T_1 data can be used for CH_3 group–proton interactions in order to determine an effective internuclear separation and that this distance can be approximated as being the distance between the proton in question and the intersection of the threefold axis of rotation and the plane containing the three protons of the CH_3 group.

5. A number of studies have been performed in which intermolecular exchange rates have been determined by NOE methods. An example that

TABLE XXI

NOE's and Relaxation Times for *cis*- and *trans*-Crotonaldehyde[71]

ij	$f_j(1)$	$T_1(C_i)$	$T_1(H_i)$	Exp[a]	Th[b]	Th[c]	Exp[d]
am	4.4	33.0	78.2	3.16	3.12	3.12	3.3
ar	0.8	—	—	3.54	3.82	3.82	4.4
ax	4.7	—	—	—	—	—	—
ma	1.5	27.2	79.0	3.54	3.12	3.12	4.0
mr	23.3	—	—	2.37	2.38	2.39	2.5
mx	1.5	—	—	—	—	—	—
ra	1.3	29.2	60.4	3.38	3.82	3.82	4.1
rm	32.8	—	—	2.33	2.38	2.38	2.35
rx	1.5	—	—	—	—	—	—
xa	32.6	46.3	26.9	2.97	2.65	2.45	2.6
xm	3.8	—	—	3.32	4.12	3.8	3.7
xr	—	—	—	2.91	2.89	2.7	2.9

ij	$f_j(1)$	$T_1(C_i)$	$T_1(H_i)$	Exp[a]	Th[b]	Th[c]	Exp[d]
am	2.4	29.7	64.0	3.66	3.13	3.13	3.7
ar	17.4	—	—	2.37	2.31	2.31	2.65
ax	−2.1	—	—	—	—	—	—
ma	2.3	28.7	91.4	3.66	3.13	3.13	3.8
mr	−2.0	—	—	—	—	—	—
mx	3.8	—	—	—	—	—	—
ra	25.7	22.5	42.1	2.37	2.31	2.31	2.45
rm	4.0	—	—	3.32	3.08	3.08	3.4
rx	0.6	—	—	—	—	—	—
xa	0.1	36.2	23.5	3.96	4.7	4.5	6.8
xm	24.2	—	—	3.04	3.10	2.85	2.7
xr	11.6	—	—	3.04	2.90	2.7	3.1

[a] Experimentally determined using Eq. (4).
[b] From optimized geometry.
[c] Measured using "centroid" model.
[d] Experimental NOE results interpreted in terms of Fig. 1 or 2.

illustrates the basic effect of intermolecular exchange on the NOE is seen in the results of Feeney and Heinrich[71] who studied phenolic compounds dissolved in $CDCl_3$ containing a trace of water. The water protons undergo a slow reversible exchange with the phenolic protons. For 2,4,6-trichlorophenol, saturation of the H_2O signal caused a complete loss of the phenolic OH signal, whereas for 2-acetyl-3-methoxyphenol the same experiment resulted in only a 15% decrease in intensity of the phenolic OH absorption. For the former, the exchange rate is such that $k > R_A'$, whereas for the latter the exchange is very slow and $k < R_A'$. Fung and Stolow[72] studied the effect of exchange on the system diphenylmethanol, **78**, and *t*-butanol in CS_2. Saturation of the OH proton resonance of **78** resulted in decoupling of the

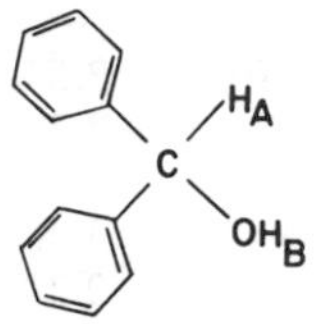

78

CH proton and complete loss of signal for the butanol OH. Saturation of the butanol OH proton caused complete loss of the resonance of the OH proton of **78** but did not decouple the CH proton. This is because the power necessary for saturation is less than that required for decoupling. In Table XXII[73] the relative intensity of the CH proton when saturating the butanol OH as a function of temperature is given. As can be seen, at 25° the dominant relaxation mechanism is the dipolar interaction whereas at $-10°$ when the CH and OH resonances are coincident, a significant contribution to the relaxation process is made by the exchange-modulated scalar coupling [Eq. (6)], as a significant decrease in intensity is observed. An excellent example of the use of transient methods for the determination of the dynamics of a system undergoing exchange can be found in the publications of Forsén and Hoffman.[74-76]

TABLE XXII

NOE Results for **77** as a Function of Temperature

Temperature (°C)	Relative intensity of H_A (%)	$\Delta\delta/J_{AB}$
25	100	38
−4	84	3.6
−7	76	2.1
−10	56	0.0
−12	74	1.8

References

[1] I. Solomon and N. Bloembergen, *J. Chem. Phys.* **25**, 261 (1956).

[2] F. A. L. Anet and A. J. R. Bourn, *J. Amer. Chem. Soc.* **87**, 5250 (1965).

[3] J. H. Noggle and R. E. Schirmer, "The Nuclear Overhauser Effect: Chemical Applications." Academic Press, New York, 1971.

[4] R. A. Bell and J. K. Saunders, *Top. Stereochem.* **7**, 1 (1972).

[5] References cited in Noggle and Schirmer[3] and Bell and Saunders.[4]

[6] R. E. Schirmer, J. H. Noggle, J. P. Davis, and P. A. Hart, *J. Amer. Chem. Soc.* **92**, 3266 (1970).

[7] R. A. Bell and J. K. Saunders, *Can. J. Chem.* **48**, 1114 (1970).

[8] F. Bloch, *Phys. Rev.* **69**, 460 (1946); F. Bloch and R. K. Wangness, *ibid.* **89**, 728 (1953).

[9] I. Solomon, *Phys. Rev.* **99**, 559 (1955).

[10] A. Abragam, "The Principles of Magnetic Resonance." Oxford Univ. Press, London and New York, 1961.

[11] A. J. Jones, D. M Grant, and K. F. Kuhlmann, *J. Amer. Chem. Soc.* **91**, 5013 (1969).

[12] J. W. ApSimon, W. G. Craig, A. Demayo, and A. A. Raffler, *Can. J. Chem.* **46**, 809 (1968); J. W. Apsimon, W. G. Craig, P. V. Demarco, D. W. Mathieson, L. Saunders, and M. Whalley, *Tetrahedron* **23**, 2357 (1920).

[13] R. A. Bell and J. K. Saunders, unpublished results; J. K. Saunders, Ph.D. Thesis, McMaster University, Hamilton, Canada, 1970.

[14] J. Horner, A. R. Dudley, and W. R. McWhinnie, *J. Chem. Soc., Chem. Commun.* p. 893 (1973).

[15] R. R. Ernst, and W. A. Anderson, *Rev. Sci. Instrum.* **37**, 93 (1966).

[16] G. N. LaMar, *J. Amer. Chem. Soc.* **93**, 1040 (1971); R. A. Freeman, K. G. Pachler, and G. N. LaMar, *J. Amer. Chem. Phys.* **55**, 4586 (1971); S. Barza and N. Engstrom, *J. Amer. Chem. Soc.* **94**, 1762 (1972); O. A. Gansow, A. R. Burke, and W. D. Vernon, *ibid.* p. 2550.

[17] F. Bloch, *Phys. Rev.* **102**, 104 (1956); B. D. Nageswaro Rao, *Advan. Magn. Resonance* **4**, 271 (1970).

[18] R. A. Bell and J. K. Saunders, *Can. J. Chem.* **46**, 3421 (1968).

[19] J. K. Saunders, R. A. Bell, C.-Y. Chen, D. B. MacLean, and R. H. F. Manske, *Can. J. Chem.* **46**, 2876 (1968).

[20] R. R. Fraser and F. J. Schuber, *Can. J. Chem.* **48**, 633 (1970).

[21] M. C. Woods, I. Miura, Y. Nakadaira, A. Terahara, M. Maruyama, and K. Nakaniski, *Tetrahedron Lett.* p. 321 (1967), and preceding 4 papers.

[22] R. T. Brown, F. Heatley, and D. Moorcroft, *J. Chem. Soc., Chem. Commun.* p. 459 (1973).

[23] G. Caron, J. Lessard, H. Beierbeck, and J. K. Saunders, in preparation.

[24] P. Yates and D. J. MacGregor, *Can. J. Chem.* **51**, 1267 (1973).

[25] J. W. ApSimon, H. Beierbeck, and J. K. Saunders, *Can. J. Chem.* **53**, 338 (1975).

[26] A. W. Brinkman, M. Gordon, R. G. Harvey, P. W. Rabideau, J. B. Stothers, and A. L. Ternay, *J. Amer. Chem. Soc.* **91**, 5912 (1972).

[27] R. A. Bell and J. K. Saunders, *Pap., 52nd CIC Conf.* 1969. Paper No. 122, p. 23 (1969).

[28] M. Gordon, W. C. Howell, C. H. Jackson, and J. B. Stothers, *Can. J. Chem.* **49**, 143 (1971).

[29] K. Tori, M. Ohtsuru, I. Horibe, and K. Takeda, *Chem. Commun.* p. 943 (1968).

[30] R. Toubiana and M.-J. Toubiana, *Tetrahedron Lett.* p. 1753 (1974).

[31] K. Takeda, K. Tori, I. Horibe, M. Ohtsuru, and H. Minato, *J. Chem. Soc.* p. 2697 (1970).

[32] N. S. Bacca and N. H. Fischer, *Chem. Commun.* p. 68 (1969).

[33] H. Koyama, M. Shiro, T. Sato, Y. Tsukuda, H. Itazaki, and W. Nagata, *Chem. Commun.* p. 812 (1967).

[34] K. Tori, *Lect., Int. Cong. Steroid. Horm., 3rd,* 1970 p. 205 (1971).

[35] J. W. ApSimon, P. Baker, J. Buccini, J. W. Hooper, and S. Macaulay, *Can. J. Chem.* **50**, 1944 (1972).

[36] J. W. ApSimon and J. K. Saunders, unpublished results.

[37] R. D. G. Cooper, P. V. Demarco, J. C. Cheng, and N. D. Jones, *J. Amer. Chem. Soc.* **91**, 1408 (1969); R. A. Archer and P. V. Demarco, *ibid.* p. 1530.

[38] D. Crowfoot, C. W. Bunn, B. W. Rodgers-Low, and A. Turner-Jones, "Chemistry of Penicillins" p. 310. Princeton Univ. Press, Princeton, New Jersey, 1949.

[39] R. A. Bell, J. W. Easton, M. B. Gravestock, J. K. Saunders, and V. Taguchi, in preparation.

[40] R. A. Hoffman and S. Forsén, *J. Chem. Phys.* **44**, 2049 (1966).

[41] R. Kaiser, *J. Chem. Phys.* **42**, 1838 (1965).

[42] T. Fukumi, Y. Arata, and S. Fujiwara, *J. Mol. Spectrosc.* **27**, 443 (1968).

[43] K. Kuhlmann and D. M. Grant, *J. Chem. Phys.* **55**, 2988 (1971).

[44] J. R. Lyerla and D. M. Grant, *J. Phys. Chem.* **76**, 3213 (1972); T. D. Alger, D. M. Grant, and R. K. Harris, *ibid.* p. 281; J. R. Lyerla and D. M. Grant, *MTP Int. Rev. Sci., Inorg. Chem. Ser.* 1 4, 155 (1972).

[45] A. J. Jones, D. M. Grant, and K. F. Kuhlmann, *J. Amer. Chem. Soc.* **91**, 5013 (1969).

[46] F. Wehrli, *Advan. Mol. Relaxation Processes* 6, 555 (1974).

[47] A. Allerhand and E. Oldfield, *Biochemistry* 12, 3428 (1973).

[48] D. Doddrell, V. Glushko, and A. Allerhand, *J. Chem. Phys.* **56**, 3683 (1972).

[49] J. C. W. Chien and W. B. Wise, *Biochemistry* 12, 3418 (1973).

[50] D. P. Miller, B. Ternai, and G. E. Maciel, *J. Amer. Chem. Soc.* **95**, 1–36 (1973).

[51] R. L. Lichter and J. D. Roberts, *J. Amer. Chem. Soc.* **94**, 2495 (1972); P. A. Cooper, R. L. Lichter, and J. D. Roberts, *ibid.* **95**, 3724 (1973).

[51a] Recently, T. K. Leipert and J. H. Noggle, *J. Amer. Chem. Soc.* **97**, 269 (1975), have demonstrated that exchange modulated scalar relaxation is not a viable mechanism for spin lattice relaxation. They suggest that the other mechanism is the spin rotation interaction which increases with increasing pH. However, no experimental proof was presented.

[52] R. L. Scholl, G. E. Maciel, and W. K. Musker, *J. Amer. Chem. Soc.* **94**, 6376 (1972).

[53] G. C. Levy, *J. Amer. Chem. Soc.* **94**, 4973 (1972); G. C. Levy, J. D. Cargioli, P. C. Juliano, and T. D. Mitchell, *ibid.* **95**, 3445 (1973).

[54] R. Martino and J. K. Saunders, *Can. J. Chem.* (submitted for publication).

[55] C. Benezra and J. K. Saunders, unpublished results.

[56] R. J. Abraham, D. F. Wileman, G. R. Bedford, and D. Greatbanks, *J. Chem. Soc., Perkin Trans.* 2 p. 1733 (1972).

[57] R. A. Bell and J. K. Saunders, *J. Chem. Soc., D,* p. 1078 (1970).

[58] I. C. Calder, P. J. Garrett, and F. Sondheimer, *Chem. Commun.* p. 41 (1967).

[59] G. Binsch, *Top. Stereochem.* 3, 97 (1968).

[60] F. A. L. Anet and A. J. R. Bourn, *J. Amer. Chem. Soc.* **89**, 760 (1967).

[61] R. A. Bell and J. K. Saunders, *Can. J. Chem.* **48**, 512 (1970).

[62] H. L. Schlafer and W. Schaffernicht, *Angew. Chem.* **72**, 618 (1960).

[63] F. A. L. Anet and M. Ahmed, *J. Amer. Chem. Soc.* **86**, 119 (1964); R. E. Klinck, D. H. Marr, and J. B. Stothers, *Chem. Commun.* p. 409 (1967).

[64] R. E. Schirmer, J. P. Davis, J. H. Noggle, and P. A. Hart, *J. Amer. Chem. Soc.* **94**, 2561 (1972).

[65] T.-D. Son, W. Guschlbauer, and M. Guéron, *J. Amer. Chem. Soc.* **94**, 7903 (1972).

[66] G. Govil and I. C. P. Smith, personal communication.

[67] J. P. Davis, *Tetrahedron* **28**, 1155 (1972); P. A. Hart and J. P. Davis, *J. Amer. Chem. Soc.* **93**, 753 (1971).

[68] P. A. Hart and J. P. Davis, *J. Amer. Chem. Soc.* **94**, 2572 (1972).

[69] R. K. Nanda, R. Tewari, G. Govil, and I. C. P. Smith, *Can. J. Chem.* **52**, 371 (1974).

[70] P. Stottart, I. D. Brown, and T. Nielson, *Acta Crystallogr., Sect. B* **18**, 6627 (1973).

[70a] R. Rowan, III, J. A. McCammon, and B. D. Sykes, *J. Amer. Chem. Soc.* **96**, 4773 (1974).

[71] J. Feeney and A. Heinrich, *Chem. Commun.* p. 295 (1966).

[72] B. M. Fung and R. D. Stolow, *Chem. Commun.* p. 257 (1967).

[73] B. M. Fung, *J. Chem. Phys.* **47**, 1409 (1967).

[74] S. Forsén and R. A. Hoffman, *J. Chem. Phys.* **40**, 1189 (1964).

[75] S. Forsén and R. A. Hoffman, *Acta Chem. Scand.* **17**, 1787 (1963).

[76] S. Forsén and R. A. Hoffman, *J. Chem. Phys.* **39**, 2892 (1963).

Molecular Structures by NMR in Liquid Crystals

6

L. LUNAZZI

I. INTRODUCTION

High-resolution nuclear magnetic resonance (NMR) spectroscopy is today a technique largely employed for the identification of molecular configurations and for the study of molecular conformations.

Essentially two kinds of parameters are obtained from the analysis of the NMR spectra: the chemical shifts (ν_i) and the spin–spin (or indirect) coupling constants (J_{ij}).[1]

These quantities should be properly described by a second rank tensor, which, for spin $\frac{1}{2}$, may have up to nine components. However, the rapid molecular tumbling that occurs in isotropic solutions makes the NMR spectrum dependent only upon the average values of these parameters, each value being proportional to the trace of the tensor itself.

The rapid molecular motions occurring in isotropic solutions are also responsible for averaging another parameter: the dipolar (or direct) coupling constant (D_{ij}). This quantity is described by a symmetric tensor with a maximum of five components; since the tensor is traceless, the average value of D_{ij} is zero and therefore, whereas the average chemical shift and spin–spin coupling still affect the NMR spectrum in an isotropic solvent, the dipolar coupling does not.

The NMR spectrum from isotropic media is thus much simpler than expected, and this is very convenient for analytical purposes; however, the information that could be extracted from the dipolar couplings (which depends on the internuclear distances) is lost.

NMR spectra in solids retain this information. Because of the absence of any averaged motion, however, both the direct couplings between the nuclei in the same molecule (intramolecular) and those between nuclei of neighboring molecules (intermolecular) do affect the spectra. This makes the number of observed transitions exceedingly large and contributes (among other effects) to the broadening of the lines in the spectra of solids. Thus, even when D_{ij} values are extracted from the spectra of the solid,[2] the accuracy is reduced by the contribution of both inter- and intramolecular dipolar couplings.

For this reason many efforts have been made to produce a situation intermediate between that of the liquid and the solid, so that only the intramolecular direct couplings could be detected without the complication arising from the intermolecular ones. To reach this goal, it is required that the molecules have a preferred orientation, in order to avoid the complete averaging that occurs in the isotropic liquids, but still retain some molecular motions to wipe out the intermolecular interactions.

Attempts to obtain NMR spectra of molecules having such partial orientation by means of applied electric fields, absorption of the molecules in stretched polymers, or inclusion in organic and inorganic lattices, although not completely unsuccessful, had so many practical limitations that there was no possibility of extended utilization. The application of an external electric field to molecules with high electric dipole moments during their observation by NMR was first performed by Buckingham and McLauchlan.[3] Their conclusions were criticized by other authors,[4] but later Hilbers and MacLean showed that it was possible to obtain NMR spectra of partially oriented nitrobenzenes in electric fields.[5]

The extent of the orientation produced by this technique, however, is very small. Also, the first experiments on molecules included in polymers which were subsequently stretched and on molecules hosted in organic crystalline lattices (e.g., urea crystallized from low molecular weight solvents that may develop cavity structures accommodating the urea molecules)[6] or in inorganic compounds (e.g., water in naturally occurring zeolites)[7] were not followed by general applications.

On the other hand, NMR spectra of molecules dissolved in liquid crystals were shown by Saupe and Englert[8] to meet the requirements of a large partial orientation and of sufficient molecular motion to give well-resolved NMR spectra that did depend upon intramolecular dipolar couplings. Accordingly, these dipolar couplings may be obtained, and the geometrical information contained therein may be extracted.

After these first experiments a large number of works have confirmed that this is the best available technique for obtaining information on molecular structures through NMR spectroscopy. We shall refer to it as the LXNMR method. Unless otherwise stated, the nucleus referred to will be the proton.

Although the method bears many more limitations than other structural techniques, it has the distinct advantage (from a chemical point of view) of allowing the determination of configurations and conformations of molecules in solutions, whereas the methods of electron diffraction (ED) and microwave spectroscopy (MW) are applicable mainly to gases, and neutron and X-ray diffraction to solids.

Because of the interest aroused from this method, a relatively large number of reviews appeared on the subject.[9-22] Some of them, however interesting, are now out of date because of the rapid development of the research, whereas others are mainly devoted to specialized audiences and are therefore not suitable for those approaching the subject for the first time. The purpose of this chapter is to give a simple presentation of this technique to those scientists who are interested in structural determinations but are not themselves specialists in NMR spectroscopy. The most recent results and developments have been included (literature has been surveyed for all of 1974) whereas only a minimum of theoretical and mathematical aspects has been introduced.*

Only information regarding molecular structures will be considered, whereas studies on the absolute sign of J couplings, orientation parameters, anisotropies of indirect couplings and chemical shifts, and determinations of nuclear quadrupole couplings will be only occasionally mentioned, although NMR in liquid crystals has proved to be a powerful tool for these kinds of investigations also.[10, 13, 20, 22]

II. THE LIQUID CRYSTALS

Whereas most organic substances melt sharply to give a clear liquid appearance, a few substances have intermediate phases between the solid and liquid. In these derivatives the solid melts to give a cloudy, viscous liquidlike state that becomes perfectly clear only at higher temperatures. This peculiar

*Some work that has appeared after the completion of the chapter is discussed in the Appendix, and some recent references are given.

state (usually referred to as the mesophase) shares some of the properties of the liquid (after all, it looks like a liquid) and some of those of the solid (particularly as regards the optical properties).[11] Liquid crystals differ from isotropic disordered liquids in that their molecules have some degree of order. However, this order is continuously destroyed and recreated in contrast to the situation in solids.

There are three types of liquid crystalline mesophase: smectic, nematic, and cholesteric. In this chapter the properties of these mesophases will be

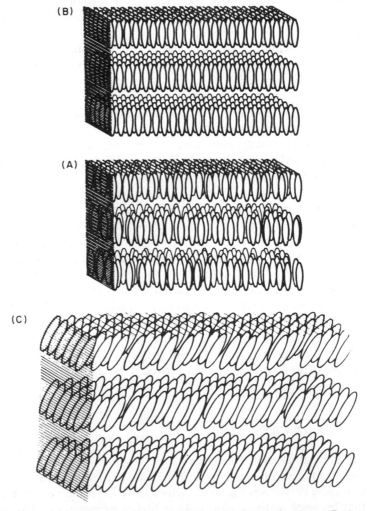

FIG. 1. Schematic picture of the local structure of smectic phases B, A, and C [from *Oesterr. Chem. Ztg.* **4**, 115 (1967) and courtesy of Dr. G. R. Luckhurst, University of Southampton, U.K.]

summarized only briefly, and only the features relevant to NMR spectroscopy will be mentioned.

A clear and concise description of these properties may be found in the articles of Saupe[11] and Luckhurst;[10] for a broader understanding of the peculiarities of liquid crystals the reader is referred to the reviews of Brown, Doane, and Neff,[23] Steinstrasser and Pohl,[23a] Kelker,[23b] as well as to the books and articles of Gray[23c,d] and de Gennes.[23e]

The molecules in the smectic phases are ordered with their longest axes all pointing in the same direction and they are also arranged in layers parallel to each other. (See Fig. 1.) Some liquid crystals may have more than one smectic phase (five and possibly even eight such phases have been detected[23a]): the three identified in Fig. 1 with the letters B, A, and C are rather often encountered and have distinct transition points between states, with slightly different degrees of order. The nematic phases have a smaller degree of order than the smectic since the molecules are aligned only with their longest axes parallel to each other and the ordered layers are no longer present. (See Fig. 2.) Some compounds have both smectic and nematic mesophases and, obviously, the latter are always obtained at a higher temperature than the former since the degree of order is reduced on going toward the completely disordered liquid state. In Table I[11,24] examples of molecules having smectic and nematic mesophases are collected.

There is also a third class of mesophases, which occurs in optically active substances. It is called cholesteric since it has been observed in derivatives of cholesterol (mostly esters). It may be regarded as a modification of the

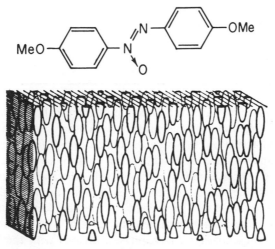

FIG. 2. Schematic picture of the local molecular distribution of a liquid crystalline substance in a nematic phase [from *Oesterr. Chem. Ztg.* **4**, 115 (1967); reprinted with permission of the copyright owner].

TABLE I

Example of Liquid Crystalline Substances Having One or More Mesophases

C$_4$H$_9$⟨⟩—N=CH—⟨⟩—CH=N—⟨⟩—C$_4$H$_9$ Terephthalbis(butylaniline) or *p*-bis(*p-n*-butylphenyliminomethyl)benzene[24]

Solid to smectic B	113°
Smectic B to smectic C	144.1°
Smectic C to smectic A	172.5°
Smectic A to nematic	199.6°
Nematic to liquid	236.5°

CH$_3$O—⟨⟩—CH=N—⟨⟩—C=CH—COOC$_2$H$_5$ Ethyl [(methoxybenzylidene)-amino]cinnamate[24]

Solid to smectic B	80°
Smectic B to smectic A	91°
Smectic A to nematic	117.6°
Nematic to liquid	138°

C$_7$H$_{15}$O—⟨⟩—N=N—⟨⟩—OC$_7$H$_{15}$ (↓O)   4,4′-bis(*n*-heptyloxy)azoxybenzene[24]

Solid to smectic C	73.1°
Smectic C to nematic	95.1°
Nematic to liquid	123.9°

C$_{12}$H$_{25}$O—⟨⟩—N=N—⟨⟩—OC$_{12}$H$_{25}$ (↓O)   4,4′-bis(*n*-dodecycloxy)azoxybenzene[11]

Solid to smectic C	82°
Smectic C to liquid	123°

CH$_3$O—⟨⟩—N=N—⟨⟩—OCH$_3$ (↓O)   4,4′-dimethoxyazoxybenzene[11]

Solid to nematic	118°
Nematic to liquid	135°

nematic mesophase in which the direction of the local (not the macroscopic) orientation is twisted along a helical path with a definite pitch: this ordering gives an optical activity to the cholesteric mesophase. (See Fig. 3.)

Since substances that exist as cholesteric mesophases may have opposite rotations of the helices, a mixture of two such substances (there is a racemic mesophase) has the same properties as a nematic mesophase.[25, 26] For instance, a mixture (1.9 to 1 by weight) of the *l*-cholesteryl chloride and *d*-cholesteryl myristate behaves as a nematic solvent.[26] Furthermore, the pitch of the helix of the cholesteric mesophase increases with the application of increasing magnetic fields, and at some critical value the pitch becomes infinite, as the helix is unwound.[27, 28] At this point the cholesteric is changed into a nematic mesophase, although this does not correspond to a

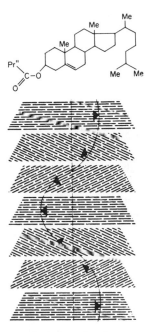

FIG. 3. Schematic picture of the molecular arrangement of a cholesteric mesophase [from *Oesterr. Chem. Ztg.* **4**, 115 (1967); reprinted with permission of the copyright owner].

phase transition since the free energy does not have a point of discontinuity at the critical field.[28]

These three mesophases (smectic, nematic, cholesteric) are usually defined as thermotropic since passages from the isotropic to the partially oriented situation occur when the temperature is changed. There is, however, another kind of mesophase which is closely related to the nematic thermotropic mesophase.[29] It is called lyotropic since it is formed by "soaps"—that is, by solutions of long-chained sulfonic acids.[30, 30a]

The transition from the isotropic to the lyotropic mesophase is made by changing the composition of the solution. The typical lyotropic nematic mesophase proposed by Lawson and Flautt[30] is obtained by mixing (by weight) sodium decyl sulfate (36%), sodium sulfate (7%), decanol (7%), and water (50%). Recently a new lyotropic mesophase has been discovered by Long and Goldstein,[31, 31a] its composition being (in wt%): potassium chloride (4.0), potassium laurate (30.0) water or deuterium oxide (60.0), and decanol (6.0). The mixture is nematic in a broad range of temperatures (10°–60°).

The orientation of the molecules within the mesophases is characterized by a vector that usually points in the direction of the longest molecular axis;

it is usually referred to as an optical axis[11] or director.[20] The direction of this vector however is not the same over a large area, but changes randomly from one region of the sample to another. We may thus consider that there is a local order (extending over many thousands of molecules) but not a macroscopic order, since the direction of the optical axis is not unique for the whole sample.

In nematic systems, however, it is possible to obtain a macroscopic order (there is a uniform orientation of the molecules of the sample in a single direction) with very simple methods. It is sufficient, for instance, to arrange each sample in a thin film between two walls; the molecules will be thus oriented through boundary effects by the walls themselves. Orientation of layers in contact with surfaces may be also obtained by rubbing the surface in a given direction. Finally, the homogeneous orientation of bulky samples may be obtained by applying magnetic fields of few thousands Gauss.[32]

This makes ESR and NMR spectroscopy, which already require magnetic fields for the detection of the transitions, particularly suited to investigations concerning liquid crystals.

Two theories have been suggested to explain the physical properties of the mesophases: the swarm theory introduced by Bose[33] and developed by Ornstein and Kast;[34] and the continuum, or distortion theory, proposed by Zocher[35] and Oseen[36] and later presented in a comprehensive form by Franck.[37]

The first theory considers the liquid crystal molecules as grouped together (in numbers of about 10^5) in swarms; within each of these the molecules are parallel to each other. The swarms have relatively short lifetimes and are dynamically destroyed and reformed. The direction of the orientation of each swarm is random and independent of those of the neighbors. Only when external forces (such as magnetic fields) are applied do they align in a uniform way. On the other hand, the continuum theory, which is now more widely used to describe liquid crystal properties, assumes that in every point there is a definite direction of molecular orientation which varies with continuity. Whereas the swarm theory predicts negligible interactions between the neighboring swarms, the continuum theory requires the existence of long-range interactions. It has been shown, however, that the two theories are equivalent when applied to the interpretation of NMR spectra.[38]

Not only are the nematic and lyotropic mesophases oriented but molecules dissolved therein take preferential directions as well. The motions of these dissolved molecules are therefore anisotropic and the intramolecular dipolar couplings may be detected by NMR spectroscopy whereas the intermolecular couplings are cancelled. Most of the work on NMR spectra of oriented molecules has been carried out using nematic (thermotropic) or lyotropic liquid crystals as solvents. Pure cholesteric solvents do not seem to allow high-resolution NMR spectra of oriented solutes to be obtained,[39] although mixtures of *d*- and *l*-cholesteric liquid crystals do orient the solute.[26] High-resolution NMR spectra using smectic mesophases as solvents have been reported only recently.[24, 40]

III. EXPERIMENTAL METHODS

The NMR spectrum of the liquid crystal itself is very difficult to detect: usually it presents itself by a few broad bands (some are many hundreds of Hz wide). This is due to the presence of a large number of nonequivalent hydrogens, which are usually present in the long hydrocarbon chains of the liquid crystal molecules. When the solvent exists as an isotropic liquid the number of peaks is not extremely high because of the large number of degenerate transitions; on the other hand, when the solvent becomes nematic the dipolar couplings remove most of the degeneracy, thus making the number of lines so large that they are no longer resolved. The introduction of deuterium atoms, by reducing the number of protons, allows the observation of NMR spectra of nematic substances with much narrower lines.[41]

The lack of resolution in the spectra of the nematic solvents, although it represents an obstacle to NMR studies of the liquid crystal itself, is very useful for studies of molecules dissolved in the mesophase. Since the solute molecules normally have a much smaller number of hydrogens, hence a smaller number of spin transitions than the solvent, they yield well-resolved NMR spectra whose sharp lines clearly appear over the bumpy background of the nematic solvent spectrum. By contrast to the situation for isotropic NMR spectra, it is not necessary to use deuterated solvents for LXNMR.

The disappearance of the resolved spectrum of the solvent on going from the isotropic to the nematic liquid is also useful from a practical point of view, since it can be used for monitoring the nematic range. The introduction of solute molecules in fact changes the known transition temperature of the liquid into the mesophase; nevertheless one can easily identify the temperature of the transformation. The NMR spectrum of the solvent (which is obviously very intense in the liquid state) almost disappears when the nematic range is reached.

NMR spectra of molecules oriented in nematic solvents can be obtained without any modification of the commercially available instruments; however, a number of factors have to be considered in order to obtain well-resolved, reproducible, and well-calibrated NMR spectra of partially oriented molecules.

A. The Solvents

In the first experiments the solvents employed usually had rather high transition temperatures so that the spectra were obtained by heating the sample in the NMR probe. Since the molecular alignment depends on the temperature,[42] the positions of the spectral lines change with temperature fluctuations. Accordingly, very stable temperature controls are required (up to $\pm$ 0.1°) to obtain accurate measurements. Also, temperature gradients within the NMR tube must be minimized to ensure uniform orientation throughout the whole sample. Such temperature stability and homogeneity is more easily achieved near room temperature. Recently a number of solvents that are nematic at room temperature have become available, thus eliminating the need for heating the sample.

Better resolution and a greater accuracy in the measurement of the line positions are thus possible. Occasionally, however, high temperatures are needed to increase the solubility of the solute. Some of the solvents employed are reported in Table II[43-52] together with examples, one for each nematogen, of work carried out in that solvent.

EBBA, MBBA (Eastman Kodak), Phase IV, and Phase V (Merck) are commercially available solvents that either are nematic at room temperature or may become nematic when solute is added: they are thus among the currently most popular solvents. Often, however, two or three different nematogens are mixed together in order to match the nematic and probe temperatures. This practice, which is now widely used, was first employed by Spiesecke and Bellion-Jourdan.[53]

Owing to the large number of lines displayed in LXNMR spectra and to the linewidths, which are larger than in normal NMR spectra, the signal to noise ratio (S/N) is usually much poorer than in isotropic spectra. Accord-

TABLE II

Examples of Work Carried Out in Various Nematogens[a]

Nematogen	Nematic range	Molecule investigated
4,4'-bis(n-Hexyloxy)azoxybenzene	80°–129°	Benzene[43]
$C_6H_{13}O$—⟨ ⟩—N=N—⟨ ⟩—OC_6H_{13} (with O below N)		
4-n-Butyl-4'-methoxyazoxybenzene (Phase IV)	16°–76°	Thienothio-phene[44]
C_4H_9—⟨ ⟩—N=N—⟨ ⟩—OCH_3[b] (with O below N)		
4-n-Butyl-4'-ethoxyazoxybenzene[b] (35%) ⎫ 4-n-Butyl-4'-methoxyazoxybenzene[b] (65%) ⎭ (Phase V)	−5° to 75°	Tropone[45]

TABLE II (*Continued*)

Nematogen	Nematic range	Molecule investigated
N-(4-Ethoxybenzylidene)-4'-*n*-butylaniline (EBBA)	36°–80°	Norbornadiene[46]

C_2H_5O—⬡—$CH=N$—⬡—C_4H_9

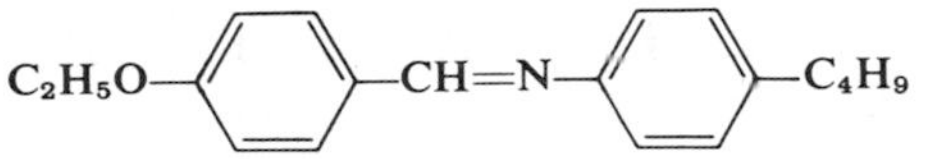		
N-(4-Methoxybenzylidene)-4'-*n*-butylaniline (MBBA)	20°–47°	Ethylene carbonate[47]

CH_3O—⬡—$CH=N$—⬡—C_4H_9

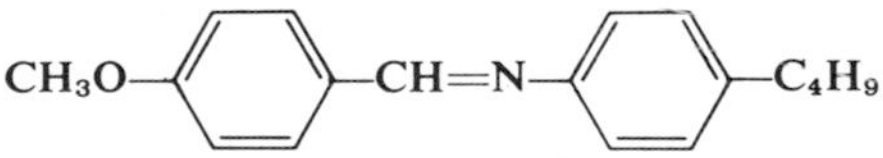		
n-Propyl *N*-(4-methoxybenzylidene)-4-amino-α-methyl-cinnamoate	54°–89°	*o*-Chloro-toluene[48]

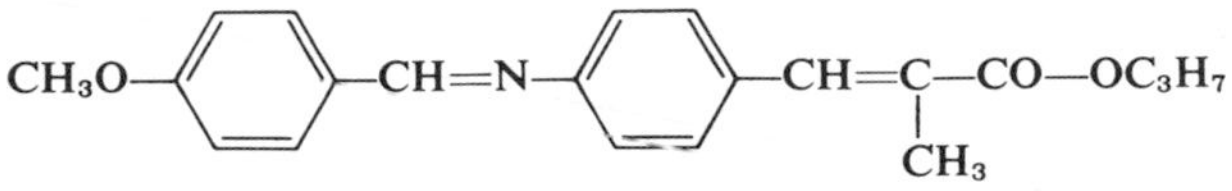

CH_3O—⬡—$CH=N$—⬡—$CH=C$—CO—OC_3H_7
               |
              CH_3

Butyl- 4-(4'-ethoxyphenoxycarbonyl)phenyl carbonate	56°–87°	*p*-Dinitroben-zene[49]

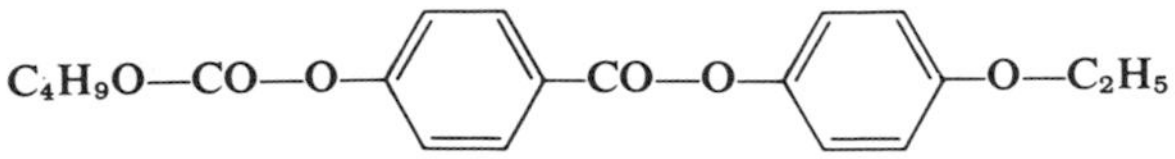

C_4H_9O—CO—O—⬡—CO—O—⬡—O—C_2H_5

Anisole- 4-azophenyl-*n*-capronate	66°–106°	*o*-Dichloroben-zene[50]

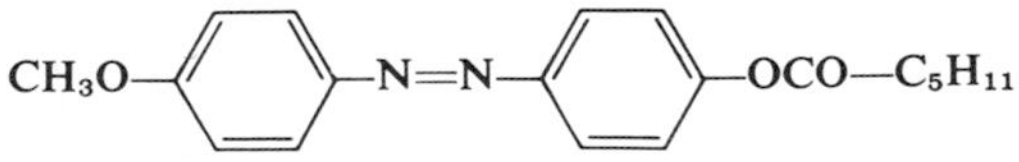

CH_3O—⬡—$N=N$—⬡—OCO—C_5H_{11}

4,4'-Bis(heptyloxybenzoyloxy)benzene	125°–206°	Bullvallene[51]

$C_7H_{15}O$—⬡—$\overset{\text{C}}{\underset{\text{O}}{\|}}$—$O$—⬡—$O$—$\overset{\text{C}}{\underset{\text{O}}{\|}}$—⬡—$OC_7H_{15}$

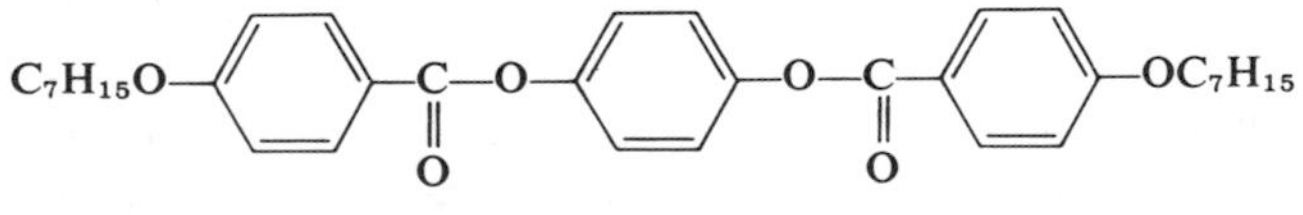		
Anisole- 4-azomethynephenyl-*n*-butanoate	50°–112°	2,5-Dimethyl-pyrazine[52]

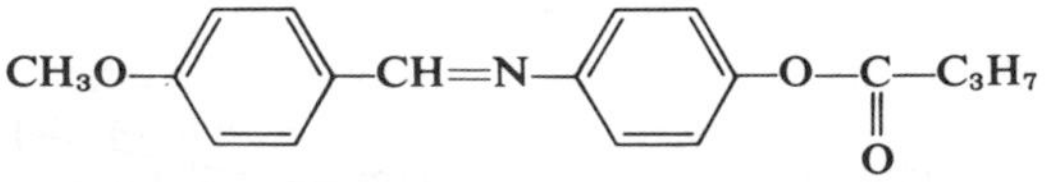

CH_3O—⬡—$CH=N$—⬡—O—$\overset{\text{C}}{\underset{\text{O}}{\|}}$—$C_3H_7$

---

[a] The name of each solvent is that used in the current literature and often is not in line with the recommended rules. The nematic range refers to that of the pure solvent: the range of the solution may therefore be much lower.

[b] The isomer with the oxygen bonded to the other nitrogen is also present.

ingly, rather concentrated solutions have to be employed: the range extends from a reported minimum of 3 mole%[54] to a maximum of 40 mole%.[55]

The melting point and the nematic–isotropic transition point are each depressed by the addition of solute to the liquid crystal so that if too much solute is added, the latter may become lower than the melting point, and the mesophase may disappear. The optimum conditions, which give an acceptable S/N ratio without destroying the nematic mesophase, thus have to be sought: the best compromises have usually been found in the range of 10–25 mole%. The advent of the time-averaging computer and, particularly, of the Fourier transform (FT) technique has reduced the need for high concentrations. However, since the solvent itself gives a background of broad lines the concentrations cannot be as low as the FT method would otherwise allow. This is because the solvent peaks are also accumulated, so that the solute spectra would appear with the intensities badly distorted due to the super-imposition of the broad solvent signals. A partial solution to this difficulty is the use of a delay between the end of the pulse and the beginning of the data acquisition[56]: in this way the broader signals may be filtered out, leaving only the sharper lines of the solute. Another technique is to take the difference spectrum between the solution and the solvent.

B. Spinning Effects

A further experimental point that has to be considered is the effect of spinning the sample.[57]

In conventional NMR spectroscopy the tube is spun about the y axis (perpendicular to the z axis of the magnetic field) in order to average and thereby reduce the field inhomogeneity experienced by the sample. Partially oriented molecules, however, have the optical axis aligned at a definite angle with respect to the magnetic field. On rotation of the sample, this angle increases,[58] and the outer lines move toward the center of the spectrum. Figure 4 shows the decrease in line separation for the proton triplet of aceto-nitrile oriented in Merck's Phase V as the rotation frequency (ω) increases from 0 to 16 revolutions per second.[55a] At a certain critical speed (usually greater than 20 rps) the spectrum becomes a complex powderlike spectrum. With these conditions it is not possible to interpret it as arising from simply

---

FIG. 4. 100 MHz FT NMR spectrum of CH_3CN oriented in a liquid crystal recorded at different spinning rates (ω). The out-of-scale signal is that of the external reference (benzene). The background line of the nematic solvent has been eliminated by computer subtraction. The overall separation of the triplet decreases from $\omega = 0$ rps to $\omega = 16$ rps, whereas the linewidth of the outer peaks increases. This is particularly evident if one compares the heights of these peaks with that of the central peak. (Courtesy of Dr. J. W. Emsley, University of Southampton.)

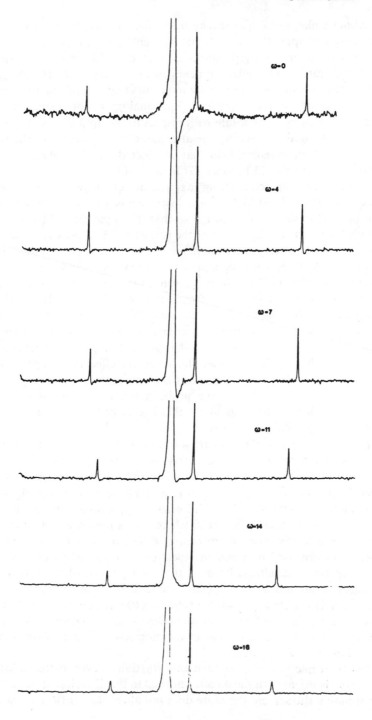

oriented molecules, as long as thermotropic nematic solutions are employed. If the sample is spun the speed must be kept less than the critical speed (3–10 rps have been employed) and must be thoroughly constant during the time of registration. This effect appears to be analogous to the effect of a change in the orientation of the solute molecules (e.g., by raising the temperature or increasing the concentration). This analogy is not correct, however, since the orientation of the solute with respect to the optical axis of the liquid crystal is unchanged;[58] it is only the alignment of the director of the solvent with respect to the magnetic field that is affected. At the critical speed the angle of alignment should become 45° instead of 0°.

However, when superconducting magnets are employed, such as those of the 220, 270, 300, and 360 MHz proton instruments, the z axis of the magnetic field is parallel to the axis of rotation so that the angle formed by the optical axis with the z axis is independent of the rotation.[10] In these cases the sample may be spun without very restricted requirements on the speed of the rotation.

Sometimes spinning the sample in conventional magnets seems to improve the spectral resolution,[59] much in the same manner as in normal isotropic solvents. The origin of such an effect is different, however, for in thermotropic nematic solutions it is due to the reduction, through averaging effects, of both thermal and concentration gradients within the sample. Indeed, the rotation of nematic solutions is on other grounds expected to increase and not reduce the linewidth.[58, 60] Often, however, the effects of temperature and concentration are dominant. But when the gradients are almost negligible, as may happen in nematics at the probe temperature, the outer lines do broaden at higher rotation speeds;[58] this is clearly visible in Fig. 4 if one compares $\omega = 16$ with $\omega = 0$ rps.[58a]

The high viscosity of the nematic solution probably affects the linewidth much more than the field inhomogeneity. As an example it may be recalled that for very similar molecules dissolved in the same nematogens at probe temperatures, almost equal linewidths (2.8 Hz) have been obtained, either in superconducting magnets (220 MHz) where sample spinning was used,[61] or in conventional magnets (100 MHz) where the sample was not spun.[62]

Often, particularly when temperatures different from the normal probe temperatures are used or when the spectrum is extremely wide, even very small temperature gradients broaden the outer pattern of the spectrum: in these cases superconducting magnets, with their merit of not requiring accurate control of the speed of sample rotation (100–150 rps are used), become useful for achieving uniform linewidths across the spectrum.[59] The proton line-widths of molecules dissolved in thermotropic nematic solvents may be kept within the 2–20 Hz range.[63]

When lyotropic systems are used the situation is even better since these liquid crystals orient with the director normal to the direction of the magnetic field: spinning the sample therefore does not affect the NMR spectra run in conventional magnets. Under these conditions resolutions equal to or better than 1 Hz have been reported.[64, 65] For this reason the accuracy of the spectral interpretation is usually higher than that obtained in systems using thermotropic mesophases, and matches that of high-resolution NMR in isotropic media. For instance, the spectral parameters of oriented benzene were determined with an accuracy (root mean square deviation) of 0.01 Hz.[31] The lyotropic mesophases however are less tolerant of large concentrations of solute molecules,[65] and only a few percent of product may be dissolved without destroying the mesophase[66-69] unless the solute is also soluble in water.[30, 64] Furthermore, it is more difficult to reach a uniform orientation throughout the sample than in thermotropic mesophases, and the samples have to be kept at controlled temperatures (often in the magnet itself) for rather long times (as much as one day).[65]

C. Methods of Recording

No additional apparatus is required for the LXNMR method, and every currently available instrument may be used without modification. Spectra may be recorded either in the field-[48] or in the frequency-swept[46] mode: external locking,[49] side-band calibration (either using the signal of the whole spectrum[70] or that of a reference signal[71]), as well as internal locking[72, 72b] have been employed for measuring the line positions. Heteronuclear locking (particularly when working in the FT mode) has proved most advantageous, and $CFCl_3$[73] or D_2O[72, 74] were successfully used as lock compounds by placement in the interannular space of two (5 and 10 mm inside diameter) coaxial tubes.

Tetramethylsilane may be used as an internal reference[75] as in isotropic solutions, but sometimes it also may be oriented by the mesophase, thus giving a triplet (due to the splitting of each hydrogen by the other two) with rather broad bands.[76] For this reason acetonitrile has been suggested[77] as a more suitable internal reference since it gives a much sharper triplet; furthermore, the three signals are better separated (1820 Hz in the quoted example) so that each of them may be conveniently used for locking. Sometimes the most intense spectral line may be used as the lock.[78] Usually however, the reference (TMS[46] or related derivatives[62]) is sealed in a capillary tube introduced into the sample containing the nematic solutions. In this way a strong, easily identified signal is provided without the complications that may arise from the orientation of the reference. Such external references may be used also in studies of the orientation of molecules since this property is affected by the introduction of even a small amount of internal reference.[57]

The spectral widths of oriented molecules usually cover a few thousand Hz, although a spectrum only 42 Hz wide has been reported.[79] Problems may thus arise because of a nonperfect linearity of the sweep over intervals that can reach even 18,000 Hz.[80] Therefore, particular care has to be taken in calibrating different regions of the same spectrum: this may be accomplished by measuring common lines present in two different overlapping sections that may be used to join the various portions of the whole spectrum in a uniform scale.[81, 82]

The advent of the FT technique has improved the accuracy of measurement since the use of a heteronuclear lock allows accurate calibration of widespread patterns, and the computerized monitoring of the line positions reduces the error. The uncertainty here is determined by the channel width employed. The short time required (even though a number of pulses are normally accumulated) almost eliminates the effects of temperature variations that may occur when the continuous wave method is employed. Even very complex and wide spectra may be obtained in few minutes.[80]

D. Decoupling Techniques

Techniques involving multiple resonance experiments can be also applied in LXNMR work. Diehl and Khetrapal have shown tickling to be effective when applied to an AB system (4,6-dichloropyrimidine).[13] The same technique was later extended to the more complex ABC spectrum of isoxazole[83] and it has been found helpful in the construction of the diagram of energy levels.

Heteronuclear spin tickling, involving irradiation of ^{14}N while protons are monitored in methylisocyanide, allowed the first experimental determination of the chemical shift anisotropy for nitrogen.[84] More recently the internuclear double resonance (INDOR) technique has also been employed to help the spectral analysis of partially oriented bicyclobutane[21] and chlorobenzene:[85] relative signs of direct with respect to indirect couplings were also obtained in the latter case. It also appears that in some liquid crystal solutions the INDOR experiment is better performed in the fast-sweep mode [analogous to the transitory selective irradiation (TSI) experiment][86] than in the customary slow-sweep mode.

By far the most interesting and fruitful LXNMR decoupling experiment is the recently achieved heteronuclear deuteron decoupling.[87-89] The potential usefulness of this method lies in the fact that it allows, in principle, the acquisition of splittings even of molecules that have otherwise unresolved spectra when oriented in nematic phases, such as molecules with a large number of protons. This is usually the case for the nematogen itself, but obviously the utility also applies to some solute molecules. So far ten is the limiting number that gives patterns sufficiently resolved to be analytically useful.[81]

A typical broad-line case is that of cyclohexane, shown in Fig. 5a.

Labeling with deuterium is not so helpful in the LXNMR method as in the

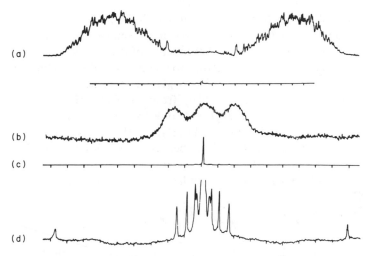

FIG. 5. NMR spectra of cyclohexane in a nematic solvent. (a) Spectrum of nondeuterated cyclohexane. The scale is given by markers spaced by 100 Hz. (b) Spectrum of cyclohexane-d_{11} without deuterium decoupling. (c) Deuteron decoupled spectrum of cyclohexane d_{11}: the triplet of trace 5(b) is reduced to a singlet. The markers are 100 Hz apart and refer to traces 5(b) and 5(d) as well. (d) Spectrum like trace 5(c) but recorded with higher sensitivity. The central signal is now out of scale and the doublets mentioned in the text are clearly evident [from *J. Chem. Phys.* **58**, 5089 (1973); reprinted with permission of the copyright owner].

normal NMR method, since the dipolar couplings are so large that even when reduced by a factor of 6.5 (corresponding to the ratio of the gyromagnetic values of protons and deuterons) they are still effective. Furthermore the spin ($= 1$) of deuterium leads to an increased number of lines because it introduces triplets instead of doublets in the spectral patterns. However, the spectrum may in principle be greatly simplified by heteronuclear decoupling at the deuteron frequency. In isotropic spectra this experiment is carried out without any great difficulty, but problems arise in oriented molecules since the deuteron resonance is now split into two halves, as much as a few kHz apart, by nuclear quadrupolar effects. It was found that irradiation at a frequency midway between the two halves induces double quantum transitions between the $+1$ and -1 spin levels of the deuterons and provides an efficient decoupling with less power than otherwise expected.[18, 21] The background theory is now well understood.[87, 88]

For example, cyclohexane-d_{11} (98% purity) was investigated in this way. The statistical distribution of deuterons in the molecules affords 80% of $C_6D_{11}H$, 18% of $C_6D_{10}H_2$, and 2% of $C_6D_9H_3$, and negligible quantities of molecules with four or more protons. The dominant species ($C_6D_{11}H$) still gave an unresolved spectrum (Fig. 5b) in which the H–C–D dipolar coupling gives a broad triplet with a coupling 6.5 times smaller than the doublet of Fig. 5a. Deuteron decoupling at the frequency of the double quantum transition reduces the whole broad spectrum to a collection of sharp lines: a single sharp line from the dominant species $C_6D_{11}H$ (Fig. 5c) and a number of sharp symmetrically disposed doublets from the $C_6D_{10}H_2$ molecules (Fig. 5d). Seven such doublets are expected, one for each of the seven possible structural isomers with formula $C_6D_{10}H_2$; at the temperature used (80°) each molecule interchanges rapidly between two equivalent chair conformations, and each of the seven $D_{1,n}$ couplings ($n = 2, \ldots, 8$) is an average between those of the two equally populated conformations:

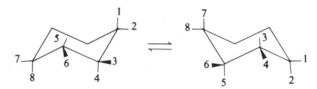

Figure 5d shows only five of the seven satellite pairs; the remaining two have been also detected by expanding the scale and improving the resolution.[89]

The largest dipolar coupling ($D_{1,2} = D_{7,8} = 1798.9 \pm 5$) corresponds to the separation of the outer lines and is due to the species with the two hydrogens in position 1,2:

The smaller coupling ($D_{1,8} = D_{2,7} = 16.0 \pm 1$) is given by the separation of the inner lines and is due to the following species:

The remaining values are as follows: $D_{1,3} = 200.0 \pm 2$, $D_{1,4} = 78.4 \pm 2$, $D_{1,5} = 316.3 \pm 2$, $D_{1,6} = 98.1 \pm 2$, and $D_{1,7} = 99.7 \pm 2$. The spectrum has therefore been interpreted despite the lack of fine structure for cyclohexane; hence, this technique should allow the analysis of rather complex molecules if deuterium can be introduced at specific sites.

IV. SPECTRAL AND STRUCTURAL ANALYSIS

A. The Hamiltonian

In order to interpret the NMR spectra of partially oriented molecules the appropriate Hamiltonian has to be solved.

The spin Hamiltonian describing the kind of spectrum with these molecules may be divided into four parts:

$$\mathscr{H} = \mathscr{H}_1 + \mathscr{H}_2 + \mathscr{H}_3 + \mathscr{H}_4 \tag{1}$$

$\mathscr{H}_1$ represents the Zeeman energy and is equal to

$$\mathscr{H}_1 = -(2\pi)^{-1} \sum_i \gamma_i (1 - \sigma_i) H_z I_{zi} \tag{2}$$

where γ_i is the gyromagnetic ratio, H_z the applied magnetic field, σ_i the shielding constant, which generates the chemical shift, and I_{zi} the z component of the angular momentum operator of the ith nuclei. The subsequent terms $\mathscr{H}_2$, $\mathscr{H}_3$, $\mathscr{H}_4$ represent the internuclear interaction energies. $\mathscr{H}_2$ is given by Eq. (3):

$$\mathscr{H}_2 = \sum_{i>j} J_{ij}(I_{xi}I_{xj} + I_{yi}I_{yj} + I_{zi}I_{zj}) \tag{3}$$

where J_{ij} is the scalar indirect coupling constant, which is invariant with respect to rotation of coordinates. This term represents the well-known through-bond interaction among nuclei in isotropic NMR spectra. The remaining terms are

$$\mathscr{H}_3 = \sum_{i>j} D_{ij}(-\tfrac{1}{2}I_{xi}I_{xj} - \tfrac{1}{2}I_{yi}I_{yj} + I_{zi}I_{zj}) \tag{4}$$

and

$$\mathscr{H}_4 = \sum_{i>j} G_{ij}(I_{xi}I_{yj} - I_{yi}I_{xj}) \tag{5}$$

Both $\mathscr{H}_3$ and $\mathscr{H}_4$ are averaged to zero in isotropic solvents, so they can usually be neglected. On the other hand, in anisotropic media such as the liquid crystalline solvents, $\mathscr{H}_3$ is not zero. $\mathscr{H}_4$ by contrast is zero even in a liquid crystalline solvent (and obviously also in an isotropic solvent) since this medium has a cylindrical symmetry about the applied magnetic field and the parameters G_{ij} are odd upon reflection of the nuclear moments in any plane parallel to the applied field.[18] However, when electric fields are applied these terms may be finite.

If nuclei with spins larger than $\tfrac{1}{2}$ are studied (such as deuterons), then an additional term $\mathscr{H}_5$ containing the quadrupolar coupling Q_i (which may become the most dominant one) also has to be taken into account. The term $\mathscr{H}_3$, which contains the D_{ij} constant, is the one we are most interested in for the LXNMR method, since the D_{ij}'s are larger than the σ_i's or J_{ij}'s.

B. The Direct Coupling

The direct coupling D_{ij} which appears in $\mathscr{H}_3$ is actually formed by two elements related through Eq. (6):

$$D_{ij} = D_{ij}^{\text{direct}} + D_{ij}^{\text{indirect}} \tag{6}$$

The prominent one (D_{ij}^{dir}) describes the interaction through space of two nuclear spins, which interact much in the same way as two macroscopic dipoles; accordingly it can be represented by simple magnetostatic relationships. D_{ij}^{ind} is due to the anisotropic components of the J_{ij} value.

Since the geometrical information is contained only in D_{ij}^{dir} (see pp. 355 and 358), it is important to know the extent of its contribution to the observable D_{ij}. There is no way of experimentally separating the contribution of D^{dir} and D^{ind}. It appears from calculations, however, that for proton–proton, proton–fluorine, and ^{13}C–proton interactions D^{ind} is negligible,[13, 90] whereas for F–F couplings it may be significant.[91–98] When it is ignored for ethylene the structural deviation from other determinations is found to be less than 1%.[93] On the other hand, in 1,1-difluorethylene the contribution of $D_{\text{FF}}^{\text{ind}}$ is so large that if the geometry of the molecule is calculated neglecting this contribution, a deviation as large as 11% with respect to the microwave geometry is observed.[93]

Additionally, Gerritsen and MacLean,[94, 95] by assuming the geometry of 1,1-difluoroethylene, calculated the contribution of $D_{\text{FF}}^{\text{dir}}$ and $D_{\text{FF}}^{\text{ind}}$ and from these values also obtained the tensor components of J_{FF}. It thus appears that the isotropic J_{FF} value (+32.5 Hz) is the average of the terms (in Hz) $J_{xx} = -720 \pm 39$, $J_{yy} = +339 \pm 39$ and $J_{zz} = +478 \pm 26$, the axes being as follows:

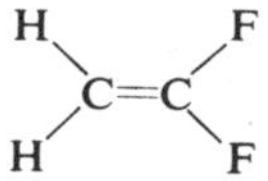

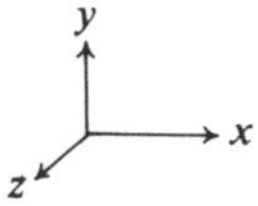

Large anisotropy values have been also observed in the F–F coupling of benzotrifluoride[97] and *trans*-difluoroethylene, for which a combination of theoretical calculations and experimental measurements shows that 5% of the anisotropic coupling arises from the indirect contribution.[98] The D^{ind} is not always so high, however, even for F–F interactions: pentafluoropyridine[99] and tetrafluoro-1,4-dithietane[69] are examples of molecules where the effect of this term can apparently be neglected.

The term D^{dir} (henceforward written D_{ij} since we will mainly be concerned with proton–proton interactions) is related to the molecular geometry by Eq. (7):

$$\frac{1}{2} D_{ij} = -P(\phi) \frac{\hbar}{2\pi} \gamma_i \gamma_j \cdot \frac{1}{2} \left\langle \frac{3\cos^2 \vartheta_{ij} - 1}{r_{ij}^3} \right\rangle \tag{7}$$

In the original treatment of Saupe[8, 11, 17] the term to the left (i.e., $D_{ij}/2$) was called B_{ij}; often however B_{ij} is called D_{ij} without taking into account the $\frac{1}{2}$ factor, so that the D_{ij} values quoted by various authors differ sometimes by a factor of 2. The other terms of the equation are the gyromagnetic ratios (γ_i and γ_j) for the two nuclei involved, the distance between the nuclei (r_{ij}) and the angle (ϑ_{ij}) between the axis connecting two nuclei of the solute molecules and the optical axis (or director) of the liquid crystalline solvent. The brackets indicate the average over the molecular motions. $P(\phi)$ represents the function

$$P(\phi) = \tfrac{1}{2}(3 \cos^2 \phi - 1) \tag{8}$$

where ϕ is the angle formed by the director of the solvent with the applied magnetic field and is therefore a measure of the orientation of the solvent. Since, however, orientation of the solvent in magnetic fields as large as those required for the NMR experiment is complete, the director is parallel to the applied field (that is, $\phi = 0$) and the $P(\phi)$ function is equal to 1. Equation (7) thus becomes

$$B_{ij} = \frac{1}{2} D_{ij} = -\frac{\hbar}{2\pi} \gamma_i \gamma_j \frac{1}{2} \left\langle \frac{3 \cos^2 \vartheta_{ij} - 1}{r_{ij}^3} \right\rangle \tag{9}$$

The term $\hbar/2\pi(\gamma_i\gamma_j) = K_{ij}$ is a constant depending upon the nature of the nuclei investigated. The values of K_{ij} in kHz Å^3 are given in Table III for some of the pairs of nuclei so far investigated. It is evident therefore that experimental couplings obtained from partially oriented spectra may provide information on internuclear distances provided the angle ϑ_{ij}, which is a measure of the orientation, can be also determined.

TABLE III

K_{ij} Values for Various Nuclei[a]

Nuclei:	H–H	H–D	H–^{13}C	H–^{19}F	^{19}F–^{19}F	^{19}F–^{13}C	^{19}F–^{31}P	^{13}C–^{13}C
K_{ij}	120.067	18.431	30.188	112.955	106.265	28.400	45.724	7.590

[a] Values are given in kHz Å^3

C. The Orientation

A more convenient definition of the alignment of an axis joining two nuclei is given by the term S_{ij} defined in Eq. (10):

$$S_{ij} = \tfrac{1}{2}\langle 3 \cos^2 \vartheta_{ij} - \delta_{ij} \rangle \tag{10}$$

where δ_{ij} is the Kronecker operator ($\delta = 1$ or 0 when $i = j$ or $i \neq j$, respectively).

In order to explain the meaning of S_{ij} better, let us consider two nuclei joined by an axis a that makes an angle ϑ_{za} with the direction z of the magnetic field, and define the orientation of the a axis with respect to z according to Eq. (11).

$$S_{aa} = \tfrac{1}{2}\langle 3\cos^2\vartheta_{za} - 1\rangle \tag{11}$$

It is evident that if the axis is completely aligned, i.e., parallel to the direction of the applied field, $\vartheta_{za} = 0$ and $S_{aa} = 1$. On the other hand, the minimum possible orientation is obtained when a is perpendicular to z, i.e., $\vartheta_{za} = \pi/2$ and $S_{aa} = -\tfrac{1}{2}$. Accordingly the term S_{aa} must lie within the limits $-\tfrac{1}{2} \leq S_{aa} \leq 1$ to have any reasonable physical meaning.

When a molecule is in an isotropic environment the average value of S_{aa} is zero, so D_{ij} becomes zero too and the "normal" NMR spectrum is obtained. In the framework of Cartesian coordinates we may indicate the axes as a, b, and c; and then we may have nine possible S_{ij} values with i or j being in turn a, b, or c. These values may be arranged in a 3×3 square matrix $\mathbf{S}$ which is called the ordering matrix. The matrix is symmetrical ($S_{ab} = S_{ba}$; $S_{ac} = S_{ca}$; $S_{bc} = S_{cb}$) and traceless ($S_{aa} + S_{bb} + S_{cc} = 0$). Only five elements therefore are required to determine the matrix: three off-diagonal and two diagonal elements.

Often, however, the coordinates chosen to define the orientation of the axis joining two nuclei are not convenient with regard to the molecular symmetry. It is possible nevertheless to relate $\mathbf{S}$ in a given coordinate system (a, b, c) to that of another system (α, β, γ) according to Eq. (12).

$$S_{\alpha\alpha} = \sum_{i,\,j=a}^{c} \cos\vartheta_{\alpha i}\cos\vartheta_{\alpha j}S_{ij} \tag{12}$$

In this way the choice of the axes can be made, taking into account the symmetry of the molecule. The existence of elements of symmetry also allows a reduction of the number of elements needed to define the orientation matrix. Therefore they will be five only in the case of an asymmetric top molecule.

As an example, a molecule like acetylene ($D_{\infty h}$ point group symmetry) needs only one element of the ordering matrix to define the orientation.[100, 101] If the axis is chosen along the HCCH bonds, this element is S_{aa} (from the angle ϑ_{za}). In the case of a molecule like pyridine (C_{2v} symmetry point group) only two elements will be required, provided that one of the axes (c in this example) is normal to the plane containing the aromatic ring and one (a for instance) is coincident with the twofold symmetry axis:

In this way S_{cc} will give the orientation of the molecular plane with respect to the z axis of the applied field and S_{aa} will define the orientation of the

TABLE IV

Ordering Matrix Elements Required to Describe the Orientation of Molecules Listed according to the Molecular Symmetry

Group symmetry	Ordering matrix elements
C_1, C_i	S_{zz}, $(S_{xx} - S_{yy})$, S_{xy}, S_{xz}, S_{yz}
C_2, C_{2h}, C_s	S_{zz}, $(S_{xx} - S_{yy})$, S_{xy}
C_{2v}, D_2, D_{2h}	S_{zz}, $(S_{xx} - S_{yy})$
C_n, C_{nh}, C_{nv} $(3 \le n \le 6)$	S_{zz}
$C_{\infty h}$	S_{zz}
D_{2d}	S_{zz}
D_n, D_{nd}, D_{nh} $(3 \le n \le 5)$	S_{zz}
D_6, D_{6h}	S_{zz}
$D_{\infty h}$	S_{zz}
S_4, S_6	S_{zz}
K_h, O, O_h, T, T_d	No elements are needed

N–C_4 axis within the plane containing the aromatic ring. When more
complex molecules are investigated it is sometimes not straightforward to
determine how many elements are required to define the orientation, and so
the requisite numbers are reported for every possible molecular symmetry
in Table IV.

The molecular orientation may be also described in an alternative way
according to Snyder's notation[102] without making use of the matrix method.
In this treatment a function $P(\vartheta, \phi)$ is defined as the probability per unit solid
angle that the applied field direction is ϑ, φ in spherical polar coordinates.
As far as partially oriented molecules in liquid crystals are concerned, $P(\vartheta, \phi)$
can be expanded as

$$P(\vartheta, \phi) = \tfrac{1}{4}\pi + C_{3z^2 - r^2} \cdot D_{3z^2 - r^2} + C_{x^2 - y^2} \cdot D_{x^2 - y^2} +$$
$$C_{xz} \cdot D_{xz} + C_{yz} \cdot D_{yz} + C_{xy} \cdot D_{xy} + \dots \qquad (13)$$

where the D terms have a mathematical expression that has the same form
as the angular part of the d-orbital wave functions:

$$D_{3z^2 - r^2} = (\sqrt{5}/8\pi)(3 \cos^2 \vartheta - 1)$$
$$D_{x^2 - y^2} = (\sqrt{15}/8\pi)(\sin^2 \vartheta \cos 2\phi)$$
$$D_{xz} = (\sqrt{15}/8\pi)(\sin 2\vartheta \cos \phi) \qquad (14)$$
$$D_{yz} = (\sqrt{15}/8\pi)(\sin 2\vartheta \sin \phi)$$
$$D_{xy} = (\sqrt{15}/8\pi)(\sin^2 \vartheta \sin 2\phi)$$

The various C terms are called motional constants and have the same meaning as the S_{ij} terms to which they are related through the following equations:

$$C_{3z^2-r^2} = \sqrt{5}\, S_{zz}$$

$$C_{x^2-y^2} = \sqrt{5/3}\,(S_{xx} - S_{yy})$$

$$C_{xz} = 2\sqrt{5/3}\, S_{xz} \qquad (15)$$

$$C_{yz} = 2\sqrt{5/3}\, S_{yz}$$

$$C_{xy} = 2\sqrt{5/3}\, S_{xy}$$

Accordingly the C values also have physical limits related to those on S; the values are listed in Table V.

The same rules as for S also apply to the choice of the C values with respect to the molecular symmetry. In this description the D_{ij} values may be related to the coordinates of the nuclei according to Eq. (16):

$$
\begin{aligned}
D_{ij} = -\frac{2}{\sqrt{5}} K_{ij} &\left\{ C_{3z^2-r^2} \left[\left\langle \frac{(\Delta z_{ij})^2}{r^5_{ij}} \right\rangle - \frac{1}{2}\left\langle \frac{(\Delta x_{ij})^2}{r^5_{ij}} \right\rangle - \frac{1}{2}\left\langle \frac{(\Delta y_{ij})^2}{r^5_{ij}} \right\rangle \right] \right. \\
&+ C_{x^2-y^2}\sqrt{3}\left[\frac{1}{2}\left\langle \frac{(\Delta x_{ij})^2}{r^5_{ij}} \right\rangle - \frac{1}{2}\left\langle \frac{(\Delta y_{ij})^2}{r^5_{ij}} \right\rangle \right] \\
&+ C_{xy}\sqrt{3}\left[\frac{(\Delta x_{ij})(\Delta y_{ij})}{r^5_{ij}} \right] + C_{xz}\sqrt{3}\left[\frac{(\Delta x_{ij})(\Delta z_{ij})}{r^5_{ij}} \right] \\
&\left. + C_{yz}\sqrt{3}\left[\frac{(\Delta y_{ij})(\Delta z_{ij})}{r^5_{ij}} \right] \right\}
\end{aligned}
\qquad (16)
$$

TABLE V

Range of Motional Constants

	Minimum	Maximum
$C_{3z^2-r^2}$	$-\dfrac{\sqrt{5}}{2} = -1.1180$	$\sqrt{5} = 2.2361$
$C_{x^2-y^2}$	$-\dfrac{\sqrt{15}}{3} = -1.2910$	$\dfrac{\sqrt{15}}{3} = 1.2910$
C_{xy}	$-\dfrac{\sqrt{15}}{2} = -1.9365$	$\dfrac{\sqrt{15}}{2} = 1.9365$
C_{yz}	$-\dfrac{\sqrt{15}}{2} = -1.9365$	$\dfrac{\sqrt{15}}{2} = 1.9365$
C_{xz}	$-\dfrac{\sqrt{15}}{2} = -1.9365$	$\dfrac{\sqrt{15}}{2} = 1.9365$

The equations given before, either in the orientation matrix or in the motional constants formalism, consist of products of orientational parameters and internuclear distances. As a consequence, a unique set of *both* orientational parameters *and* distances cannot be obtained from the D_{ij} values; but a solution can be always obtained for an arbitrary r_{ij} value provided that the S_{ij} (or the motional constants) are modified accordingly. Since one does not usually know the molecular orientation by independent techniques with sufficient accuracy, only ratios of distances, instead of absolute distances, can be determined. This requires the knowledge (or a reasonable guess) of at least one absolute distance from an independent source in order to determine the value of the other r_{ij}. In other words one could say that molecular shapes rather than absolute structures may be determined.

D. The Effects of the Vibrational Motions

A further point has to be carefully considered, i.e., which kind of distance is determined? The equations employed allow us to obtain the term $\langle r^{-3} \rangle$ averaged over the vibrational motions. Therefore the distance that is determined is $\langle r^{-3} \rangle^{-1/3}$. If vibrational motions were not present this distance would be equal to those obtained by microwave spectroscopy ($\langle r^{-2} \rangle^{-1/2}$), electron diffraction ($\langle r \rangle$) and neutron diffraction ($\langle r^{3/2} \rangle^{2/3}$). Since the vibrational motions do exist, however, the various averages are not equal, in principle, and discrepancies should be observed among distances determined by different techniques.

For diatomic molecules the average distance is given by Eq. (17)

$$\langle r^n \rangle^{1/n} = \langle r \rangle - \tfrac{1}{2}(1 - n)r_e\langle \xi \rangle^2 \tag{17}$$

where r_e is the equilibrium distance and $\xi = (r - r_e)/r_e$. As an example, a C—H bond distance, determined by electron diffraction as $\langle r \rangle$ is 1.1338 Å, whereas $\langle r^{-2} \rangle^{-1/2}$ is 1.1307 Å from microwave spectroscopy, the equilibrium value (r_e) being 1.1198 Å. As a consequence, in a neutron diffraction experiment we would expect to obtain 1.1401 Å (corresponding to the average $\langle r^{3/2} \rangle^{2/3}$) and in an LXNMR determination we expect 1.1280 Å (corresponding to $\langle r^{-3} \rangle^{-1/3}$). One can see that the LXNMR data are expected to be closer to those determined by microwave spectroscopy than to those obtained from other techniques.

Recently some attention[18, 103-109] has been paid to the problem of correcting the r values for the effects of vibrational motions. Since however these corrections are relatively small they are worth applying only when other major sources of error and approximation have been eliminated. In the model suggested by Snyder and Meiboom[18] the direct dipolar coupling D_{ij} is expressed as a function of the internal molecular motions described by the coordinates $q_1, q_2, \ldots, q_n$. The D_{ij} values were expanded in a Taylor series about the values corresponding to the equilibrium geometry, neglecting the terms higher than quadratic. The following expression was eventually obtained:

$$D_{ij} = D_{ij}^{eq} + \sum_k \left(\frac{\delta D_{ij}}{\delta q_k}\right)^{eq} \bar{q}_k + \frac{1}{2} \sum_k \left(\frac{\delta^2 D_{ij}}{\delta q_k^2}\right)^{eq} \bar{q}_k^2$$

When the above equation was applied to benzene, they found that the deviation of D_{ij} with respect to the equilibrium value D_{ij}^{eq} expressed as a percentage is rather small as far as proton–proton distances are concerned. The deviation is in fact -0.3% for the D values between ortho protons, and 0.1% for meta as well as para protons. By contrast, the deviation is rather relevant (-5.5%) when ^{13}C–H couplings are considered. This means that whereas vibrational effects may be neglected for the H–H distances since they are of the same order of magnitude as the experimental errors, they are important when directly bonded atoms, such as carbon and hydrogen, are concerned. Because of the relationship between D_{ij} and r_{ij}, if the 5.5% reduction of D_{CH} due to vibrations is not taken into account, the C—H bond distance will be 1.8% too large with respect to the H—H distance, which is almost unaffected by the correction. Experimentally the D_{CH} values of benzene were matched assuming a C—H bond distance of 2.5% larger than the values determined by other techniques; the agreement with the calculated 1.8% deviation is quite satisfactory.

Subsequently Lucas[106–108] proposed the use of an average structure derived only from harmonic vibrational corrections. The correction, which has to be applied to the direct couplings, was derived within the framework of normal-coordinate analysis. This method has been successfully applied to the elimination of discrepancies observed in the interpretation of the structure of cyclopropane[107] by means of the LXNMR method.

The benzene molecule has also been studied by this approach,[109] and it has been confirmed that vibrational corrections are important only when C—H bonds are studied, and that they are negligible when distances between hydrogens are under investigation. Indeed in some cases (such as cyclopentadiene[110]) the agreement with microwave determinations is better when vibrational corrections are *not* taken into account: other sources of error are probably more important than neglect of the vibrational analysis. In this particular example, for instance, the MW geometry is presumably not very accurate.

E. The Interpretation of the Spectra

The interpretation of LXNMR spectra can be achieved by solving the Hamiltonian [Eq. (1)]. The kinds of calculations required are similar to those needed to solve the Hamiltonian for isotropic NMR spectra. However, in

addition to chemical shifts [Eq. (2)] and J_{ij} couplings [Eq. (3)], the D_{ij} couplings have also to be considered [Eq. (4)]. Some additional differences have
also to be taken into account which depend upon the molecular orientation.
They may be better understood by examination of some simple examples. Let
us consider a molecule in which there are two chemically equivalent hydrogens
such as 2,5-dibromothiophene:

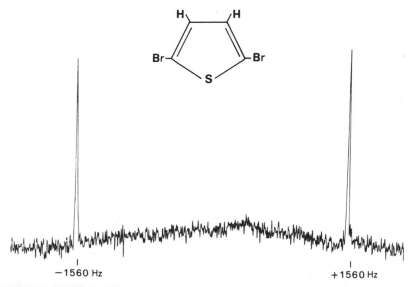

In the isotropic NMR spectrum we have only a single line, since the two
nuclei are magnetically equivalent and thus have the same chemical shift.
The J_{HH} coupling does not affect the spectrum in this situation and therefore
its value cannot be determined unless the satellite signals due to ^{13}C are
investigated. On the other hand, the D_{ij} value *does* affect the spectrum even
though the nuclei have the same chemical shift, and as a result the LXNMR
spectrum of two equivalent nuclei is a doublet instead of a singlet (see Fig. 6).
The spectral notation, which was A_2 for the isotropic spectrum, becomes AA'
in the case of oriented molecules. The solution of the Hamiltonian and the
subsequent energy diagram allows an understanding of the relation between
the experimental splitting (which in this case is 3120 Hz) and the D_{ij} value
(see Table VI). The separation between the two lines with nonzero intensity is

FIG. 6. 60 MHz NMR spectrum of 2,5-dibromothiophene in EBBA at probe
temperature.

TABLE VI

Transition Diagram, Frequencies, and Intensities
for an Oriented Two-Spin System, A_2

Transition	Frequency	Intensity
$1S_0-S_1$	$\nu_0 + J + \frac{1}{4}D$	0
$2S_0-S_1$	$\nu_0 + \frac{3}{4}D$	2
$S_{-1}-2S_0$	$\nu_0 - \frac{3}{4}D$	2
$S_{-1}-1S_0$	$\nu_0 - J - \frac{1}{4}D$	0

therefore $\frac{3}{2}D$ or (since $B = D/2$) $3B$. The chemical shift (ν_0) is given by the middle point of the doublet, whereas J does not affect the spectrum and thus cannot be determined in this case.

Since the distance between the two hydrogens in this molecule is approximately 2.64 Å we may determine the degree of orientation of the interprotonic axis H-3-H-4.

$$D_{3,4} = \pm \frac{2}{3}(3120) = -K_{ij}/r^3 (3 \cos^2 \vartheta - 1) = -120067/(2.64)^3$$
$$(3 \cos^2 \vartheta - 1)$$

Therefore $S_{3,4}$ is equal to $\pm 0.159_5$ and $\vartheta = 48° 28'$ or $61° 35'$. Since, however, we do not know whether D_{ij} is positive or negative, we do not know whether $S_{3,4}$ is -0.159_5 or $+0.159_5$. This uncertainty could be lifted in an absolute way only if one of the two values exceeded the lower physical limits of S (i.e., $|S| > 0.5$). Since this does not happen in the present case the conclusion remains ambiguous. In the case of a molecule with three chemically equivalent nuclei, we again observe a difference between the spectra from the isotropic and nematic phases. The first gives a single line whereas in the second each hydrogen is split by the other two, thus leading to a triplet with a 1:2:1 intensity ratio. Such an example is reported in Fig. 7, which gives the 60 MHz spectrum of 1,3,5-trinitrobenzene dissolved in MBBA.

It can be shown also in this case that the D_{HH} coupling is equal to two-thirds of the line separation (487.5 Hz in this case). Since this molecule belongs to the D_{3h} point group only one element is required to describe the orientation, provided its direction is taken along the threefold symmetry axis.

If the interprotonic distance is assumed (a reasonable value is 4.30 Å), then not only can the orientation of the interprotonic axis be derived (as in

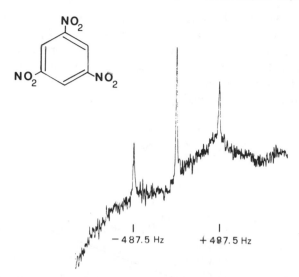

FIG. 7. 60 MHz NMR spectrum of 1,3,5-trinitrobenzene in a nematic solvent (MBBA) at probe temperature (31°).

the previous example) but also the orientation of the whole molecule. The orientation of the interprotonic axis (here S_{xx}) is given by Eq. (11) and is equal to ± 0.1076 since the separation between the lines is 487.5 Hz. Again the sign of D and hence of S_{xx} cannot be determined from the spectrum. The orientation parameters in the xy plane are equally probable because of the symmetry, i.e., $S_{xx} = S_{yy}$. Since $S_{xx} + S_{yy} + S_{zz} = 0$, it turns out that $S_{zz} = -2S_{xx}$; therefore $S_{zz} = \mp 0.2152$.

It is not possible to decide definitely whether S_{zz} is positive or negative, however, a large number of experiments indicate that flat molecules orient preferentially with the molecular plane along the applied magnetic field. Accordingly, the z axis, which is normal to the aromatic plane, is preferentially perpendicular to the magnetic field and the S_{zz} value is most likely negative.

It is evident that in the simple cases presented so far we cannot obtain geometrical information, since only one D_{ij} coupling is available. Similarly, for molecules containing only two or three interacting nuclei we cannot provide information about the molecular structure by this method. When, however, the number of nuclei is four or more and the molecule also has a sufficiently high symmetry, it is possible to obtain structural information as well as orientational parameters. The two conditions must be fulfilled simultaneously in order to have a sufficient number of independent D_{ij} values (each corresponding to an independent equation) and a small enough number of orientational parameters (unknowns) to allow the determination of some additional unknowns (geometrical parameters). For instance, a four-spin system such as AB_3, which does not have axial (C_3) symmetry, does not satisfy both the required conditions, since it only has two D_{ij} values (D_{AB} and D_{BB}) and the lack of C_3 symmetry means that two or more S values are needed to describe the orientation. Accordingly no direct coupling is left for structural determination.

For example, the spectrum of 2,3,4,6-tetrachloroanisole

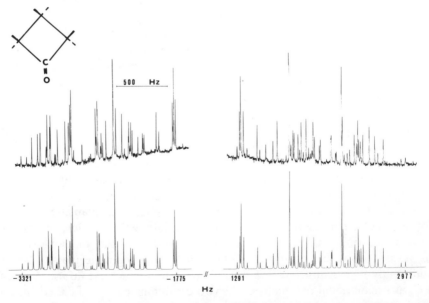

has been interpreted[111] and the two D_{ij} values have been obtained (666 and 56 Hz for D_{BB} and D_{AB}, respectively), but no structural information could be deduced. On the contrary when C_3 symmetry is present (as in the case, for instance, of propyne, $CH_3C\equiv CH$) only one S value is sufficient to define the molecular orientation and structural data can be obtained. Actually, it

FIG. 8. Experimental (top) and computer-simulated (bottom) spectra at 100 MHz of cyclobutanone in a nematic solvent (Phase IV) at probe temperature. [from *J. Chem. Soc., Perkin Trans. 2* p. 1908 (1973): reprinted with permission of the copyright owner].

can be shown that the ratio of the D_{ij} values in a homonuclear AB_3 system with C_3 symmetry is given by

$$\frac{D_{AB}}{D_{BB}} = \left(\frac{r_{BB}}{r_{AB}}\right)^3 \left[\left(\frac{r_{BB}}{r_{AB}}\right)^2 - 2\right]$$

In the case of propyne the ratio of interprotonic distances is[112]

$$R/r = 2.353 \pm 0.005$$

The result is in good agreement with the microwave data.[113]

It is obvious that when the spectra of molecules with three, four, or more nuclei have to be analyzed, the D_{ij} values cannot be obtained from line separations as easily as shown in the two simple examples given. The spectral analysis is then carried out by means of computer programs derived from those already employed in the spectral analysis of NMR spectra in isotropic solutions. The most widely used is that called LAOCOONOR[30a, 114] which is a generalization of the well known LAOCOON program of Castellano and Bothner-By.[115] In its original version it could handle up to six nuclei but it has since been extended to accommodate up to eight nuclei.

TABLE VII

Spectral Parameters of Cyclobutanone Partially Oriented in the Nematic Phase (in Hz)[a]

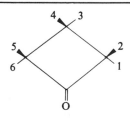

$\nu_1 = -283.16 \pm 0.05$	
$\nu_3 = -192.60 \pm 0.07$	
$J_{1,2} = -17.60$	$J_{1,5} = 4.09$
$J_{1,3} = 10.05$	$J_{1,6} = -3.19$
$J_{1,4} = 6.33$	$J_{3,4} = -10.92$
$D_{1,2} = +3027.69 \pm 0.10$	$D_{1,5} = -207.47 \pm 0.12$
$D_{1,3} = -568.69 \pm 0.07$	$D_{1,6} = -40.21 \pm 0.11$
$D_{1,4} = +8.19 \pm 0.08$	$D_{3,4} = +3045.95 \pm 0.18$

[a] The spectrum has been taken in the internally locked, frequency-swept mode at 100 MHz using a capillary of hexamethyldisiloxane as reference. The theoretical spectrum has been plotted assuming a Lorentzian line shape with a width at half height of 2.8 Hz. The program used is the LAOCOONOR: the J values were assumed equal to those of the isotropic spectrum.

Another kind of program is called LEQUOR: in this case the concept of magnetic equivalence has been introduced;[116] it allows calculations even with ten nuclei[81] provided that the spin system has a number of totally equivalent nuclei (for instance methyl groups). Usually these programs have a plotting subroutine so that the theoretically interpreted spectrum can be displayed. As an example, the spectrum of the six spin system of cyclobutanone is shown in Fig. 8. The parameters used in its interpretation are given in Table VII.

In order to have a good understanding of the possible applications of the LXNMR method in determining molecular shapes, let us discuss an example of a rather usual case, that is, the AA'BB' system with C_{2v} symmetry, specifically tellurophene.[117]

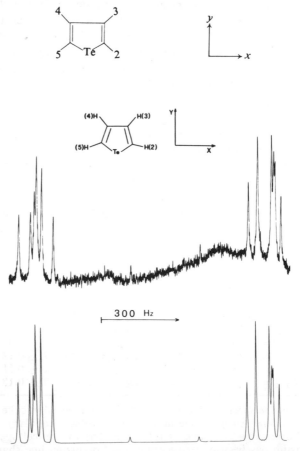

FIG. 9. Experimental (top) and computer-simulated (bottom) 60 MHz spectra of tellurophene in Phase IV at probe temperature [from *Mol. Phys.* **27**, 257 (1974): reprinted with permission of the copyright owner].

The spectrum (taken at 60 MHz) is given in Fig. 9. Twenty-four lines are expected to have a nonzero intensity: fourteen of them were detected experimentally.[13] The computer simulation allowed reproduction of the experimental lines within the accuracy of 1 Hz. It can be seen that only lines with an intensity larger than the arbitrary value 0.1 can be observed over the noise. The root mean square deviation is 0.5 Hz (Table VIII). The parameters used for the spectral interpretation are given in Table IX.

The relation between the interprotonic distances and the experimental couplings is as follows for an AA'BB' system with C_{2v} symmetry:

$$r_{2,5}/r_{3,4} = [D_{3,4}/D_{2,5}]^{1/3}$$

$$D_{2,5}(r_{2,3}/r_{3,4})^5 - D_{2,4}[(D_{3,4}/D_{2,5})^{1/3} + (r_{2,3}/r_{3,4})^2]^{5/2} = - D_{3,4}[D_{3,4}/D_{2,5}]^{1/3}$$

$$(r_{2,4}/r_{3,4})^2 = [D_{3,4}/D_{2,5}]^{1/3} + [r_{2,3}/r_{3,4}]^2$$

TABLE VIII

Computed and Experimental Frequencies (in Hz) of the Tellurophene Spectrum Relative to the Average Frequency of the Outer Lines (Downfield Is Negative)

Experimental frequency	Calculated frequency	Deviation	Intensity
—	−778.70	—	0.01_5
−505.5	−505.70	0.20	2.04
−459.6	−460.63	1.03	1.97
$−445.6_5$	−445.56	−0.09	2.07
−436.5	−436.19	−0.31	3.82
−415.5	−415.56	0.06	3.81_5
−372.0	−371.54	− 0.46	1.97
—	−172.73	—	0.03
—	−163.20	—	0.01_5
—	−161.53	—	0.04
−73.5	−73.06	−0.44	0.18_5
—	−70.96	—	0.03
—	92.25	—	0.03
—	133.80	—	0.09
—	160.70	—	0.04
—	181.34	—	0.03
193.5	193.93	−0.43	0.17
379.8	380.16	−0.36	1.97
414.0	414.73	−0.73	4.00
446.7_5	466.93	−0.18	3.75
477.0	476.30	0.70	1.98
482.3	481.92	0.38	1.97
505.5	504.87	0.63	1.88
—	777.86	—	0.08

TABLE IX

Spectral Parameters in Hz Employed in the Interpretation
of the Tellurophene Spectrum[a]

$\nu_2 - \nu_3 = -74.7 \pm 1.5$	$J_{2,3} = +6.58 \pm 0.44$
$D_{2,3} = -581.5 \pm 0.3$	$J_{2,4} = +1.12$
$D_{2,4} = -25.0_5 \pm 0.2_5$	$J_{2,5} = +1.82$
$D_{2,5} = +17.6 \pm 0.9$	$J_{3,4} = +3.76$
$D_{3,4} = +109.6 \pm 0.8_5$	

[a] $J_{2,3}$ has been iterated, whereas the remaining J values
have been kept equal to those determined in isotropic
solution.

Accordingly the three interprotonic ratios can be obtained without any
knowledge of the orientation:*

$$r_{2,3}/r_{3,4} = 0.982$$
$$r_{2,4}/r_{3,4} = 1.675$$
$$r_{2,5}/r_{3,4} = 1.840$$

If a distance between any two protons is assumed, the remaining distances as
well as the elements of the ordering matrix may be calculated. For the latter
depend upon D and r according to the equations:

$$S_{xx} = -\frac{2\pi^2}{h\gamma^2} D_{3,4} r^3_{3,4}$$

$$S_{yy} = -\left(\frac{2\pi^2}{h\gamma^2}\right) r^3_{2,3}\left\{D_{2,3} - \left(\frac{D_{3,4}}{4}\right)\left(\frac{r_{2,3}}{r_{3,4}}\right)^{-5}\left[\left(\frac{D_{3,4}}{D_{2,5}}\right)^{1/3} - 1\right]^2\right\} \times$$
$$\left\{1 - \frac{1}{4}\left(\frac{r_{2,3}}{r_{3,4}}\right)^{-2}\left[\left(\frac{D_{3,4}}{D_{2,5}}\right)^{1/3} - 1\right]^2\right\}^{-1}$$

The distance $r_{3,4}$ is taken equal to 2.538 Å, a value deduced from microwave
investigations of tellurophene, selenophene, and thiophene.[118] The other
interprotonic distances thus become $r_{2,3} = 2.493$, $r_{2,4} = 4.251$, and $r_{2,5} =$
4.670. The ordering matrix elements are $S_{xx} = -0.00746$ and $S_{yy} = 0.04759$,
and, consequently $S_{zz} = -0.04013$. The z axis is expected (see p. 363)
to be directed toward the direction perpendicular to the magnetic field and
thus S_{zz} has a negative value. The absolute signs for the D_{ij}'s can therefore
be deduced (see Table IX). It should be observed that $J_{2,3}$ has been iterated
also and its sign (positive) was found to be opposite to that of $D_{2,3}$ (negative).
Thus, LXNMR spectroscopy is an efficient method for the determination of
the signs of J_{HH} values in an absolute way, provided that there is sufficient

*Different weights for the D_{ij} values have been used since the error of $D_{3,4}$ and $D_{2,5}$
are about three times larger than those of $D_{2,3}$ and $D_{2,4}$.

information about the orientational behavior of the molecule being investigated.

When molecules have more than four interacting nuclei it is rather cumbersome to calculate the ratio of interprotonic distances from the experimental couplings since the analytical expressions, although available,[13] are not very simple. Computer programs have been thus developed[48, 71, 77] to obtain the best set of protonic coordinates matching the experimental couplings (SHAPE,[48] GEOFIT,[71] YMODI,[77] and SPIRAL[119]). The most widely used is the program SHAPE, developed by Diehl and co-workers,[48] by which, with a number of assumptions of some geometrical data, the coordinates and the orientational parameters are refined until the minimum deviation between experimental and computer D_{ij} values is reached. An accurate method for the determination of the errors is also included and the D_{ij} couplings may be differentially weighted according to their experimental uncertainty.

An example of such an application is the work of Veracini, Bucci, and Barili[49] on benzonitrile.

The 60 MHz spectrum taken at probe temperature in 4-n-butyl-4′-methoxy-azoxybenzene is given in Fig. 10. The interprotonic ratios obtained from the SHAPE program are given in Table X. These values are in good agreement with the corresponding data obtained from microwave investigations.[120]

TABLE X

Experimental Couplings and Interprotonic Distances of Benzonitrile[a]

$\nu_2 - \nu_1 =$	27.4 ± 0.5	
$\nu_3 - \nu_1 =$	37.3 ± 0.5	
$D_{1,2} =$	-3165.5 ± 0.2	$r_{1,2}/r_{3,4} = 0.996 \pm 0.003\ (0.994)$
$D_{1,3} =$	-426.2 ± 0.7	$r_{1,3}/r_{3,4} = 1.729 \pm 0.005\ (1.725)$
$D_{1,4} =$	-50.3 ± 0.9	$r_{1,4}/r_{3,4} = 1.998 \pm 0.004\ (1.997)$
$D_{1,5} =$	$+100.9 \pm 0.3$	$r_{1,5}/r_{3,4} = 1.735 \pm 0.004\ (1.734)$
$D_{2,3} =$	-392.4 ± 0.2	$r_{2,4}/r_{3,4} = 1.730 \pm 0.003\ (1.732)$
$D_{2,4} =$	$+101.4 \pm 0.2$	$r_{1,5}/r_{1,2} = 1.742 \pm 0.004\ (1.744)$
$C_{3z^2 - r^2} =$	0.440 ± 0.001	
$C_{x^2 - y^2} =$	-0.167 ± 0.001	

[a] The microwave ratios are given in parentheses. The chemical shifts and direct couplings are in Hz.

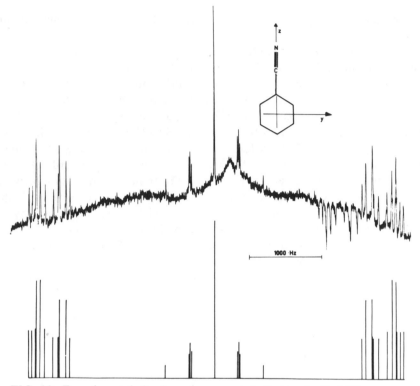

FIG. 10. Experimental (top) and computed (bottom) 60 MHz spectra of benzonitrile partially oriented in a nematic solvent (Phase IV) at probe temperature [from *Mol. Phys.* **23**, 59 (1972): reprinted with permission of the copyright owner].

By using an assumed interprotonic distance (in this case $r_{3,4} = 2.478$ Å, taken from microwave spectroscopy) they determined the actual distances and the orientational parameters.

V. EXAMPLES OF STRUCTURAL DETERMINATIONS

In this section a number of examples of the applications of LXNMR spectroscopy to structural determinations is reported. The first part will deal with molecules having a rigid structure and the second part will be concerned with molecules in which internal molecular motions may be expected.

A. Rigid Molecules

For classification purposes, we shall present here the molecules divided in two classes: hydrocarbons and heterocyclic compounds.

Among the molecules which do not have internal motions rigid hydrocarbons have been investigated most: the results obtained for some of the more significant molecules are reported here.

1. Hydrocarbons

a. *Ethylene.*[121] The spectrum of ethylene has been investigated by MacLean and co-workers[121]:

The spectral parameters (in Hz) they obtained by interpreting the spectrum in hexyloxyazoxybenzene are as follows:

$$D_{gem} \pm 480 \qquad J_{gem} \pm 2.5$$
$$D_{cis} \mp 740.5 \qquad J_{cis} \pm 11.7$$
$$D_{trans} \mp 199.6 \qquad J_{trans} \pm 19.1$$

Since two matrix elements are needed, only one structural parameter can be obtained. The authors chose the ratio of the cis–gem protonic distances, which was found to be

$$r_{cis}/r_{gem} = 1.3198$$

Electron diffraction, infrared, and Raman data give very close values for this ratio (1.320 ± 0.020, 1.330 ± 0.016, and 1.327 ± 0.0409, respectively).

b. *Acetylene.*[101] As mentioned before, this molecule has only one D_{HH} coupling so that structural information cannot be obtained. However in a sample enriched in ^{13}C we may expect three couplings from species (a) and four from species (b):

$$H—^{13}C≡^{12}C—H \qquad H—^{13}C≡^{13}C—H$$
$$\quad 1 \quad\;\; 3 \quad\;\; 4 \quad\;\; 2 \qquad\quad 1 \quad\;\; 3 \quad\;\; 4 \quad\;\; 2$$
$$\qquad (a) \qquad\qquad\qquad (b)$$

By investigating such enriched samples Spiesecke[101] obtained the following $B_{ij} = D_{ij}/2$ values (in Hz):

	(a)	(b)
B_{12}	81.87	80.85
B_{13}	616.37	617.94
B_{23}	62.46	62.50
B_{34}	—	103.71

TABLE XI

Distance Ratios of Acetylene Obtained with
Various Methods (See Text) (distances in Å)

Ratio	$r_e{}^a$	$r_e - \delta$	r_{NMR}
$r_{1,3}/r_{2,3}$	0.4682	0.4720	0.4662
$r_{1,3}/r_{1,2}$	0.3189	0.3229	0.3216
$r_{1,3}/r_{3,4}$	0.8808	—	0.8771
$r_{2,3}/r_{1,2}$	0.6811	0.6840	0.6844
$r_{2,3}/r_{3,4}$	1.880	1.865	1.876
$r_{1,2}/r_{3,4}$	2.761	2.728	2.727

^{a} r_e is the equilibrium distance.

The distance ratios obtained from derivative (b) are compared in Table XI
with the equilibrium distances r_e obtained by other techniques (column 1)
and also with r_e values corrected for the Bastiansen–Morino shrinkage
effect (δ) (column 2).

   c. *Cyclopropane.*[103] Snyder and Meiboom not only obtained the D_{HH}
values but also determined the D_{CH} couplings by examining the proton
satellite spectra from ^{13}C in natural abundance.[103]

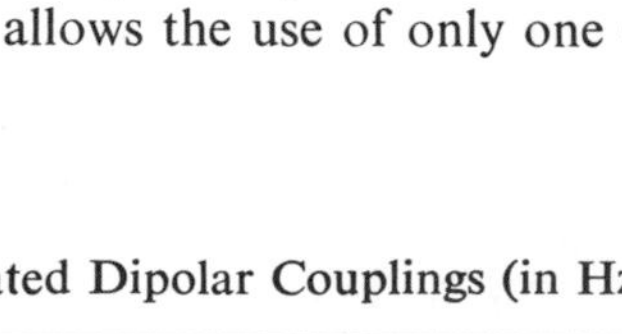

Owing to the D_{3h} molecular symmetry, the choice of the z axis perpendicular
to the plane of the ring allows the use of only one orientational parameter.

TABLE XII

Experimental and Calculated Dipolar Couplings (in Hz) for Cyclopropane

	Experimental	Calculated	
		Without vibrational corrections	With vibrational corrections
$D_{1,2}$	−194.51	−194.21	−194.54
$D_{1,4}$	+974.00	+974.05	+974.00
$D_{1,5}$	+4.35	+5.50	+4.28
$D_{C,1}$	+651.78	+651.79	+651.78
$D_{C,3}$	−33.95	−33.94	−33.91

By assuming a C—C bond length of 1.510 Å, the value from electron diffraction studies, the authors determined a C—H bond length of 1.123 Å and an HCH angle of 114.4°. Refinement of these data taking into account the effect of molecular vibrations by Lucas[107] gave a C—H bond length of 1.0807 ± 0.0009 Å, an HCH angle of 115.16° ± 0.04°, and $C_{3z^2 - r^2}$ = −0.05679 ± 0.00015.* This is one of the few cases where the correction for molecular vibrations substantially improved the fit between experimental and computed D_{ij} values, as shown in Table XII.

d. Allene.[122] An LXNMR spectrum of allene was obtained by Sackmann.[122] Both the D_{HH} and D_{CH} values were determined from the main proton spectrum and the ^{13}C satellite patterns.

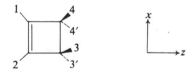

The values in Hz are as follows: $D_{1,2}$ = 1356.7, $D_{1,3}$ = 248.0, $D_{1,5}$ = 357.0, $D_{1,6}$ = −348.0, and $D_{1,7}$ = −114.0. The D_{2d} symmetry allows the use of only one orientational parameter and the geometry was calculated with the assumption of either the C—H (1.0816 Å) or the C—C bond distance (1.3116 Å), each taken from electron diffraction measurements. The results do not agree perfectly with the electron diffraction data, although the deviations are not very large. For instance the HCH angle was determined to be 117.9°, as opposed to the value 118.5° from electron diffraction.

e. Cyclobutene.[77]

The spectrum of cyclobutene in the nematic phase gave the following results in Hz:

$D_{1,2}$	−122.8	$D_{3,3'}$	+801.52
$D_{1,3}$	−32.84	$D_{3,4}$	−183.99
$D_{1,4}$	−52.54	$D_{3,4'}$	−7.34

The NMR data were found to be in agreement with C_{2v} symmetry, corresponding to the planar structure of cyclobutene. The ratios of the determined interprotonic distances turned out to be very close to those obtained by

*In the original work[107] this value is incorrectly quoted as S_{zz}.

microwave spectroscopy in the gaseous phase. The data are as follows, with the MW ratios in parentheses:

$$r_{1,2}/r_{1,4} \quad 0.99\ (0.98)$$
$$r_{1,2}/r_{3,4} \quad 1.14\ (1.14)$$
$$r_{1,2}/r_{3,3'} \quad 1.58\ (1.59)$$

If $r_{1,2} = 2.83$ Å, the elements of the ordering matrix are $S_{zz} = 0.0156$ and $S_{xx} = 0.0232$.

f. Benzocyclopropene.[123]

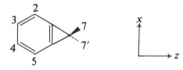

The spectral analysis leads to the following interprotonic distance ratios[123] (the corresponding data of benzene are given in parentheses):

$r_{2,3}/r_{3,4}$	1.028 (1.000)	$r_{2,7}/r_{3,4}$	1.669
$r_{2,4}/r_{3,4}$	1.771 (1.732)	$r_{8,7}/r_{3,4}$	2.214
$r_{2,5}/r_{3,4}$	2.081 (2.000)	$r_{7,7'}/r_{3,4}$	0.760

The shape of the aromatic ring is distorted with respect to benzene; in particular, the two angles adjacent to the three-membered ring are reduced $(-\delta)$ and the others increased $(+\delta)$.

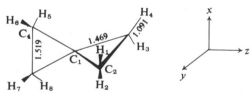

g. Spiropentane.[124]

This molecule consists of two cyclopropane rings sharing a carbon atom; accordingly, the two rings are expected to lie in orthogonal planes with D_{2d} symmetry. Therefore (see Table IV) only one orientational parameter is required. The NMR spectrum proved that this symmetry could account for the experimental data.

By using the C—C or the C—H bond lengths from electron diffraction data (see structure) the authors determined the angles HCH (115° 13' or 115° 21'), C_1C_2H (117° 48' or 117° 43') and C_3C_2H (117° 49' or 117° 48').

h. Norbornadiene.[46]

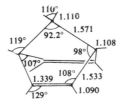

A 220 MHz spectrometer was used in the study of this rather complicated eight-spin system.[46] Among various possible structures obtained by electron diffraction only one agreed with the NMR data. The geometrical results were thus obtained by combining some of the electron diffraction data with those derived from the NMR spectrum.

The conclusions reported in the formula represent a very good example of the possible achievement of the LXNMR method as a complementary technique in structural determination.

i. Phenylacetylene.[125, 126]

This is an example of a structure which has been investigated only by means of LXNMR spectroscopy. After the first studies of Diehl and Khetrapal[125] the problem has been reinvestigated and the structural determination completed by Dereppe, Arte, and Van Meerssche.[126] The ratios of the interprotonic distances from the two investigations are given in Table XIII. A large number of variously substituted benzenes has also been investigated: the data can be found in Diehl and Khetrapal,[13, 50, 128] Veracini et al.,[49] Yim and Gilson,[75, 131] Barili and Veracini,[127] Bulthuis et al.,[129] and de Kievut and de Lange.[130]

j. Naphthalene.[132]

An LXNMR spectrum of this very complicated eight-spin system gave more than 300 lines at 100 MHz; 210 of them were used in the spectral analysis and were matched with an accuracy of 1.0 Hz. The significant interprotonic distances were found to be in general agreement with the results of neutron diffraction study in the solid state.[133] The data are collected in Table XIV: the expected D_{2h} symmetry was observed, so that only two orientational parameters were required. From an assumed $r_{1,2}$ bond distance of 2.468 Å, the orientational parameters were also determined: $S_{zz} = +0.1912 \pm 0.0004$ and $S_{xx} = -0.0393 \pm 0.0001$. As one could have expected, the molecule is primarily oriented in the direction of the longest molecular axis.

TABLE XIII

Structural Parameters of Phenylacetylene from NMR Studies in the Nematic Phase

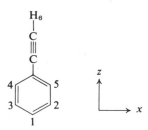

Ratio	Ref. 125	Ref. 126 $(r_{2,4} = 2.481$ Å, fixed)
$r_{1,2}/r_{1,6}$	—	0.333 ± 0.001
$r_{1,4}/r_{1,6}$	—	0.570 ± 0.002
$r_{2,6}/r_{1,6}$	—	0.880 ± 0.001
$r_{4,6}/r_{1,6}$	—	0.578 ± 0.001
$r_{4,5}/r_{2,3}$	1.0 (fixed)	0.978 ± 0.002
$r_{2,4}/r_{2,3}$	0.584 ± 0.008	0.568 ± 0.001
$r_{3,4}/r_{2,3}$	1.158 ± 0.008	1.141 ± 0.002
$r_{1,2}/r_{2,3}$	0.575 ± 0.005	0.578 ± 0.001
$r_{1,4}/r_{2,3}$	1.02 ± 0.02	0.989 ± 0.003
$r_{1,6}/r_{2,3}$	2.91 ± 0.02	3.051 ± 0.003

TABLE XIV

Geometrical Data for the Protons of Naphthalene

Ratio	NMR[132]	Neutron diffraction[133]
$r_{2,3}/r_{1,2}$	1.004 ± 0.002	1.002 ± 0.003
$r_{1,3}/r_{1,2}$	1.742 ± 0.002	1.736 ± 0.005
$r_{1,8}/r_{1,2}$	1.005 ± 0.001	0.992 ± 0.003

2. Heterocyclic Compounds

A few examples selected from the more recent works will be given of structures determined for heterocyclic derivatives.

a. Five-Membered Aromatics. The molecules having general formula

$$X = NH \quad \text{(pyrrole)}$$
$$= O \quad \text{(furan)}$$
$$= S \quad \text{(thiophene)}$$
$$= Se \quad \text{(selenophene)}$$
$$= Te \quad \text{(tellurophene)}$$

have been all examined by NMR spectroscopy in the nematic phase; their interprotonic ratios thus give a relatively homogeneous set of data. As far as orientational parameters are concerned, those obtained in thermotropic nematogens cannot be compared with those obtained in lyotropic solvents because of the differences in the orienting properties of these media.

The interprotonic ratios obtained are reported in Table XV; the comparison with the microwave data (when available) appears to be quite satisfactory. Also, the differences between structural data obtained for the same molecule in the two kinds of mesophases are rather small and should be attributed to experimental error.

The proton spectrum of pyrrole differs from those of the other five-membered rings because of the additional couplings with the hydrogen of the NH group. The spectrum of pyrrole in its naturally occurring isotopic composition however cannot be interpreted, since the quadrupole moment of ^{14}N makes

TABLE XV

Ratios of Interprotonic Distances Determined by LXNMR for Five-Membered Heterocyclics[a]

Ratio	Pyrrole T^{73}	Furan T^{134}	Furan L^{135}	Thiophene $T^{136,137}$	Thiophene L^{135}	Selenophene L^{67}	Tellurophene T^{117}
$r_{2,3}/r_{3,4}$	1.000	0.98	0.985	0.995	0.985	0.988	0.982
	(1.000_3)		(0.994)		(0.998)	(0.995)	
$r_{2,4}/r_{3,4}$	1.597	1.56	1.567	1.653	1.638	1.663	1.675
	(1.598_7)		(1.571)		(1.648)	(1.668)	
$r_{2,5}/r_{3,4}$	1.550	1.47	1.487	1.745	1.714	1.788	1.840
	(1.555_4)		(1.485)		(1.719)	(1.791)	

[a] Columns T refer to experiments in thermotropic nematogens, columns L in lyotropics; the values in parentheses are those obtained by microwave spectroscopy in the gaseous phase.

the spectral lines too broad to be analyzed. However the use of a 95% ^{15}N-enriched pyrrole allowed Randall and co-workers[73] to overcome this difficulty, since ^{15}N has spin $\frac{1}{2}$ and therefore no quadrupolar effects. In this way the couplings due to proton–^{15}N interactions were also detected and the distances between protons and nitrogen were determined. In addition to the distance ratios quoted in Table XV,[67, 73, 117, 134–137] five more data were determined and are reported in Table XVI; the corresponding microwave results are in parentheses. It can be seen that this molecule offers one of the best examples of agreement between microwave and NMR data, since the deviations between the ratios determined with the two techniques never exceed 0.005.

TABLE XVI

Additional Internuclear Distance Ratios Obtained for ^{15}N-Pyrrole

Ratio	LXNMR	Microwave
$r_{1,2}/r_{3,4}$	0.924	0.924_8
$r_{1,3}/r_{3,4}$	1.540	1.544
$r_{1,6}/r_{3,4}$	0.363	0.367_4
$r_{2,6}/r_{3,4}$	0.787	0.789_0
$r_{3,6}/r_{3,4}$	1.203	1.202_7

At this point it is worth mentioning the study of a molecule formed by two condensed thiophene rings, namely thieno[2,3-b]thiophene.[44] The distance ratios experimentally determined are compared with those obtained by building up the molecule from two thiophene rings, under the assumption that the ring geometry is not affected by the condensation (Table XVII). Although differences were observed in the two cases, the practice of using the known geometry of simpler molecules to give indications on the geometry of more complex derivatives does not seem to be very dangerous, as the deviations were found to be 6% or less.

b. *Pyridine and Aza Heterocyclics.* Diehl and co-workers first examined the LXNMR spectrum of pyridine[138] and compared the structure obtained in this way with early microwave data.[139, 139a] The agreement between these results is quite good,[138] even though the use of more accurate microwave measurements[139b] showed slightly larger discrepancies (see Table XVIII). Recently Schumann and Price obtained a FT spectrum of pyridine enriched in ^{15}N (see Fig. 11).[140]

TABLE XVII

Ratios of Interprotonic Distances of Thieno[2,3-*b*]-thiophene from LXNMR and a Model Formed from Two Condensed Thiophene Rings

Ratio	Experimental	Computed
$r_{2,3}/r_{3,4}$	0.79_8	0.85
$r_{2,4}/r_{3,4}$	1.66_3	1.70
$r_{2,5}/r_{3,4}$	2.12_9	2.16

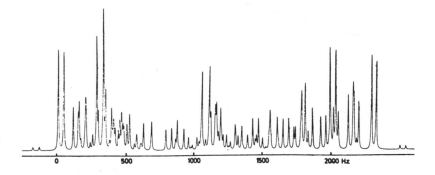

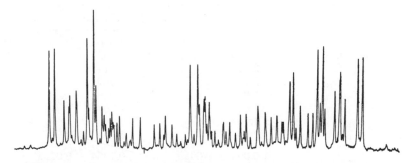

FIG. 11. Experimental (bottom) and computer-simulated (top) 90 MHz FT NMR spectrum of ^{15}N-pyridine partially oriented in a nematic phase [from *Angew. Chem., Int. Ed. Engl.* **12**, 930 (1973): reprinted with permission of the copyright owner].

TABLE XVIII

Interprotonic Distances of Pyridine and Perfluoropyridine

Ratio	Pyridine			Perfluoropyridine
	NMR[138]	NMR[140]	MW[139b]	NMR[99]
$r_{2,6}/r_{3,5}$	0.962	0.958	0.955	0.958
$r_{2,3}/r_{3,5}$	0.579	0.576	0.577	0.571
$r_{2,5}/r_{3,5}$	1.139	1.136	1.135	1.133
$r_{3,4}/r_{3,5}$	0.587	0.582	0.582	0.558
$r_{2,4}/r_{3,5}$	1.009	0.997	0.997	0.983

As mentioned previously in the case of pyrrole, this isotopic substitution also allows the determination of proton–nitrogen dipolar couplings and of the corresponding distances. The protonic distances obtained from this source are also collected in Table XVIII, and match very well those obtained by microwave.[139b] This agreement shows that the small discrepancies of the early determination were due to the experimental uncertainty and could be

TABLE XIX

Molecular Shape of Quinoxaline and Phthalazine as Determined by NMR Spectra in the Nematic Phase[a]

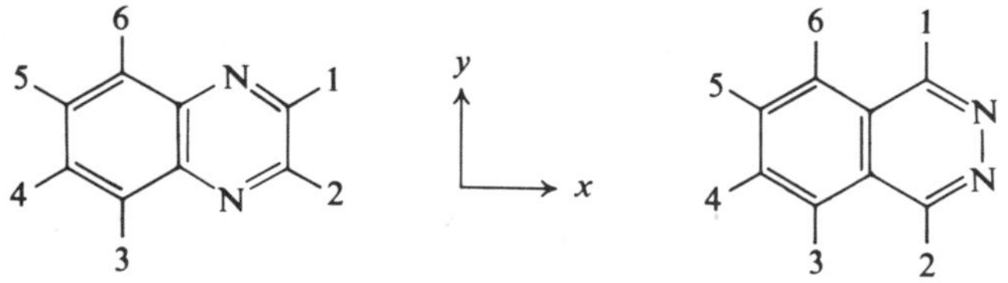

Ratio	Quinoxaline	Phthalazine
$r_{1,2}/r_{4,5}$	1.013	1.966
$r_{1,3}/r_{4,5}$	2.375	2.258
$r_{1,4}/r_{4,5}$	2.87	2.417
$r_{1,5}/r_{4,5}$	2.69	1.969
$r_{1,6}/r_{4,5}$	1.89	1.046
$r_{3,4}/r_{4,5}$	1.01	1.008
$r_{3,5}/r_{4,5}$	1.75	1.748
$r_{3,6}/r_{4,5}$	2.04	2.038
S_{xx}	$+0.094 \pm 0.005$	$+0.1864 \pm 0.0033$
S_{yy}	$+0.0260 \pm 0.0001$	-0.0161 ± 0.0002

[a] The orientational parameters were obtained assuming $r_{4,5} = 2.481$ Å.

overcome by means of the more sophisticated instruments now available. The determination of the geometry of pyridine and deuteropyridines with inclusion of the effects of vibrational motions has also been made by Emsley et al.[141] It is worth also mentioning that the ratios deduced from pentafluoropyridine[99] are, as expected, not very dissimilar to those of pyridine: the major deviation is observed for the $r_{3,4}/r_{3,5}$ distance ratios.

A study of the structure of pyridine in the presence of a lanthanide shift reagent showed that the complex formation does not significantly affect the geometry of the ring.[72] The complex of pyridine with bromine[72a] was also investigated, and on the basis of the orientational behavior, a structure has been suggested in which bromine is linked to the nitrogen along the C_2 axis.

Molecules containing two nitrogens in six-membered rings have also been studied and their shapes determined.[142] Good results have been obtained as well in the investigation of two diazanaphthalenes, quinoxaline,[143] and phthalazine.[143a] The data are given in Table XIX. Additionally, the spectrum of 2,7-naphthyridine has been investigated—the structural parameters are given in Table XX.[144] Again the deviations between the experimental inter-

TABLE XX

Experimental Inteprotonic Ratios of 2,7-Naphthyridine Compared with Those "Computed" from Two Joined Pyridine Rings[a]

Ratio	Experimental	Computed	Deviation
$r_{1,2}/r_{2,3}$	1.642 ± 0.006	1.656	0.85%
$r_{1,3}/r_{2,3}$	1.963 ± 0.010	1.967	0.20%
$r_{1,4}/r_{2,3}$	2.223 ± 0.010	2.217	0.27%
$r_{1,5}/r_{2,3}$	2.339 ± 0.005	2.341	0.08%
$r_{1,6}/r_{2,3}$	1.020 ± 0.003	1.012	0.78%
$r_{2,4}/r_{2,3}$	1.975 ± 0.003	1.949	1.31%
$r_{2,5}/r_{2,3}$	2.723 ± 0.003	2.710	0.47%
$r_{3,4}/r_{2,3}$	1.065 ± 0.004	1.033	3.00%

S_{xx} 0.1726 ± 0.0018
S_{zz} −0.0239 ± 0.0006

[a] The orientational parameters given have been calculated on the assumption that $r_{2,3}$ is equal to the corresponding distance in pyridine (2.4833 Å).

protonic ratios and those one would have expected by "building up" the molecule from two pyridine rings[139b] are relatively small.

B. Molecules with Possible Internal Motions

Mainly because this technique determines molecular shapes rather than absolute distances, it seems well suited to the investigation of conformational problems in nonrigid molecules. The complications arising from the internal motions are quite large however, and many studies have been performed at various stages of approximation. In this section some of the results obtained in the field will be presented according to the following criteria:

1. Molecules where the motion can be described essentially as a free rotational process, for which there is no need to take into account preferred conformations.
2. Molecules which do have preferred conformers which are interconverting but are equivalent.
3. Molecules where two or more inequivalent interconverting isomers may be expected; cases where only one conformer was actually observed will be presented first, followed by cases where the presence of more than one conformer was inferred.

1. Free Rotation

This group of investigations concerns molecules containing methyl groups that were shown to possess such a low energy barrier to the internal motion that a free rotational model (i.e., a model which assumes no barrier) could not be distinguished, as far as NMR studies are concerned, from models taking into account a curve of probability distribution of the rotamers. One of the most detailed studies in this sense is that of Diehl and co-workers on toluene.[116]

They interpreted this eight-spin system with the program LEQUOR and, assuming the H-2–H-3 distance equal to 2.481 Å as in benzene, determined the geometry of the aromatic ring.

They showed that the coupling $D_{1,6}$ (-201.42 ± 0.15 Hz) was almost insensitive to the assumption of a free rotation or of a stable preferred conformation, since the calculated value changed by only ± 1.6 Hz. This insensitivity has been explained with the fact that in the case of a sixfold potential, such as that of methyl in toluene, the probability distribution almost approaches the free rotation distribution and the technique is therefore insensitive to the effects of such small barriers. It could be inferred nevertheless that the potential minimum is likely to exist when any one of C—H bonds lies in a plane perpendicular to the benzene ring. Actually microwave studies have determined the barrier as being 0.014 kcal/mole. The ratio of the average distance of the methyl protons from the ring protons was determined as $r_{6,7}/r_{3,6} = 0.3052 \pm 0.0004$, and the intermethyl distance $r_{6,7}$ was found to be 1.779 ± 0.008 Å.

Analogous conclusions were reached in the case of the 3,5-dichlorotoluene:[145] again the computed $D_{1,6}$ coupling (-326.38 ± 0.14 Hz) differed by only 1.1 Hz from the free rotational model and a model with a potential of 1 kcal/mole, and by only 2.2 Hz when the limiting case of an infinite potential (stable conformer) was considered. Since the barrier to the rotation is expected to be much smaller than 1 kcal/mole, the free rotation model can be safely used as an approximation. The LXNMR method was shown to be effective in determining the energy barrier only when the barrier is larger than 0.5–1 kcal/mole. Even the more hindered 2,6-dichlorotoluene[146] was analyzed successfully with a free rotational approach. The following analogous cases were thus treated with a free rotational model: 4-methylpyridine (γ-picoline),[147] 2,5-dimethylpyrazine,[52] tetrachloro-p-xylene,[54] 2,5-dichloro-p-xylene,[56] p-xylene,[148] and trifluoromethylbenzene.[97]

A similar case, 2-propene enriched in ^{13}C, has been also studied.[149] The authors examined the problem by comparing the free rotational model with a model having different weights of staggered and eclipsed rotamers.

(a)   (b)

The free rotational model was used with the introduction of a variable number (n) of equipopulated rotamers with conformations differing by angles of $120/n$. For $n \geq 4$ the best fit reached its optimum when the deviation between the couplings was 0.37 Hz. By assuming a nonequal ratio of staggered (a) and eclipsed (b) conformations the best fit was improved (0.014 Hz) when the eclipsed conformer was 63% and the staggered 37%. However, both models fit the experimental values within the error limits and the improvement does not seem to have a physical meaning.[149] The assumption of the existence of a single conformation, either (a) or (b) also does not give very large deviation: surprisingly however, the eclipsed conformation fitted [the root mean square (rms) error was 0.68 Hz] the experiment better than the staggered (rms error = 1.48 Hz).

The structure deduced for each model (unequal population of rotamers and free rotation) agrees reasonably well with the microwave data (Table XXI).

TABLE XXI

Interatomic Distances for 2-Propene (in Å)^a

Distance	Microwave	LXNMR (two rotamers)	LXNMR (free rotation)
$r_{1,2}$	2.431	2.466 ± 0.010	2.455 ± 0.058
$r_{1,3}$	3.072	3.096 ± 0.007	3.093 ± 0.043
$r_{1,4}$	3.100	3.109 ± 0.005	3.104 ± 0.029
$r_{1,5}$	2.580	2.572 ± 0.003	2.565 ± 0.024
$r_{1,7}$	1.090	1.114 ± 0.004	1.100 ± 0.025
$r_{2,3}$	1.861	1.917 ± 0.009	1.911 ± 0.055
$r_{2,4}$	3.726	3.725 ± 0.005	3.726 ± 0.017
$r_{2,5}$	4.135	4.138 ± 0.006	4.135 ± 0.024
$r_{2,7}$	2.113	2.139 ± 0.006	2.134 ± 0.027
$r_{3,4}$	2.435	2.362 ± 0.004	2.379 ± 0.027
$r_{3,5}$	3.509	3.462 ± 0.003	3.477 ± 0.028
$r_{3,7}$	2.111	2.103 ± 0.004	2.107 ± 0.026
$r_{4,5}$	1.758	1.770	1.776
$r_{4,7}$	2.147	2.128 ± 0.001	2.128 ± 0.001
rms deviation (Å) from MW		0.032	0.026

^{a} The values of the LXNMR measurements are based on the fixed value of the distance H-4–H-5 that allows the best fitting with the MW data.

When two rotational isomers are considered the problem arises as to how to handle the orientational parameters for the two conformers involved. In this case average values were used: the problem will be discussed in more detail in Section V, B, 3.

2. Molecules with a Preferred Conformation

This section deals with molecules where only one possible conformation is present and different conformational isomers are not expected. In these cases the NMR of oriented molecules in liquid crystals proved to be a powerful tool in deciding, for instance, whether planar or nonplanar conformations were the most stable situation in a nonrigid molecule.

a. Tropone.[45, 150] This molecule has been studied by two groups of workers which independently reached the same conclusion.[45, 150]

It was found that the preferred conformation of this molecule is the planar one. Furthermore, the model of a rigid structure matched the experimental dipolar couplings quite well so that the existence of relevant ring puckering vibrations around the planar conformer could be ruled out. A slight bond alternation was also observed, which means that the molecule is not a regular heptagon. The alternation, however, is not as large as expected for pure single and double bonds.[150] Accordingly, canonical structures of the type (a), which would give pure bond alternation, and (b), which would correspond to an heptagonal shape, both contribute to the molecular geometry.

(a)   (b)

The experimental interprotonic ratios are compared with those expected for a regular heptagon (in parentheses in Table XXII).[45]

TABLE XXII

Experimental Interprotonic Ratios Found for Tropone[a]

$r_{1,3}/r_{1,2}$	1.828 ± 0.006	(1.802)
$r_{1,4}/r_{1,2}$	2.274 ± 0.008	(2.247)
$r_{1,5}/r_{1,2}$	2.264 ± 0.008	(2.247)
$r_{1,6}/r_{1,2}$	1.812 ± 0.008	(1.802)
$r_{2,3}/r_{1,2}$	1.029 ± 0.006	(1.000)
$r_{2,4}/r_{1,2}$	1.832 ± 0.007	(1 802)
$r_{2,5}/r_{1,2}$	2.276 ± 0.008	(2.247)
$r_{3,4}/r_{1,2}$	1.010 ± 0.004	(1.000)

[a] Values in parentheses are those expected for a regular heptagon

b. *Cyclopentadiene*.[110, 151, 152] Three research groups deduced the structure of cyclopentadiene almost simultaneously. The problem mainly consisted in deciding whether the molecule was planar or bent and also whether relevant vibrational modes involving ring bending were important:

As in the case of tropone, the molecule was found to be essentially planar with negligible contributions of ring bending modes. Actually the assumption

TABLE XXIII

Interprotonic Distance Ratios of Cyclopentadiene Obtained from LXNMR and MW Data[110]

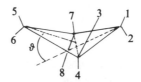

Ratio	LXNMR	MW
$r_{1,3}/r_{1,2}$	1.510 ± 0.009	1.476 ± 0.18
$r_{1,4}/r_{1,2}$	2.253 ± 0.009	2.164 ± 0.21
$r_{3,4}/r_{1,2}$	1.460 ± 0.022	1.460 ± 0.17
$r_{4,5}/r_{1,2}$	1.526 ± 0.009	1.518 ± 0.02
$r_{3,5}/r_{1,2}$	2.439 ± 0.022	2.375 ± 0.18
$r_{3,6}/r_{1,2}$	2.501 ± 0.009	2.447 ± 0.18

of a rigid planar structure matched the experimental D_{ij} values with a rms deviation of only 0.29 Hz.[110] A comparison between the structural parameters obtained from LXNMR[110] and MW is given in Table XXIII.

c. Cyclobutane.[153, 154] A very interesting case is that of cyclobutane, which was first studied by Meiboom and Snyder.[153] They found that the spin Hamiltonian has a symmetry D_{4h} as expected for a planar structure, but also showed that this was due to a rapid interconversion of the molecule between the two nonplanar forms, each having a D_{2d} symmetry.

They were also able to determine the bending angle ($\vartheta = 27°$). In their study the authors assumed that only the two equivalent bent structures were populated and no allowance was made for possible ring puckering vibrations. In other words, the barrier to the ring interconversion was assumed to be quite large. Despite this approximation the agreement with other structural determinations is quite acceptable, as electron diffraction gave 35°, infrared spectra in the solid state gave 37° ± 6° and infrared in the gaseous phase 33.3° ± 0.5°. Microwave investigations of 1,1-difluorocyclobutane also indicated the value of 27°.

Recently Cole and Gilson[154] reinvestigated the problem by introducing a potential function taking into account the ring puckering and methylene

TABLE XXIV

Experimental D Values (Hz) of Cyclobutane and Their Deviations Relative to Values Computed according to Various Models

	Experimental	Rigid planar $(\varDelta)$	Rigid nonplanar $(\varDelta)$	Puckering $(\varDelta)$	Puckering plus rocking $(\varDelta)$
$D_{1,2}$	$+1256.01$	-0.15	-0.06	-0.04	-0.04
$D_{1,3}$	-221.21	$+16.35$	-0.25	$+2.41$	-0.27
$D_{1,4}$	$+3.73$	-0.09	-0.86	-0.12	-0.80
$D_{1,5}$	-130.46	-46.47	-0.00	-4.30	$+0.08$
$D_{1,6}$	-30.10	-7.69	-0.71	-1.87	-0.25
rms		22.30	0.51	2.36	0.40
$C_{3z^2 - r^2}$		-0.06411	-0.07898	-0.07835	-0.07838

rocking vibrational effects. The results and those of Meiboom and Snyder[153] are summarized in Table XXIV. It can be seen that a planar structure is completely unacceptable since the computed D values give exceedingly large rms deviations. The rigid nonplanar structure matches the experimental data quite well, whereas the introduction of a ring puckering potential function leads to worse results. The use of both ring puckering and methylene rocking slightly improves the agreement with the experimental data, but the difference is too small to be considered significant. Within the framework of the latter model the angle was determined as 29°.

d. 2,5-Dihydrofuran.[155] A very accurate spectrum of this compound has been obtained, and therefore the small deviations observed between experimental and computed couplings for a planar rigid model indicate the existence of ring puckering vibrations. The rms deviation for the planar structure is only 1.30 Hz (as opposed to the 22 Hz of cyclobutane); nevertheless, the introduction of a potential function for the ring puckering allowed the cancellation of this small discrepancy, the deviation being eventually 0.06 Hz (see Table XXV). The puckering angle was found to be about 16°. The overall geometry of this molecule was obtained by making a number of assumptions, namely that $r_{C=C}$ is 1.350 Å, r_{CO} is 1.430 Å and that the CCO angle is bisected by the direction of the C—H bond of methylene. On this basis the data reported below were obtained (see Appendix).

TABLE XXV

Computed and Experimental D Values (Hz) for 2,5-Dihydrofuran

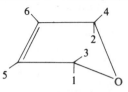

	Experimental	Rigid planar	Ring puckering
$D_{1,2}$	−154.11	−153.69	−154.09
$D_{1,3}$	+2215.33	+2215.37	+2215.33
$D_{1,4}$	−41.44	−39.99	−41.39
$D_{1,5}$	−232.53	−231.92	−232.50
$D_{1,6}$	−83.51	−86.24	−83.64
$D_{5,6}$	−362.19	−362.03	−362.18
rms deviation		1.30	0.06

e. Trimethylene Oxide, Sulfide, and Ketone (Cyclobutanone).[61, 62, 78] These molecules have been investigated by three different groups and represent good examples of the sensitivity of the LXNMR method for deciding whether the structures are rigid or not.

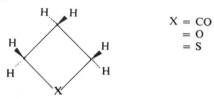

The dipolar couplings of cyclobutanone (X = CO) were satisfactorily matched with the assumption of a rigid planar structure, and, therefore, the ring puckering effects should be almost negligible (the rms deviation between experimental and computed values is 0.73 Hz).[62] By contrast, the discrepancy between experimental and computed data is larger when rigid planar structures are assumed for trimethylene oxide (X = O) and sulfide (X = S). The deviations are 1.85 Hz[78] or 1.64 Hz[61] for X = O, and 2.9 Hz[71] or 3.1 Hz[62] for X = S. It seems clear that in both cases the population of puckered conformations cannot be neglected and therefore the rings cannot be considered rigidly planar. Although it is not correct to assume a model of only two rapidly interconverting bent structures in order to evaluate the angle of puckering, such an assumption nevertheless led to an improvement of the best fit and indicated that the angles between the CCC and CXC planes may cover the range 2°–20° for trimethylene oxide[78] and 10°–26° for methylene sulfide.[62, 71] For larger angles the errors become exceedingly large. It was also found that the bending is expected to occur both along the S–C-β and C-α–C-α directions in the case of the sulfide.[62]

The ring puckering effects have been studied in a more appropriate way for trimethylene oxide by the use of a potential function (symmetrical) for the population distribution. A better fit for those couplings that are affected by the motion was then obtained. Interestingly, these conclusions are essentially the same as those obtained by microwave studies in the gaseous phase. In cyclobutanone (X = CO) the barrier to the puckering was found to be so low (0.022 kcal/mole)[156] that the molecule is essentially planar. In the case of trimethylene oxide the barrier is twice as large (0.043 kcal/mole)[156a] but the energy at the zero-point vibration is still above the maximum of the potential curve; thus the molecule may still be considered planar although the vibrational effects are expected to be more detectable than in cyclobutanone. Accordingly, the error is larger when the rigid structure is assumed in the liquid crystal study. Finally, in trimethylene sulfide (X = S) the barrier is much larger (0.783 kcal/mole)[157] and therefore the molecule may be considered to approach the limiting case of two equivalent bent conformations.

f. Cyclooctatetraene.[158] The LXNMR spectrum of this sample has been studied over a wide temperature range ($-35°$ to $+170°$) and linewidth variations were observed and interpreted. The spectral interpretation at low temperature allowed the authors to assign a D_{2d} symmetry to the molecule (corresponding to a "tub" conformation) and to exclude the C_{4v} (planar) and D_4 (crown) possibilities. (See below; + and − refer to positions above and below the plane, respectively.)

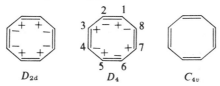

Again the authors made use of the deuteron spin decoupling method reported for cyclohexane[89] to help in the spectral analysis. The experimental couplings and those calculated on the basis of a tub shaped molecule are reported in Table XXVI.

When the temperature is raised progressively ($0°$–$100°$) the spectral lines first broaden selectively and then sharpen again (see Fig. 12). This effect has been attributed to the bond shift phenomenon which averages the D_{ij} couplings.

TABLE XXVI

Experimental and Computed D_{ij} Values (Expressed as $D_{ij}/\sum D_{ij})^a$ for Deuterated Cyclooctatetraene[158]

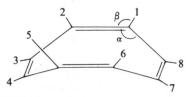

	Experimental	Best fit (tub conformation)
$D_{1,2}$	−1.236	−1.241
$D_{1,3} = D_{1,7}$	−0.0796	−0.0795
$D_{1,4}$	−0.0660	−0.0647
$D_{1,5}$	−0.115	−0.113
$D_{1,6}$	−0.159	−0.158
$D_{1,8}$	+0.735	+0.736

^{a} The data refer to the spectrum taken at 25°.

Accordingly, instead of the six different D_{ij} values observed at low temperature only four average values (D's) are appropriate when the process is fast. The relationship between the values obtained at high temperature (fast exchange region) with those obtained at low temperature (slow exchange region) is as follows:

$$D'_{1,2} = D'_{1,8} = \tfrac{1}{2}(D_{1,2} + D_{1,8})$$
$$D'_{1,3} = D'_{1,7} = D_{1,3} = D_{1,7}$$
$$D'_{1,4} = D'_{1,6} = \tfrac{1}{2}(D_{1,4} + D_{1,6})$$
$$D'_{1,5} = D_{1,5}$$

Luz and Meiboom[158] determined the activation energy for the bond shift process to be $\Delta E = 10.9 \pm 1.0$ kcal/mole. Finally, using the bond length from the electron diffraction studies and assuming that the fragment CC(H)C is planar, they determined the angles α and β to be 128.1° and 117.6° respectively.

g. Biphenyl-like Molecules.[80, 159] The biphenyl molecule is planar in the solid state (X-ray diffraction) but twisted by 42° in the gaseous state (electron diffraction). In order to obtain reliable information on the conformation in solution, symmetrically substituted biphenyls were investigated by the LXNMR method, since biphenyl itself gives unresolved lines in nematic solvents.

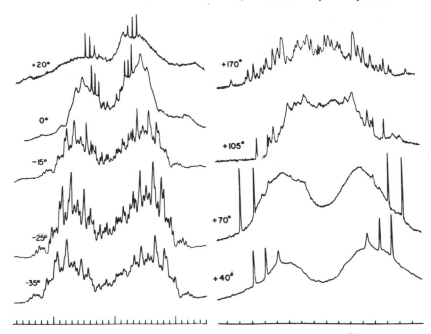

FIG. 12. 100 MHz NMR spectrum of cyclooctatetraene in different nematic phases at various temperatures. The scale at the left-hand side is spaced at 100 Hz and refers to each of the five spectra; that at the right hand side is spaced at 50 Hz for the top spectrum and 100 Hz for the three lower spectra [from *J. Chem. Phys.* **59**, 1077 (1973): reprinted with permission of the copyright owner].

In a first approach[159] Lunazzi and co-workers examined 3,3′,5,5′-tetra-chlorobiphenyl, the spectrum for which is shown in Fig. 13. This molecule has D_2 symmetry unless the angle of twist (θ) is 90°; therefore (see Table IV), two ordering matrix elements are needed for measuring the orientation. On the other hand, only one element (S_{zz}) is required if $\theta = 90°$ (D_{2d} symmetry) or if free rotation occurs (effective symmetry $D_{\infty h}$). From Table XXVII it can be seen that the experimental dipolar couplings cannot be reproduced by the free rotation or planar models; accordingly, the molecule must exist in a twisted conformation with θ smaller than 90°. The authors assumed the geometry of the biphenyl skeleton as determined from X-ray studies and calculated the dipolar couplings as a function of θ. Since only one dipolar coupling is sensitive to the angle between the two rings, the rms deviation is expected to be zero.

For this reason no allowance could be made for a potential function that included the barrier to the internal motions: the model thus assumed that the molecules populate only the most stable conformation, an assumption that is correct only when the barrier to the internal rotation is quite high.

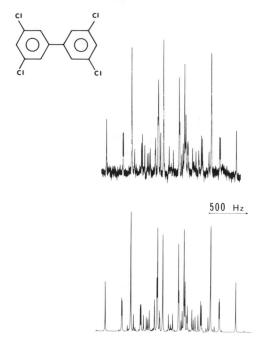

FIG. 13. Experimental (top) and computed (bottom) 100 MHz spectrum of 3,3′,5,5′-tetrachlorobiphenyl in 4,4′-bis(n-hexyloxy)azoxybenzene at 78° [from *J. Chem. Soc., Perkin Trans.* 2 p. 1396 (1973): reprinted with permission of the copyright owner].

TABLE XXVII

Experimental Dipolar Couplings (Hz) of 3,3′,5,5′-Tetrachlorobiphenyl Compared with Those Calculated Assuming Various Conformational Models[159]

	Experimental	Twisted (34° 20′)	Planar	Free rotation
$D_{2,2'}$	-224.3 ± 0.4	-224.3	-268.2	-236.1
$D_{2,4}$	-330.6 ± 0.4	-330.2	-120.1	-80.2
$D_{2,4'}$	-125.6 ± 0.4	-126.8	-26.2	-38.3
$D_{2,6}$	-259.8 ± 0.7	-259.8	-311.9	$+64.5$
$D_{4,4'}$	-35.0 ± 0.8	-34.8	-3.8	-12.6
rms error		0.6	109.4	187.7

The results obtained in this way indicate $\theta = 34°$, quite close to the value obtained for the vapor phase.

The problem was further investigated by using 4,4'-dichlorobiphenyl, in which the larger number of dipolar couplings allowed an estimate of the energy involved in the interconversion to be obtained.[80] Seven dipolar couplings are available: four of them ($D_{2,3}$, $D_{2,5}$, $D_{2,6}$, $D_{3,5}$) were employed to determine the benzene ring shape and the orientation whereas the other three ($D_{2,2'}$, $D_{2,3'}$, and $D_{3,3'}$), which depend upon θ, were used to study the conformation.

A potential function that included a barrier due to a nonbonded interaction (planar form) of 2.0 kcal/mole and one due to conjugation (orthogonal form) of 3.3 kcal/mole as suggested by theoretical work[160] was used, and the twisting angle in the nematic phase was found to be $37.7° \pm 6.7°$. The intermolecular potential due to the effects of the liquid crystalline solvents was estimated (4.5 ± 1.0 kcal/mole). Neglect of the potential function led, in this case, to a value of $31.7° \pm 0.3°$ for θ.

This kind of investigation has been also extended to 4,4'-bipyridyl.[161]

When a rigid model is assumed θ is found to be 29.35°. However, the deviations between experimental and computed D_{ij}'s are larger than the experimental errors. This discrepancy has been eliminated by introducing the effects of vibration: θ is almost unaffected by the correction (29.58°).

h. *Benzaldehyde*[162] *and Phenol.*[163] One major problem in studying molecules having internal motions is the scale of times (and the corresponding rates) involved. There are three different times (and rates) which have to be considered: (1) the lifetime of the nuclear spin (relaxation rate), (2) the time required for the molecule to reorient in the liquid crystal, and (3) the time of the dynamic processes that the molecule undergoes (internal rotation, inversion, ring puckering, etc.). The first time is quite long (about 10^{-3} second) and it is usually responsible for averaging the D_{ij} values, which, on the basis of the interprotonic distances, should be different. As an example, let us consider benzaldehyde.

H_6 O
 \ //
  C
5   1

4   2

3

(a)

O H_6
 \\ /
  C
5   1

4   2

3

(b)

z

$\longrightarrow x$

$D_{1,6}$ (for instance) should be different from $D_{5,6}$ if the rotation of the -CHO group (rate 3) is slower than the relaxation rate (1). Experimentally, $D_{5,6}$ is equal to $D_{1,6}$; therefore, the interconversion of (a) and (b) is faster than the relaxation rate.

The relative order of magnitude of the times mentioned in (2) is much more difficult to ascertain since no accurate measurements of the reorientation times of molecules in liquid crystals are available. The values have been estimated to lie in the range 10^{-8}–10^{-12} second. Accordingly if the rotational process has a quite low energy barrier (as in cases of methyl rotations[145-149] or ring inversions[61, 62, 78]) little doubt is left that the intramolecular process happens in a time shorter than the time of molecular reorientation. When the energy is higher it is much more difficult to assess the times involved. The existence of an internal motion (3) shorter or longer than the reorientation rate (2) affects the molecular symmetry and therefore the number of orientational parameters one has to use. Benzaldehyde, for instance, would only have a plane of symmetry if reorientation is faster than internal rotation, and therefore three S values would be needed. On the other hand, if reorientation is slower than internal motion, the fast (a) $\rightleftharpoons$ (b) interconversion would dynamically create an effective C_{2v} symmetry, for which only two S values are needed. In principle a comparison between the two possibilities would allow a differentiation: however, it is evident that a model having one more parameter is expected to give better results *per se*. It is thus difficult to decide whether the better agreement is merely a mathematical process or has a real physical meaning.

In the case of benzaldehyde, the model with a dynamic C_{2v} symmetry gave results not significantly better than those for the model in which reorientation is assumed to be faster than interconversion, despite the fact that the S_{xz} term, although assumed to be finite, nevertheless became negligible within the experimental error upon iteration. The results obtained for benzaldehyde are reported in Table XXVIII.

In the case of phenol[163] however, the geometry obtained using three S values (tantamount to assuming that internal motion is slower than re-

TABLE XXVIII

Coordinates (Å) of the Hydrogen Atoms 1, 5, and 6 in Benzaldehyde[a]

	(a)	(b)	(c)
x_1	2.15 ± 0.04	2.20 ± 0.19	2.01 ± 0.09
z_1	3.72 ± 0.03	3.79 ± 0.11	3.64 ± 0.11
x_5	-2.05 ± 0.02	-2.05 ± 0.15	-2.25 ± 0.14
z_5	3.71 ± 0.03	3.63 ± 0.09	3.76 ± 0.14
x_6	-0.76 ± 0.52	-1.22 ± 0.09	-1.25 ± 0.12
z_6	5.73 ± 0.09	5.80 ± 0.02	5.79 ± 0.03
S_{xx}	-0.0133 ± 0.0007	-0.0135 ± 0.0004	-0.0135 ± 0.0004
S_{zz}	0.1746 ± 0.0026	0.1744 ± 0.0014	0.1744 ± 0.0015
S_{xz}	0.062 ± 0.127	—	—

[a] Key to letters: (a) assuming three parameters; (b) and (c) the two solutions obtained in the case of two orientational parameters.

orientation) gave better results than the model employing two S values, as one would expect because of the additional parameter introduced. The result is quite surprising in view of the fact that the barrier to the rotation in phenol is expected to be lower and therefore (given a similar frequency factor) the internal motion is expected to be even faster. However, the orientational parameters of phenol indicate the existence of strong hydrogen bonding with the nematic solvent, and this fact would affect the rates of both internal rotation and reorientation to a different extent. Furthermore neglect of the potential curve and other assumptions about the behavior of the -OH motion may affect the results. It seems, therefore, that the problem of the times involved in the processes (2) and (3) has yet to be clarified.

i. Ortho-Substituted Toluenes.[48, 81, 164] One of the most accurate investigations so far obtained in conformational studies by means of LXNMR concerns the ortho-substituted toluenes.

Owing to the large number of nonequivalent dipolar couplings (10 values), quite accurate information was obtained on the geometry of these molecules, particularly in *o*-chlorotoluene (X = Cl).[48] The rotational mode of the methyl group was studied using a threefold potential function of the type

$$V = \frac{V_3}{2}(1 - \cos 3\phi)$$

A comparison was also made between the limiting cases of free rotation ($V_3 = 0$) and two stable conformers: eclipsed (a) and staggered (b) ($V_3 = \infty$).

$$\text{(a)} \quad \overset{H}{\underset{H}{\diagdown}}\text{---}\overset{|}{C}\text{—H---Cl} \qquad \text{(b)} \quad \text{---H—}\overset{H}{\underset{H}{\diagup}}\overset{|}{C}\text{---Cl}$$

For free rotation rms deviation from the experimental couplings of 0.25 Hz was obtained: the rigid eclipsed conformation (a) gave a deviation of 0.8 Hz whereas for the rigid staggered conformation (b) the deviation was 0.4 Hz. The inclusion of the potential function gives a zero deviation when the rotational barrier is 1.2 kcal/mole, the staggered conformation still being the more stable species. As one can see from the relatively small range of deviations involved (the worst is only 0.8 Hz), the accuracy of the determination of the potential barrier is quite low; the estimated error is ± 0.6 kcal/mole.

The uncertainty of these energies is shown by the absence of significant variations in V_3 for X = Br and X = I,[164] although the increasing bulk of the halogen atoms should give variations in the potential. The reported values are 0.89 ± 0.55 and 1.80 ± 0.58 kcal/mole in the two cases, respectively. However, since no other conformational studies were available on these molecules the importance of the LXNMR method in assessing the preferred conformation (staggered in all the three cases) and the order of magnitude of the barrier is quite evident. It is also worth mentioning that theoretical calculations[165] predicted a barrier of 1.77 kcal/mole and a staggered conformation for o-chlorotoluene.

TABLE XXIX

Dipolar Couplings (in Hz) of o-Xylene Oriented in the Nematic Phase

$D_{1,2} = -696.78 \pm 0.07$	$D_{2,3} = -311.83 \pm 0.11$
$D_{1,3} = -85.03 \pm 0.08$	$D_{2,5} = -49.52 \pm 0.07$
$D_{1,4} = -40.70 \pm 0.13$	$D_{2,8} = -90.19 \pm 0.08$
$D_{1,5} = -52.89 \pm 0.06$	$D_{5,6} = +899.74 \pm 0.03$
$D_{1,8} = -460.15 \pm 0.05$	$D_{5,8} = -88.88 \pm 0.03$

The interpretation of the spectrum of *o*-xylene molecule (X = CH$_3$)[81] was a considerable achievement since this ten-spin system displayed 678 lines: 582 of them were matched with an accuracy of 1.0 Hz. The D_{ij} values obtained are reported in Table XXIX.

The potential function used to describe the motion of this molecule is quite complex. It contains both a threefold and sixfold potential (V_3 and V_6 respectively); it also takes into account the barrier (V_a) that the coupled rotation experiences when the two methyls rotate in the same direction (slipping geared wheel) and the barrier (V_g) when the methyls rotate in opposite direction (as for two wheels in contact). If V_a is the dominant term and V_a/V_g is large the system might be considered as existing in a "gearing" situation. Although the number of parameters is too high to be determined with the available dipolar couplings, the authors concluded that V_3 is the largest term and the potential due to the coupling between the motion of the methyl groups is negligible; therefore, no gearing occurs in this system. They suggested a value for the threefold potential V_3 of 2.0 kcal/mole, whereas $V_a = 0.29$, $V_g = 0.37$, and $V_6 \simeq 0$ kcal/mole. Analogous conclusions were also reached for the similar case of *cis*-2-butene,[166]

$$CH_3 \quad CH_3$$
$$\diagdown \quad \diagup$$
$$C{=}C$$
$$\diagup \quad \diagdown$$
$$H \quad\quad H$$

for which the geometry was determined on the assumption of a simple threefold potential taken from microwave studies (0.75 kcal/mole).[167]

3. Molecules with Possible Conformational Isomers

Because absolute distances cannot be obtained, the LXNMR method is less satisfactory than other techniques. However, it may be very effective when the choice between two or more possible conformations has to be made. In this case an accurate determination of the absolute distances is not usually required, but the type of conformer has to be determined. The method additionally has a considerable advantage: it allows the determination in solution, whereas most of the other structural determinations are obtained in solid or gaseous phase.

For conformational problems we may distinguish two cases:

1. Only one conformation is actually present out of the two or more that can in principle be expected.

2. A preferred conformation is not dominant but an equilibrium between two or more conformers has to be considered.

The first case can be handled in the same way as the cases of rigid (or flexible) molecules and has been found to be successful in deciding, for instance, between syn or anti conformers. In practice, if the assumption of a single conformation allows satisfactory reproduction of the experimental dipolar couplings, there is little doubt that this is by far the most stable species and the presence of other conformers can be neglected. A number of examples of this type will be presented.

On the contrary, when the presence of a single conformational isomer does not allow the fitting of the experimental data, a model has to be made which includes an equilibrium between two or more conformers. The problems connected with the approximations required to approach these cases will be also mentioned.

a. Butadiene.[71] This case represents a good example of a molecule that could exist, in principle, either in the cis or in the trans conformation.

trans   cis

Electron diffraction (ED) studies indicate that the molecule exists as a trans isomer in the gaseous phase,[168] although suggestions have been advanced of a possible equilibrium between the trans and a gauche skewed conformation.[169]

Liquid crystal NMR work confirmed that in solution the trans is the only form observed. The agreement with the ED measurement is quite satisfactory, although differences of 1–2% may be inferred for the C—H bond length. As a whole the NMR data seem to indicate that the molecule has a slightly more elongated shape than expected. The possibility of a nonplanar structure could be rejected since deviations from the planarity as low as 2° lead to unacceptable deviations between computed and experimental couplings.

b. Salicylaldehyde.[170] Although hydrogen bonding is expected to strongly favor conformer (a), this molecule could in principle exist, at least to a certain extent, also in conformations (b), (c), and (d).

(a)   (b)   (c)   (d)

The distances involving either H-5 or H-6 and the ring protons (H-1, H-2, H-3, H-4), however, were found to be in agreement only with a structure of type (a), without any evidence of substantial contributions from other conformers (see Table XXX).

TABLE XXX

Ratios of Interprotonic Distances Determined for Salicylaldehyde: $r_{2,3}$ Was Assumed Equal to 2.481 Å.

$r_{1,2}/r_{2,3} = 0.989 \pm 0.003$	$r_{2,6}/r_{2,3} = 2.066 \pm 0.010$
$r_{1,3}/r_{2,3} = 1.712 \pm 0.005$	$r_{3,4}/r_{2,3} = 0.986 \pm 0.003$
$r_{1,4}/r_{2,3} = 1.949 \pm 0.009$	$r_{3,5}/r_{2,3} = 1.851 \pm 0.008$
$r_{1,5}/r_{2,3} = 2.163 \pm 0.011$	$r_{3,6}/r_{2,3} = 2.206 \pm 0.011$
$r_{1,6}/r_{2,3} = 1.371 \pm 0.008$	$r_{4,5}/r_{2,3} = 0.925 \pm 0.005$
$r_{2,4}/r_{2,3} = 1.707 \pm 0.004$	$r_{4,6}/r_{2,3} = 1.744 \pm 0.011$
$r_{2,5}/r_{2,3} = 2.320 \pm 0.008$	$r_{5,6}/r_{2,3} = 1.308 \pm 0.009$

$S_{xx} = 0.0161 \pm 0.0003$ $S_{yy} = 0.0786 \pm 0.0010$ $S_{xy} = 0.0318 \pm 0.0005$

A better fit was obtained if a slight deviation from the planarity was assumed for the six-membered ring formed, through hydrogen bonding by CO and OH. The deviation should be about 10°–15° for both HCO and OH, in agreement with an out-of-plane structure having the aldehyde and hydroxyl hydrogens both on the same side of the ring.[170] In order to investigate this possibility the authors made the reasonable assumption that the interconversion between the two identical nonplanar forms with the hydrogens either below or above the aromatic ring plane would take place fast enough to dynamically recreate the same symmetry as for the planar form. Therefore instead of the five S values required for a molecule without any symmetry, only three values could be used.

c. Bisisoxazoles.[171, 172] Two structural isomers have been examined, the 3,3′-bisisoxazole (**1**) and the 5,5′-bisisoxazole (**2**). Both these molecules may exist as cis or trans rotational isomers, because of easy rotation around the single bond joining the two five-membered rings.

In each case there are only four hydrogen atoms, so that only four D_{ij} values can be obtained; the geometrical information one can derive thus depends upon the number of orientational parameters which have to be used. The authors discussed the following models: (i) free rotation, (ii) single conformation, and (iii) equilibrium between cis and trans isomers.

i. Free rotation. In this case, owing to the effective symmetry which is dynamically created, only one orientational parameter is required. The model however could be unambiguously rejected since the results were internally inconsistent.[172]

ii. Single conformation. If a single conformation is present it could be the cis, the trans, or a twisted isomer. In the case of cis only two S values are required (C_{2v} symmetry) and the ratio of the interprotonic distances can thus be obtained. However the ratios determined under this assumption for both **1** and **2** disagree completely with the shape predicted for a cis isomer on the basis of X-ray studies and other structural determinations of the isooxazole ring.

If the trans isomer is present, then three orientational parameters are required (see Table IV); accordingly, no geometrical information can be directly obtained. However, if the position of the hydrogen atoms within the five-membered ring is assumed from other sources, then there are only three unknowns (the orientational parameters) and four equations (D_{ij} values). If the computed dipolar couplings match the experimental ones satisfactorily, then the existence of the trans conformation may be proved.

In the case of the 3,3′-bisisoxazole (**1**) the trans conformation was found to satisfy the experimental data,[171] whereas in the case of the 5,5′-bisisoxazole (**2**) neither the cis nor the trans conformations did.[172] For the 5,5′-derivative (**2**) the trial with a single nonplanar isomer also led to negative results. This is because although for a certain twisting angle the experimental D_{ij} values were matched (there are four equations and four unknowns, i.e., three orientational parameters and the twisting angle), the orientational parameters were found to exceed the allowed limits[172] (see Table V), thus denying any physical meaning to this hypothesis.

iii. Equilibrium between cis and trans isomers. Finally, the assumption was made that in the case of 5,5′-bisisoxazole (**2**) both isomers were present in different amounts. Two limiting possibilities are considered: either the internal rotation is faster than the molecular reorientation in the liquid crystal, or the internal rotation is slower than reorientation. With the first hypothesis, three average values for the orientational parameters have to be used; for the second, three for the trans and two for the cis isomer (a total of five) should be employed. In the latter case, unless additional approximations are used (as discussed in some of the following examples), the problem cannot be solved. In the framework of the first hypothesis the authors found that a rapid equilibrium between $22.5 \pm 5\%$ of the cis isomer and $77.5 \pm 5\%$ of the trans accounts for the experimental data. Theoretical calculations[172] are not in disagreement.

d. Furan and Thiophene Aldehydes. The first of these examples concerns the determination of the conformation of 2-furan[173] and 2-thiophene[174] carboxaldehyde.

XO-cis   XO-trans

$$X = O, 3$$
$$X = S, 4$$

Whereas in the case of **3** low-temperature NMR studies showed that both cis and trans rotational isomers are present, the OO-cis being the more stable,[175, 176] no clear information had been obtained on the preferred rotational isomer in the case of thiophene aldehyde in solution. In both cases the four protons give six D_{ij} couplings, and because of the symmetry both the XO-cis and XO-trans isomers require three orientational parameters. The spectrum of 2-thiophenecarbaldehyde (**4**) is reported in Fig. 14. Its interpretation led the authors to conclude that only the SO-cis isomer was present.[174]

The basic geometry of the thiophene ring and the bond lengths involved in the carbonyl group were assumed from those of similar molecules. The only structural parameter that was allowed to change was the C-3–C-2–CHO angle. Thus in both the SO-cis and the SO-trans form there are six experimental data and four unknowns, which leave the problem overdetermined. In Table XXXI the values of the dipolar couplings computed for the SO-cis and the SO-trans conformations are compared with the experimental data: the SO-cis couplings agree within the experimental error, whereas the SO-trans definitely does not.

Lunazzi and Veracini[174] also proved that a free rotational model is incorrect and established that the uncertainty about the coplanarity of CHO and thiophene is less than $10°$; they showed furthermore that theoretical

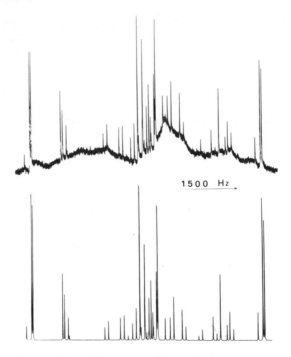

FIG. 14. Experimental and computed 100 MHz spectrum of 2-thiophene-carboxaldehyde in Phase IV at probe temperature [from *J. Chem. Soc., Perkin Trans. 2* p. 1739 (1973); reprinted with permission of the copyright owner].

TABLE XXXI

Experimental Dipolar Couplings (Hz) of Oriented 2-Thiophenecarboxaldehyde Compared with Those Obtained Assuming the SO-cis or the SO-trans Conformation

D values	Experimental	SO-cis	SO-trans
CHO–H-3	-2371 ± 5	-2370.7	-1988.8
CHO–H-4	-340 ± 4	-344.0	-763.9
CHO–H-5	-130 ± 7	-127.6	-465.6
H-3–H-4	-1653 ± 4	-1653.6	-1670.6
H-3–H-5	-9 ± 2	-9.8	$+526.1$
H-4–H-5	$+364 \pm 4$	$+364.0$	$+232.7$
rms deviation (Hz)		2	352

calculations favor the existence of the SO-cis conformation solely. An MW study also gives SO-cis as the more stable conformer.[177]

As far as the furan derivative **3** is concerned, neither the OO-cis, nor the OO-trans form gave a reasonable fit of the couplings, in agreement with the low-temperature NMR observations.[175, 176] Even with the basic geometry from the MW method, there are not sufficient experimental values to determine the ratio of the isomers. The authors thus assumed that instead of six orientational parameters, only three could be used. This assumption may be justified in two ways: (1) either the reorientation is slower than the internal rotation and therefore the three pairs of orientational parameters are physically averaged to three values, or (2) the reorientation is faster than the internal motion, but each of the three parameters of one isomer is sufficiently close to one of the others that the use of a single term instead of a pair does not affect the result too strongly. The ratio of the two isomers obtained in this way (60% OO-cis and 40% OO-trans) is in surprisingly good agreement with the relative amount obtained by Dahlqvist and Forsén,[175] if the variation of the isomer ratio with temperature, according to the Boltzmann distribution, is taken into account.[173] Owing to the mathematical implications, however, the agreement might be due to a fortuitous coincidence.

The different conformational behavior of thiophene and furan has been also explained, by means of CNDO calculations, with the presence of an oxygen–oxygen repulsion in the furan derivative that would allow the existence of a certain amount of the OO-trans rotamer. The absence of such a repulsion in the thiophene analog **4** is responsible for the existence of the SO-cis isomer alone.[174]

Even in the case of 2,5-thiophenedicarboxaldehyde (**5**) it was shown that a single conformer is not present but an equilibrium has to be considered between the possible rotational isomers.[178]

(a)   (b)   (c)

5

By using similar approximations it was shown that the trans-trans isomer (c) is not present at all and the equilibrium is restricted to the isomers (a) and (b). The cis-cis isomer (a) is expected to be the more stable, its amount being in the range 70–80% depending upon the approximation used. A parallel ESR study on the corresponding radical anion[179] did prove that the rotational isomer (c) is absent and that (a) is the more stable species. Dipole moment investigations on the same molecule[180] also agree with the NMR method.

It seems therefore that, despite the quite coarse approximations that have to be employed, the results of the conformational studies agree, at least from a qualitative point of view, with the conclusions of other techniques. The furan analog (2,5-furandicarboxaldehyde) has been studied. The results are reported in Huckerby[181] and Bucci et al.[182]

e. *Bithienyl Derivatives.* The 2,2'-bithienyl **6** was shown to be planar in the solid state and to exist solely in the SS-trans form.[183]

SS-cis   SS-trans

X = H, **6**; = Cl, **7**; = Br, **8**; = NO$_2$, **9**

On the other hand the ESR spectrum of the radical anion in solution[184] did show that two unequally populated rotamers are present, although it was not possible to identify the preferred conformer. Electron diffraction measurements in the gaseous phase also seem to indicate that two isomers (probably not planar) are present;[185] theoretical calculations predict the existence of two planar isomers, the SS-trans being more stable than SS-cis by 0.46 kcal/mole.[186]

In a preliminary investigation carried out on 5,5'-disubstituted derivatives (**7**, **8**, **9**), Veracini, Macciantelli, and Lunazzi showed[187] that LXNMR could unambiguously exclude the free rotation model as well as the existence of a single rotational isomer, that was either planar (cis or trans) or twisted. By assuming the basic structure of 2,2'-bithienyl as well as the approximations on the orientational parameters previously mentioned,[172, 178] they reached the conclusion that an equilibrium between the two planar forms, SS-cis and SS-trans (the latter in larger amount), could account for the experimental data. Subsequent NMR investigations on the unsubstituted derivative **6** confirmed that a cis-trans equilibrium is present.[188, 189] An exact determination of the relative weight of the two forms was not possible;[188, 189] however the use of assumptions leading to a reduced number of orientational parameters indicates that the SS-trans is the most stable species ($\sim$ 70%).[188]

An attempt has also been made to estimate the barrier to the rotation, although resulting from an approximate treatment the value obtained (5 $\pm$ 2 kcal/mole)[188] seems quite reasonable (cf. the theoretical prediction of 4 kcal/mole [186]). Preliminary studies on the structural isomer 3,3'-bithienyl in liquid crystalline medium also indicated the existence of two unequally populated cis-like and trans-like isomers.[190]

Other cases have been reported where conformational preferences and interconversion barriers have been studied.[191-193] It seems therefore that NMR spectroscopy in liquid crystalline solvents may become a quite useful technique in studying the properties of molecules with internal motions.

APPENDIX

A number of papers published after the completion of this chapter contain material relevant to the subject of LXNMR. Pertinent sources are given below.

Further investigations on cyclobutanone and trimethylene sulfide have been reported.[194,195] Swinton[194] found large deviations between computed and experimental D_{ij} values, assuming rigid structures for both cyclobutanone and trimethylene sulfide. The first case may be due to experimental errors since cyclobutanone has been observed to react with the nematogen which was employed (EBBA) and the spectrum changes with time.[62] Actually, the deviations reported[194] for a rigid planar structure are much smaller when an inert nematic solvent is employed.[62] On the other hand, the deviation observed for the trimethylene sulfide confirms that the molecule does not have a rigid planar structure.

However, Cole and Gilson[195] discovered that inclusion of the potential function for ring puckering leads to an even worse result, and the inclusion of methylene rocking does not improve the situation. These authors then used a third set of orientational parameters to reduce the error. It is likely however that the motions of this molecule are complex and that complete vibrational corrections would be required in addition to the corrections for puckering and rocking effects.

2,5-Dihydrofuran has been studied[196,197,197a] and a slightly different geometry than that given by Cole and Gilson[155] has been proposed. Planar rigid structures have been found to match the experimental couplings with quite good accuracy (i.e., within a few tenths of a Hz).[197,197a]

Both ethylene sulfide and oxide have been reinterpreted by making use of ^{13}C satellites.[198] Acetylene labeled with ^{13}C has been also reinvestigated,[1 a more accurate vibrational analysis has been carried out, and a peculia feature concerning the orientation of this molecule has been discovered.[1 Pyridine and pyridine N-oxide have been studied in lyotropic mesophase an differences between the two six-membered rings have been detected.[200]

Benzonitrile,[201] 1,4-dichloro- and dibromobenzene,[202,202a] acetalde hyde,[203] cyanopropyne,[204] and methyl fluoride[205,205a] have been studiec using both proton and ^{13}C dipolar couplings.

Vibrational corrections have been applied to norbornadiene[206] and tc ^{13}C-labeled ethylene.[207]

Further works on 1,4-difluorobenzene[208] and 1,2-difluoroethene[209] have been published as well as studies of rigid molecules such as 1-chloronaph-thalene[210] and coumarine,[211] which have not previously been investigated. Organometallic compounds have been also receiving considerable attention.[212–218]

Among flexible molecules, p-dioxene was found to be nonplanar with a twisting angle of 29°.[219] A quite odd result was found for 2,2'-bipyrimidine[220] in that the more favorable conformations were suggested to be planar and twisted by 90°; both were found to be equally probable.

This result might depend upon the insensitivity of the inter-ring couplings to the twisting angle, so that a meaningful determination of this parameter is not possible. Actually the presence of an inter-ring coupling between the hydrogens ortho to the second ring (which obviously are not present in 2,2'-bipyrimidine) is almost essential to determine the conformation with reasonable accuracy.[80] Beside the conclusions reached by Niederberger et al.,[80] d'Annibale et al.,[159] and Emsley et al.,[161] a recently studied substituted biphenyl[221] also confirms that biphenyl-like molecules are twisted by angles of about 30°.

The usefulness of LXNMR spectroscopy in deciding among various possible conformers has been confirmed by the recent results of Veracini and co-workers. They determined that the only conformation of 2,5-pyridine-dicarboxaldehyde is the NO-trans-trans:[222]

Investigations on technical aspects of LXNMR with particular regard to double resonance,[141, 203, 223] the effect of sample spinning,[224] and double quantum transitions are in progress.[225] Finally a comprehensive and up-to-date book on the various aspects of the NMR spectroscopy in liquid crystalline solvents has been recently published by Emsley and Lindon.[226]

Acknowledgments

The author wishes to thank Dr. E. E. Burnell (University of British Columbia), Dr. D. Canet (University of Nancy, France), Prof. P. Diehl (University of Basel), Dr. J. W. Emsley (University of Southampton), Dr. D. F. R. Gilson (McGill University), Prof. J. H. Goldstein (Emory University), Prof. H. Gunther (University of Cologne), Dr. C. W. Haigh (University of Wales, Swansea), Dr. C. L. Khetrapal (Raman Institute,

Bangalore), Dr. C. A. de Lange (Vrije Universiteit, Amsterdam), Dr. M. C. McIvor (Imperial Chemical Industries, Runcorn), Dr. S. Meiboom (Bell Laboratories, Murray Hill, New Jersey), Prof. E. W. Randall (Queen Mary College, London) and Dr. A. J. Rest (University of Cambridge), for making manuscripts available in advance of publication.

The author is also indebted to Drs. J. W. Emsley and G. R. Luckhurst (University of Southampton) and Dr. C. A. Veracini (University of Pisa) for helpful comments and suggestions. Many thanks are also due to one of the editors, Prof. E. W. Randall for his encouragement and advice.

The support of the "Laboratorio dei composti del carbonio" of the Italian CNR (Ozzano, E.) and a NATO grant is acknowledged.

References

[1] J. W. Emsley, J. Feeney, and L. H. Sutcliffe, eds., "High Resolution NMR Spectroscopy," Vol. 1. Pergamon, Oxford, 1963.

[2] R. E. Richards and T. S. Smith, *Trans. Faraday Soc.* **47**, 1261 (1951); R. E. Richards, *Quart. Rev., Chem. Soc.* **10**, 480 (1956).

[3] A. D. Buckingham and K. A. McLauchlan, *Proc. Chem. Soc., London* p. 144 (1963).

[4] R. E. J. Sears and E. L. Hahn, *J. Chem. Phys.* **45**, 2753 (1966); **47**, 348 (1967).

[5] C. W. Hilbers and C. MacLean, *Chem. Phys. Lett.* **2**, 445 (1968); **7**, 587 (1970); in "NMR Basic Principle and Progress" (P. Diehl, E. Fluck, and R. Kosfeld, eds.), Vol. 7, p. 1. Springer-Verlag, Berlin and New York, 1972.

[6] Z. M. ElSaffar, *J. Chem. Phys.* **45**, 4643 (1966); D. Lawton and H. N. Powell, *J. Chem. Soc., London,* p. 2339 (1958).

[7] P. Averburch, P. Ducros, and X. Pare, *C. R. Acad. Sci.* **250**, 322 (1960); K. A. McLauchlan, "NMR in Chemistry." Academic Press, New York, 1966.

[8] A. Saupe and G. Englert, *Phys. Rev. Lett.* **11**, 462 (1963); G. Englert and A. Saupe, *Z. Naturforsch. A* **19**, 172 (1964).

[9] A. D. Buckingham and K. A. McLauchlan, in "Progress in NMR Spectroscopy" (J. W. Emsley, J. Feeney, and L. H. Sutcliffe, eds.), Vol. II, p. 63. Pergamon, Oxford, 1967.

[10] G. R. Luckhurst, *Quart. Rev., Chem. Soc.* **22**, 179 (1968).

[11] A. Saupe, *Angew. Chem. Int. Ed. Engl.* **7**, 97 (1968).

[12] S. Meiboom and L. C. Snyder, *Science* **162**, 1337 (1968).

[13] P. Diehl and C. L. Khetrapal, in "NMR Basic Principle and Progress" (P. Diehl, E. Fluck, and R. Kosfeld, eds.), Vol. 1, p. 1. Springer-Verlag, Berlin and New York, 1969.

[14] L. C. Snyder and S. Meiboom, *Mol. Cryst. Liquid Cryst.* **7**, 181 (1969).

[15] P. J. Black, K. D. Lawson, and T. J. Flautt, *Mol. Cryst. Liquid Cryst.* **7**, 201 (1969).

[16] R. Brière, *Comm. Energ. At. Publ. CEA–BIB*, 167 (1969).

[17] A. Saupe, in "Magnetic Resonance" (C. K. Coogan et al., eds.), p. 339. Plenum, New York, 1970.

[18] S. Meiboom and L. C. Snyder, *Accounts Chem. Res.* **4**, 81 (1971).

[19] D. H. Whiffen, *Chem. Brit.* **7**, 57 (1971).

[20] J. Bulthuis, C. W. Hilbers, and C. MacLean, *MTP Int. Rev. Sci., Phys. Chem. Ser. One* **4**, 201 (1972).

[21] S. Meiboom, R. C. Hewitt, and L. C. Snyder, *Pure Appl. Chem.* **32**, 251 (1972).

[22] P. Diehl and P. M. Henrichs, *in* "Specialistic Periodical Reports of the Chemical Society: Nuclear Magnetic Resonance" (R. K. Harris, ed.), Vol. 1 p. 321 Chem. Soc., London, 1971; P. Diehl and W. Niederberger, *ibid.* Vol. 3, Chapter 11, p. 368 (1973).

[23] G. H. Brown, J. W. Doane, and V. D. Neff, "A Review of Structural and Physical Properties of Liquid Crystals," CRC Monosci. Ser. Butterworth, 1972. *Advan. Liquid Cryst.* **1** (1975).

[23a] R. Steinstrasser and L. Pohl, *Angew. Chem., Int. Ed. Engl.* **12**, 617 (1973).

[23b] H. Kelker, *Mol. Cryst. Liquid Cryst.* **21**, 1 (1973).

[23c] G. W. Gray, "Molecular Structure and the Properties of Liquid Crystals." Academic Press, New York, 1962.

[23d] G. W. Gray, *Mol. Cryst. Liquid Cryst.* **21**, 161 (1973).

[23e] P. G. de Gennes, "The Physics of Liquid Crystals." Oxford Univ. Press (Clarendon), London and New York, 1974.

[24] Z. Luz and S. Meiboom, *J. Chem. Phys.* **59**, 275 (1973). S. Meiboom and Z. Luz, *Mol. Cryst. Liquid Cryst.* **22**, 143 (1973).

[25] M. G. Friedel, *Ann. Phys.* **18**, 273 (1922); M. R. Caro, *C. R. Acad. Sci.* **251**, 1139 (1960).

[26] S. Sackmann, S. Meiboom, and L. C. Snyder, *J. Amer. Chem. Soc.* **89**, 5981 (1967).

[27] G. Durand, L. Leger, F. Rondelez, and M. Veyssie, *Phys. Rev. Lett.* **22**, 227 (1969); R. B. Meyer, *Appl. Phys. Lett.* **14**, 208 (1969).

[28] G. R. Luckhurst and H. J. Smith, *Mol. Cryst. Liquid Cryst.* **20**, 319 (1973).

[29] V. Luzzati, H. Mustacchi, and A. Skoulies, *Discuss. Faraday Soc.* **25**, 43 (1958).

[30] K. D. Lawson and T. J. Flautt, *J. Amer. Chem. Soc.* **89**, 5489 (1967).

[30a] P. J. Black, K. D. Lawson, and T. J. Flautt, *J. Chem. Phys.* **50**, 542 (1969).

[31] R. C. Long, Jr., *J. Magn. Resonance* **12**, 216 (1973).

[31a] R. C. Long, Jr. and J. H. Goldstein, to be published.

[32] W. Maier, *Phys. Z.* **45**, 285 (1944).

[33] E. Bose, *Phys. Z.* **10**, 32 (1909).

[34] S. Ornstein and W. Kast, *Trans. Faraday Soc.* **29**, 31 (1933).

[35] H. Zocher, *Phys. Z.* **28**, 790 (1927).

[36] C. W. Oseen, *Trans. Faraday Soc.* **29**, 883 (1933).

[37] F. C. Franck, *Discuss. Faraday Soc.* **25**, 19 (1958).

[38] G. R. Luckhurst, *Mol. Cryst. Liquid Cryst.* **2**, 363 (1967).

[39] E. Sackmann, S. Meiboom, and L. C. Snyder, *J. Amer. Chem. Soc.* **90**, 2184 (1968).

[40] C. S. Yannoni, *J. Amer. Chem. Soc.* **91**, 4611 (1969).

[41] J. C. Rowell, W. D. Phillips, L. R. Melby, and M. Panar, *J. Chem. Phys.* **43**, 3442 (1965); J. Charvolin, P. Manneville, and B. Deloche, *Chem. Phys. Lett.* **23**, 345 (1973).

[42] H. C. Longuett-Higgins and G. R. Luckhurst, *Mol. Phys.* **8**, 613 (1964).

[43] A. Saupe, *Z. Naturforsch. A* **20**, 572 (1965).

[44] C. A. Boicelli, A. Mangini, L. Lunazzi, and M. Tiecco, *J. Chem. Soc., Perkin Trans.* 2 p. 599 (1972).

[45] J. W. Emsley and J. C. Lindon, *Mol. Phys.* **25**, 641 (1973).

[46] E. E. Burnell and P. Diehl, *Can. J. Chem.* **50**, 3566 (1972).

[47] M. A. Raza and L. W. Reeves, *J. Magn. Resonance* **8**, 222 (1972).

[48] P. Diehl, P. M. Henrichs, and W. Niederberger, *Mol. Phys.* **20**, 139 (1971).

[49] C. A. Veracini, P. Bucci, and P. L. Barili, *Mol. Phys.* **23**, 59 (1972).

[50] P. Diehl and C. L. Khetrapal, *Mol. Phys.* **15**, 201 (1968).
[51] C. S. Yannoni, *J. Amer. Chem. Soc.* **92**, 5237 (1970).
[52] D. Canet and P. Granger, *C. R. Acad. Sci.* **276**, 315 (1973).
[53] H. Spiesecke and J. Bellion-Jourdan, *Angew. Chem. Int. Ed. Engl.* **6**, 450 (1967).
[54] D. Canet and P. Granger, *C. R. Acad. Sci.* **272**, 1345 (1971).
[55] J. Russell, *Org. Magn. Resonance* **4**, 433, (1972).
[56] D. Canet and R. Price, *J. Magn. Resonance* **9**, 35 (1973).
[57] P. Diehl and C. L. Khetrapal, *Mol. Phys.* **14**, 283 (1968).
[58] J. W. Emsley, J. C. Lindon, G. R. Luckhurst, and D. Shaw, *Chem. Phys. Lett.* **19**, 345 (1973).
[58a] J. W. Emsley, personal communication.
[59] A. D. Buckingham, E. E. Burnell, and C. A. de Lange, *Mol. Phys.* **15**, 285 (1958).
[60] F. M. Leslie, G. R. Luckhurst, and H. J. Smith, *Chem. Phys. Lett.* **13**, 368 (1972).
[61] K. C. Cole and D. F. R. Gilson, *J. Chem. Phys.* **56**, 4362 (1972).
[62] A. d'Annibale, L. Lunazzi, G. Fronza, R. Mondelli, and S. Bradamante, *J. Chem. Soc., Perkin Trans.* 2 p. 1908 (1973).
[63] D. Bailey, A. D. Buckingham, and A. J. Rest, *Mol. Phys.* **26**, 233 (1973).
[64] L. W. Reeves, J. M. Riveros, and J. A. Vanin, *Mol. Phys.* **25**, 9 (1973).
[65] S. A. Barton, M. A. Raza, and L. W. Reeves, *J. Magn. Resonance* **9**, 45 (1973).
[66] I. J. Gazzard, *Mol. Phys.* **25**, 469 (1973).
[67] K. I. Dahlqvist and A. B. Hornfeldt, *Chem. Scripta* **1**, 125 (1971).
[68] J. M. Dereppe, J. P. Morisse, and M. Van Meerssche, *Org. Magn. Resonance* **3**, 583 (1971).
[69] R. C. Long, Jr. and J. H. Goldstein, *J. Chem. Phys.* **54**, 1563 (1971).
[70] P. F. Swinton and G. Gatti, *J. Magn. Resonance* **8**, 293 (1972).
[71] A. L. Segre and S. Castellano, *J. Magn. Resonance* **7**, 5 (1972).
[72] I. M. Armitage, E. E. Burnell, M. B. Bunn, L. D. Hall, and R. B. Malcolm, *J. Magn. Resonance* **13**, 167 (1974).
[72a] C. A. Veracini, M. Longeri, and P. L. Barili, *Chem. Phys. Lett.* **19**, 592 (1973).
[73] J. M. Briggs, E. J. Rahkmaa, and E. W. Randall, *J. Magn. Resonance* **17**, 55 (1975).
[74] N. Zumbulyadis and B. P. Dailey, *Mol. Phys.* **26**, 777 (1973).
[75] C. T. Yim and D. F. R. Gilson, *Can. J. Chem.* **46**, 2783 (1968).
[76] L. C. Snyder and S. Meiboom, *J. Chem. Phys.* **44**, 4057 (1966).
[77] W. Herrig and H. Gunther, *J. Magn. Resonance* **8**, 284 (1973).
[78] C. L. Khetrapal, A. C. Kunwar, and A. Saupe, *Mol. Phys.* **25**, 1405 (1973).
[79] C. S. Yannoni, *J. Chem. Phys.* **51**, 1682 (1969).
[80] W. Niederberger, P. Diehl, and L. Lunazzi, *Mol. Phys.* **26**, 571 (1973).
[81] E. E. Burnell and P. Diehl, *Mol. Phys.* **24**, 489, (1972).
[82] J. M. Dereppe, J. Degelaen, and M. Van Meersche, *J. Mol. Struct.* **17**, 225 (1973).
[83] B. M. Fung and M. J. Gerace, *J. Chem. Phys.* **53**, 1171 (1970).
[84] C. S. Yannoni, *J. Chem. Phys.* **52**, 2005 (1970).
[85] D. Canet, E. Haloui, and H. Nery, *J. Magn. Resonance*, **10**, 121 (1973).
[86] R. A. Hoffman, B. Gestblom, and S. Forsén, *J. Chem. Phys.* **39**, 486 (1963).
[87] J. S. Emsley, J. C. Lindon, and J. Tabony, *J. Chem. Soc., Faraday Trans.* 2 **69**, 10 (1973); *Mol. Phys.* **26**, 1485 and 1499 (1973).
[88] L. C. Snyder and S. Meiboom, *J. Chem. Phys.* **58**, 5096 (1973).

89 R. C. Hewitt, S. Meiboom and L. C. Snyder, *J. Chem. Phys.* **58**, 5089 (1973).
90 P. K. Bhattacharyya and B. P. Dailey, *J. Chem. Phys.* **59**, 3737 (1973).
91 L. C. Snyder and E. W. Anderson, *J. Chem. Phys.* **62**, 3336 (1965).
92 A. D. Buckingham, E. E. Burnell, and C. A. de Lange, *Mol. Phys.* **16**, 299 (1969).
93 H. Spiesecke and A. Saupe, *Mol. Cryst. Liquid Cryst.* **6**, 287 (1970).
94 J. Gerritsen and C. MacLean, *Mol. Cryst. Liquid Cryst.* **12**, 97 (1971).
95 J. Gerritsen and C. MacLean, *J. Magn. Resonance* **5**, 44 (1971).
96 C. W. Haigh and S. Sykes, *Chem. Phys. Lett.* **19**, 571 (1973).
97 J. Degelaen, P. Diehl, and W. Niederberger, *Org. Magn. Resonance* **4**, 721 (1972).
98 G. J. den Otter, C. MacLean, C. W. Haigh, and S. Sykes, *J. Chem. Soc., Chem. Commun.* p. 24 (1974).
99 J. W. Emsley, J. C. Lindon, and S. R. Salman, *J. Chem. Soc., Faraday Trans.* 2 p. 1343 (1972).
100 G. Englert, A. Saupe, and J. P. Weber, *Z. Naturforsch A* **23**, 152 (1968).
101 H. Spiesecke, *in* "Liquid Crystals and Ordered Fluids" (J. F. Johnson and R. S. Potter, eds.), p. 123. Plenum, New York, 1970.
102 L. C. Snyder, *J. Chem. Phys.* **43**, 4041 (1965).
103 L. C. Snyder and S. Meiboom, *J. Chem. Phys.* **47**, 1480 (1967).
104 J. Bulthuis and C. MacLean, *Chem. Phys. Lett.* **7**, 242 (1970).
105 J. Bulthuis and C. MacLean, *J. Magn. Resonance* **4**, 148 (1971).
106 N. J. D. Lucas, *Mol. Phys.* **22**, 147 (1971).
107 N. J. D. Lucas, *Mol. Phys.* **22**, 233 (1971).
108 N. J. D. Lucas, *Mol. Phys.* **23**, 825 (1972).
109 P. Diehl and W. Niederberger, *J. Magn. Resonance* **9**, 495 (1973).
110 J. W. Emsley, J. C. Lindon, D. S. Stephenson, and M. C. McIvor, *Mol. Phys.* **28**, 93 (1974).
111 G. Englert and A. Saupe, *Z. Naturforsch A* **20**, 1401 (1965).
112 G. Englert and A. Saupe, *Mol. Cryst.* **1**, 503 (1966); E. Haloui and D. Canet, *Org. Magn. Resonance* (to be published).
113 C. C. Costain, *J. Chem. Phys.* **29**, 864 (1958).
114 P. Diehl, C. L. Khetrapal, and H. P. Kellerhals, *Mol. Phys.* **15**, 333 (1968).
115 S. Castellano and A. A. Bothner-By, *J. Chem. Phys.* **41**, 3863 (1964).
116 P. Diehl, H. P. Kellerhals, and W. Niederberger, *J. Magn. Resonance* **4**, 352 (1971).
117 A. d'Annibale, L. Lunazzi, F. Fringuelli, and A. Taticchi, *Mol. Phys.* **27**, 257 (1974).
118 B. Bak, D. Christensen, L. Hansen-Nygaard, and J. Rastrup-Andersen, *J. Mol. Spectrosc.* **7**, 58 (1962); R. D. Brown and J. G. Crofts, *Chem. Phys.* **1**, 217 (1973).
119 A. Jones, *Comput. J.* **13**, 301 (1970).
120 B. Bak, D. Christensen, W. B. Dixon, L. Hansen-Nygaard, and J. Rastrup-Andersen, *J. Chem. Phys.* **37**, 2027 (1962).
121 W. Bovée, C. W. Hilbers, and C. MacLean, *Mol. Phys.* **17**, 75 (1969).
122 E. Sackmann, *J. Chem. Phys.* **51**, 2984 (1969).
123 J. B. Pawliczek and H. Gunther, *J. Amer. Chem. Soc.* **93**, 2050 (1971).
124 A. D. Buckingham, E. E. Burnell, and C. A. de Lange, *Mol. Phys.* **17**, 205 (1969).
125 P. Diehl and C. L. Khetrapal, *Org. Magn. Resonance* **1**, 467 (1969).
126 J. M. Dereppe, E. Arte, and M. Van Meerssche, *J. Cryst. Mol. Struct.* (in press).

[127] P. L. Barili and C. A. Veracini, *Chem. Phys. Lett.* **8**, 229 (1971).

[128] P. Diehl and C. L. Khetrapal, *Mol. Phys.* **15**, 633 (1968).

[129] J. Bulthuis, J. Gerritsen, C. W. Hilbers, and C. MacLean, *Rec. Trav. Chim. Pays-Bas* **87**, 417 (1968).

[130] W. de Kievut and C. A. de Lange, *Chem. Phys. Lett.* **22**, 378 (1973).

[131] C. T. Yim and D. F. R. Gilson, *Can. J. Chem.* **47**, 1057 (1969).

[132] J. M. Dereppe, J. Degelaen, and M. Van Meerssche, *J. Mol. Struct.* **17**, 225 (1973).

[133] G. C. Pawley and E. A. Yeats, *Acta Crystallogr.*, *Sect. B* **25**, 2009 (1969).

[134] P. Diehl, C. L. Khetrapal, and H. P. Kellerhals, *Helv. Chim. Acta* **51**, 529 (1968).

[135] R. C. Long, Jr., S. L. Baughcum, and J. H. Goldstein, *J. Magn. Resonance* **7**, 253, (1972); see also Dereppe *et al.*[68]

[136] P. Diehl, C. L. Khetrapal, and U. Lienhard, *Can. J. Chem.* **46**, 2645 (1968).

[137] P. Diehl, C. L. Khetrapal, and U. Lienhard, *Mol. Phys.* **14**, 465 (1968).

[138] P. Diehl, C. L. Khetrapal, and H. P. Kellerhals, *Mol. Phys.* **15**, 333 (1968).

[139] B. Bak, L. Hansen, and J. R. Anderson, *J. Chem. Phys.* **22**, 2013 (1954).

[139a] C. W. N. Cumper, *Trans. Faraday Soc.* **54**, 1261 and 1266 (1958).

[139b] B. Bak, L. Hansen-Nygaard, and J. Rastrup-Andersen, *J. Mol. Spectrosc.* **2**, 361 (1958).

[140] C. Schumann and R. Price, *Angew. Chem., Int. Ed. Engl.* **12**, 930 (1973).

[141] J. W. Emsley, J. Lindon, and J. Taboni, *J. Chem. Soc., Faraday Trans. 2* **71**, 579, (1975).

[142] R. C. Long, Jr., K. R. Long, and J. H. Goldstein, *Mol. Cryst. Liquid Cryst.* **21**, 299 (1973); R. C. Long, Jr. and J. H. Goldstein, *ibid.* **22**, 137 (1973).

[143] C. L. Khetrapal and A. C. Kunwar, *Mol. Cryst. Liquid Cryst.* **15**, 363 (1972).

[143a] C. L. Khetrapal, A. Saupe, and A. C. Kunwar, *Mol. Cryst. Liquid Cryst.* **17**, 121 (1972).

[144] R. Danieli, L. Lunazzi, and C. A. Veracini, *J. Chem. Soc. Perkin Trans. 2* (in press).

[145] P. Diehl, H. P. Kellerhals, and W. Niederberger, *J. Magn. Resonance* **3**, 230 (1970).

[146] P. Diehl, C. L. Khetrapal, W. Niederberger, and P. Partington, *J. Magn. Resonance* **2**, 181 (1970).

[147] C. L. Khetrapal and A. Saupe, *J. Magn. Resonance* **9**, 275 (1973).

[148] D. Canet, *C. R. Acad. Sci.* **276**, 807 (1973); D. Canet and J. Barriol, *Mol. Phys.* **27**, 1705 (1974).

[149] L. F. Williams and A. A. Bothner-By, *J. Magn. Resonance* **11**, 314 (1973).

[150] C. A. Veracini and F. Pietra, *J. Chem. Soc., Chem. Commun.* p. 1262 (1972).

[151] C. A. Veracini, M. Guidi, M. Longeri, and A. M. Serra, *Chem. Phys. Lett.* **24**, 99 (1974).

[152] H. Gunther, W. Herrig, and J. B. Pawliczek, *Z. Naturforsch. B* **29b**, 104 (1974).

[153] S. Meiboom and L. C. Snyder, *J. Chem. Phys.* **52**, 3857 (1970).

[154] K. C. Cole and D. F. R. Gilson, *J. Chem. Phys.* **60**, 1191 (1974).

[155] K. C. Cole and D. F. R. Gilson, *Can. J. Chem.* **52**, 281 (1974).

[156] L. H. Scharpen and V. W. Laurie, *J. Chem. Phys.* **49**, 221 (1968).

[156a] S. I. Chan, J. Zinn, and D. W. Gwinn, *J. Chem. Phys.* **34**, 1319 (1961).

[157] D. O. Harris, H. W. Harrington, A. C. Luntz, and W. D. Gwinn, *J. Chem. Phys.* **44**, 3467 (1966).

[158] Z. Luz and S. Meiboom, *J. Chem. Phys.* **59**, 1077 (1973).

[159] A. d'Annibale, L. Lunazzi, C. A. Boicelli, and D. Macciantelli, *J. Chem. Soc., Perkin Trans. 2* p. 1396 (1973).

[160] I. Fischer-Hjalmars, *Tetrahedron* **19**, 1805 (1963).
[161] J. W. Emsley, D. S. Stephenson, J. C. Lindon, L. Lunazzi, and S. Pulga, *J. Chem. Soc. Perkin Trans 2* p. 1541 (1975).
[162] P. Diehl, P. M. Henrichs, and W. Niederberger, *Org. Magn. Resonance* **3**, 243 (1971).
[163] P. Diehl and P. M. Henrichs, *Org. Magn. Resonance* **3**, 791 (1971).
[164] P. Diehl, P. M. Henrichs, W. Niederberger, and J. Vogt, *Mol. Phys.* **21**, 377 (1971).
[165] J. F. Yan, F. A. Momany, and H. A. Scheraga, *J. Amer. Chem. Soc.* **92**, 1109 (1970).
[166] E. E. Burnell and P. Diehl, *Org. Magn. Resonance* **5**, 137 (1973).
[167] S. Kondo, Y. Sakuray, E. Hirota, and Y. Morino, *J. Mol. Spectrosc.* **34**, 231 (1970).
[168] A. Almenningen, O. Bastiansen, and M. Traetteberg, *Acta Chem. Scand.* **12**, 1221 (1958); W. Haugen and M. Traetteberg, *Acta Chem. Scand.* **20**, 1726 (1966); K. Kuchitsu, T. Fukuyama, and Y. Morino, *J. Mol. Struct.* **1**, 463 (1967).
[169] E. B. Reznikova, V. I. Tulin, and V. M. Tatevskii, *Opt. Spectrosc. (USSR)* p. 364 (1962); A. L. Segre, L. Zetta, and A. di Corato, *J. Mol. Spectrosc.* **32**, 296 (1969).
[170] P. Diehl and P. M. Henrichs, *J. Magn. Resonance* **5**, 134 (1971).
[171] P. Bucci, P. F. Franchini, A. M. Serra, and C. A. Veracini, *Chem. Phys. Lett.* **8**, 421 (1971).
[172] P. Bucci and C. A. Veracini, *J. Chem. Phys.* **56**, 1290 (1972).
[173] P. L. Barili, L. Lunazzi, and C. A. Veracini, *Mol. Phys.* **24**, 673 (1972).
[174] L. Lunazzi and C. A. Veracini, *J. Chem. Soc. Perkin Trans. 2* p. 1739 (1973).
[175] K. I. Dahlqvist and S. Forsén, *J. Phys. Chem.* **69**, 4062 (1965).
[176] K. I. Dahlqvist and A. B. Hornfeldt, *Tetrahedron Lett.* p. 3837 (1971).
[177] J. F. Bertran, E. Ortiz and L. Ballester, *J. Mol. Struct.* **17**, 161 (1973).
[178] L. Lunazzi, G. F. Pedulli, M. Tiecco, and C. A. Veracini, *J. Chem. Soc., Perkin Trans. 2* p. 755 (1972).
[179] L. Lunazzi, G. F. Pedulli, M. Tiecco, C. Vincenzi, and C. A. Veracini, *J. Chem. Soc., Perkin Trans. 2* p. 751 (1972).
[180] H. Lumbroso and P. Pastour, *C. R. Acad. Sci.* **261**, 1279 (1965); H. Lumbroso, D. M. Bertin, M. Robba, and B. Roques, *ibid.* **262**, 36 (1966).
[181] T. N. Huckerby, *Tetrahedron Lett.* p. 3497 (1971).
[182] P. Bucci, C. A. Veracini, and M. Longeri, *Chem. Phys. Lett.* **15**, 396 (1972).
[183] G. J. Visser, G. J. Heeres, J. Wolters, and A. Vos, *Acta Crystallogr., Sect. B* **24**, 467 (1968).
[184] P. Cavalieri d'Oro, A. Mangini, G. F. Pedulli, P. Spagnolo, and M. Tiecco, *Tetrahedron Lett.* p. 4179 (1969).
[185] A. Almenningen, O. Bastiansen, and P. Svendsas, *Acta Chem. Scand.* **12**, 1671 (1958).
[186] V. Galasso and N. Trinajstic, *Tetrahedron* **28**, 4419 (1972).
[187] C. A. Veracini, D. Macciantelli, and L. Lunazzi, *J. Chem. Soc., Perkin Trans. 2* p. 751 (1973).
[188] P. Bucci, M. Longeri, C. A. Veracini, and L. Lunazzi, *J. Amer. Chem. Soc.* **96**, 1305 (1974).
[189] C. L. Khetrapal and A. C. Kunwar, *Mol. Phys.* **28**, 441 (1974).
[190] L. Lunazzi, F. Salvetti, and C. A. Veracini, to be published.
[191] I. J. Gazzard and N. Sheppard, *Mol. Phys.* **21**, 169 (1971).

[192] J. Bulthuis, J. Van Den Berg, and C. MacLean, *J. Mol. Struct.* **16**, 11 (1973).
[193] J. Courtieu and Y. Gounnelle, *Org. Magn. Resonance* **6**, 11 (1974).
[194] P. F. Swinton, *J. Magn. Resonance* **13**, 304 (1974).
[195] K. C. Cole and D. F. R. Gilson, *Mol. Phys.* **29**, 1749 (1975).
[196] J. Courtieu and Y. Gounelle, *Mol. Phys.* **28**, 161 (1974).
[197] P. F. Swinton, *J. Mol. Struct.* **22**, 221 (1974).
[197a] D. G. de Kowalewski and V. J. de Kowalewski, *J. Mol. Struct.* **23**, 203 (1974).
[198] E. Haloui and D. Canet, *J. Mol. Struct.* **24**, 85 (1975).
[199] P. Diehl, S. Sykora, W. Niederberger, and E. E. Burnell, *J. Magn. Resonance* **14**, 268 (1974).
[200] C. L. Khetrapal, A. C. Kunwar, and A. C. Patankar, *J. Magn. Resonance* **15**, 219 (1974).
[201] J. P. Jacobsen and K. Schaumburg, *Mol. Phys.* **28**, 1644 (1974).
[202] E. E. Burnell, P. Diehl, and W. Neiderberger, *Can. J. Chem.* **52**, 151 (1974).
[202a] E. E. Burnell and M. A. J. Sweeney, *Can. J. Chem.* **52**, 3565 (1974).
[203] J. W. Emsley, J. C. Lindon, and J. Tabony, *J. Chem. Soc. Faraday Trans 2*, **71**, 586 (1975).
[204] E. Haloui and D. Canet, *Chem. Phys. Lett.* **26**, 261 (1974).
[205] P. K. Bhattacharyya and B. P. Dailey, *J. Magn. Resonance* **13**, 317 (1974).
[205a] E. E. Burnell, J. R. Council, and S. E. Ulrich, *Chem. Phys. Lett.* **31**, 395 (1975).
[206] J. W. Emsley and J. C. Lindon, *Mol. Phys.* **29**, 531 (1975).
[207] P. Diehl, S. Sykora, and E. Wullschleger, *Mol. Phys.* **29**, 305 (1975).
[208] G. J. denOtter, W. Heijser, and C. MacLean, *J. Magn. Resonance* **13**, 11 (1974).
[209] G. J. denOtter and C. MacLean, *Chem. Phys.* **3**, 119 (1974).
[210] P. Diehl and J. Vogt, *Organic Magn. Resonance* **6**, 33 (1974).
[211] E. Cappelli, A. di Nola, and A. L. Segre, *Mol. Phys.* **27**, 1385 (1974).
[212] D. Bailey, A. D. Buckingham, M. C. McIvor, and A. J. Rest, *J. Organometall. Chem.* **61**, 311 (1973).
[213] J. P. Yesinowski and D. Bailey, *J. Organometall. Chem.* **65**, 270 (1974).
[214] I. R. Beattie, J. W. Emsley, and R. M. Sabine, *J. Chem. Soc., Faraday Trans. 2* **70**, 1356 (1974).
[215] J. W. Emsley and J. C. Lindon, *Mol. Phys.* **28**, 1373 (1974).
[216] I. R. Beattie, J. W. Emsley, J. C. Lindon, and R. M. Sabine, *J. Chem. Soc. Dalton Trans.* p. 1264 (1975).
[217] J. D. Kennedy and W. McFarlane, *J. Chem. Soc., Chem. Commun.* p. 595 (1974).
[218] A. D. Buckingham, *5th Int. Symp. Magn. Resonance*, p. 1 (1974).
[219] C. A. de Lange and K. J. Peverelli, *J. Magn. Resonance* **16**, 159 (1974).
[220] J. Courtieu, Y. Gounelle, C. Duret, P. Gonord, and S. K. Kan, *Org. Magn. Resonance* **6**, 622 (1974).
[221] L. Lunazzi and D. Macciantelli, *Gazz. Chim. Ital.* **105**, 657 (1975).
[222] P. L. Barili, M. Longeri, and C. A. Veracini, *Mol. Phys.* **28**, 1101 (1974).
[223] P. Diehl and W. Niederberger, *J. Magn. Resonance* **15**, 391 (1974).
[224] J. W. Emsley and J. C. Lindon, *Mol. Phys.* **28**, 1253 (1974).
[225] H. Wennerstrom, N. O. Persson, and B. Lindman, *J. Magn. Resonance* **13**, 348 (1974).
[226] J. W. Emsley and J. Lindon, "NMR Spectroscopy Using Liquid Crystal Solvents." Pergamon, Oxford, 1975.

Author Index

Numbers in parentheses are reference numbers and indicate that an author's work is referred to although his name is not cited in the text. Numbers in italics show the page on which the complete reference is listed.

A

Abe, H., 122(95), 129, *152*
Abragam, A., 273, *331*
Abraham, E. P., 24(104), 42(104), *84*
Abraham, K. M., 244(161), *266*
Abraham, R. J., 313(56), *332*
Abrahamson, E. W., 195(156, 157), *201*
Abrahamsson, S., 94(23), 98, 102(32, 33), 103, 104(33), *149, 150*
Abramson, F. P., 112(84, 85, 86), 113(86), *151*
Adams, A., 57(213), *88*
Adams, R., 195(156), *201*
Adlercreutz, H., 147(182), *154*
Agadzhanyan, Ts. E., 22(98), *84*
Agarwal, K. L., 4(30), 25(30), 43(194), 49(194), 50(30, 194), 51(30), 52(30), 70(30), *81, 87*
Ahmed, M., 317(63), *333*
Aimura, A., 25(111), 49(111), 51(111), *84*
Airey, W., 244(195), 247(185), 248(192, 194, 195), 249(195), *266, 267*
Alakhov, Yu. B., 2(12), 3(12), 4(12), 5(12), 9(76), 14(12), 20(93), 21(93), 22(93), 25(76, 93), 26(93, 121), 28(93, 123), 37(123, 151), 41(12, 93, 123, 167), 43(93, 167), 49(167), *80, 83, 84, 85, 86*
Alchalel, A., 195(160, 161), *201*
Aldanova, N. A., 2(12), 3(12), 4(12), 5(12, 60), 14(12), 28(123), 37(60, 123, 151), 41(12, 167), 43(167), 49(167), 59(225), *80, 82, 85, 86, 88*
Alford, A. L., 107(54), 136(54), *150*
Alger, T. D., 307(44), *332*

Ali, A. W., 163(28), *198*
Allen, L. C., 173(66), *199*
Allerhand, A., 309(47, 48), *332*
Almenningen, A., 398(168), 404(185), *412*
Altamura, M. R., 41(187), 42(187), 61(187), *87*
Ambler, R. P., 4(34), 10(34), 12(34), 21(34), 25(34), 43(34), 53(34), 54(34), 59(34), 70(34), 73(34), *81*
Amoss, M., 25(109, 115), 51(109), *84*
Anderson, D. N., 130, 131(125), 132(125), *152*
Anderson, D. W. W., 223(93), 241(93), 249(93, 205), 250(93, 205), 253(205), *264, 267*
Anderson, E. W., 354(91), *410*
Anderson, J. H., 204(5), 207(5), 209(5), 211(5), 212(5), *261*
Anderson, J. R., 378(139), *411*
Anderson, J. W., 253(235), 255(235), *268*
Anderson, W. A., 280(15), *331*
Andersson, B. A., 11(82), 41(82), *83*
Andersson, C. O., 35(140), 39, *85*
Andreotti, R. E., 41(187), 42(187), 61(187), *87*
Anet, F. A. L., 271, 289, 314(60), 317(63), *331, 332, 333*
Aoyagi, T., 40(174), 41(174, 175), *87*
Aplin, R. T., 14(86, 87), 36(86), 41(86, 87, 181, 182), 42(86), 43(86, 181), 44(86), *83, 87*
Applebury, M. L., 195(162), *201*
ApSimon, J. W., 278(12), 294(25), 301, *331, 332*
Arata, Y., 305(42), *332*
Archer, R. A., 302(37), *332*

Subject Index*

* Letters after a compound indicate the technique applied, viz., ESR = electron spin resonance, FP = flash photolysis, LXNMR = NMR in liquid crystals, MS = mass spectrometry, NMR = nuclear magnetic resonance, NOE = nuclear Overhauser effect.

A 6
B 7
C 8
D 9
E 0
F 1
G 2
H 3
I 4
J 5